New Studies in Weak Arithmetics

CSLI *Lecture Notes Number 211*

New Studies in Weak Arithmetics

Edited by

Patrick CÉGIELSKI
Charalampos CORNAROS
Costas DIMITRACOPOULOS

PUBLICATIONS
Center for the Study of
Language and Information
STANFORD, CALIFORNIA

PRESSES UNIVERSITAIRES
DU PÔLE DE RECHERCHE
ET D'ENSEIGNEMENT
SUPÉRIEUR PARIS-EST

Center for the Study of Language and Information
Leland Stanford Junior University
and
Presses universitaires du Pôle de Recherche et d'Enseignement Supérieur de Paris-Est
Printed in the United States
17 16 15 14 13   1 2 3 4 5

Library of Congress Cataloging-in-Publication Data

New studies in weak arithmetics / edited by Patrick Cegielski, Charalampos Cornaros, Costas Dimitracopoulos.
pages cm. – (CSLI lecture notes ; 211)
Includes bibliographical references and index.
ISBN 978-1-57586-723-6 (pbk. : alk. paper) –
ISBN 978-1-57586-724-3 (electronic : alk. paper)
1. Number theory. 2. Mathematics–Philosophy. 3. Logic, Symbolic and mathematical. I. Cegielski, Patrick, 1954– editor of compilation. II. Cornaros, Charalampos, 1964– editor of compilation. III. Dimitracopoulos, Costas, editor of compilation. IV. Title: Weak arithmetics.

QA241.N465 2013
512.7–dc23

2013030651
CIP

∞ The acid-free paper used in this book meets the minimum requirements of the American National Standard for Information Sciences—Permanence of Paper for Printed Library Materials, ANSI Z39.48-1984.

CSLI was founded in 1983 by researchers from Stanford University, SRI International, and Xerox PARC to further the research and development of integrated theories of language, information, and computation. CSLI headquarters and CSLI Publications are located on the campus of Stanford University.

CSLI Publications reports new developments in the study of language, information, and computation. Please visit our web site at
http://cslipublications.stanford.edu/
for comments on this and other titles, as well as for changes and corrections by the author and publisher.

Contents

II Some Problems in Logic and Number Theory, and their Connections

III Models of Arithmetic

Foreword

LUC HITTINGER[1]

I am very proud that the publishing collaboration between Stanford University, more precisely CSLI Publications, and the University Paris 12 (*Pôle de Recherche et d'Enseignement Supérieur Paris-Est*), continues. After the first Lecture Notes (number 196 in 2010) devoted to Studies in Weak Arithmetics and two more volumes (numbers 190 and 194 in 2011) of translation in French of select papers by Donald Knuth, the famous computer scientist who has inspired much work in our laboratory LACL (*Laboratoire d'Algorithmique, Complexité et Logique*), a second Lecture Notes devoted to Studies in Weak Arithmetics is being published. This time the volume contains papers based on lectures given during JAF31—the thirty first meeting of the *Journées sur les Arithmétiques Faibles* conferences that began in Lyon in 1990.

Although one of the successors of the original University of Paris, created in 1100, we like to consider our university as born in 1970 with the Faculty of Medicine and the Institute of Technology. At that time the aim was towards professional students, which continues to be our first objective. Fundamental science was essentially nonexistent in the initial plan—no masters in Mathematics or Theoretical Computer Science was offered, unlike what happened at many other universities in France and around the world. However, we had to adapt to the needs of our 32,000 students and our Laboratories of Mathematics (LAMA) and Theoretical Computer Science (LACL) were created; LAMA and LACL developed quickly and are now widely recognized by the international community of researchers.

[1] President of the University Paris-Est Créteil Val-de-Marne

Studies in Weak Arithmetics.
Patrick Cégielski, Charalampos Cornaros,
Costas Dimitracopoulos.

Congratulations from Samos

PETROS PARIANOS[1]

On behalf of all the residents of Samos I extend my best wishes on the occasion of the "thirty first Weak Arithmetics Days" Conference which is hosted by the Department of Mathematics of the School of Sciences of the University of the Aegean. It give us great pleasure that the Conference offers us the presentation of the latest results in the area of mathematics, which leads to the contemporary and rapid evolution of this important field of science.

Samos has the blessing to have had the University of the Aegean since 1985. We are proud because during these years it has won great recognition and reputation for the quality of its teaching and research.

The University of the Aegean contributes to the growth of the level of education of the residents of Samos and with its collaborations with all the local institutions it aims to the cultural, intellectual and economic development in our area.

I congratulate all those who contributed towards the realization of this Conference as well as all its participants.

[1]Karlovassi deputy mayor, Samos Island

Studies in Weak Arithmetics.
Patrick Cégielski, Charalampos Cornaros,
Costas Dimitracopoulos.

Introduction

PATRICK CÉGIELSKI,[1] CHARALAMPOS CORNAROS,[2]
COSTAS DIMITRACOPOULOS[3]

The conference series, *Journées sur les Arithmétiques Faibles* (*Weak Arithmetics Days*, JAF), began in June 1990 when the first meeting was held at École Normale Supérieure in Lyon, France. That meeting has been followed by thirty more meetings, during which researchers from countries all over the world presented work and discussed research ideas concerning *Weak Arithmetics*, that is, the research area concerning, roughly speaking, the application of logical methods to Number Theory and related topics.

The latest meeting of the series, JAF31, took place from May 30 until June 1, 2012, on Samos, the island of Pythagoras, one of the most renowned mathematicians of all times. JAF31 was dedicated to the memory of Alan R. Woods, a colleague who died on December 11, 2011, having made significant contributions to the area of Weak Arithmetics and having participated is several meetings of the series.

The Steering Committee of the JAF series decided that a volume be published, containing papers based on lectures delivered during the meeting on Samos, as well as Alan's Ph.D. thesis, parts of which remained unpublished despite having motivated a tremendous amount of work by many distinguished researchers. While preparing the volume, which we undertook with enthusiasm, we invited more contributions by

[1]LACL, EA 4219, Université Paris-Est Créteil, France cegielski@u-pec.fr
[2]University of Aegean, Karlovassi, Greece kornaros@aegean.gr
[3]University of Athens, Greece cdimitr@phs.uoa.gr

Studies in Weak Arithmetics.
Patrick Cégielski, Charalampos Cornaros,
Costas Dimitracopoulos.

colleagues who had not been able to participate in JAF31 but wished to honor the memory of Alan R. Woods. After a proposal of Roman Kossak, we decided to also include in the volume another unpublished Ph.D. thesis that had contributed significantly to progress in the area of Weak Arithmetics, namely the thesis of Hamid Lesan, who died in 2006.

All papers published in the present volume were refereed by (at least) two referees each. We are grateful to the following colleagues, for having kindly helped us with the refereeing of the papers: Andrés Cordon-Franco, David Fernandez Duque, Alex Esbelin, Benedetto Intrigila, Francisco Félix Lara Martín, Vassilis Paschalis and Albert Visser. We are also grateful to Vassilis Paschalis, for having undertaken the typesetting of A. Woods' thesis and for having offered us valuable technical support.

We would like to express our gratitude to Alan's son, Robert, and daughter, Sarah, for their kind permission to include his thesis in this volume and for providing us with information concerning his life and career; we are also grateful to Hamid's wife, Lynda, and daughters, Sara and Leila, for granting us their kind permission to include his thesis and to George Wilmers, Hamid's thesis supervisor, for having prepared a note on Hamid's life and work.

We are grateful to Ch. Cornaros, E. Tachtsis and their team in Karlovassi, for organizing a memorable conference, and to all the sponsors of the meeting—University of Aegean, University of Athens, Regional Government of North Aegean (Mr. A. Yiakalis, Governor, and Mr. Th. Papatheofanous, Deputy Governor), Municipality of Samos (Mr. S. Thanos, Mayor, and Mr. P. Parianos, Deputy Mayor), Municipal Department of Karlovassi (Mr. S. Makris, Chairman), DOPONAS, ALPHA BANK, Energiaki Samou SA (Mr. Nikos Elenis), Mr. Evangelos Mytilinaios, Samaina Hotels, AB Shop & Go, DMGRAPHICS, EKTYPON and TO SXEDIO—for having provided generous financial and other support, without which the meeting would not have been so successful. Finally, we would like to express our deep gratitude to the Université Paris Est-Créteil, for having generously covered the cost of production of the present volume, and to Marie-Annick Le Traon (Université Paris Est Créteil–IUT de Sénart-Fontainebleau) for the design of the cover.

Internet site:

`http://lacl.fr/jaf/`

contains further information on “Journées sur les Arithmétiques Faibles”.

Part I

Contributions

1

A few questions concerning consistency and conservativeness

ZOFIA ADAMOWICZ[1]

Abstract: In the paper we give a survey of results concerning the provability of various consistency statements in weak arithmetics and we pose a problem. In our survey we focus on the Π_1-conservativeness question, namely the question of Π_1-conservativeness between theories $I\Delta_0 + \Omega_i$ (these theories are recalled in section 2). We consider those consistency statements which are helpful in proving non conservativeness results.

1 Introduction

Weak fragments of arithmetic have been studied since the eighties of the former century. One of the basic weak fragments is the theory $I\Delta_0$. Some variants of it are the theories $I\Delta_0+\Omega_i$ and Buss's theories S_i, T_i, S_i^n, T_i^n, where the most interesting are S_2, T_2, S_2^n, T_2^n. One of the crucial problems in the area is the problem of the collapse of the hierarchy of theories S_2^n, T_2^n. A weaker property than collapse is conservativeness. The most interesting is Π_1 or $\forall\Sigma_n^b$-conservativeness. The main known result in this respect is the $\forall\Sigma_{n+1}^b$-conservativeness of S_2^{n+1} over T_2^n.

A criterion for a theory to be called *weak* is that it does not prove the totality of the exponential function. A non weak theory is considered as *strong*.

As for strong theories a lot of results about non conservativeness

[1]Institute of Mathematics of the Polish Academy of Sciences, Śniadeckich 8, P.O.B. 21, 00-956 Warszawa, Poland, Z.Adamowicz@impan.pl

Studies in Weak Arithmetics.
Patrick Cégielski, Charalampos Cornaros,
Costas Dimitracopoulos.

are known. For instance $I\Sigma_{n+1}$ is not Π_1-conservative over $I\Sigma_n$. Also, $I\Delta_0 + \exp$ is Π_1-conservative neither over $I\Delta_0$ nor over any $I\Delta_0 + \Omega_i$, [14]. However, it is open whether any of $I\Delta_0 + \Omega_i$ is Π_1-conservative over any other $I\Delta_0 + \Omega_k$, for a $k < i$.

In all the cases in which the Π_1-conservativeness question is settled, it is settled via a consistency statement. For instance, the counterexample to the Π_1-conservativeness of $I\Sigma_{n+1}$ over $I\Sigma_n$ is the consistency of $I\Sigma_n$.

For, the consistency of $I\Sigma_n$ is provable in $I\Sigma_{n+1}$, while, in view of Gödel's theorem, it is not provable in $I\Sigma_n$. The provability of the consistency of $I\Sigma_n$ in $I\Sigma_{n+1}$ is usually shown via provability of the cut free consistency of $I\Sigma_n$, the Gentzen, Tableau or Herbrand one. On the other hand, using Gödel's theorem and the availability of cut elimination in $I\Sigma_n$, this cut free consistency of $I\Sigma_n$ is not provable in $I\Sigma_n$.

In the case of the non-conservativeness of $I\Delta_0 + \exp$ over $I\Delta_0$ or over $I\Delta_0 + \Omega_i$, the situation is more delicate. Here the Hilbert style consistency and the cut free consistency behave in different ways. The Hilbert kind of consistency cannot be an example for non conservativeness since $I\Delta_0 + \exp$ does not prove the Hilbert style consistency of $I\Delta_0$ (and even of PA^- and of Robinson's arithmetic Q) [14]. On the other hand, the cut free consistency can be such an example. The cut free consistency of $I\Delta_0 + \Omega_i$ is provable in $I\Delta_0 + \exp$, [14]. For some time it has been unknown whether it is unprovable in $I\Delta_0 + \Omega_i$. That is, it was unknown whether the second Gödel incompleteness theorem holds for arithmetics $I\Delta_0 + \Omega_i$ for a cut free provability predicate. In [7] and [3] the second incompleteness theorem for $I\Delta_0 + \Omega_i$ was proved for $i \geq 2$, for the Herbrand notion of consistency and in [2] this result was proved for $I\Delta_0 + \Omega_1$ for the Tableau notion of consistency.

As a matter of fact the non-Π_1-conservativeness of $I\Delta_0 + \exp$ over $I\Delta_0 + \Omega_i$ can be witnessed also by some other sentences and it was proved already in [14]. The proof is simple for a variant of the Herbrand consistency of $I\Delta_0 + \Omega_i$, namely for a sentence expressing the Herbrand consistency of $I\Delta_0 + \Omega_i$ together with the set of true $\exists\Pi_1^b$ sentences. This variant is provable in $I\Delta_0 + \exp$ and is easily unprovable in $I\Delta_0 + \Omega_i$. We study in detail this variant of consistency below in sections 2,3 and 4.

Kołodziejczyk showed in [10] a strengthening of the results of [7] and [3]. He proved that there is a finite fragment S of $I\Delta_0 + \Omega_1$ such that no theory $I\Delta_0 + \Omega_i$ proves the Herbrand consistency of S. Thus, if one wants to distinguish the theories $I\Delta_0 + \Omega_i$ it cannot be done by means of provability of Herbrand consistency. One should consider

Herbrand proofs restricted to some definable cuts of a given model of a bounded arithmetic.

For theories still weaker than $I\Delta_0 + \exp$, cut free provability can be shown provable when restricted to some cuts. Restricting the consistency predicate to some definable initial segment was already considered in the literature, for instance by P. Pudlák in [12]. He proved that as far as the Hilbert consistency predicate is concerned many interesting restrictions to a cut still satisfy the Gödel independence phenomenon, however as far as the Herbrand or Genzen predicate is concerned for many natural cuts this is not the case. For instance $I\Delta_0$ proves the cut free consistency of itself and related theories $I\Delta_0 + \Omega_i$ prove cut free consistency of $I\Delta_0$ and of themselves when the consistency is restricted to some cut of iterated logarithms [12]. A similar approach was studied carefully in [9] for Herbrand consistency. There are also negative results, $I\Delta_0 + \Omega_i$ do not prove cut free consistency of themselves when the consistency is not restricted to the deep enough cut of iterated logarithms [3, 9, 10, 15]. For a most careful and fine study and for the discussion of positive and negative results the reader is referred to [9]. There are analogous results for Buss's theories T_2^n and S_2^n, where induction is restricted, although these theories behave more regularly and more is known [3, 10].

Let us focus on the problem of the Π_1-conservativeness between theories $I\Delta_0 + \Omega_i$.

In the special case it is the question:

Is $I\Delta_0 + \Omega_2$ Π_1-conservative over $I\Delta_0 + \Omega_1$?

We would like to ask a new question, modifying the above. This is done in section 5. Instead of Π_1-conservativeness we consider a kind of Δ_0-conservativeness. Possibly our Δ_0-conservativeness may be a weaker property, where a positive answer is more likely.

2 Some definitions recalled

First we shall recall some results.

Definition 1. *Recall:*

$$\omega_0(x) \doteq x^2,$$
$$\omega_1(x) = 2^{(\log x)^2},$$
$$\omega_2(x) = 2^{2^{(\log \log x)^2}},$$
$$\omega_{i+1}(x) = 2^{\omega_i(\log x)}.$$
$$\Omega_i : \ \forall x \exists y \ y = \omega_i(x).$$

We can restate the conservativeness problem as follows:

Is there a Π_1 *sentence* ξ *such that*

$$I\Delta_0 + \Omega_2 \vdash \xi$$

and

$$I\Delta_0 + \Omega_1 \nvdash \xi?$$

A good candidate for ξ roughly is:

$$Cons^I(I\Delta_0 + \Omega_1 + \text{the set of true } \Sigma_1 \text{ sentences}),$$

where $Cons^I(\cdot)$ is some predicate expressing consistency relativized to some definable initial segment I.

A good candidate for I is the log log log cut. Let us denote it by $\log^3$. Why should we consider the $\log^3$ cut? We have

$$I\Delta_0 + \Omega_1 \nvdash \forall x \exists y\ y = \omega_1^{\log^3 x}(x)$$

and

$$I\Delta_0 + \Omega_2 \vdash \forall x \exists y\ y = \omega_1^{\log^3 x}(x).$$

For the following formula is easy to be checked by induction:

$$\omega_1^n(x) = 2^{(\log x)^{2^n}},$$

for $n \geq 1$. Hence,

$$\omega_1^{\log^3 x}(x) = 2^{(\log x)^{2^{\log^3 x}}} = 2^{(\log x)^{\log \log x}} = 2^{2^{(\log \log x)^2}} = \omega_2(x).$$

There is a number of non-conservativeness results.

Consider the sentence:

$$HCons^{\log^3}(I\Delta_0 + \Omega_1 + \text{the set of true } \Sigma_1 \text{ sentences}),$$

where $HCons$ denotes the Herbrand consistency.

Before formulating it precisely, let us indicate why it should be unprovable in $I\Delta_0 + \Omega_1$ and why it can be provable in $I\Delta_0 + \Omega_2$.

Suppose

$$I\Delta_0 + \Omega_1 \vdash HCons^{\log^3}(I\Delta_0 + \Omega_1 + \text{the set of true } \Sigma_1 \text{ sentences}).$$

Let $T^{\#} \subseteq \Sigma_1$ be maximal consistent with $I\Delta_0 + \Omega_1$. Let $M \models I\Delta_0 + \Omega_1 + T^{\#}$. Then in M, for each Σ_1 sentence ϕ,

$$\phi \text{ iff } HCons^{\log^3}(I\Delta_0 + \Omega_1 + \text{the set of true } \Sigma_1 \text{ sentences} + \phi).$$

For, if ϕ is true, then in view of

$$HCons^{\log^3}(I\Delta_0 + \Omega_1 + \text{the set of true } \Sigma_1 \text{ sentences}),$$

$$HCons^{\log^3}(I\Delta_0 + \Omega_1 + \text{the set of true } \Sigma_1 \text{ sentences} + \phi)$$

holds.

Conversely, if the right hand side holds, then ϕ is consistent with

$$I\Delta_0 + \Omega_1 + \text{the set of true } \Sigma_1 \text{ sentences},$$

whence $\phi \in T^{\#}$, by maximality. But this is a Π_1 truth definition for Σ_1 sentences, contradiction.

3 Herbrand Consistency

Now let us indicate that $I\Delta_0 + \Omega_2$ may prove

$$HCons^{\log^3}(I\Delta_0 + \Omega_1 + \text{the set of true } \Sigma_1 \text{ sentences}).$$

Before doing it, we have to roughly formalize our notion of consistency. This is a way of formalizing the Herbrand consistency. We adopt some ideas from [10].

Let φ be a sentence of the form

$$\exists x_1 \forall y_1 \ldots \exists x_k \forall y_k \varphi'(x_1, y_1, \ldots, x_k, y_k),$$

where φ' is open. The Herbrandization of φ, $He(\varphi)$ is defined as

$$\exists x_1, \ldots, x_k \varphi'(x_1, f_1(x_1), \ldots, x_k, f_k(x_1, \ldots, x_k)),$$

where f_i are new function symbols, called *Herbrand functions.*

The Herbrand Theorem states that φ is provable iff a finite disjunction of formulas of the form

$$\varphi'(t_1, f_1(t_1), \ldots, t_k, f_k(t_1, \ldots, t_k)),$$

where t_i are terms of the language extended by adding Herbrand functions, is a propositional tautology. The above disjunction is called a *Hebrand proof* of φ.

Note that Herbrandization is dual to Skolemization, i.e. $He(\varphi)$ is the negation of the Skolemization of $\neg\varphi$. Hence, a finite disjunction of formulas of the form

$$\varphi'(t_1, f_1(t_1), \ldots, t_k, f_k(t_1, \ldots, t_k)),$$

is not a propositional tautology iff a finite conjunction of formulas of the form

$$\neg\varphi'(t_1, f_1(t_1), \ldots, t_k, f_k(t_1, \ldots, t_k))$$

is satisfiable in the propositional calculus. Now, the function symbols $f_1, \ldots, f_k$ play the role of Skolem functions.

We shall say that a theory T is *Herbrand consistent* if there is no Herbrand proof of $0 = 1$ in T, i.e. if every conjunction ψ of axioms of T of the form

$$\forall x_1 \exists y_1 \ldots \forall x_j \exists y_j \forall x_{j+1} \exists y_{j+1} \ldots \forall x_k \exists y_k \psi'(x_1, y_1, \ldots, x_k, y_k),$$
$$\psi'(t_1, f_1(t_1), \ldots, t_j, f_j(t_1, \ldots, t_j), \ldots, t_k, f_k(t_1, \ldots, t_k))$$

is satisfiable in the propositional calculus. The Skolem function f_j now will be denoted by (see [4]):

$$s^{\exists y_j \forall x_{j+1} \exists y_{j+1} \ldots \psi'(t_1, s^{\exists y_1 \forall x_2 \exists y_2 \ldots \psi'(t_1, y_1, \ldots, x_k, y_k)}, \ldots, t_j, y_j, x_{j+1}, y_{j+1}, \ldots, x_k, y_k)}.$$

This satisfiability can be formulated in terms of evaluations.

Given a tuple of variables $\langle x_1, \ldots, x_m \rangle$, the collection of *simple atomic formulae* over $\langle x_1, \ldots, x_m \rangle$ consists of $x_i = x_j$, $x_i < x_j$, $x_i + x_j = x_l$, $x_i \cdot x_j = x_l$ etc. Basically, simple atomic formulae are those which would still be considered atomic if the vocabulary were relational.

Consider a sequence Λ of closed terms of the extended language.

Given Λ, the collection $\mathcal{S}(\Lambda)$ of *simple atomic sentences over* Λ consists of all sentences obtained by substituting terms from Λ for variables in simple atomic formulas.

An evaluation over Λ is supposed to be a function p which assigns truth values to formulae in $\mathcal{S}(\Lambda)$ in such a way that the assignment can be suitably extended to more complex formulae based on just Λ and p.

Given Λ, an *evaluation* on Λ is any function $p : \mathcal{S}(\Lambda) \longrightarrow \{0, 1\}$ which maps every axiom of equality over $\mathcal{S}(\Lambda)$ to 1. We shall write $p \models \psi$ if $p(\psi) = 1$.

An evaluation can be extended in an obvious way to arbitrary open formulas whose terms are in Λ. We shall say that $p \models T$, for a theory T, if $p \models \psi$ for any ψ obtained by substituting the terms from Λ into the open part of the Skolemization of some axiom of T.

Let ψ be a sentence. For any k, define the formula $HCons^k(T + \psi)$ as:

> *Given any Λ which does not contain terms of depth greater than k, there exists an evaluation p on Λ such that $p \models T + \psi$.*

By $HCons^I(T + \psi)$ we mean

> *For every k in I, given any Λ which does not contain terms of depth greater than k, there exists an evaluation p on Λ such that $p \models T + \psi$.*

By $HCons^I(T + \Sigma_1 \text{ truth})$ we mean

> *For every k in I, every true $\psi \leq k$, $\psi \in \Sigma_1$, given any Λ which does not contain terms of depth greater than k, there exists an evaluation p on Λ such that $p \models T + \psi$.*

How to prove in $I\Delta_0 + \Omega_2$ the sentence

> *For every k in $\log^3$, every true $\psi \leq k$, $\psi \in \Sigma_1$, given any Λ which does not contain terms of depth greater than k, there exists an evaluation p on Λ such that $p \models I\Delta_0 + \Omega_1 + \psi$.*

In a model M of $I\Delta_0 + \Omega_2$, given $k \in \log^3 M$, ψ true in M and $\Lambda = \langle t_1, \ldots, t_k \rangle$ we may construct a sequence $H_\Lambda = \langle h_{t_1}, \ldots, h_{t_k} \rangle$ of true interpretations of terms in Λ in M. We construct the sequence H_Λ by constructing $\langle h_{t_1}, \ldots, h_{t_j}\rangle$ by induction on $j \leq k$. In particular, the term s^ψ is interpreted as the witness for the unbounded existential quantifier for ψ and terms of depth k of the form $s^{\exists\omega_1^k(s^\psi)}$ – more exactly $s^{\exists\omega_1(s^{\exists\omega_1(\ldots\exists\omega_1(s^\psi)\ldots)})}$ – are interpreted as the value of ω_1 iterated k times on the witness for ψ. Those values exist in M because of Ω_2.

Then, having H_Λ we define an appropriate evaluation on Λ, whose truth values follow truth in H_Λ.

All this would work perfectly, however there are some problems. First, to formulate $HCons^{\log^3}(I\Delta_0 + \Omega_1 + \Sigma_1 \text{ truth})$ we need to say "*for every Σ_1 sentence $\psi \leq k$, whenever ψ is true, then given any Λ which does not contain terms of depth greater than k, there exists an evaluation p on Λ such that $p \models T + \psi$*". Thus, we need to be able to say "*ψ is true*". Secondly, to be able to perform induction on $j \leq k$ to construct $\langle h_{t_1}, \ldots, h_{t_j}\rangle$, we need a truth definition for Δ_0 formulas. We can get round these problems in two ways. One is considering $\exists\Pi_1^b$ instead of Σ_1 and T_2^n instead of $I\Delta_0 + \Omega_1$ and T_3^n instead of $I\Delta_0 + \Omega_2$.

Recall: $T_3^n = I\Sigma_n^b + \Omega_2$, $T_2^n = I\Sigma_n^b + \Omega_1$ (up to the language), $I\Sigma_n^b$ a finitely axiomatizable fragment of $I\Delta_0$.

Following the first approach we get the following result

$$T_3^n \textit{ is not } \forall\Sigma_1^b \textit{ conservative over } T_2^n, \text{ [11]}.$$

The witness is the sentence $HCons^{\log^3}(T_2^n + \exists\Pi_1^b \text{ truth})$.

The second approach is instead of restricting $I\Delta_0 + \Omega_1$ to T_2^n, adding something to $I\Delta_0 + \Omega_1$.

4 Self reproduction

The idea is adding to $I\Delta_0 + \Omega_1$ the sentence $HCons^{\log^3}(I\Delta_0 + \exists\Pi_1^b -$ truth) and considering the theory

$$I\Delta_0 + \Omega_1 + HCons^{\log^3}(I\Delta_0 + \exists\Pi_1^b \text{ truth}).$$

However, now the sentence $HCons^{\log^3}(I\Delta_0 + \exists\Pi_1^b \text{ truth})$ does no longer express the consistency of the above theory, since it does not express the consistency of $I\Delta_0$ plus itself. We want to modify it in such a way that it would almost express the consistency of $I\Delta_0$ plus itself.

If $M \models HCons^I(T)$, this allows us to construct a new model M' of T, like in the arithmetized completeness theorem, however M' may no longer satisfy $HCons^I(T)$ (like in the arithmetized completeness theorem).

We may consider a sentence expressing the iteration over I of $HCons^I(T)$. We shall denote it by $HCons^I(T + it)$. This idea was first introduced in [1], it was used in [8] and then it has been refined, carefully studied and used in [6]. The reader is referred mainly to [6] for the study of this idea.

If $M \models HCons^I(T + it)$, this allows us to construct a new model M' of T in which $HCons^I(T + it)$ is satisfied at the initial segment of M' which is cofinal with M. If T is for instance $I\Delta_0 + \Omega_1$, we may cut M' to this segment and obtain a model of $I\Delta_0 + \Omega_1 + HCons^I(T + it)$. Thus $HCons^I(T + it)$ has the *self reproducing property.*

Now let us follow the second approach.

The approach is considering $\exists\Pi_1^b$ instead of Σ_1 and

$$I\Delta_0 + \Omega_1 + HCons^{\log^3}(I\Delta_0 + \exists\Pi_1^b \text{ truth} + it)$$

instead of $I\Delta_0 + \Omega_1$.

We get the following result, [1]:

$I\Delta_0 + \Omega_2 + HCons^{\log^3}(I\Delta_0 + \exists\Pi_1^b$ truth $+ it)$ is not Π_1-conservative over $I\Delta_0 + \Omega_1 + HCons^{\log^3}(I\Delta_0 + \exists\Pi_1^b$ truth $+ it)$.

The witness is the sentence $HCons^{\log^3}(I\Delta_0 + \Omega_1 + \exists\Pi_1^b \text{ truth} + it)$.

5 A new question

Consider now the language $L(\underline{a})$ with an additional constant $\underline{a}$ and the theory $I\Delta_0 + \Omega_1 + \zeta(\underline{a})$, where ζ is a fixed bounded formula. Let $\Delta_0(\underline{a})$ denote the set of bounded sentences of $L(\underline{a})$ having $\underline{a}$ as their bound.

We shall say that $I\Delta_0 + \Omega_2 + \zeta(\underline{a})$ is $\Delta_0(\underline{a})$-conservative over $I\Delta_0 + \Omega_1 + \zeta(\underline{a})$ if for every $\Delta_0(\underline{a})$ sentence $\xi(\underline{a})$, if $I\Delta_0 + \Omega_2 + \zeta(\underline{a}) \vdash \xi(\underline{a})$ then $I\Delta_0 + \Omega_1 + \zeta(\underline{a}) \vdash \xi(\underline{a})$.

Question *Is there a Δ_0 formula ζ such that the theory $I\Delta_0 + \Omega_1 + \zeta(\underline{a})$ is consistent and $I\Delta_0 + \Omega_2 + \zeta(\underline{a})$ is $\Delta_0(\underline{a})$-conservative over $I\Delta_0 + \Omega_1 + \zeta(\underline{a})$?*

The initial segment which is naturally related to this question is $I_a = \{x : \ \omega_1^x(a) \text{ exists}\}$.

And the natural candidate for non Π_1-conservativeness (for an arbitrary ζ) could be

$$HCons^{2I_a}(I\Delta_0 + \Omega_1 + \Delta_0(\underline{a})).$$

There is a chance for the provability of this sentence in $I\Delta_0 + \Omega_2$ and unprovability in $I\Delta_0 + \Omega_1$.

Why provability?

First observe:

$I\Delta_0 + \Omega_2 \vdash I_a = 2I_a$.

For, we have

$$\omega_1^x(a) = \omega_1^{\log^3 a+x}(4) = \exp^3(\log^3 a + x).$$

Thus, the equality $I_a = 2I_a$ takes the form:

$$\forall x\big(\log^3 a + x \in \log^3 \Rightarrow \log^3 a + 2x \in \log^3\big).$$

Now observe:

$I\Delta_0 + \Omega_2 \vdash I_a \supseteq \log^3$.

For, let $x \in \log^3$. Then $x = \log^3 y$ for a y. Note that $\log^3 a \in I_a$. For, if $y \leq a$, then $\omega_1^{\log^3 y}(a) \leq \omega_1^{\log^3 a}(a) = \omega_2(a)$. If $y > a$, then $\omega_1^{\log^3 y}(a) \leq \omega_1^{\log^3 y}(y) = \omega_2(y)$. Hence $\omega_1^x(a)$ exists.

Hence, if $\log^3 = 2\log^3$, which holds under Ω_2, then $I_a = 2I_a$.

Let $M \models I\Delta_0 + \Omega_2$.

If we want to prove

$$HCons^{2I_a}(I\Delta_0 + \Omega_1 + \Delta_0(\underline{a}))$$

in $I\Delta_0 + \Omega_2$, we can show the existence of an evaluation of depth x for $x \in 2I_a$ by evaluating all terms of depth x occurring in it in M. There is a chance for it since $\omega_1^x(a)$ exists in M.

The above is not possible in $I\Delta_0 + \Omega_1$, since $\omega_1^x(a)$ may not exist for $x \in 2I_a$. This is an indication that

$$HCons^{2I_a}(I\Delta_0 + \Omega_1 + \Delta_0(\underline{a}))$$

may be unprovable in $I\Delta_0 + \Omega_1$.

However, note that possibly

$$I\Delta_0 + \Omega_1 \vdash HCons^{I_a}(I\Delta_0 + \Omega_1 + \Delta_0(\underline{a})).$$

We have no conjecture for an answer of our question. For theories $I\Delta_0 + \exp + \zeta(\underline{a})$ and $I\Delta_0 + \Omega_i + \zeta(a)$ for some ζ there is no $\Delta_0(\underline{a})$-conservativeness. However we do not know whether there is a ζ for which conservativeness holds.

As for weak theories, if $I\Delta_0 + \Omega_2$ is not Π_1-conservative over $I\Delta_0 + \Omega_1$, then again for some ζ there is no $\Delta_0(\underline{a})$-conservativeness. However the question itself may have a positive answer. May be a suitable choice of ζ could give $\Delta_0(\underline{a})$-conservativeness. Looking for such a ζ may give new insight to the conservativeness question.

A related question *Is there a Δ_0 formula ζ such that the theory $T_3^n + \zeta(\underline{a})$ is $\Delta_0(\underline{a})$-conservative over $T_2^n + \zeta(\underline{a})$?*

6 Some other witnesses

Now we drop Σ_1 truth and we consider classical consistency statements, namely sentences of the form:

$$HCons^{\log^3}(I\Delta_0 + \Omega_1).$$

Is this sentence provable in $I\Delta_0 + \Omega_1$?

Provably in $I\Delta_0 + \Omega_1$,

$$\exp^3(k) = \omega_1^k(4)$$

for each $k \in \log^3$.

Hence, the term $\omega_1^k(4)$ can be evaluated in any model of $I\Delta_0 + \Omega_1$ for $k \in \log^3$.

We have

$$I\Delta_0 + \Omega_i \vdash HCons^{\log^{i+3}}(I\Delta_0 + \Omega_i).$$

Open are:

$$I\Delta_0 + \Omega_i \vdash HCons^{\log^{i+2}}(I\Delta_0 + \Omega_i)?$$

In particular,

$$I\Delta_0 + \Omega_1 \vdash HCons^{\log^3}(I\Delta_0 + \Omega_1)?$$

However,

$$I\Delta_0 + \Omega_i \not\vdash HCons^{2\log^{i+2}}(I\Delta_0 + \Omega_i),$$

in particular,

$$I\Delta_0 + \Omega_1 \not\vdash HCons^{2\log^3}(I\Delta_0 + \Omega_1).$$

Another approach

Instead of restricting $I\Delta_0 + \Omega_1$ to T_2^n, add something to $I\Delta_0 + \Omega_1$.

For instance, consider

$$I\Delta_0 + \Omega_1 + HCons^{\log^3}(I\Delta_0 + \Omega_1).$$

This is not yet a theory we can work with. We have to introduce self reproduction again.

We have:

$$I\Delta_0 + \Omega_2 + HCons^{\log^3}(I\Delta_0 + \Omega_1) \vdash HCons^{2\log^3}(I\Delta_0 + \Omega_1 + it)$$

and

$$I\Delta_0 + \Omega_1 + HCons^{\log^3}(I\Delta_0 + \Omega_1 + it) \not\vdash HCons^{2\log^3}(I\Delta_0 + \Omega_1 + it).$$

Hence

$$I\Delta_0 + \Omega_2 + HCons^{\log^3}(I\Delta_0 + \Omega_1 + it)$$

is not Π_1-conservative over

$$I\Delta_0 + \Omega_1 + HCons^{\log^3}(I\Delta_0 + \Omega_1 + it)$$

and the witness is the sentence $HCons^{2\log^3}(I\Delta_0 + \Omega_1 + it)$.

For a further study of the above mentioned results and methods we refer the reader to [1, 3, 7, 4, 5, 6, 9, 10].

References

[1] Adamowicz Z., A contribution to the end-extension problem and the Π_1 conservativeness problem, *Annals of Pure and Applied Logic*, vol. 61, 1993, pp. 3–48.

[2] Adamowicz Z., *On tableaux consistency in weak theories*, preprint 618, Institute of Mathematics, Polish Academy of Sciences, 2001.

[3] Adamowicz Z., Herbrand Consistency and Bounded Arithmetic, *Fundamenta Mathematicae*, vol. 171, 2002, pp. 279–292.

[4] Adamowicz Z. and Kołodziejczyk L., Well-behaved principles alternative to bounded induction, *Theoretical Computer Science*, vol. 322, 2004, pp. 5–16.

[5] Adamowicz Z. and Kołodziejczyk L., Partial Collapses of the Σ_1 Complexity Hierarchy in Models for fragments of Bounded Arithmetic, *Annals of Pure and Applied Logic*, vol. 145, 2007, pp. 91–95.

[6] Adamowicz Z., Kołodziejczyk L. and Zbierski P., An application of a reflection principle, *Fundamenta Mathematicae*, vol. 180, 2003, pp. 139–159.

[7] Adamowicz Z. and Zbierski P., On Herbrand consistency in weak arithmetic, *Archive for Mathematical Logic*, vol. 40, 2001, pp. 399–413.

[8] Adamowicz Z. and Zbierski P., On Complexity reduction of Σ_1 Formulas, *Archive for Mathematical Logic*, vol. 42, 2003, pp. 45–58.

[9] Adamowicz Z. and Zdanowski K., Lower bounds for the provability of Herbrand consistency in weak arithmetics, *Fundamenta Mathematicae*, vol. 212, 2011, pp. 191–216.

[10] Kołodziejczyk L., On the Herbrand notion of consistency for finitely axiomatizable fragments of bounded arithmetic theories, *Journal of Symbolic Logic*, vol. 71, 2006, pp. 624–638.

[11] Krajíček J. and Takeuti G., On induction free provability, *Annals of Mathematics and Artificial Intelligence*, vol. 6, 1992, pp. 107–126.

[12] Pudlák P., Cuts, consistency statements and interpretations, *Journal of Symbolic Logic*, vol. 50, 1985, pp. 423–441.

[13] Salehi S., *Herbrand Consistency in Arithmetics with Bounded Induction*, PhD Dissertation, Institute of Mathematics, Polish Academy of Sciences, 2002.

[14] Wilkie A. J. and Paris J. B., On the scheme of induction for bounded arithmetical formulas, *Annals of Pure and Applied Logic*, vol. 35, 1987, pp. 261–302.

[15] Willard D., How to extend the semantic tableaux and cut-free versions of the second incompleteness theorem to Robinson's arithmetic Q, *Journal of Symbolic Logic*, vol. 67, 2002, pp. 465–496.

2

In memoriam of Alan Robert Woods

PATRICK CÉGIELSKI,[1] DENIS RICHARD[2]

Some of the works inspired
(especially in the French community of Logicians and Number Theorists) by his seminal thesis

Abstract: We recall a first result of Alan Woods: the undecidability of the first-order theory $Th(\mathbb{N}, S, \perp)$, where $\perp$ denotes the binary relation of coprimeness, with two different proofs. The question of the $(S, \perp)$-definability of the full first-order arithmetic leads to the famous Erdös-Woods Conjecture (EWC) (There is some $k \in \mathbb{N}$ such that every natural number x is determined uniquely by the sequence $S_0, S_1, \ldots, S_k$ of sets of prime numbers defined by $S_i = \{ p \mid p \textit{ divides } x + i \}$). In case EWC is false, then the a-b-c conjecture, Hall's conjecture, Hall-Schinzel's conjecture, and Lang-Waldschmidt's conjecture are false (Michel Langevin). After Alan Woods, and except for EWC, the situations (mutual definability, decidability) of many (not all!) reasonable natural theories $Th(\mathbb{N}, R_1, R_2)$ – where R_1 is a subset of the relation of addition and R_2 a subset of the relation of divisibility – were investigated (Alexis Bès, Patrick Cégielski, Françoise Maurin, Denis Richard) with some surprising results. Concerning the theory of addition and a predicate, for primes essentially, nothing was known before Alan, in a joint work with C. G. Jockush and P. T. Bateman, proved under Schinzel's hypothesis (a polynomial generalization

[1]LACL, EA 4219, Université Paris-Est Créteil, France, cegielski@u-pec.fr
[2]Université d'Auvergne, denis.richard@udamail.fr

Studies in Weak Arithmetics.
Patrick Cégielski, Charalampos Cornaros, Costas Dimitracopoulos.

of the twin-primes conjecture) that $Th(\mathbb{N}, +, \mathbb{P})$ is undecidable. Patrick Cégielski, Denis Richard, and Maxim Vsemirnov have shown an absolute result: the undecidability of $Th(\mathbb{N}, +, n \mapsto n \times f(n))$, where f is a good approximation of $n \mapsto p_n/n$ and p_n is the $(n{+}1)$-th prime.

Alan also proved that the class RUD of all rudimentary (i. e. Δ_0-definable) sets of positive integers is contained in the class consisting of the spectra $S_\Phi = \{|M| \; such\ that\ M\ is\ finite\ and\ M \models \Phi\}$ of those sentences Φ having only graphs as models. Malika More, Alex Esbelin, and Frédéric Olive gave a toolbox for investigating rudimentary predicates and also extended results to second order languages.

The name of Alan is attached to another very well-known concept: the Erdös-Woods numbers, which are the positive integers k such that there is a closed interval of integers $[a, a+k]$ with the property that every integer in the interval has a factor in common with either a or $a+k$. Results on these numbers are due to M. Bienkowski, Patrick Cégielski, D. L. Dowe, François Heroult, M. Korzeniowki, K. Lorys, N. Lygeros, Denis Richard, Maxim Vsemirnov, and others.

Although sure of not being at all exhaustive in speaking of the works of Alan Robert Woods, we add some results and interesting open questions about the p-destinies (Francis Nézondet) within $< \mathbb{N}, S, \perp >$.

1 Undecidability of successor and coprimality

In the *Open Days in Model Theory and Set Theory* held in Jadwisin in the Karpacz Mountains near Warsaw in September 1981, Jeff Paris gave a lecture in which he mentioned the thesis of Alan Woods, quoting one of Alan's results, the proof that $I\Delta_0 + \Delta_0 - PHP \vdash$ *Bertrand's theorem.* Denis Richard met Jeff for discussing about his lecture and two of his own results on questions appearing in the thesis of Julia Robinson, namely the $(\leq, \perp)$-definability of the whole arithmetic and the undecidability of $Th(\mathbb{N}, S, \perp)$, where S is the successor function. Jeff told him that one of his PhD students, namely Alan, had also proved those two results. Denis was glad that Jeff could know his arguments to prove the undecidability result just mentioned and compare them with those of Alan. He was at that time at the University Claude Bernard of Lyon, and organized a visit of Alan in order to give some lectures on his thesis. In fact, Alan visited Lyon in the following winter and they were able to compare their independent proofs of the undecidability of $Th(\mathbb{N}, S, \perp)$ and $Th(\mathbb{N}, =, S, \perp)$.

Alan's Proof. Alan introduces infinitely many equivalence relations such as

$x \approx y$ iff $\nu(x) = \nu(y)$, where $\nu(a)$ is the cardinal of the set of prime divisors of a

$x \sim y$ iff x and y have the same set of prime divisors

$x \sim_n y$ iff $(x \sim y) \wedge (x+1) \sim (y+1) \wedge \ldots \wedge (x+n) \sim (y+n)$

For $x \neq 0$, let us denote by $\overline{x}$, $\overline{x}^n$, and $\overline{\overline{x}}$, respectively, the $\sim$, $\sim_n$, and $\approx$ equivalence classes of which x is a representative.

Let $\overline{\overline{\mathbb{N}}} = \{\overline{\overline{x}} \mid x \in \mathbb{N}\backslash\{0\}\}$, $\overline{\mathbb{N}} = \{\overline{x} \mid x \in \mathbb{N}\backslash\{0\}\}$, $\overline{\mathbb{N}}^n = \{\overline{x}^n \mid x \in \mathbb{N}\backslash\{0\}\}$.

Alan proves $< \mathbb{N}, =, +, \times >$ is a model isomorphic to $< \overline{\mathbb{N}}, =, +, \times >$ under the mapping ν^{-1}. Then he proves that, considered as predicates on $\mathbb{N}$, the expressions $x = y$, $x+y = z$, and $x \times y = z$ are $(S, \perp)$-definable and hence there is an effective way of transforming any sentence Φ of the language $L(=, +, \times)$ into an equivalent $L(S, \perp)$ sentence Φ^* asserting that

$$< \overline{\mathbb{N}}, =, +, \times > \models \Phi^*.$$

This result, of course, permits to deduce the undecidability of theories $Th(\mathbb{N}, S, \perp\ )$ and of $Th(\mathbb{N}, =, S, \perp)$ from the known undecidability of the full first order arithmetic. Also the existence of these formulas reduces the problem of the $(S, \perp)$-definability of multiplication to the question of whether ν^{-1} is $(S, \perp)$-definable.

Through some lemmas of definability using the Chinese Remainder Theorem, Alan proves the $(S, \perp)$-definability of relations $\overline{\overline{x}} \leq \overline{\overline{y}}$, $\overline{\overline{x}} + \overline{\overline{y}} = \overline{\overline{z}}$, and $\overline{\overline{x}} \times \overline{\overline{y}} = \overline{\overline{z}}$.

Richard's proof. Instead of "Richard's proof", maybe it could be better to say "proof as a corollary of the following result, due to G.D Birkhoff and H.S. Vandiver–Zsigmondy–Carmichael".

Theorem (ZBV-theorem). *If x and y are coprime positive integers such that $x > y \geq 1$ then, for every $n > 0$, there exists at least a prime divisor of $x^n - y^n$ which divides none of the of $x^m - y^m$ for $0 < m < n$ (such divisors are called* primitive divisors *of $x^n - y^n$) except in the following cases*:

$n = 1, x - y = 1$ (*since there is no prime divisor*)

$n = 2, x + y = 2^\alpha$ (*for $\alpha > 0$*)

$n = 6, x = 2, y = 1$.

Actually, it is not difficult to give a $(S, \perp)$-definition of $x = n$, for every n in $\mathbb{N}$, of $\mathbb{P}(x)$ (meaning x is prime), of the relation "*p is prime and x is a power of p different from 1*", of the relation "*x and y are powers of (a priori different) primes and are equal*", of set-theoretic relations of union, intersection, and complementation on the supports (the *support* of an integer a is the set of the prime divisors of a), and of the relation "$x \neq 1$ *and* x *is a power of some prime and* q *is a primitive divisor of* $x - 1$".

Let p be some fixed prime. We denote by $p^{\mathbb{N}}$ the set of powers of the prime p, by $p^\alpha \oplus_p p^\beta = p^{\alpha+\beta}$ a binary function of addition and by $p^\alpha \otimes_p p^\beta = p^{\alpha\times\beta}$ a binary function of multiplication in $p^{\mathbb{N}}$. We also denote by $=_p$ the restriction to $p^{\mathbb{N}}$ of the equality relation. It is obvious that the structure $< p^{\mathbb{N}}, =_p, \oplus_p, \otimes_p >$ is isomorphic to the usual model $< \mathbb{N}, =, +, \times >$. Then we have proved that the structure $< p^{\mathbb{N}}, =_p, \oplus_p, \otimes_p >$ is $(S, \perp)$-definable in $\mathbb{N}$.

As a by-product, we get the fact that the $(S, \perp)$-definability of any map $n \mapsto p^n$, for any fixed prime p, would be sufficient for giving the $(S, \perp)$-definability of multiplication.

In fact, the two proofs have their advantages. On the one hand, Alan does not need any deep result in Number Theory (just the Chinese Remainder Theorem) and proves by using quotient structures the transfer of undecidability from full arithmetic to the weak first order $(S, \perp)$-arithmetic. On the other hand, the ZBV-theorem provides a powerful tool for investigating $Th(\mathbb{N}, S, \perp)$ and $Th(\mathbb{N}, =, S, \perp)$. The coding Denis Richard uses to transfer undecidability is explicit and, in case of a negative answer, we shall know in terms of Number Theory, exactly where the $(S, \perp)$-definability fails.

The previous result straightforwardly shows that:

(i) $z = x \times y$ is $(S, \perp)$-definable in $\mathbb{N}$

(ii) $z = x + y$ is $(S, \perp)$-definable in $\mathbb{N}$

(iii) $x \leq y$ is $(S, \perp)$-definable in $\mathbb{N}$

(iv) $x = y$ is $(S, \perp)$-definable in $\mathbb{N}$

(v) For any fixed prime p, the map $n \mapsto p^n$ is $(S, \perp)$-definable in $\mathbb{N}$.

Statements (i) to (iv) appear at first in the thesis of Alan [35], (v) in [29].

2 Erdös-Woods Conjecture (EWC)

Erdös-Woods Conjecture. *There is some* $k \in \mathbb{N}$ *such that every natural number* x *is determined uniquely by the sequence* S_0, S_1, ..., S_k *of sets of prime numbers defined by* $S_i = \{ p \mid p \text{ divides } x + i\}$.

Let $Supp(a)$ denote the set of prime divisors of a. Consider the following conjectures of Number Theory.

Oesterlé-Masser's conjecture. (Also called **a-b-c conjecture**) *For all* $(a, b, c) \in (\mathbb{N}^*)^3$ *such that* $a + b = c$ *and for every real* r, *there exists an effectively computable constant* $K(r)$ *such that*

$$\Pi_{p \in Supp(abc)}\, p > K(r)^{1-\epsilon(a)}$$

with $\epsilon(r)$ *tends to 0 when* r *tends to infinity.*

Hall's conjecture. *If* $x^3 \neq y^2$ *then there exists an effectively computable constant* C *such that*

$$|x^3 - y^2| > [C.Max(x^3, y^2)]^{1/6}$$

Hall-Schinzel's conjecture. *If* $x^m \neq y^n$ *then there exists an effectively computable constant* C *such that*

$$|x^m - y^n| > [Max(x^m, y^n)]^C$$

Michel Langevin proved in 1988 [20] that the three previous conjectures are false in case EWC fails. Moreover every conjecture which is false if Hall's conjecture is false becomes false in turn if EWC is. This is the case for Lang-Waldschmidt's conjecture (which gives lower bounds of linear forms of logarithms) and for Vojta's conjecture about abelian varieties.

Obviously, a positive solution to one of the conjectures above would give a positive solution to EWC and Alan was right when he gave the title of his thesis [35]. Denis Richard [31] was able to prove, after some discussion with A. Schinzel, that if the a-b-c conjecture is true then the equality of supports (sets of prime divisors) leads to the equality of numbers

$$Supp[x^{2^n} - 1] = Supp[y^{2^n} - 1] \Rightarrow x = y.$$

Before leaving EWC, we must note that Maxim Vsemirnov was able to extend the EWC to a conjecture on polynomial rings. Parallel to the second proof using the so-called ZBV-theorem, Maxim Vsemirnov developed an investigation of Carmichaël numbers useful for coding in the logical approach of EWC [34].

3 Definability, Decidability, and Undecidability within theories of $\mathbb{N}$ in various first-order languages

After Alan's thesis, a lot of PhD students and experienced researchers investigate the questions first opened by Julia Robinson, and then revisited by Alan and members of the team of Denis Richard.

Here is an (non-exhaustive) abstract table giving some of the results on decidability found between 1980 and 2000, where $M_1 \to M_2$ means the structure M_2 is definable in the structure M_1. The theories of the structures of the first four lines are undecidable and decidable in the two last lines.

$$
\begin{array}{ccccc}
 & & <\mathbb{N},=,+,\times> & & \\
 & & \uparrow\downarrow & & \\
<\mathbb{N},\perp,<> & \rightarrow & <\mathbb{N},|,<> & \rightarrow & <\mathbb{N},=,\times,<> \\
[35] & \leftarrow & & & \\
\uparrow (\downarrow ?) & & \uparrow\downarrow & & \uparrow\downarrow \\
<\mathbb{N},\perp,S> & \rightarrow & <\mathbb{N},|,S> & \rightarrow & <\mathbb{N},=,\times,S> \\
[35, 28] & \leftarrow ? & [32] & \leftarrow & \\
\uparrow \neg \downarrow & & \uparrow\downarrow & & \uparrow\downarrow \\
<\mathbb{N},\perp,<_{\Pi}> & \rightarrow & <\mathbb{N},|,<_{\Pi}> & \rightarrow & <\mathbb{N},=,\times,<_{\Pi}> \\
[4] & & [4] & & [4] \\
\uparrow \neg \downarrow & & \uparrow \neg \downarrow & & \uparrow \neg \downarrow \\
<\mathbb{N},\perp,<_{P_2}> & \rightarrow & <\mathbb{N},|,<_{P_2}> & \rightarrow & <\mathbb{N},=,\times,<_{P_2}> \\
[4] & & [4] & & [4] \\
 & & & & \uparrow \neg \downarrow \\
 & & & & <\mathbb{N},=,\times,<_{\mathbb{P}}> \\
 & & & & [21] \\
 & & & & \uparrow \neg \downarrow \\
 & & & & <\mathbb{N},=,\times,(\overline{n})_{n\in\mathbb{N}}> \\
 & & & & [33]
\end{array}
$$

4 Results about addition and primes

Before 1981, almost nothing was known about the association of the operation of addition and the predicate of being a prime. Of course, there is the famous theorem of Schnirelmann on the fact that every integer is a sum of less than 27 (nowadays 19) primes and the theorem of Vinogradov which reduces asymptotically this sum to three primes. Such results show, at least, the depth and the interest of $Th(\mathbb{N},+,\mathbb{P})$.

The first results in this area were conditional results under a conjecture of Schinzel. Let SH denote *Schinzel's hypothesis*, i.e. the following generalization of the twin primes conjectures:

(SH) *Let $f_1(x), f_2(x), \ldots, f_n(x)$ irreducible polynomials over $\mathbb{Z}$, each having a positive leading coefficient. If there is no prime p such that $p|f_1(x)f_2(x)\ldots f_n(x)$ then there are infinitely many integers x such that $[f_1(x) \in \mathbb{P} \wedge f_2(x) \in \mathbb{P} \wedge \ldots \wedge f_n(x) \in \mathbb{P}]$.*

Let **(DC)** denote *Dickson's conjecture*, which is (SH) restricted to linear polynomials. In 1981, Alan proved

$$(\mathrm{DC}) \Rightarrow Th_{\forall}(\mathbb{N},+,\mathbb{P}) \textit{is decidable.}$$

In 1983, Belyakov and Martyanov extended this result of Alan by showing

$$(DC) \Rightarrow Th_{\forall}(\mathbb{Z}, +, \mathbb{P}, 1) \textit{ is decidable.}$$

In 1993, Alan, T. Bateman, and C.G. Jockush [2] showed, using arguments from automata theory, a conditional result on a subtheory of $Th(\mathbb{N}, +, \mathbb{P})$, namely

$$(DC) \Rightarrow Th_2(\mathbb{N}, S, \mathbb{P}) \textit{ is decidable.}$$

So that multiplication is $(+, \mathbb{P})$-definable and hence

$$(SH) \Rightarrow Th(\mathbb{N}, S, \mathbb{P}) \textit{ is undecidable.}$$

Maurice Boffa briefly proves in the one-page paper [3] that

$$(SH) \Rightarrow Th(\mathbb{N}, +, \{ \text{ primes of the form } an+b \text{ for } a \perp b\}) \textit{ is undecidable.}$$

Patrick Cégielski, Yuri Matiyasevich, Denis Richard, and Maxim Vsemirnov decided to investigate on unconditional results around the same theory $Th(\mathbb{N}, +, \mathbb{P})$. In 1996 [10], they proved that

$$Th(\mathbb{N}, n \mapsto p^n, R) \text{ for } R \in \{\times, |, \perp\} \textit{ are undecidable.}$$

Now remember that

$$p_n = nlog(n) + n(log(log(n) - 1) + O(nlog(log(n))/log(n)).$$

In [13], they proved the following unconditional theorem

The three theories $Th(\mathbb{N}, +, n \mapsto nlog(n) + n(log(log(n) - 1))$, $Th(\mathbb{N}, +, n \mapsto p_n, n \mapsto rem(p_n, n))$, and $Th(\mathbb{N}, +, n \mapsto nf(n))$ are undecidable, for any function f "close" to the integer part (floor) of p_n/n.

Below, PP denotes the set of products of two primes, X the set of primary numbers (powers of a prime number), and $<_A$ the restricted natural order of $\mathbb{N}$ to A for a part A of $\mathbb{N}$.

R. Villemaire and Véronique Bruyère showed in 1992 that if $V_k(x)$ is the valuation of an integer for the (k+1)-th prime then $Th(\mathbb{N}, +, V_k)$ is decidable but $Th(\mathbb{N}, +, V_k, V_h)$ is undecidable for k and h multiplicatively independent integers. Alexis Bès [1] completed these results by showing that V_h can be replaced by any set of n-tuples h-recognizable by a finite automaton.

To complete this part of the problems on addition and primes first opened by Alan, we must mention the result of Françoise Maurin [21] showing that $Th(\mathbb{N}, <_P, \times)$ is decidable so that $Th(\mathbb{N}, <_P, \mathbb{P})$ is decidable although $Th(\mathbb{N}, <_{PP}, \mathbb{P})$ and $Th(\mathbb{N}, <_X, \mathbb{P})$ are undecidable [4].

5 Rudimentary predicates, Spectra, and Rudimentary languages of first and second order

Alan Woods, in chapter 2 of his thesis, considered the problem of defining addition and multiplication on the natural numbers by first order bounded quantifier formulas involving only the predicate of coprimeness ($x \perp y$) and the natural order predicate ($\leq$).

He also proved as an application that the class RUD of all rudimentary (i.e. Δ_0-definable) sets of positive integers is contained in the class consisting of the spectra $S_\Phi = \{|M| \textit{ such that } M \textit{ is finite and } M \models \Phi\}$ of those sentences Φ having only graphs as models.

The equality between these classes holds if and only if S_Φ is rudimentary for every first order sentence Φ.

An analogous result holds for partial orderings instead of graphs. Alan noted that if S_Φ is rudimentary for every first order sentence Φ then we have NP $\neq$ co-NP. It is worth noting that Alan was often concerned with key-questions.

This part of Alan's thesis inspired a lot of logicians and computer scientists. In the team of Denis Richard, Alex Esbelin and Malika More stubbornly worked to make some links with *Grzegorczyk small classes* (Esbelin), to investigate *binary spectra by explicit polynomial transformations of graphs* (More), to give tools for recognizing rudimentary predicates (Esbelin and More) and also to extend results to second order languages (More and Frédéric Olive).

In his thesis [17], Alex Esbelin studied the problem of equality of RUD with the first class $(E^0)^*$ containing the projections, the constant functions, the successor function and closed under composition and bounded recursion. He investigated two specific cases of $(E^0)^*$:

1) the over-rudimentary classes, the associated functions of which contain RUD (it is possible somehow counting in all these classes) and he has found results of derecursivation;
2) the sub-rudimentary classes, the associated classes of relations which are strictly contained in RUD and he solved the problem of derecursivation for these classes. Roughly speaking, derecursivation consists in finding the complete definability of the values of a function or of a relation from the variables and the parameters occurring in these without using the history of the computation, hence without referring to recursion.

The question of whether a given primitive recursive relation is rudimentary is in some cases difficult and is related to several well-known open questions in theoretical computer science. In [16], Alex Esbelin and Malika More present systematic tools to study this question, with various applications. Namely, they give rudimentary definitions of various classical spectra. One of them gives a sufficient condition for the collapsing of the first classes of Grzegorczyk's hierarchy. Also they prove there exist one-to-one functions with rudimentary graph that provide representations of various sets of spectra into RUD. Their results are obtained via arithmetical coding of finite structures by a transformation of the first-order sentences into rudimentary formulas describing the same semantic properties.

In [22], Malika More and Frédéric Olive make conspicuous the equivalence between three notions coming, respectively, from the theory of recursive functions and relations, from complexity of computing and, finally, from (finite) model theory. On one hand, we know rudimentary languages are characterized by the linear hierarchy. On the other hand, we can prove this complexity class corresponds to second-order monadic logic with addition. The authors bring to light the deep links between those distant topics by providing a straightforward logical characterization of the rudimentary languages and also a result of representation of the second-order logic within those languages. (They use arithmetical tools for doing that).

6 Erdös-Woods numbers

About the works of Alan, Gordon Royle said in *SymOmega* on the 18th of December 2011: "*Mathematically, his name is attached to at least one concept the Erdös-Woods numbers are the integer numbers k such that there is a closed interval of integers $[a, a+k]$ with the property that every integer in the interval has a factor in common with either a or $a+k$*".

As we have already seen, many interesting problems in number theory emerge from the thesis of Alan Woods [35]. The most famous of them is now known as *Erdös-Woods conjecture*, after its publication in the book [19] of Richard Guy.

Erdös-Woods conjecture. *There exists an integer k such that integers x and y are equal if and only if, for $i = 0, \ldots, k$, integers $x+i$ and $y+i$ have same prime divisors.*

This problem is a source of an active domain of research.

In relation with this problem, Alan Woods had conjectured [35, p. 88] that for any ordered pair (a, d) of natural numbers, with $d \geq 3$, there exists a natural number c such that $a < c < a + d$ and c is coprime with a and with $a + d$. In other words

$$\forall a, \forall d > 2, \exists c\, [a < c < a + d \wedge a \perp c \wedge c \perp a + d]$$

Very quickly, he realized the conjecture was false, by finding the counterexample (2184, 16) (published in [14]). In 1989, David Dowe [14] proved that there exist infinitely many such numbers d. We call such numbers d *Erdös-Woods numbers*.

The main aim of the paper [9] is to prove that the set of Erdös-Woods numbers is recursive.

A second aim is to give the first values of Erdös-Woods numbers and to show there are a lot of natural open problems concerning these numbers.

Notation Let us denote by $NoCoprimeness(a, d)$ the property

$$\forall c\,[a < c < a + d \rightarrow \neg(a \perp c) \vee \neg(c \perp a + d)]$$

Let us begin with some remarks.

Remarks

(1) The relation $NoCoprimeness(a, d)$ is recursive: it is easy to write a program to check whether an ordered pair belongs to it.

(2) The set $\{(a, d) \mid NoCoprimeness(a, d)\}$ is infinite: we know an element (2184, 16) of this set and it is easy to check that for every $k \neq 0$, the ordered pair $(2184, 2184 + k.2.3.5.7.11.13)$ is also an element of this set.

(3) The two unary relations defined by

$$ExtremNoCoprime(a)\ iff \exists d\ NoCoprimeness(a, d)$$

$$AmplitudeNoCoprime(d)\ iff \exists a\ NoCoprimeness(a, d)$$

projections of $NoCoprimeness(a, d)$, are recursively enumerable: it is easy to write a program to list elements of these sets (but not in natural order, unfortunately).

First elements of *AmplitudeNoCoprime* Theoretical reasons allow us to implement an algorithm (in language C) to compute the first elements of the set *AmplitudeNoCoprime*. The algorithm is not very efficient, but it allows to test quickly the first six hundred integers. We obtain the beginning of the set *AmplitudeNoCoprime*:

16; 22; 34; 36; 46; 56; 64; 66; 70; 76; 78; 86; 88; 92; 94; 96; 100; 106;112; 116;118; 120; 124; 130; 134; 142; 144; 146; 154; 160; 162; 186;190; 196; 204; 210; 216; 218; 220; 222; 232; 238; 246; 248; 250; 256;260; 262; 268; 276; 280; 286; 288; 292; 296; 298; 300; 302; 306; 310;316; 320; 324; 326; 328; 330; 336; 340; 342; 346; 356; 366; 372; 378.

On odd elements of *AmplitudeNoCoprime* Dowe has found [14] an infinite subset of *AmplitudeNoCoprime*, every element being even. He conjectures every element of *AmplitudeNoCoprime* is even. Marcin Bienkowski, Mirek Korzeniowski, and Krysztof Lorys, from Wroclaw University (Poland), have found the counterexamples $d = 903$ and 2545 by computation, then a general method to generate many other examples: 4533, 5067, 8759, 9071, 9269, 10353, 11035, 11625, 11865, 13629,

15395, ... Nik Lygeros, from Lyon 1 University (France), has independently found the counterexample $d = 903$, making precise the related extremity:

a = 9 522 262 666 954 293 438 213 814 248 428 848 908 865 242 615 359 435 357 454 655 023 337 655 961 661 185 909 720 220 963 272 377 170 658 485 583 462 437 556 704 487 000 825 482 523 721 777 298 113 684783 645 994 814 078 222 557 560 883 686 154 164 437 824 554 543 412 509 895 747 350 810 845 757 048 244 101 596 740 520 097 753 981 676 715 670 944 384 183 107 626 409 084 843 313 577 681 531 093 717 028 660 116 797 728 892 253 375 798 305 738 503 033 846 246 769 704 747 450 128 124 100 053 617.

Lygeros found other values ($d = 907$ and 909), proving that the previous sufficient condition is not necessary. Also he discovered that the solution $d = 903$ is an old result from Erdös and Seldfridge [15].

On even squares of *AmplitudeNoCoprime* We see, in scanning the above list, that every even square except 4 appears at the beginning. However $676 = 26 \times 26$, $1156 = 34 \times 34$ and $1024 = 32 \times 32$ are not Erdös-Woods numbers.

On prime elements of *AmplitudeNoCoprime* An Erdös-Woods number may be a prime number as 15 493 and 18 637 show.

Open problems

The above list of first elements of *AmplitudeNoCoprime* suggests a great number of open problems, curiously similar to problems for the set of primes.

We may implement a program to compute, for an Erdös-Woods number d, the smallest associated extremity a. Numerical experiments suggest that $2 \perp a + 1$ whenever the amplitude d is even, hence 2 divides a. Is it a general property?

The solution $(a, 903)$, with the a found by Nik Lygeros, shows it is not the case for d odd.

Open problem 1. (**Even extremity for even amplitude**) *For an even d, is every element a such that $NoCoprimeness(a, d)$ even?*

We may note we have a great number of twin Erdös-Woods numbers among the first elements of *AmplitudeNoCoprime*: 34 and 36, 64 and 66, 76 and 78, 86 and 88, 92 and 94, ...

Open problem 2. (**Infinity of twin Erdös-Woods numbers**) *There exists an infinity of integers d such that d, $d + 2$ belong to AmplitudeNoCoprime.*

Indeed we also have a sequence of three consecutive even Erdös-

Woods numbers (as 92, 94, 96), even four consecutive ones (as 216, 218, 220, 222).

Open problem 3. (**Polignac's conjecture**) *For any integer k, there exists an even integer d such that $d, d+2, d+4, \ldots, d+2.k$ belong to AmplitudeNoCoprime.*

Nik Lygeros has searched segments of consecutive natural numbers which are not Erdös-Woods numbers. He has found long such segments.

Open problem 4. *There exist segments of any length without elements of AmplitudeNoCoprime:*

$$\forall k, \exists e \; [e, e+1, \ldots, e+k \notin \mathit{AmplitudeNoCoprime}]$$

Passing from patterns in *AmplitudeNoCoprime* to complexity, we may remark that the algorithm we have given to decide whether an integer belongs to *AmplitudeNoCoprime* is worst than exponential. It is interesting to improve it if it is possible.

Open problem 5. (**Complexity of** *AmplitudeNoCoprime*) *To which complexity classes does AmplitudeNoCoprime belong?*

Open problem 6. *Find a lower bound for AmplitudeNoCoprime.*

Also we may ask questions à la Vallée-Poussin–Hadamard, passing from the complexity to the density of the set *AmplitudeNoCoprime*. Let us denote by $d(n)$ the cardinality of the set

$$\{d \leq n \mid AmplitudeNoCoprime(d)\}.$$

Open problem 7. *Find a (simple) function f such that $d(n) \sim f(n)$.*

Open problem 8. *Is the density of AmplitudeNoCoprime linear? In a more precise statement*

$$d(n) = O(n)?$$

The last open problems we suggest concern Logic, more precisely Weak Arithmetics. Problems of definability and decidability are important: Presburger's proof of decidability for the elementary theory of $<\mathbb{N}, +>$ implies that a set $X \subseteq \mathbb{N}$ is definable in $<\mathbb{N}, +>$ iff X is ultimately periodic. The negative solution given by Matiyasevich to Hilbert's Tenth Problem relies on the fact that the exponential function is existentially definable in $<\mathbb{N}, +, \times>$. In the same way, the following problems deserve consideration.

Open problem 9. *Is the theory $Th(\mathbb{N}, NoCoprimeness, R)$ decidable? where R is some relation or function to specify. (Addition + is an interesting candidate).*

At the opposite, we may search for undecidability.

Open problem 10. *Is the theory $Th(N, +, AmplitudeNoCoprime)$ def-complete (i.e. is multiplication definable in the underlying structure)?*

7 Is Nézondet's method a beginning to an answer to EWC?

Via the whole theory of p-destinies, Francis Nézondet proved a surprising result which is too complicated to discuss here. Roughly speaking, Nézondet constructs in a functional language (not relational) a non-standard structure, the theory which is very like that of the standard model and in which EWC is false.

Last open problems

Using p-destinies, Marcel Guillaume, Jilei Yin and Denis Richard have tried to construct an algorithm for deciding $Th_3(\mathbb{N}, S, \perp)$. It turns out that two sets are the keys of this construction, namely

$$X = \{a \in \mathbb{N} \mid \exists b \, (Supp(a) \subseteq Supp(a) \wedge Supp(b) \neq Supp(a) \\ \wedge \neg(b-1 \perp a) \wedge \neg(b+1 \perp a)\}$$

$$Y = \{a \in \mathbb{N} \mid \exists b \, (Supp() \subseteq Supp(a) \wedge Supp(b) \neq Supp(a) \\ \wedge (b-1 \perp a) \wedge \neg(b+1 \perp a)\}$$

It can be proved that $a \in Y \cap (2\mathbb{N}+1)$ iff $\exists p \in Supp(a), \exists q \in Supp(a)$ ($ord_p(q)$ *is even*), where $ord_p(q)$ is the order of the element q within the group $(\mathbb{Z}/q\mathbb{Z})*$.

The construction of the essential 3-destiny of $Th_3(N, S, \perp)$ depends on the answer (yes or no) to the following nine questions

1) $\exists a = (2^n - 1) \wedge |Supp(a)| = 2 \wedge a \notin Y$
   Answer: YES with $a = 2^{227} - 1$ or $a = 2^{269} - 1$.
2) $\exists a = 2(2^n - 1) \wedge |Supp(a)| = 3 \wedge a \notin Y$
   Answer: YES with $a = 2(2^{269} - 1)$.
3) $\exists a = (2^n - 1) \wedge |Supp(a)| \geq 3 \wedge a \notin Y$
   Answer: OPEN
4) $\exists a = (2^n + 1) \wedge n \in 2\mathbb{N} \wedge |Supp(a)| = 2 \wedge a \notin Y$
   Answer: OPEN
5) $\exists a = (2^n + 1) \wedge n \in 2\mathbb{N} \wedge |Supp(a)| \geq 3 \wedge a \notin Y$
   Answer: OPEN
6) $\exists a = (2^n - 1) \wedge n \in 2\mathbb{N} + 1 \wedge |Supp(a)| \geq 3 \wedge a \in Y \setminus X$
   Answer: YES with $2^{25} - 1 = 31.601.1801$
7) $\exists a = (2^n - 1) \wedge n \in 2\mathbb{N} \wedge |Supp(a)| \geq 3 \wedge a \notin X$
   Answer: NO

8) $\exists a = (2^n + 1) \wedge n \in 2\mathbb{N} + 1 \wedge |Supp(a)| \geq 3 \wedge a \notin X$
   Answer: OPEN

9) $\exists a = (2^n + 1) \wedge n \in 2\mathbb{N} \wedge |Supp(a)| \geq 3 \wedge a \notin Y \backslash X$
   Answer: OPEN

Conclusion

We believe neither in paradise nor in hell. But as Sartre puts it in *Les mots*, authors write with the aim of some kind of survival. In the case of mathematicians, we cannot survive only by our theorems and works, but also, and above all, by our questions. Questions of Alan Robert Woods are so pertinent, deep and difficult that we guess he will last very long.

References

[1] Alexis Bès, *Problèmes de définissabilité arithmétique, Triangles de Pascal, arithmétique de Presburger et arithmétique de Skolem*, Thèse de doctorat, Université Paris 7, 1996.

[2] P.T. Bateman, C.G. Jockusch, and A.R. Woods, Decidability and undecidability of theories with a predicate for the primes, *The Journal of Symbolic Logic*, vol. 58, 1993, pp. 672–687.

[3] Maurice Boffa, More on an undecidability result of Bateman, Jockusch and Woods, *The Journal of Symbolic Logic*, vol. 63, 1998, p. 50.

[4] Alexis Bès and Denis Richard, Undecidable Extensions of Skolem Arithmetic, *The Journal of Symbolic Logic*, vol. 63, 1998, pp. 379–401.

[5] R. Balasubramian, T.N. Shorey, and Michel Waldschmidt, On the maximal length of two sentences of consecutive integers with the same prime divisors, *Acta Mathematica Hungarica*, vol. 54, 1989, pp. 225–236.

[6] Patrick Cégielski, Definability, decidability and complexity, *Annals of Mathematics and Artificial Intelligence*, vol. 111, 1996, pp. 311–341.

[7] Patrick Cégielski, Decidability and p-destinies, *Journal of Mathematical Sciences*, vol. 130, 2005, pp. 4620–4623.

[8] Patrick Cégielski and Malika More, Foreword – Weak Arithmetics, *Theoretical Computer Science - B*, vol. 322, 2004, pp. 1–3.

[9] Patrick Cégielski, François Heroult, and Denis Richard, On the amplitude of intervals of natural numbers whose every element has a common prime divisor with at least an extremity, *Theoretical Computer Science*, vol. 303, 2003, pp. 53–62.

[10] Patrick Cégielski, Yuri Matiyasevich, and Denis Richard, Definability and decidability issues in extensions of the integers with the divisibility predicate, *The Journal of Symbolic Logic*, vol. 61, 1996, pp. 515–540.

[11] Patrick Cégielski and Denis Richard, Indécidabilité de la théorie des entiers naturels munis d'une énumération des premiers et de la divisibilité, *Comptes Rendus de l'Académie des Sciences*, Série 1, Mathématique, vol. 315, 1992, pp. 1431–1434.

[12] Patrick Cégielski, Jean-Pierre Raoult, and Denis Richard, Colloque National sur les Recherches en IUT, *Quadrature*, vol. 24, 1996.

[13] Patrick Cégielski, Denis Richard, and Maxim Vsemirnov, On the additive theory of prime numbers, *Fundamenta Informaticae*, vol. 81, 2007, pp. 83–96.

[14] D. L. Dowe, On the existence of sequences of co-prime pairs of integers, *Journal of the Australian Mathematical Society*, Series A, vol. 47, 1989, pp. 84–89.

[15] P. Erdös and J. L. Selfridge, Complete prime subsets of consecutive integers, *Proceedings of the Manitoba Conference on Numerical Mathematics*, 1971, p. 114.

[16] Henri-Alex Esbelin and Malika More, Rudimentary relations and primitive recursion: a toolbox, *Theoretical Computer Science*, vol. 193, 1998, pp. 129–148.

[17] Henri-Alex Esbelin, *Relations rudimentaires et petites classes de fonctions recursives*, Thèse de doctorat, Université d'Auvergne, 1994.

[18] O. Eterevski and Maxim Vsemirnov, On the number of prime divisors of higher-order Carmichael numbers, *Fibonacci Quarterly*, vol. 42, 2004, pp. 141–148.

[19] Richard K. Guy, *Unsolved Problems in Number Theory*, Springer, Berlin, 1981, xviii + 161 pp.

[20] Michel Langevin, Autour d'un problème d'Erdös et Woods, (Dédié au professeur Paul Erdös pour son 75-ième anniversaire), 1988 in

U.A. 763 *Problèmes Diophantiens*, I.H.P. 11, rue P. et M. Curie 75231 Paris cedex 05.

[21] Françoise Maurin, The theory of integer multiplication with order restricted to primes is decidable, *The Journal of Symbolic Logic*, vol. 62, 1997, pp. 123–130.

[22] Malika More and Frédéric Olive, Rudimentary languages and second-order logic, *Mathematical Logic Quarterly*, volume 43, 1997, pp. 419–426.

[23] Malika More, *Définissabilité dans les graphes finis et en arithmétique bornée*, Thèse de doctorat, Université d'Auvergne, 1994.

[24] Malika More, Investigation of Binary Spectra by Explicit Polynomial Transformations of Graphs, *Theoretical Computer Science*, vol. 124, 1994, pp. 221–272.

[25] Francis Nézondet, Indépendance de la conjecture d'Erdös sur une théorie arithmétique, unpublished paper, presented at JAF (*Journées sur les Arithmétiques Faibles*), Paris, 1996.

[26] Francis Nézondet, *p-destinées et applications aux structures arithmétiques avec successeur et coprimarité*, Thèse de doctorat, Université d'Auvergne, 1997.

[27] Frédéric Olive and Malika More, *Caractérisation de RUD par la logique monadique*, unpublished presentation at JAF (*Journées sur les Arithmétiques Faibles*), Paris, 1996.

[28] Jean-François Pabion and Denis Richard, Synonymy and re-interpretation for some sublanguages of Peano arithmetic, *Open Days in Model Theory and Set Theory*, W. Guzicki, W. Marek, A. Pelc, and C. Rauszer, eds, Proceedings of a conference held September 1981 at Jadwisin, near Warsaw, Poland, Leeds University Press.

[29] Denis Richard, La théorie sans égalité du successeur et de la coprimarité des entiers naturels est indécidable. Le prédicat de primarité est définissable dans le langage de cette théorie, *Comptes Rendus de l'Académie des Sciences*, Paris, t. 294 (25 janvier 1982).

[30] Denis Richard, All arithmetical sets of powers of primes are first-order definable in terms of the successor function and the coprimeness predicate, *Discrete Mathematics*, vol. 53, 1985, pp. 221–247.

[31] Denis Richard, Equivalence of some questions in mathematical logic with some conjectures in number theory, *Number Theory and applications*, Richard Mollin ed., NATO ASI Series, Series C, Mathematical and Physical Sciences - Vol. 265, 1988, pp. 529–545.

[32] Julia Robinson, Definability and decision problems in arithmetic, *The Journal of Symbolic Logic*, vol. 14, 1949, pp. 98–114. Reprinted in Feferman, Solomon ed., *The Collected Works of Julia Robinson*, American Mathematical Society, 1996, 338 p.

[33] Thoralf Skolem, Über einige satzfunktionen in der Arithmetik, Videnskabselskabets i Kristiana Skriften, I. Matematisk-naturwidenskabelig Klasse No. 7, Oslo, 1930. Reprinted in *Selected Works in Logic*, ed. J. E. Fendstad, Universitets-forlag, Oslo, 1970, pp. 281–306. French translation by Claude Richard, 1979.

[34] Maxim Vsemirnov, The Woods-Erdös conjecture for polynomial rings, *Annals of Pure and Applied Logic*, vol. 113, 2002, pp. 331–344.

[35] Alan Robert Woods, *Some problems in logic and number theory, and their connections*, Ph.D. Thesis, University of Manchester, 1981. Part II of this volume.

3

Degrés Ludiques : une introduction

CHRISTOPHE CHALONS, JEAN-PIERRE RESSAYRE[1]

1 Introduction

Le préordre et les degrés ludiques, introduits en 1999 dans Chalons [10], sont à la Mécanique Quantique ce que le préordre et les degrés de Tukey sont à la Topologie : les degrés ludiques classifient les expériences de Mécanique Quantique comme les degrés de Tukey classifient les espaces topologiques. Par ailleurs les degrés de Tukey se plongent dans les degrés ludiques, qui enrichissent encore toute la problématique des premiers.

Les résultats ont été complétés en 2012 par Anatole Khélif [7] et Chalons [11]. La présente introduction à ce domaine donne les idées et résultats de base, donne ou résume leurs preuves et renvoie aux travaux cités pour des compléments.

Les deux auteurs expriment leur infinie gratitude à Anatole Khélif. Un grand merci également aux éditeurs scientifiques, dont la diligence a permis au manuscrit initial d'être pris en considération malgré ses défauts formels. Et à Olivier Finkel dont l'aide et la patience inépuisable ont sauvé le second auteur du naufrage informatique.

Les Degrés Ludiques sont un peu éloignés des Arithmétiques Faibles, mais le mélange de sujets réalisé par la présentation des premiers dans le cadre des JAF est souhaitable par principe : il faut lutter contre le cloisonnement des approches et domaines mathématiques. De manière plus concrète voici un rapprochement entre le présent article et

[1]Université Paris 7, ressayre@logique.jussieu.fr

Studies in Weak Arithmetics.
Patrick Cégielski, Charalampos Cornaros,
Costas Dimitracopoulos.

les Arithmétiques Faibles. Un de leurs thèmes est la recherche de la plus faible méta-théorie permettant de formaliser un théorème ou un domaine donné des mathématiques : la même question se pose avec (en place du domaine mathématique) une partie de la Physique telle que la Mécanique Quantique ("MQ"). Une réponse partielle est que la théorie WKL_0 introduite par Friedmann suffit à formaliser une grande partie de la MQ (étant entendu que, dans la pratique, ce serait très fastidieux). Mais le résultat principal de cet article (Théorème 62) pointe vers une exception de taille. En effet ce théorème est étroitement lié à la MQ (voir Sections 4 et 5), mais la seule preuve du théorème que nous connaissions nécessite l'axiome du choix ainsi que l'Hypothèse du Continu ou, à sa place, le lemme de Ketonen. Ce qui nous fait passer d'une Arithmétique WKL_0 bien plus faible que celle de Peano à une Théorie des ensembles sans comparaison plus forte.

Il nous paraît donc une bonne idée de conjuguer les deux approches, et nous remercions les organisateurs des JAF, pour leur invitation qui en offre l'occasion.

1.1 2-Réductions

Tout part de la notion de 2-réduction entre des relations binaires :

Définition 1. *Soient* E, F, E', F' *des ensembles et* $R \subset E \times F$, $S \subset E' \times F'$ *deux relations binaires.*

Une 2-réduction de R *à* S *est un couple de fonctions* $(h(x), g(x, y))$ *tel que* $\forall x \forall y\ [S(h(x), y) \Rightarrow R(x, g(x, y))]$

• Les domaines de x et y ainsi que les ensembles de départ et d'arrivée des fonctions h et g doivent vérifier les conditions évidentes pour que la définition ait un sens : puisque x apparaît comme premier argument de R, nous voulons $x \in E$; et puisque x est l'argument de h nous voulons aussi E comme domaine de h. De proche en proche nous voulons finalement : $x \in E$, $y \in F'$, h va de E vers E' et g va de $E \times F'$ vers F.

• De telles conditions - à supposer ou à démontrer, suivant les cas - seront sous-entendues et laissées systématiquement aux soins du lecteur: étant manifestement nécessaires pour que les raisonnements exposés fassent sens, elles sont induites par une gymnastique intellectuelle utile et facile ; de sorte que leur omission est souhaitable pour aérer l'exposé.

• Formellement, une relation binaire est un triplet (E, F, R) avec $R \subset E \times F$; mais par abus nous dirons : "une relation $R \subset E \times F$".

• On écrit parfois $R \leq S$ pour "il existe une 2-réduction de R à S".

Commentaire heuristique.

• Une première interprétation intuitive de la 2-réduction sera utile dans le contexte des Sections 3 (degrés de Tukey) et 4 (degrés ludiques) : on suppose disposer de "téléphones" qui transmettent les messages en les déformant - si on émet un message $x \in E$, le message reçu $y \in F$ peut différer de x. Un tel téléphone T est dit **R-garanti** si chaque fois qu'on a donné x et que T a transmis y, on est sûr au moins que $R(x, y)$ est vrai ; autrement dit, par exemple si R est l'identité, T est R-garanti si sa transmission est toujours exacte... Alors une 2-réduction h, g de R à S donne un moyen de fabriquer un téléphone T' qui est R-garanti à partir d'un téléphone T qui est S-garanti : x étant donné, T' applique T au message $h(x)$; comme T est S-garanti, le message reçu y satisfait $S(h(x), y)$. Alors $R(x, g(x, y))$ est vrai, de sorte que si $g(x, y)$ est le message transmis par T', T' est R-garanti. La 2-réduction de R à S désigne donc ce moyen de construire des téléphones R-garantis à partir de téléphones S-garantis.

• L'alinéa qui précède se sert des mots "intuitif" et "heuristique" ; dès que l'un de ces mots est utilisé cela signifie pour nous deux choses : a) l'alinéa ne fait pas partie des définitions, preuves et théorèmes de notre article - si on l'enlève ou qu'on en saute la lecture, le texte qui reste est mathématiquement auto-suffisant. b) L'alinéa est là pour le lecteur qui souhaite compléter les développements mathématiques par des considérations intuitives.

• Comme la pratique (a+b) sera systématique par la suite, elle pourrait brouiller la frontière entre le texte mathématique auto-suffisant et les commentaires heuristiques et/ou intuitifs. Mais voici la délimitation précise du texte proprement mathématique : i) il comprend d'abord les définitions et les énoncés de lemmes, propositions et théorèmes ; ce sont exactement les phrases *en italique* de l'article. ii) Il comprend d'autre part les preuves ; chacune commence par "Preuve", et le symbole ⊣ en délimite la fin.

• Certaines de nos considérations intuitives pourraient être formalisées, et leur complexe statut épistémologique pourraît être précisé ; mais nous ne le ferons pas. Au lecteur qui nous reprocherait cette timidité, nous présentons trois excuses : 1, correctement fait, cela poserait de gros problèmes de taille de l'article. 2, dans le cadre du présent article cela n'apporterait aucun résultat de plus. 3, nous proposons au lecteur de chercher une définition mathématique de la notion de "téléphone R-garanti" : "R-garanti", pas de problème, mais... ce "téléphone", qu'est-ce à part la relation R elle-même ? Pour l'étude de la MQ, il conviendrait que ce soit : "un dispositif de transmission d'information qui existe ou pourrait exister, dans un Monde Physique qui lui même existe ou pourrait exister". Chapeau à celui qui formalisera cette notion !

Une autre interprétation de la notion de 2-réduction est la suivante : pour décider si $\exists y R(x, y)$ est vrai, il faut toute une recherche d'un "témoin de $\exists y R(x, y)$", c'est-à-dire d'un $y \in E$ tel que $R(x, y)$; la fonction g dispense de cette recherche, si on dispose d'un témoin r pour $\exists y' S(h(x), y')$. Car $g(x, r)$ est alors un témoin de $\exists y R(x, y)$.

Autrement dit, $R \leq S$ signifie : "si on a des témoins pour $\exists y' S$, alors (grâce aux deux fonctions de la 2-réduction) il s'en déduit des témoins pour $\exists y R$". Cette deuxième façon de voir la 2-réduction intervient en Complexité Algorithmique ; dans ce contexte, la notion primordiale de réduction est celle de 1-réduction : une 1-réduction de l'ensemble (ou relation unaire) X à l'ensemble Y est une fonction f telle que $X = f^{-1}(Y)$; pour que cette définition fasse sens, il convient que f aille d'un domaine E contenant X vers un co-domaine E' contenant Y. Alors l'application f ramène le problème de décider (pour $x \in E$) si $x \in X$ au problème de décider (pour $y \in E'$) si $y \in Y$.

Quand on demande à la fonction f d'être effective, cette notion de 1-réduction est au coeur de la Complexité Algorithmique ; et le cœur de ce cœur est le cas où on exige que f soit calculable en temps déterministe polynomial : la 1-réduction en temps polynomial amène à distinguer la classe P, la classe NP et les ensembles NP-complets ; elle nous guide vers les problèmes centraux de la Complexité Algoritmique tels P=?NP. Mais la 2-réduction s'immisce par la même occasion dans le sujet ; en effet les mille et un résultats de NP-complétude qui montrent l'importance pratique du problème P=?NP se démontrent tous en utilisant la 2-réduction. Car chaque fois qu'on a besoin de 1-réduire l'un à l'autre deux ensembles X, Y de la classe NP, on met X, Y sous la forme $\exists y R, \exists y S$ - où R, S sont des relations binaires de la classe P - et on construit explicitement une 2-réduction (h, g) de R à S en temps déterministe polynomial, telle que $\exists y R(x, y) \Rightarrow \exists y S(h(x), y)$; on conclut que h 1-réduit X à Y. En effet, supposons $h(x) \in Y$, d'où un y tel que $S(h(x), y)$; alors $R(x, g(x, y))$ donc $x \in X$. Inversement supposons $x \in X$ donc $\exists y R(x, y)$, d'où un y tel que $S(h(x), y)$, par hypothèse sur h ; alors $R(x, g(x, y))$ donc $x \in X$.

Par ce biais, la 2-réduction s'est invitée au cœur de l'Informatique; mais on peut aussi étudier la 2-réductibilité en temps polynomial pour elle-même. Toutefois notre intérêt pour cette étude est tempéré :

- certes, elle pose des problèmes mathématiques et informatiques profonds (i)

- mais ces problèmes ont tendance à être **trop** difficiles pour faire l'objet d'une théorie qui avance rapidement (ii)

Fait à l'appui de (i+ii) : notons $Vrai$ la relation binaire toujours vraie, et $Div(x, y)$ la relation "y est un diviseur de x, non trivial excepté si x est premier" ; la factorisation des entiers est faisable en temps déterministe polynomial si et seulement si Div est 2-réductible à $Vrai$ en temps polynomial.

On touche ainsi à des problèmes fondamentaux en théorie des nombres et en cryptographie ; et Div n'est qu'une seule des relations binaires de P à comparer aux autres (pas seulement à $Vrai$) par 2-réductibilité. Ainsi, dans ce cadre, l'étude de la 2-réductibilité est un peu marginalisée par sa difficulté, puisque un seul cas particulier d'une question parmi beaucoup d'autres excède déjà les capacités actuelles de la Science.

Nous allons donc étudier d'autres cadres où l'étude de la 2-réductibilité est plus accessible. Chacun de ces cadres est défini en fixant une classe bien choisie Φ de réductions $(h(x), g(x, y))$: pour R, S relations binaires on pose $R \leq_\Phi S$ si R se réduit à S par une de ces Φ-réductions. Dans tout cas intéressant (et sous des conditions évidentes), $\leq_\Phi$ est un préordre ; on appellera Φ-relations ses éléments. La relation d'équivalence associée est notée $\equiv_\Phi$; ses classes d'équivalences sont appelées les degrés de Φ-réductibilité ou Φ-degrés, elles sont munies du préordre induit, noté de la même façon. Et il y a un degré minimum noté 1_Φ : celui de "la" relation toujours vraie (de n'importe quel domaine non vide). Il y a un degré maximum noté ∞_Φ : c'est le Φ-degré de la relation toujours fausse.

Aux notations qui précèdent il convient d'ajouter la distinction entre le cas dit **externe**, où si une relation $R \subset E \times F$ est une Φ-relation alors toute relation $R' \subset E \times F$ en est une également ; et le cas **interne**, à l'opposé. Considérons, par exemple, le cas $\Phi = P$, comme ci-dessus : les Φ-réductions sont tous les couples (h, g) avec h, g calculables en temps polynomial. Il est alors naturel de se placer dans un cas interne en prenant comme P-relations seulement les relations binaires de la classe P ou celles de la Hiérarchie en temps polynomial, par exemple. Mais désormais tous les cadres Φ considérés seront externes.

Remarque.

- Le degré ∞_Φ est constitué par toutes les Φ-relations non totales ; en effet, si $\forall y \neg S(a, y)$, prendre pour h la fonction constante a et pour fonction g n'importe quelle fonction avec les bons ensembles de départ et d'arrivée permet de construire une 2-réduction (a, g) de n'importe quelle relation $R \in \Phi$ à S. Et les cadres Φ qui nous intéressent ne sont

pas idiots au point de ne pas inclure une Φ-réduction de cette forme.
• Le degré 1_Φ est constitué par toutes les relations $R \in \Phi$ qui sont totales et possèdent une fonction de choix (c'est-à-dire f tel que $\forall x R(x, f(x))$), qui est une Φ-fonction – on appelle ainsi toute fonction f faisant partie d'une Φ-réduction (h, g).

Nous passons maintenant en revue les cas remarquables de préordre $<_\Phi$ qui seront étudiés par la suite. On écrira $\leq, \infty, 1$ pour $\leq_\Phi, \infty_\Phi, 1_\Phi$ quand il n'y a pas risque de confusion.

Exemple trivial. : Φ est la collection de toutes les 2-réductions. Alors le préordre associé $\leq$ a pour degrés $\infty, 1$ uniquement.

Cet exemple pose la question préliminaire de rendre $\leq_\Phi$ non triviale sans que son étude se perde dans les nuages. Un moyen est d'abandonner toute exigence d'effectivité des Φ-réductions, mais de restreindre celles-ci par des conditions "**algébriques**", en un sens ou un autre. C'est ce que font les deux exemples suivants, dont le premier définit le préordre de Tukey et le second le préordre ludique.

1.2 Exemple de la Section 3 : préordre et degrés de Tukey

C'est le préordre noté $\leq_T$ qui prend pour Φ-réductions tous les couples de la forme $(h(x), g(y))$. Autrement dit g ne dépend plus que de son **deuxième** argument. Le préordre de Tukey est déjà ancien [13] ; Tukey l'a introduit pour comparer finement les divers espaces topologiques.

Commentaire.

• On peut aussi le voir en termes intuitifs de téléphones R-garantis – comme déjà décrit pour le cas général de 2-réduction $h(x), g(x, y)$ de R à S, mais en supposant $g = g(y)$. Notez que pour rendre naturelle cette supposition restrictive on peut se placer dans le cadre de la Relativité et imaginer que le récepteur de chaque téléphone est très éloigné de son émetteur ; de sorte qu'il n'est pas possible de disposer en même temps du message émis x et du message reçu y. En particulier une fonction $g(x, y)$ n'est pas disponible, mais seulement : $g(y)$ (disponible à proximité du "récepteur téléphonique" qui reçoit y) ; ou $h(x)$ (disponible à proximité de "l'émetteur téléphonique" qui émet x).
• Clairement, le degré de Tukey de S est une mesure de la quantité d'information que permet de faire passer un téléphone S-garanti. Apparemment la réduction de Tukey n'était pas vue comme ça avant Chalons [10].

Sur le préordre de Tukey il y a une structure supplémentaire canonique définie plus loin : opérations de somme et de produits, ainsi que de borne supérieure et inférieure d'une famille de degrés. Ces opérations sont des plus naturelles, et pourtant l'existence des deux dernières n'était pas connue avant [10]. Le fait tient au point de vue uniquement topologique qui fut longtemps celui sur les degrés de Tukey ; car la définition des bornes supérieure et inférieure devient évidente une fois qu'on remplace la perspective topologique par celle de transmission d'information (telle qu'on vient de l'expliquer au moyen des "téléphones R-garantis").

La collection des cardinaux munis de leur ordre et des opérations de somme et produits se plonge canoniquement dans le préordre de Tukey: le plongement envoie le cardinal κ sur le degré noté I_κ de la relation $x = y$ $(x, y < \kappa)$. Et les degrés de Tukey, comme les cardinaux, mesurent des quantités – mais cette fois des quantités **d'information** ; c'est le point de vue que développera la Section 3.

Il y a aussi une opération de "dualité" sur les degrés de Tukey, qui est essentiellement celle de la complémentation (ou négation) d'une relation R :

Définition 2. *Soit $R \subset E \times F$, E et F ensembles quelconques ; on note R^{-1} et on appellera duale de R la relation* $\{(y, x) : \neg R(x, y)\}$.

C'est bien la négation, à la permutation des arguments x, y près ; cette permutation est nécessaire pour que le préordre de Tukey vérifie la propriété "$R < S$ implique $S^{-1} < R^{-1}$". Car si h, g est une 2-réduction de Tukey de R à S, grâce à la permutation de x, y et à l'équivalence entre "$A \longrightarrow B$" et sa contraposée "$\neg B \longrightarrow \neg A$", on a (g, h) comme réduction de S^{-1} à R^{-1} : de $S(h(x), y) \Rightarrow R(x, g(y))$ se déduit $\neg R(x, g(y)) \Rightarrow \neg S(h(x), y)$ soit $R^{-1}(g(x), y) \Rightarrow \neg S^{-1}(x, h(y))$.

On obtient ainsi que le degré de Tukey de R^{-1} ne dépend que de celui de R : $R \leq S \leq R$ implique $S^{-1} \leq R^{-1} \leq S^{-1}$.

1.3 Exemple de la Section 4 : degrés ludiques

Le préordre $\leq_L$ des degrés ludiques va au-delà du préordre de Tukey en permettant à nouveau des fonctions de réduction $g = g(x, y)$; mais en contrepartie il demande à h, g d'être des fonctions **locales** (une terminologie empruntée à la Mécanique Quantique, comme expliqué plus loin dans la "Parenthèse heuristique").

Ce caractère "local" est la condition de type algébrique qui sert à restreindre les fonctions de réduction de manière à donner naissance à un préordre dont l'étude soit ni triviale ni hors de portée. On va considérer des 2-réductions au moyen de fonctions $(h(x), g(x, y))$ dont les arguments x, y sont des κ – *uplets* pour un cardinal $\kappa > 0$. On dira que κ est l'**uplicité** de h, g (pour ne pas confondre avec leur arité qui est 1 ou 2).

Définition 3.

- *On dit qu'une fonction $f(x)$ est d'***uplicité** *κ si son ensemble de départ et d'arrivée sont des produits cartésiens de longueur κ.*
- *Cette fonction f est dite* **locale** *s'il existe des fonctions $f_i, i < \kappa$ telles que $f(x) = (f_i(x_i))_{i<\kappa}$. Ces notions sont étendues à une fonction de deux variables g en exigeant que $g(x, y) = (g_i(x_i, y_i))_{i<\kappa}$.*
- *Le* **préordre ludique** *$\leq_\kappa$ est le préordre $\leq_\Phi$ lorsque les Φ-réductions sont tous les couples $(h(x), g(x, y))$ de fonctions locales d'uplicité κ.*
- *κ est appelé l'***uplicité** *du préordre $\leq_\kappa$, et les* **degrés ludiques d'uplicité** *κ sont les classes d'équivalences modulo $\equiv_\kappa$.*
- *$\leq_L$ est le préordre réunion de tous les $\leq_\kappa$, dont les classes d'équivalences sont tous les degrés ludiques.*

Les degrés de Tukey se plongent dans ces degrés ludiques, qui enrichissent les degrés de Tukey d'une manière intéressante pour les topologues et théoriciens des ensembles, notamment. D'autre part, les degrés ludiques sont des objets mathématiques naturels dans le cadre de la Mécanique Quantique et constituent un outil pour l'approfondissement de cette dernière.

Si $\kappa = 1$ toute fonction d'uplicité κ est locale ; alors $\leq_\kappa = \leq_1$ est le préordre à deux éléments de l'Exemple trivial ci-dessus. Mais $\leq_\kappa$ a une classe propre de degrés, dès que $\kappa > 1$. En effet, on a vu plus haut que les degrés de Tukey forment une classe propre puisque s'y plonge la classe des Cardinaux ; or on indique ci-dessous un plongement simple et canonique du préordre de Tukey dans $\leq_2$.

Plongement des degrés de Tukey dans les degrés ludiques.- Il envoie la relation binaire $R \subset E \times F$ sur la relation $L_R \subset E^2 \times F^2$ définie par : $(x_0, x_1)L_R(y_0, y_1)$ si et seulement si ($y_0 = x_0$ et $R(x_0, y_1)$). (Notez que la variable x_1 n'apparait pas dans la formule qui définit L_R : x_1 ne sert que pour donner formellement l'uplicité 2 à la relation L_R) De cette manière on plonge le préordre de Tukey en préservant aussi ses opérations de somme, produit, borne supérieure et inférieure – si on étend canoniquement ces opérations.

Remarque 4. Le degré trivial 1 dans le cas de Tukey est constitué de toutes les relations R telles que $\exists b \forall x R(x, b)$; dans le cas ludique $\leq_2$, il comprend toutes les relations R qui admettent une fonction de choix **locale** : $\forall x_0, x_1\ (x_0, x_1)R(g_0(x_0), g_1(x_1))$. Mais le plongement de Tukey dans $\leq_2$ envoie bien 1 sur 1, et ∞ sur ∞.

En revanche on a clairement un problème pour préserver les bonnes propriétés de l'opération de dualité R^{-1} par ce plongement : pour montrer que $R \leq S$ implique $S^{-1} \leq R^{-1}$ dans le cas Tukey, en partant d'une réduction $(h(x), g(y))$ de R à S il suffisait de prendre $(g(x), h(y))$ comme réduction de S^{-1} à R^{-1} ; or dans le cas ludique, les deux fonctions h, g d'une réduction de R à S n'ont pas le même nombre d'arguments donc ne peuvent être permutées de cette manière. Mais une définition plus astucieuse de R^{-1} dans le cas ludique va rétablir la situation : appliquée à $\leq_2$, elle permet au plongement ci-dessus de préserver la dualité. (Elle préserve donc aussi des opérations telles que la somme duale, définie par $R +^{-1} S := (R^{-1} + S^{-1})^{-1}$)

La définition qui suit de la "dualité de Chalons" est donnée pour un cadre Φ au départ quelconque ; on appelle Φ-relations les relations dans le domaine de $\leq_\Phi$.

Définition 5. *Soit $R \subset E \times F$ une Φ-relation ; on note R^{-1} et on appellera duale de R la relation R^{-1} définie par : $R^{-1}(f, x)$ si et seulement si $\neg R(x, f(x))$ – où x varie sur E, f sur les Φ-fonctions : $E \mapsto F$.*

Sous des hypothèses naturelles sur Φ – qui sont vérifiées dans le cas ludique – on obtient pour $\leq := \leq_\Phi$:

Théorème 6. *$R \leq S$ implique $S^{-1} \leq R^{-1}$, et $(R^{-1})^{-1} \equiv R$.*

Le plongement ci-dessus : $R \mapsto L_R$ des degrés de Tukey dans les $<_2$-degrés préserve la dualité en plus des autres opérations, grâce à cette redéfinition de R^{-1} mais à la condition supplémentaire d'opérer une permutation des arguments de L_R.

Nous n'avons pas encore parlé de la Section 2, destinée à introduire la dualité de Chalons en l'étudiant dans un cadre un peu plus plus simple que celui des degrés ludiques ; il est temps de le faire.

1.4 Exemples de la Section 2 : préordres d'uplicité 1

Les deux Exemples précédents sont obtenus en mettant des restrictions de type algébrique sur la collection de toutes les Φ-réductions ; il y a un autre type de restriction fructueuse sur Φ : les restrictions ensemblistes. Elles consistent à restreindre les Φ-fonctions à des fonctions finies, ou effectives, ou définissables en divers sens. Ou à restreindre les Φ-réductions **plus fortement** que les Φ-relations, comme dans le Cas 1 ci-dessous.

Cas 1 Soit V un modèle fixé de ZFC (qui est un ensemble, pas une classe propre) ; prenons pour Φ la collection de tous les couples (h, g) tels que $h, g \subset V$. On se place dans le cas externe, donc si par exemple V est dénombrable, alors les V-réductions sont en nombre dénombrable pendant que les V-relations ont la puissance du continu ; ainsi le préordre $\leq_V$, loin d'être trivial comme $\leq_1$, a un continu de degrés. On verra que le théorème de dualité $R \equiv (R^{-1})^{-1}$ est vrai dans ce cadre $\leq_V$, pour la dualité de Chalons.

Cas 2 Une deuxième source fructueuse de restrictions ensemblistes (sur les Φ-réductions) consiste à employer des conditions de calculabilité ou définissabilité pour définir Φ. Pour mémoire, nous l'avons déjà fait en considérant le cas où Φ est l'ensemble des fonctions sur ω calculables en temps polynomial ; nous avions noté que ce cas est plutôt trop difficile. Ici signalons un autre défaut: la dualité de Chalons ne fait pas bon ménage avec l'effectivité (des fonctions de réductions) ; certes elle n'est pas incompatible avec des propriétés de définissabilité, mais celles que nous connaissons sont très loin de l'effectivité.

Pour examiner cela, écrivons $\leq_\omega$ pour $\leq_\Phi$ quand toutes les Φ-réductions sont des couples $(h(x), g(x,y))$ tels que chaque argument de h, g varie sur l'un des trois ensembles ω, ${}^\omega\omega$ et ${}^\omega({}^\omega\omega)$; de la sorte $(R^{-1})^{-1}$ est définie pour toute relation R sur ω. Pour que de plus $R \leq (R^{-1})^{-1}$ soit toujours vrai, il suffit que la classe des Φ-fonctions ait quelques propriétés de clôture simples : par composition, évaluation et opération λ. En effet nous obtiendrons une 2-réduction (H, G) de R à $(R^{-1})^{-1}$ qui est assez simple et canonique pour exister dans Φ sous ces seules conditions.

Mais pour que inversement $(R^{-1})^{-1} < R$ soit toujours vrai, il faut payer plus cher: nous obtiendrons une 2-réduction (σ, τ) de $(R^{-1})^{-1}$ à R où (σ, τ) sont des fonctionnelles Σ^1_2-définissables ; et à part cela nous n'avons que des 2-réductions non effectives (car basées sur une fonction de choix donc sur l'axiome du même nom). Or entre définissabilité Σ^1_2 et effectivité, il existe plus qu'une marge ! Cette distance illustre la difficulté de trouver des cadres Φ qui concilient intéret théorique avec solvabilité pratique. Toutefois ce cas Σ^1_2 de dualité offre deux consolations pour la non-effectivité ; la première est inattendue : les réductions (H, G) et σ, τ sont indépendantes de la relation R – ne dépendant que de son domaine ! La seconde consolation est que ces réductions conservent l'une des conséquences importantes de l'effectivité : leur **absoluité** - au sens du théorème de Schoenfield selon lequel sont absolues les formules Σ de théorie des ensembles à paramètres héréditairement dénombrables ; et les formules Σ^1_2 d'Arithmétique.

L'introduction plus haut de $\leq_T$ et $\leq_L$ a donné une première motivation pour les degrés ludiques : enrichir la perspective (topologique et de théorie des ensembles) fournie par les degrés de Tukey. Mais ceci ne donne pas encore la motivation principale des degrés ludiques.

Pour expliquer cette motivation il convient de dépasser le cadre des mathématiques pures, et d'utiliser – en complément et d'une manière heuristique – celui de la Physique.

1.5 Parenthèse heuristique : motivation des degrés ludiques par la Mécanique Quantique

Rappelons l'heuristique donnée en 1.1. de la 2-réduction de $R \subset E \times F$ à $S \subset E' \times F'$: on appelle téléphone R-garanti un dispositif qui à toute entrée $x \in E$ fait correspondre une sortie y telle que la relation $R(x, y)$ vaut toujours entre l'entrée et la sortie. Alors une 2-réduction f, g de R à S est un moyen de fabriquer un téléphone T qui est R-garanti à partir de n'importe quel téléphone U qui est S-garanti : T applique le téléphone U au message $h(x)$; comme U est S-garanti, le message y reçu satisfait $S(h(x), y)$. Alors $R(x, g(x, y))$ est satisfait, donc T est bien R-garanti si on veille à ce que $g(x, y)$ soit le message transmis par T.

Cette heuristique de la 2-réduction s'adapte entièrement au présent cas ludique, où R est une relation d'uplicité $\kappa \geq 1$ et h, g sont des fonctions locales ; car il suffit d'imaginer qu'un téléphone R-garanti T est constitué de κ **"combinés indépendants"** au sens suivant : si $x = (x_i)_{i<\kappa}$ est le message en entrée de T, le i-ième combiné ne reçoit que l'entrée x_i et sa sortie y_i ne dépend que de x_i. De cette manière, pour qu'une 2-réduction h, g de R à S permette de fabriquer un téléphone R-garanti T à partir de tout téléphone U qui est S-garanti, obligatoirement h, g doivent être des fonctions locales, autrement dit on ne considère que les 2-réductions permises par la relation $R <_\kappa S$.

Désormais nous ne regardons que le **cas fini** des degrés ludiques : relations binaires sur les domaines constitués d'ensembles finis d'uplicité finie. Et nous examinons ce cadre ludique dans divers Mondes Physiques possibles

1.5.1 Degrés ludiques et Relativité

Plaçons-nous dans le cadre de la Physique Relativiste (sous-entendu: **non** Quantique). Dans ce cadre restrictif, quelles sont les relations R telles qu'il est possible de "fabriquer" un téléphone R-garanti ? (Les guillemets autour de "fabriquer" attirent l'attention sur le fait que cette notion est initialement floue et que le sens en sera précisé par la réponse à la question).

Remarque 7. Dans le cadre de la Physique (**seulement**) Relativiste et pour toute relation binaire R représentant un degré ludique fini il y a équivalence entre (i) et (ii) :

(i) il existe un téléphone R-garanti (on peut le "fabriquer")

(ii) R admet une fonction de choix locale – c'est-à-dire

$$g = (g_0(x_0), g_1(x_1), g_2(x_2))$$

telle que $R(x, g(x))$ est toujours vrai.

Autrement dit le seul degré ludique de téléphones fabricables par la Relativité est le degré **trivial** 1_L.

(ii) implique (i) - Cela va de soi parce que toute fonction de domaine fini peut être réalisée ou simulée par un automate à mémoire finie, tout bêtement ; et un tel automate est toujours physiquement réalisable (dans le monde réel, si son cardinal est suffisamment petit ; et dans le monde de la Physique **théorique** Newtonienne, sinon).

(i) implique (ii) - Ici il faut préciser ce qu'on exige toujours d'un téléphone R-garanti : a) il doit fonctionner même si ses "combinés" sont placés à distance arbitraire les uns des autres ; b) une fois que l'entrée x_i du i-ième combiné est inscrite pour tout $i <$ uplicité de R, instantanément la sortie y_i doit apparaître sur tous ces combinés également; c) et bien entendu $R(x, y)$ doit valoir. On sait que les axiomes de la Relativité à eux seuls ne permettent aucune transmission d'information ou action instantanée à distance. Donc si T est R-garanti par la Relativité, la sortie y_i transmise par son i-ième combiné ne dépend que de x_i : il existe un ensemble $Y_i \subset F$ tel que les i-ièmes sorties possibles sont tous les éléments y_i de Y_i ; soit $f = (f_i(x_i))_i$ une fonction telle que $i <$ uplicité de R implique : pour tout $x_i \in X_i$ $f_i(x_i)$ est défini et $f_i(x_i) \in Y_i$. Alors f est la fonction de choix locale de R demandée.

Remarque 8. La relation $R \leq_n S$ implique que à partir d'un téléphone S-garanti T ("à n combinés indépendants") on peut "fabriquer" (dans le monde de la Physique newtonienne) un téléphone T' qui est R-garanti.

Notez que T n'est pas supposé "relativiste" : il se peut que la relation S qu'il garantit soit de degré non trivial ; c'est seulement le bricolage de T' à partir de T qui est relativiste et même newtonien. T, lui, peut être comme le "téléphone GHZ" ci-dessous, qui est "fabriquable" par la MQ. En fait et bien au-delà, rien n'empêche T de garantir une relation mathématique non effective sur un domaine infini. Et dans ce cadre élargi on a affaire à toute la structure des n-degrés ludiques.

1.5.2 Degrés ludiques dans un Monde Quantique

Mais que deviennent les remarques ci-dessus si la Relativité est remplacée par la MQ ? Dans ce cadre il existe une relation R pour laquelle on peut "fabriquer" un téléphone R-garanti, mais qui n'admet pas de fonction de choix locale – autrement dit, R est de degré ludique non trivial. Concrètement, c'est un téléphone qui fonctionne de manière instantanée même quand ses "combinés" sont loins les uns des autres ; mais qui **n'existerait pas** dans un monde régi uniquement par la Relativité ou la Physique Newtonienne...

En voici un exemple frappant, d'uplicité 3. Dans [9] est défini un dispositif expérimental que nous appellerons "téléphone quantique GHZ"; ce "téléphone" comporte trois "combinés", et le i-ième combiné ($i < 2$) a une entrée $x_i \in E$ et une sortie $y_i \in F$, où E, F sont des ensembles avec deux éléments – nous convenons que $E := \{+, -\}, F := \{vert, rouge\}$. De plus un combiné donné ne réagit en allumant une de ses deux lampes que si sur **chacun** des trois combinés une touche a été appuyée (à trois instants qui n'ont pas besoin d'être simultanés) ; notez qu'alors trois lampes s'allument : une par combiné. Notez que ces conditions disent exactement que le téléphone GFZ est R^*-garanti pour une certaine relation $R^* \subset \{+, -\} \times \{vert, rouge\}$. Mais le Fait suivant ajoute qu'il s'agit d'une relation R^* bien précise :

Fait 9. Si les trois touches (du téléphone GHZ) appuyées sont +, alors le nombre de lampes rouges allumées est pair (0 ou 2) ; si deux touches appuyées sont + et une est -, alors le nombre de lampes rouges allumées est impair (1 ou 3).

Autrement dit, le téléphone GHZ est R^*-garanti pour la relation $R^*(x, y)$ qui est vérifiée si et seulement si l'un des cas suivants est réalisés : i) $x = (+, +, +)$ et y comporte un nombre pair de "rouge" ; ii) x comporte un seul "-" et y comporte un nombre impair de "rouge" ; ii) x comporte au plus un "+" (et y est quelconque). Ce Fait est à la fois prédit par la Mécanique Quantique (à partir de la nature du dispositif expérimental), **et** vérifié expérimentalement (par GHZ justement).

Proposition 10. *Ce "téléphone" n'est pas local : il n'existe pas de fonctions $g_i, i < 3$ telles que le i-ème combiné satisfasse à chaque expérience g_i (touche appuyée)= couleur de la lampe qui s'allume.*
Autrement dit, le degré ludique de la relation R^* est non trivial.

Preuve. Nous démontrons que la relation R^* n'admet pas de fonction de choix locale $g = (g_0(x_0), g_1(x_1), g_2(x_2))$.

Sinon, les fonctions g_i étant données, posons $x_i = 1$ si $g_i(+) =$ vert, $x_i = -1$ si $g_i(+) =$ rouge, $y_i = 1$ si $g_i(-) =$ vert, $y_i = -1$ si $g_i(-) =$ rouge (en somme $x_i =$ "indicatrice" de la couleur qui s'allume quand on appuie sur la touche + du i-ème combiné et $y_i =$ "indicatrice" de la couleur qui s'allume quand on appuie sur la touche -). Appliquons le Fait précédent, c'est-à-dire le résultat de l'expérience GHZ, quand celle-ci est faite avec + + - , c'est-à-dire touches + appuyées sur les deux premiers combinés, touche - sur le troisième ;

de la sorte un nombre impair de lampes rouges s'allume, et cela entraîne que le produit $x_0x_1y_2$ vaut -1. Pour le même genre de raisons (mais en invoquant le Fait avec d'autres touches appuyées), le produit $x_0x_1y_2\ x_0y_1x_2\ y_0x_1x_2$ vaut $(-1)^3 = -1$; et le produit $y_0y_1y_2$ vaut 1. Or du simple fait que les nombres x_i, y_i valent tous 1 ou -1, on a toujours $x_0x_1y_2x_0y_1x_2\ y_0x_1x_2 = x_0^2x_1^2x_2^2\ y_0y_1y_2 = y_0y_1y_2$; donc "-1=1", une contradiction qui prouve la Proposition. ⊣

Remarque 11.

• Le téléphone GHZ est un exemple d'uplicité 3 de degré ludique non trivial "fabriqué" par la MQ (on dira : **degré FMQ**) ; c'est à dire un dispositif physique qui (de manière théorique dans le cadre de la MQ, et/ou expérimentalement) réalise un téléphone R-garanti pour une relation $R >_L 1_L$. Car Fait 9 et Proposition 10 peuvent s'énoncer : il existe un 3-degré ludique non trivial et dont le caractère FMQ, en plus d'être démontrable par la MQ, est un fait expérimental.

• D'autres effets quantiques expérimentaux et/ou théoriques (Aspect, Conway-Specker, etc.) donnent lieu de manière similaire à d'autres degrés ludiques non triviaux, pour des uplicités commençant à $n = 2$. Et à partir de ces degrés, les opérations de somme, produits, bornes et dualité déduisent une infinité d'autres degrés non triviaux qui sont FMQ.

• Actuellement le seul effet quantique connu qui ne se laisse pas exprimer par l'existence d'un "téléphone R-garanti" (où R est de degré ludique non trivial) est celui de "l'inspection non explosive d'une bombe", voir Elitzur-Vaidman [8].

• Mais il s'en faut de beaucoup que ces degrés ludiques obtenus par la Mécanique Quantique épuisent tous les degrés ludiques, même en se restreignant à ceux qui sont (d'uplicité et de domaine) finis : par exemple, dès qu'une relation R d'uplicité 2 est le graphe d'une fonction, si cette fonction n'est pas locale on peut montrer qu'il n'existe pas de téléphone quantique R-garanti ; autrement dit, R n'est pas un "effet quantique" – R n'est pas FMQ.

• Puisque les degrés ludiques finis vont largement au-delà des effets quantiques, il se pose la question de caractériser, parmi les degrés ludiques finis, tous les degrés (théoriquement) FMQ. (La solution est indiquée sans preuve plus loin)

• Il se pose aussi la question de savoir combien de degrés quantiques existent : si un degré quantique est non trivial – par exemple celui de GHZ – est-ce que par somme et produits de ce degré on obtient tous les autres ? (La réponse est non). Il y a quantité d'autres questions intuitives et intuitivement intéressantes sur la Mécanique Quantique qui deviennent formulables rigoureusement à l'aide du cadre des degrés ludiques (et qui sinon sont trop floues pour qu'on puisse y répondre. . .).

Cette Remarque termine notre "Parenthèse heuristique" (ou "Sous-section 1.5"). Malgré la motivation de l'étude des degrés ludiques par la Physique qu'elle a esquissée, la thèse est purement mathématique. C'est pourquoi on a développé les considérations mathématiques bien avant cette Parenthèse. Et la suite continue dans l'esprit mathématique, sauf pour la Conclusion de la thèse (ou "Section 5"), qui revient sur l'interprétation (certains diront la spéculation) Physique.

1.5.3 Un peu de souplesse

La formalisation des degrés de 2-réductibilité a tendance – pour avoir des définitions simples – à cloisonner à l'excès les diverses notions qui en résultent. Mais il y a diverses astuces pour décloisonner sans toucher aux définitions. Nous traitons ici un seul exemple : si on a un Φ-degré représenté par une relation R de domaine $E \times F$ et s'il serait utile de considérer formellement R comme étant de domaine plus grand $E' \times F'$, que faire pour ne pas changer le degré de R ? La réponse est que R a une extension canonique R' de domaine $E' \times F'$ telle que R' a même Φ-degré que R ; cette réponse est valable avec la même définition de R' pour tous les exemples de cadre Φ examinés en Section 1.

Voici R' : $R'|E \times F = R$; $R'(x, y)$ est faux pour tous $(x, y), x \in E, y \in (F' - F)$; et $R'(x, y)$ est vrai pour tous $(x, y), x \in (E' - E), y \in E$. On a $R \leq_\Phi R'$ au moyen d'une réduction h, g qui est la même dans tous nos cadres Φ. Car $R(h(x), y) \Longrightarrow R'(x, y)$ est toujours vrai si on pose $x \in E \Longrightarrow h(x) = x$, $x \in (E' - E) \Longrightarrow h(x) = a$ où a est n'importe quel élément fixé de E. On a aussi $R \leq_\Phi R'$ (donc $R \equiv_\Phi R'$), mais cette fois-ci la 2-réduction de R' à R va dépendre du cadre Φ. On traite ici le cadre ludique d'uplicité 2 (les autres uplicités et autres cadres étant laissés au lecteur) : supposant $F = Y^2, F' = Y'^2$ on a $R'(x, y) \Longrightarrow R(x, g(y))$ pour tous $x \in E, y \in F'$ – en posant $g(y) = (g_1(y_1), g_2(y_2)))$ où chaque $g_i(y_i)$ vaut y_i si $y_i \in Y$ et vaut $b \in Y$ quelconque fixé si $y_i \in (Y' - Y)$. En effet $g(y) = y$ si $y \in F$ auquel cas $R'(x, y) \Longrightarrow R(x, y)$ et $R(x, y) \Longleftrightarrow R(x, g(y))$ est vrai ; et sinon $R'(x, y)$ est faux puisque $x \in E$, auquel cas $R'(x, y) \Longrightarrow R(x, g(y))$ est vrai aussi.

Conclusion de la Section

Pour conclure cette Introduction (ou Section 1), il est bon de remarquer que la théorie des degrés ludiques comprendra ici deux parties de nature et d'esprit différents:

- d'un côté, la classe de tous les degrés infinis est en liaison avec la structure des degrés de Tukey, donc avec la Topologie et la Théorie des ensembles. Cette partie infinie utilise toute la force de la Théorie des Ensembles - y compris celle des “axiomes de grand cardinal”, qui vont bien au-delà des axiomes usuels de Zermelo-Fraenkel. (Pour ce qui est des bases de Théorie des Ensembles que nous supposons connues, voir [1, 2, 12].)

- de l’autre côté, une partie restreinte des degrés finis est liée, comme on l’a vu, à la Mécanique Quantique.

À cet égard un parallèle entre la théorie des degrés ludiques et la détermination des jeux infinis nous paraît donner une intuition convenable : d’un coté, les degrés ludiques infinis sont dans le même esprit que la détermination des jeux infinis quelconques ; de l’autre côté, les degrés ludiques finis sont dans le même esprit que la détermination des jeux infinis arbitrés par un automate avec condition d’acceptation de Büchi ou un automate à compteurs ou à pile (et autres “machines” théoriques finies).

En effet du premier côté, l’étude des degrés ludiques infinis est dans le même esprit que la détermination des jeux boréliens, analytiques voire projectifs : non seulement la force des axiomes de ZF y est utile, mais les développements les plus intéressants font même appel à des cardinaux de plus en plus grands allant au-delà de ZF, par exemple les “dièses”. Et de l’autre côté, chacun des deux domaines modélise l’extension jusqu’à l’infini potentiel de choses qui existent matériellement dans le monde réel : expériences quantiques dans un cas, processus informatiques infinis dans le deuxième.

2 Préordres ludiques d’uplicité 1

Le préordre ludique $\leq_1$ est trivial parce qu’on prend pour fonctions de réduction toutes les fonctions possibles entre les domaines concernés. Cette trivialisation cesse dans un cadre Φ qui impose aux fonctions de réduction des conditions restrictives – telles que la définissabilité. D’où (en faisant varier Φ) divers préordres d’uplicité 1 et cependant non triviaux, intéressants à divers titres ; le chapitre en expose quelques-uns en les utilisant pour introduire et tester la dualité de Chalons.

2.1 Le lemme de l’étoile

Ce lemme est le cœur de la preuve du théorème de dualité de Chalons : résultats du type $(R^{-1})^{-1} \equiv_\Phi R$ dans des cadres Φ appropriés. Il implique le théorème diagonal de Cantor : “pas d’injection de ${}^E F$ dans F” ; car si $L : {}^E F \longrightarrow F$, il donne toute une fibre $L^{-1}(a)$ qui est en bijection avec F.

C'est plus fort que Cantor puisque ce dernier demande seulement à la fibre $L^{-1}(a)$ d'avoir au moins deux éléments.

Et c'est un résultat d'existence alors que le théorème de Cantor est un résultat de **non** existence. Le lemme de l'étoile peut aussi être vu comme combinant l'axiome du choix avec la détermination des jeux de longueur 2, laquelle est plus un axiome qu'un théorème :

$$\forall x \exists y\ R(x,y)\ ou\ \exists x \forall y\ \neg R(x,y).$$

En fait nous démontrons 3 lemmes : le premier est le lemme de l'étoile sous axiome du choix qui rend la preuve élémentaire. Le deuxième lemme atteste qu'en un certain sens, le lemme de l'étoile dit quelque chose de maximal ; le troisième est une forme du lemme de l'étoile sans axiome du choix.

Lemme 12. *Soient E, F des ensembles quelconques, et $L : ({}^E F) \to E$, une application de ${}^E F$ dans E, où ${}^E F$ est l'ensemble des applications de E dans F. Alors il existe $a \in E$ tel que $\forall y \in F \exists f \in {}^E F : L(f) = a \& f(a) = y$.*

Preuve. Dans le cas général, on utilise AC (l'axiome du choix) : supposons que la conclusion soit fausse, et pour chaque x choisissons un $g(x) \in F$ qui en témoigne: $\forall f : L(f) = x \to f(x) \neq g(x)$. Prenant $x := L(g)$ et $f := g$ on aboutit à une contradiction. ⊣

Lemme 13. *Soit maintenant $T \subseteq {}^E F$ vérifiant la conclusion du lemme, i.e. : $\forall L : T \to E \exists a \in E \forall y \in F \exists f \in T : L(f) = a \& f(a) = y$. Alors $T = {}^E F$.*

Preuve. Soit $g \notin T$. Prendre alors $L(f) :=$ un x tel que $f(x) \neq g(x)$. ⊣

On peut ré-énoncer la conclusion du premier lemme comme suit : il existe une famille de fonctions $f_y, y \in E$ telle que

$$*)\ \forall y[f_y(a) = y \& L(f_y) = a].$$

Sous une hypothèse supplémentaire, le troisième lemme renforce les propriétés de cette famille, tout en se passant de l'axiome du choix :

Lemme 14. *Dans la situation du premier lemme ajoutons la condition que F est muni d'un bon ordre noté $<$; et disons qu'une famille de fonctions $g_i \in {}^E F$ où i parcourt un ensemble ordonné $(I, <)$ est nulle part décroissante si*

$$(**)\ i < j \in I\ implique\ \forall x \in E : g_i(x) \leq g_j(x).$$

Alors il existe une famille $f_y, y \in F$ vérifiant (*) et qui est de plus nulle part décroissante.

Preuve. Par récurrence sur l'ordinal α, pour tout $x \in F$ on définit $A_\alpha(x) \subset F, g_\alpha(x) \in F$ par $A_\alpha(x) := \{t \in F : \forall\beta < \alpha L(g_\beta) = x \Rightarrow g_\beta(x) \neq t\}$ et tant que aucun des ensembles $A_\alpha(x)$ n'est vide, $g_\alpha(x) := \min A_\alpha(x)$. Remarquez que pour chaque x fixé, $A_\alpha(x)$ est non-décroissant quand α augmente, et la suite des fonctions g_α est nulle part décroissante au sens (**). De plus si $L(g_\alpha) = a$ et $A_\alpha(a)$ est non vide alors $A_{\alpha+1}$ est strictement inclus dans $A_\alpha(a)$; donc tant que les ensembles $A_\alpha(x), x \in F$ sont tous non vides, l'un d'eux décroît strictement quand on passe de α à $\alpha+1$. Pour des raisons de cardinalité il finit par exister un ordinal γ (de même cardinal que F) et un a tels que $A_\gamma(a) = \emptyset$. Notez que chaque fois que $A_\alpha(a)$ diminue en passant de α 'a $\alpha + 1$, c'est de son plus petit élément ; de $A_\gamma(a) = \emptyset$ résulte alors $\forall y \in F \exists \beta^y < \gamma y = g_{\beta^y}(a)$. Soit α^y le premier ordinal$\geq \beta^y$ tel que $L(g_{\alpha^y}) = a$ ($\alpha^y < \gamma$ sinon $A_\gamma(a)$ en serait encore à contenir l'élément y) ; $g_{\alpha^y}(a) = y$ et (*) est satisfait en posant $f_y := g_{\alpha^y}$. $\dashv$

2.2 Application du lemme de l'étoile à la dualité pour la 2-réduction

Soit V un modèle quelconque des axiomes de ZFC. On s'intéresse ici aux relations binaires $R \subset E \times F, S \subset E' \times F'$ avec $E, F, E', F' \in V$, mais sans requérir $R, S \in V$. On pose $R \leq_V S$ – presque toujours noté $\leq$ – s'il existe une 2-réduction (f, g) de R à S avec $f, g \in V$, et $R \equiv S$ si $R \leq S \leq R$. C'est le cadre $\Phi = V$ dans le cas externe. Notez que le cas interne où $R, S \in V$ est trivial, car par axiome du choix dans V le préordre n'a plus que deux classes d'équivalence : celle constituée par toutes les relations S qui sont totales, et celle des relations non totales. Dans le cas externe qui est désormais le notre, la classe d'équivalence de R représente en un sens le "degré d'externalité de R au-dessus de V". En particulier, si V est un modèle transitif de la théorie des ensembles, notez que $R \leq_V S$ implique $R \in V[S]$ mais que la réciproque est fausse : le pré-ordre $R \leq_V S$ ressemble à la relation $R \in V[S]$ mais en bien plus fin. Mais nous étudions ce cadre pour une toute autre raison : il est parfait pour exposer avec simplicité les preuves et idées de la dualité de Chalons, qui serviront en Section 4.

Voici les principales propriétés de la relation $R \leq S$:

Proposition 15.

- $R < S$ *est un préordre*
- $R < S \Longleftrightarrow S^{-1} < R^{-1}$ *; et* $(R^{-1})^{-1} \equiv R$

Autrement dit l'opération $R \mapsto R^{-1}$ *est partout définie, décroissante et involutive sur les classes de l'équivalence associée au préordre.*

Preuve. La preuve de chaque énoncé A se fera en deux étapes ; la première étape est la preuve de A faite pour le seul cadre $\Phi = V$. La deuxième étape se donne un cadre **arbitraire** Φ ; elle énonce l'ensemble P de toutes les propriétés de V qui ont été utilisées par la première étape – de sorte que l'énoncé A est démontré vrai pour toute collection Φ vérifiant P.

- $\leq$ est un préordre
  - Reflexivité : $R \leq R$ s'obtient en prenant $h(x) := x, g(x,y) := y$.

Notez que l'argument vaut dans n'importe quel cadre Φ dès que sa collection de fonctions de Φ-réductions (entre produits cartésiens de domaines) contient les projections. Par la suite, cette propriété de Φ sera toujours sous-entendue.

  - Transitivité : supposons $R \leq S \leq T$ et soient $(h,g),(l,f)$ les 2-réductions correspondantes ; ainsi $S(h(x),y) \longrightarrow R(x,g(x,y)))$ et $T(l(x),y) \longrightarrow S(x,f(x,y))$, pour toutes valeurs permises de x,y. On obtient donc $T(l(h(x)),y) \longrightarrow S(h(x),f(h(x),y))$ et $S(h(x),f(h(x),y)) \longrightarrow R(x,g(x,f(h(x),y)))$.

Ainsi $T \leq S$ au moyen des réductions $(l(h(x)),g(x,f(h(x),y))$.
Notez que la preuve vaut dans Φ dès que sa collection de fonctions est close par composition (ce qu'on sous-entend désormais).

- $R \leq S$ implique $S^{-1} \leq R^{-1}$ – Supposons $\forall x,yS(h(x),y) \longrightarrow R(x,g(x,y))$ (2-réduction de R à S) ; nous voulons H,G tels que $\forall y,fR^{-1}(H(f),y) \longrightarrow S^{-1}(f,G(f,y))$ (2-réduction de S^{-1} à R^{-1}). Autrement dit, nous voulons pour toute valeur permise de la variable f que :

$$\neg R(y,H(f)(y)) \longrightarrow \neg S(G(f,y),f(G(f,y)));$$

soit :

$$S(G(f,y),f(G(f,y))) \longrightarrow R(y,H(f)(y)).$$

En prenant $G(f,y) := h(y)$, on a donc $f(G(f,y)) = f(h(y))$ et en prenant $H(f)(y) := g(y,f(h(y))$, cet énoncé souhaité devient $S(h(y),f(h(y)) \longrightarrow R(y,g(y,f(h(y))$, qui résulte de l'hypothèse :

$$\forall x,yS(h(x),y) \longrightarrow R(x,g(x,y)).$$

Notez que l'argument vaut dans Φ dès que R^{-1} y est défini et que sa collection de fonctions de réduction comporte la fonctionnelle H : ${}^{E'}F' \longrightarrow {}^{E}F$ ci-dessus (ce qu'on sous-entend désormais ; notez que la fonctionnelle G est une projection, donc est dans Φ par sous-entendu précédent). R^{-1} est défini pour tout $R \subset E \times F$, pourvu que ${}^{E}F := \{f; f : E \longrightarrow F\}$ soit l'un des domaines de Φ ; notez qu'il peut s'agir d'un cas aussi bien interne que externe de la notion ${}^{E}F$; c'est-à-dire que dans sa définition ci-dessus, on peut supposer que f varie seulement dans Φ, mais aussi que f varie sur le "vrai" ensemble ${}^E F$.

• $R \leq (R^{-1})^{-1}$ – Supposons $R \subset E \times F$; il faut trouver H, G tels que $(R^{-1})^{-1}(H(x), f) \longrightarrow R(x, G(x, f))$ pour tous $x \in E, f \in {}^E F$ dans V ; autrement dit, $\neg R^{-1}(f, H(x)(f)) \longrightarrow R(x, G(x, f)$, ou encore

$$R(H(x)(f), f(H(x)(f))) \longrightarrow R(x, G(x, f).$$

Définissons H, G par $H(x)(f) := x, G(x, f) := f(x)$ et c'est gagné ! Notez que l'argument vaut dans Φ dès que sa collection de fonctions de réduction est close par λ-abstraction : $F(x, y) \hookrightarrow \lambda y F(x, y)$ et contient la fonctionnelle G, c'est-à-dire l'application d'évaluation : $E \times ({}^E F) \longrightarrow F$.

• Reste le point le plus instructif : $(R^{-1})^{-1} \leq R$. On doit trouver σ, τ de sorte que: $R(\sigma(L), y)$ implique $(R^{-1})^{-1}(L, \tau(L, y))$ (pour tous $L \in {}^E F \cap V$ et tout $y \in F$) ; en appliquant la définition de S^{-1} pour $S = R^{-1}$ puis pour $S = R$ cette implication s'écrit:

$$R(\sigma(L), y) \longrightarrow R(L(\tau(L, y)), \tau(L, y)(L(\tau(L, y)))) \qquad (*)$$

On va construire σ, τ qui vérifient même la réciproque de (*), donc l'équivalence entre les deux formules de cette implication. C'est le lemme de l'étoile qui permet ce tour de passe-passe – le lemme n'est rien d'autre que ce tour de passe-passe, en fait.
Nous l'appliquons à l'extérieur de V, en utilisant ZFC : étant donné $L \in {}^E F$, choisir a tel que $\forall y \in E \exists f_y L(f_y) = a \& f_y(a) = y$ (a existe d'après le lemme de l'étoile) ; poser $\sigma(L) := a$. Choisir ensuite la famille $(f_y), y \in E$ et poser $\tau(L, y) := f_y$.
Pour vérifier (*) et sa réciproque il suffit de ramener à la même formule $R(a, y)$ l'hypothèse et la conclusion de cette implication ; s'agissant de l'hypothèse, c'est immédiat puisque par définition $a = \sigma(L)$. Reste à ramener à $R(a, y)$ la conclusion de (*) ; en substituant dedans f_y à $\tau(L, y)$ on la ramène d'abord à $R(L(f_y), f_y(L(f_y))$; formule qui se ramène à $R(a, y)$ en substituant a à $L(f_y)$ et y à $f_y(a)$ – ce qui est permis par définition de $\sigma(L)$ et $\tau(L, y)$. $\dashv$

Nous avons prouvé tout le lemme de dualité pour Φ sous les conditions suivantes :

- Φ contient les projections et fonctions constantes, est clos par composition, évaluation et λ-abstraction.
- pour toute fonctionnelle $L \in \Phi$, $L : {}^E F \longrightarrow E$, le lemme de l'étoile est vérifié par une constante $a = \sigma(L)$ et par des fonctions $f_y = \tau(L, y)$ où σ, τ appartiennent tous deux à la collection des Φ-fonctions.

Remarque 16. Il est facile de voir que la collection des fonctions locales d'uplicité κ possède ces propriétés pour chaque cardinal κ ; moyennant ce fait la démonstration ci-dessus prouve le théorème de dualité pour le préordre ludique $\leq_\kappa$ – ce qui nous servira à la section 4.

2.3 Dualité et définissabilité

Écartons d'abord la possibilité d'avoir le résultat $(R^{-1})^{-1} \equiv R$ dans un cadre effectif tel que $\Phi =$ les fonctions calculables. Un problème est rédhibitoire : les preuves du lemme de l'étoile ne sont pas effectives, de sorte que les fonctions qui en résultent ne sont pas calculables même si la fonctionnelle L l'était.

Dans ces conditions, on va s'occuper seulement d'avoir pour la dualité un cadre non trivial qui soit non pas effectif, mais du moins définissable. Soit Ω un ordinal infini et écrivons $\leq_\Omega$ pour $\leq_\Phi$ quand les Φ-réductions sont tous les couples $(h(x), g(x,y))$ tels que :

- chaque argument de h, g est l'un des trois ensembles $\Omega, {}^\Omega\Omega$ et ${}^\Omega({}^\Omega\Omega)$ (de la sorte $(R^{-1})^{-1}$ est définie pour toute relation R sur Ω) (i)

- h, g sont Σ-définissables en le paramètre Ω (ii).

Le préordre $\leq_\Omega$ est non trivial pour raison de cardinalité ; et la Proposition 15 reste vraie quand $\leq_\Omega$ remplace $\leq_V$. Car le présent Φ vérifie les propriétés (1+2) énumérées à la fin des preuves de la sous-section précédente.

Pour le points 1. c'est clair : les fonctions Σ à paramètre Ω sont closes par composition et comprennent les projections, les λ-abstractions et les fonctions d'évaluation : $(f, x) \hookrightarrow f(x)$. Reste le point 2. ; pour le montrer ici on utilise la forme lemme 14 du lemme de l'étoile, lorsque $E = F = \Omega$ muni de son bon ordre, et L est une fonctionnelle (pas nécessairement définissable) : ${}^\Omega\Omega \longrightarrow \Omega$.

Il est clair que la famille $g_\alpha, \alpha < \gamma$ fournie par la preuve du lemme 14 est Σ-définissable en le paramètre Ω (à partir de L, de manière uniforme). C'est pareillement clair pour l'élément a de cette preuve, si nous précisons que c'est le premier faisant l'affaire, dans l'ordre sur Ω ; et alors, la famille $(f_y^L), y \in \Omega$ fournie par la preuve est Σ-définissable en le paramètre Ω. La 2-réduction (σ, τ) qui en résulte est donc bien permise par $\leq_\Omega$

Ainsi la dualité pour la 2-réduction est vérifiée dans ce cadre où les réductions sont en un sens définissables et (si Ω est dénombrable) absolues. Notez que pour $\Omega = \omega$, la collection des Ω-réductions est celle des fonctions et fonctionnelles Δ^1_2 (cf [12]).

3 Degrés de Tukey : une unité de mesure de l'information

Les degrés ludiques sont le vrai objet de ce travail mais, dans la mesure où ils enrichissent les degrés de Tukey et où ces derniers constituent une ossature déjà très riche et étudiée, il semble nécessaire de consacrer une section aux degrés de Tukey. Elle est réduite, dans l'esprit : "trop brefs extraits d'un riche panorama". Pour la compléter le lecteur est renvoyé à [14] et Chalons [10]. Rappelons que $R \leq_T S$ ssi $\exists h, g \forall x, y [h(x) S y \Rightarrow x R g(y)]$.

Les degrés de Tukey des relations suivantes peuvent être identifies aux cardinaux : posons $I_\kappa :=$ relation $x = y$ sur le domaine κ^2. Le degré de I_1 est le degré trivial – celui des relations binaires R telles que $\exists y \forall x \; xRy$; et c'est celui des relations R telles qu'un téléphone R-garanti ne transmet aucune information. Autrement dit le cardinal de Tukey I_1 correspond au cardinal ensembliste 0. Et toute la suite est décalée de même manière : le cardinal κ correspond à $I_{1+\kappa}$, qui peut transmettre exactement κ informations.

Autres exemples :

- être capable de transmettre un réel x à ϵ près, c'est disposer d'un téléphone R-garanti, où $R := \{(x, y) : |x - y| < \epsilon\}$. Mais bizarrement, le degré de Tukey de R ne dépend pas de ϵ dès que $\epsilon > 0$.
- Soit κ un cardinal quelconque. On note D_κ le degré de Tukey de la relation suivante : $R := \{(a, b) : a \in \kappa, b \in \kappa, a \leq b\}$ Voir [14] ou plus bas pour d'autres exemples.

3.1 Comparaison entre cardinaux et "cardinaux de Tukey" dans le cas fini

Commentaire heuristique. Le préordre et les degrés de Tukey, c'est en somme une approche phénoménologique de l'univers mathématique V. Au lieu qu'on ait directement accès aux objets mathématiques de V, il y a une seule sorte de "phénomène observable" : toutes les (classes d'équivalences des) relations binaires auxquelles donnent lieu ces objets. Et il s'agit de se faire une idée du monde mathématique "en soi" qu'est V, à partir uniquement de ces relations binaires – dont la donnée est trop restreinte pour permettre de décrire complètement même les objets mathématiques les plus basiques. Par exemple, le cardinal κ (ou un ensemble de κ éléments distincts) n'est pas dans le monde de Tukey ; et tout ce qu'il y a pour en rendre compte est une classe d'équivalence de relations binaires dont la plus canonique est I_κ. Cette idée va être illustrée par la comparaison de toutes les relations binaires R avec les

$I_n, n < \omega$.

Définition 17.

- *R est n-filtrante ssi pour tout $x_1, ..x_n$ $\exists y \forall i$ $x_i R y$*
- *R est n-majorée s'il existe $y_1, \dots y_n$ tels que $\forall x \exists i$ $x R y_i$. Cet ensemble $y_1, \dots y_n$ est alors appelé une n-majoration de R*

Proposition 18.

1. *$I_2 \leq R$ si et seulement si R n'est pas 2-filtrante.*
2. *$R \leq I_2$ si et seulement si R est 2-majorée.*
3. *$R \equiv I_2$ si et seulement si R est 2-majorée mais pas 2-filtrante.*

Preuve.

1. Soit $h : i \mapsto x_i$ et $g : y \mapsto g(y)$ une 2-réduction de I_2 à R ; on ne peut pas avoir en même temps $R(x_0, y)$ et $R(x_1, y)$ car cela entraînerait $0 = g(y) = 1$. Donc R n'est pas 2-filtrant.
   Réciproquement, supposons R non 2-filtrant et soient x_0, x_1 qui en témoignent : ils n'ont pas de R-majorant commun. En posant $h(0) = x_0, h(1) = x_1$ et g quelconque on obtient une 2-réduction de I_2 à R.
2. Soit $h(x), g(y)$ une 2-réduction de R à I_2 : on a $h(x) = y \longrightarrow R(x, g(y))$; on en tire une 2-majoration de R en posant $g(0) = y_0, g(1) = y_1$.
   En sens inverse on fixe un 2-majorant y_0, y_1 de R, et on pose $h(x) = 0$ si $R(x, y_0)$ et $h(x) = 1$ sinon (auquel cas $R(x, y_1)$). Alors en posant $g(0) = y_0, g(1) = y_1$ on a bien $h(x) = y \longrightarrow R(x, g(y))$.
3. Le dernier item est conséquence des deux premiers. $\dashv$

En remplaçant 2 par n dans la preuve de "$R \leq I_2$ si et seulement si R est 2-majorée", on obtient sa généralisation à tout n : $R \leq I_n$ si et seulement si R est n-majorée.

En revanche, pour passer de 2 à n dans "$I_2 \leq R$ si et seulement si R n'est pas 2-filtrante" il y a un détour par un fait facile à voir :

Fait 19. $I_2 = I_2^{-1}$ (On utilise pour $R = I_2$ la dualité R^{-1} introduite pour le cadre de Tukey en Section 1).

Moyennant ce fait, "$I_2 \leq R$ si et seulement si R n'est pas 2-filtrante" se traduit aussi par : "$I_2^{-1} \leq R$ si et seulement si R n'est pas 2-filtrante". Sous cette forme, le résultat passe de 2 à tout n : "$I_n^{-1} \leq R$ si et seulement si R n'est pas n-filtrante", avec preuve semblable au cas n=2. L'opération de dualité inversant le préordre de Tukey, cela s'écrit aussi "$R^{-1} \leq I_n$ si et seulement si R n'est pas n-filtrante"; d'où encore

"$R \leq I_n$ si et seulement si R^{-1} n'est pas n-filtrante", et finalement la proposition générale :

Proposition 20.

- *$I_n \leq R$ si et seulement si R^{-1} n'est pas n-filtrante.*
- *$R \leq I_n$ si et seulement si R est n-majorée.*
- *$R \equiv I_n$ si et seulement si R est n-majorée mais R^{-1} pas n-filtrante.*

Un premier corollaire est que les "cardinaux de Tukey" finis (c'est-à-dire les $R \equiv I_n$) ne sont pas un segment initial des degrés de Tukey, contrairement aux cardinaux finis pour tous les cardinaux. En effet il est facile de construire R n-majorée mais telle que R^{-1} ne soit pas n-filtrante.

Dans ce cas $R \leq I_n$ est vrai mais $\exists p R \equiv I_p$ est faux, d'après la proposition ci-dessus ; et le degré de Tukey de R – bien que borné pour $\leq_T$ par un cardinal de Tukey fini – n'en est pas un lui-même.

Un deuxième corollaire montrera que dans le cas fini, l'analogue de Tukey de l'arithmétique s'éloigne de l'arithmétique habituelle ; mais il faut déjà introduire cette arithmétique analogue.

3.2 Les opérations sur les degrés de Tukey

Une opération déjà vue et facile à comprendre dans le cas des degrés de Tukey est la suivante :

Définition 21. $R^{-1} := \{(x, y) : \neg y R x\}$

Ses propriétés sont :

Proposition 22. *$(R^{-1})^{-1} = R$, autrement dit elle est involutive. De plus elle est décroissante, i.e. $R \leq S$ entraîne que $S^{-1} \leq R^{-1}$.*

Voir la preuve en Section 1.

En particulier $1^{-1} = \infty$, où ∞ est le degré des relations R telles que $\exists x \forall y \neg(x R y)$. Il est plus grand que n'importe quel autre degré de Tukey.

Définition 23. *Soient $R_i \subset E_i \times F_i$ $(i \in I)$ des relations binaires regardées comme des degrés de Tukey. La somme des degrés des R_i est le degré de la relation R suivante : $(x, i)R(y, i)$ est défini si $i \in I, x \in E_i, y \in F_i$; degrés des R_i est le degré de la relation R et $(x, i)R(y, i)$ est vrai si de plus x R_i y.*

Définition 24. *Le produit des degrés des $R_i, i \in I$ est le degré de la relation S suivante : fSg est défini si $f \in \Pi_{i \in I} E_i, g \in \Pi_{i \in I} F_i$; fSg est vrai si de plus $\forall i$ $f(i) R_i g(i)$.*

Proposition 25. *Les opérations précédentes ont un sens : elles sont compatibles avec la relation $\equiv_T$.*

La démonstration en est laissée au lecteur.

Pour ces opérations, les degrés de Tukey se comportent comme les entiers et les cardinaux :

Proposition 26. $\prod_i a_i \geq_T a_j$ *et de même pour la somme.*

La démonstration en est laissée au lecteur.

La proposition qui suit n'est pas seulement une définition :

Proposition 27. *La borne supérieure des degrés des $R_i, i \in I$ est le degré de la relation R telle que $(x, i)Rf$ est défini si $i \in I$, $x \in E_i$, f a pour domaine I, $\forall i\ f(i) \in F_i$; et $(x, i)Rf$ est vrai si de plus $xR_if(i)$.*

Preuve. Supposons que toutes les R_i sont $\leq S$ et soient f_i, g_i qui en témoignent. Montrons que $\sup_i R_i \leq S$. Le couple de fonctions $(x, i) \mapsto f_i(x), y \mapsto \lambda i g_i(y)$ représente une réduction de $\sup_i R_i$ à S. $\dashv$

Un deuxième corollaire montre que les degrés de Tukey n'étendent pas les cardinaux de la même façon que les rationnels étendent les entiers.

Corollaire 28. *La structure des rationnels $(Q, \leq, \times)$ ne se plonge pas dans celle des degrés de Tukey $(T, \leq_T, \times)$. En effet,* 1.5 *n'y aurait pas d'image.*

Preuve. $1.5 \times 1.5 = 2.25 \geq 2$ donc l'image de 1.5 devrait être un degré de Tukey R telque $\neg R \geq_T I_2$ mais aussi $R \times R \geq_T I_2$. Or un tel Tukey n'existe pas. $\dashv$

Commentaire.

- La morale de cette comparaison entre cardinaux finis et "cardinaux de Tukey" finis, c'est i) une bonne intuition de la relation entre degrés de Tukey et cardinaux consiste à considérer que les premiers sont la phénoménologie des seconds dans un monde où les phénomènes observables se limitent à des communications R-garanties pour des relations R binaires quelconques - en quotientant ces relations par $\equiv_T$; ii) mais il faut regarder de près, car le détail des différences et ressemblances entre le "monde en soi" ensembliste et le monde de Tukey s'avère plein de subtilités. Et cette morale tirée à propos de cardinaux finis est générale, valant pour la classe propre des cardinaux et des degrés quelconques.
- Il existe une foule d'approches algébriques qui restreignent fortement le langage utilisé pour décrire les objets mathématiques auxquels on s'intéressent ; mais presque toujours cette restriction a pour effet qu'une métathéorie faible suffit pour formaliser (au sens prouver) l'approche algébrique en question. Ici la situation est différente : i) les "téléphones

de Tukey" sont un moyen de transmission – donc d'expression – très limité ; ii) mais la métathéorie la plus appropriée pour étudier la structure des degrés de Tukey est la plus forte qui soit (de sorte que sa formalisation est toujours incomplète et incertaine) : ZF + grands cardinaux. Il en est de même pour les degrés ludiques ; cette situation (i+ii) propre à nos "phénoménologies" contraste aussi bien avec celle des approches algébriques qu'avec celle des Arithmétiques Faibles.

Le résultat suivant illustre encore l'analogie de comportement entre degrés de Tukey et cardinaux. C'est l'équivalent du lemme cardinal de Koenig en ce qui concerne les degrés de Tukey :

Théorème 29. *Étant donnés des degrés de Tukey a_i, b_i qui sont tels que $\sum a_i \geq \prod b_i$ il existe au moins un i tel que : $a_i \geq b_i$.*

On utilise pour ça le lemme de l'étoile (de la Section 2).

Preuve du théorème. Si la relation $R_i \subset E_i \times E'_i$ représente le degré a_i et $S_i \subset F_i \times F'_i$ représente b_i, alors nous avons supposé $S \leq R$ - où $R := \sum_i R_i$, $S := \prod_i S_i$; il y a donc une réduction de Tukey de la relation fSg à la relation $(x,i)R(y,i)$.

Nota Bene : les symboles f, g, x, i, y utilisés sont repris de notre définition ci-dessus de la somme et du produit de degrés ludiques ; cette définition précisait les domaines attribués à ces symboles – de telle manière que, par exemple, f parcourt $\Pi_{i \in I} F_i$. (Rappel de la convention posée dès la sous-section 1.1 : le lecteur prend en charge la vérification qu'en prenant ces domaines précis, la preuve que nous donnons est exacte).

La réduction consiste en une application : $f \mapsto (x(f), i(f))$, et une autre application : $(y,i) \mapsto g(y,i)$ telles que pour tous f, y, i on a l'implication $(x(f), i(f))R(y,i) \Rightarrow fSg(y,i)$. En revenant à la définition de R, S l'implication se détaille en :

$$(*)\ [\ (x(f), i)R_i(y,i) \textbf{ et } i(f) = i\ ] \Rightarrow \forall j \in I\ f(j)S_j g(y,i)(j)$$

(observez que $g(y,i)$ a pour valeur une fonction de domaine I). C'est le moment d'appliquer le lemme de l'étoile quand la fonction L est $i(.) : f \mapsto i(f)$ (qui va de $\Pi_{i \in I} F_i$ vers I ainsi que nous avons pointé dans le Nota Bene plus haut). Le lemme nous donne $k \in I$ tel que $\forall x \in E \exists f_x \in ({}^I E)\ i(f_x) = k$ et $f_x(k) = x$. Pour montrer $b_k \leq a_k$ (ce qui achève la preuve), il reste à construire une réduction h, G de S_k à R_k : assurer $\forall x, y\ h(x)R_k y \Rightarrow xS_k G(y)$; posons $G(y) = g(y,k)(k)$ et appliquons la propriété (*) ci-dessus avec $f = f_x$ et $j = k$: nous obtenons $(x(f_x)R_k y$ et $i(f_x) = k \Rightarrow f_x(k)S_k G(y)$. Comme $i(f_x) = k$ et $f_x(k) = x$ sont vrais par application du lemme, il reste à poser

$h(x) = x(f_x)$ pour avoir fait tout coller ensemble. ⊣

3.3 Un crochet par la Topologie

À titre d'illustration des possibilités de la théorie de Tukey dans le domaine qui en est le berceau, on va associer à chaque espace topologique un degré de Tukey qui peut être pensé comme le "degré de non-compacité" de cet espace. Ce sera un des nombreux cas où l'opération borne supérieure permet de faire référence à un seul degré là où on avait besoin d'une famille auparavant. Précisons que la notion de compacité retenue est celle qui ne demande pas à l'espace d'être séparé.

Définition 30. *Soit E un espace topologique. Soit W un recouvrement ouvert de E. On note T_W la relation binaire $\{(x, U)/x \in U \in W\}$. On note incompact(E) la borne supérieure des degrés de Tukey de T_W quand W parcourt tous les recouvrements ouverts de E.*

Nous pouvons comparer les compacités de deux espaces en demandant *incompact*(E) $\leq$ *incompact*(F) :

Proposition 31. *E est compact si et seulement si incompact(E) $\leq$ incompact(F) pour tout espace non compact F.*

Preuve. Soit E un espace compact et F un espace non compact, montrons que le degré (d'incompacité) de E est le plus petit. Cela revient à montrer que pour tout recouvrement ouvert V de E il en existe un W de F tel que $T_V \leq_T T_W$. Il suffit de le montrer quand V est fini ; car si V' est quelconque et V est extrait de V' alors $T_V \leq_T T_W$ entraîne facilement $T_{V'} \leq_T T_W$.

Disons qu'un sous-ensemble Y de l'espace F est W-indépendant si $x \neq y \in X$ implique l'existence de $U, V \in W$ tels que $x \in U - V, y \in V - U$. Soit X un sous-ensemble maximal V-indépendant de E ; son cardinal est borné par 2^n où n est le cardinal de V. Soit W un recouvrement ouvert de F sans sous-recouvrement fini, et soit Y un sous-ensemble W indépendent maximal de F ; Y étant infini donc plus grand que 2^n nous pouvons fixer une injection h de X dans Y.

Pour étendre h à E tout entier notons $x \equiv y$ si x, y appartiennent aux mêmes ouverts V_j de V ; pour $x \in E$, s'il existe $y \in X, y \equiv x$ posons $h(x) = h(y)$. Sinon nous posons $h(x) = b$ où b est quelconque mais fixé dans F. Reste à définir une fonction g de W dans V qui avec h témoigne que $T_V \leq_T T_W$, c'est-à-dire $h(x) \in W_i \in W$ implique $x \in g(W_i) \in V$. C'est possible sous réserve de modifier W . soit W' l'ensemble de tous les ouverts W^j qui sont union finie d'intersections finies d'éléments de W et tels que $h(V_j \cap X) = W^j \cap h(X)$ pour un élément $V_j \in V$; W' est un recouvrement de F parce que la réunion pour tout j des $W^j \cap h(X)$ étant égale à $h(X)$, la réunion des W_j vaut F. Alors on peut poser $g(W^j) = V_j$ pour témoigner de $T_V \leq_T T_{W'}$. $\dashv$

Remarque. Entre espaces discrets finis donc tous deux compacts, le degré d'incompacité varie comme le cardinal de l'espace, ou encore comme l'entier n tel que tout recouvrement ouvert a un sous-recouvrement de cardinal $\leq n$. Nous verrons dans la section sur les degrés ludiques une autre façon de 'mesurer' la compacité d'un espace topologique ; elle nous permettra une autre comparaison des compacités respectives de deux espaces... même s'ils sont tous les deux compacts.

3.4 Inoffensivité

Cette notion utilise celle de jeu à information parfaite $[M, A]$ entre un joueur 1 et un joueur 2 : A est un sous-ensemble de X^ω pour un X fixé ; X est appelé l'ensemble des coups possibles et A ensemble gagnant du joueur 1. En commençant par J1 les deux joueurs choisissent en alternance les éléments successifs d'une suite $s \in X^\omega$, cette suite est la partie jouée et elle est gagnée par J1 si et seulement si $s \in A$. M désigne l'opération Match qui, à un couple de stratégies S_1, S_2 (c'est-à-dire à deux manières de jouer les coups, la première pour J1 et la deuxième pour J2), associe l'ω-suite

$$M(S_1, S_2) =< S_1(<>), S_2(<>, S_1(<>)), \ldots >$$

obtenue en faisant jouer le joueur 1 avec S_1 contre le joueur 2 avec S_2.

Et $M(S_1, S_2) \in A$ signifie que la stratégie de J1 bat celle de J2. Il est sous-entendu que la notation $M(a, b)$ comprend le cas où a (ou b, ou les deux) sont des ω-suites de coups inconditionnels : des éléments de X^ω, et non des stratégies dépendant des coups de l'adversaire. Ce sous-entendu est permis par le fait que les éléments de X^ω s'identifient avec les stratégies inconditionnelles ; il sera utilisé dans la définition de relation inoffensive, quand l'opération Match a des éléments de X^ω pour arguments.

Voici cette définition :

Définition 32. *Soit R une relation binaire sur X^ω ; R est inoffensive si et seulement si pour tout jeu $[M, A]$ tel qu'il existe F, G vérifiant*

$$\forall y, z \in X^\omega [F(y) R z \Rightarrow M(G(z), y) \in A],$$

le joueur 1 a une stratégie gagnante au jeu $[M, A]$ (autrement dit,

$$\exists a \forall x \; M(a, x) \in A).$$

Commentaire.

• L'idée qui rend intuitive cette définition est celle de communication – brouillée – avec le futur ; dans cette idée le joueur 1 possède un téléphone R-garanti dont l'émetteur est situé dans le futur : un accolyte de J1 se trouve déjà au stade temporel où toute la partie infinie a été jouée entre J1 et J2. Et par le téléphone, l'accolyte envoie y (suite des coups joués par J2) à J1 ; ou plus généralement il envoie $F(y)$, F étant n'importe quelle fonction sur X^ω convenue avant la partie entre J1 et le téléphoniste du futur. De son côté J1 a la capacité de "décoder" le message reçu au moyen de n'importe quelle fonction fixée G sur X^ω.

• Si $F = id = G$ et si ce téléphone depuis le futur a la capacité de transmission à l'identique – autrement dit si R est la relation identité $x = y$ sur X^ω, notée Id_{X^ω} – alors J1 jouerait en sachant d'avance tous les coups que jouera J2. De sorte que J2 ne pourrait gagner une partie que s'il est même capable de la gagner en annonçant son jeu avant le début de la partie. Mais il est bien connu que les communications depuis le futur sont mal garanties, et en général, R ne garantit pas cette transmission parfaite...

• Justement la définition dit que R est inoffensif quand ce téléphone est assez défectueux pour n'apporter aucune aide réelle à J1 : quand un téléphone R-garanti depuis le futur ne permet à J1 de gagner toutes les parties d'un jeu $[M, A]$ que si J1 y parvient aussi en se passant du téléphone – si J1 a une stratégie gagnante sans téléphone ni accolyte.

Soit maintenant une relation binaire R sur un domaine quelconque et qu'on regarde comme un degré de Tukey ; on dit que R et son degré sont inoffensifs s'il existe une relation binaire S de même degré que R, dont le domaine est de la forme X^ω et qui est inoffensive. Remarquer qu'alors toute autre relation R_1 de même degré et avec un domaine de la forme X_1^ω est inoffensive aussi. De sorte qu'être ou ne pas être inoffensif ne dépend finalement que du degré de Tukey de R.

La preuve de cette $\equiv_T$-invariance de l'inoffensivité a besoin de la possibilité – que la définition accorde à J1 et à son accolyte – d'appliquer aux messages émis et reçus entre eux n'importe quelles fonctions F et G sur X^ω qu'ils se sont fixées. C'est ce qui donne le pas à cette définition précise de l'inoffensivité sur d'autres définitions, auxquelles l'idée de communication depuis le futur fait d'abord songer.

Supposons donc $R_1 \leq_T R$: nous avons h, g tels que $R(h(x), y)$ implique $R_1(x, g(y))$; soient F, G les fonctions qui satisfont l'hypothèse de l'inoffensivité de R : $F(y)Rz \Rightarrow M(G(z), y) \in A$. On peut l'écrire :

$$h(h^{-1}oF(y))Rz \Rightarrow M(G(z), y) \in A;$$

ce qui implique l'hypothèse de l'inoffensivité de R_1 avec $h^{-1}oF$ à la place de F, et Gog à la place de G. Ainsi l'hypothèse de l'inoffensivité :

> H_R:= "il existe F, G tels que le téléphone R-garanti permet à J1 de gagner toutes les parties du jeu $[M, A]$"

se transfère de R à R_1 chaque fois que $R_1 \leq R$. En échangeant R avec R_1 on conclut que si $R \equiv_T R_1$ alors $H_R \iff H_{R_1}$. D'où le résultat voulu en appliquant la tautologie : $B \iff C$ implique $(B \Rightarrow F) \iff (C \Rightarrow F)$.

3.5 Exemples

• La relation $y = z \quad : y, z \in X^\omega$ (autrement dit, Id_{X^ω}) est non-inoffensive (on dira offensive), dès que X a au moins deux éléments ; car il est facile de construire un ensemble gagnant A pour lequel J1 n'a pas de stratégie gagnante, mais le joueur 2 n'a aucune suite de coups inconditionnellement gagnante : si $y \in X^\omega$ il existe $G(y) \in X^\omega$ tel que $M(G(y), y) \in A$. Posons $F(y) = y$; avec ces valeurs de F, G l'hypothèse de l'inoffensivité de la relation Id_{X^ω} est vérifiée. Or la conclusion en est fausse par choix de A : J1 n'a pas de stratégie gagnante.

• On va voir que le degré I_ω, lui, est inoffensif.

Soit R une relation $\leq_T I_\omega$: on a h, g tels que $h(x) = n \Rightarrow R(x, g(n))$. En posant $g(n) := y_n$ on a une ω-suite telle que $\forall x \bigvee_n R(x, y_n)$ (autrement dit R est "ω-majorée"). Supposons l'hypothèse de l'inoffensivité de R : $\forall y, z R(F(y), z) \Rightarrow (G(z), y) \in A$; il reste à montrer que J1 possède une stratégie gagnante. On peut coder les stratégies pour les deux joueurs dans le jeu $[M, A]$ par des éléments de X^ω, et l'opération Match s'applique à des stratégies, pas seulement à des parties. De sorte que l'hypothèse de l'inoffensivité de R reste vraie quand $y, F(y)$ varient sur les stratégies pour J1 et $z, G(z)$ sur les stratégies pour J2.

R étant ω-majorée, il existe alors une ω-suite de stratégies (z_n) telle que $\forall y \bigvee_n R(y, z_n)$, et par notre hypothèse $\forall y \bigvee_n M(z_n, y) \in A$. Autrement dit chaque stratégie pour J2 est battue par au moins une des stratégies z_n ; en vue d'une preuve par l'absurde, supposons que J1 n'a pas de stratégie gagnante – ce qui contredit l'inoffensivité. Le lemme suivant entraîne alors l'existence d'une stratégie pour J2 qui gagne contre tous les z_n – une contradiction ; et l'inoffensivité de R est démontrée par l'absurde.

Lemme 33. *Dans un jeu où l'un des joueurs (mettons J1) n'a pas de stratégie gagnante, pour toute famille de ω stratégies pour J1, il existe une stratégie pour J2 qui les bat toutes.*

Pour prouver ce lemme, on a besoin d'un fait qui est amusant :

Fait 34. Supposons dans un jeu G que J1 a une stratégie telle que pour la battre il faut obligatoirement passer par une position désespérée pour J2. C'est-à-dire une position telle que à partir d'elle, J1 a une stratégie gagnante pour finir la partie. Alors J1 a une stratégie gagnante au jeu G.

Preuve du lemme. Soit $s_{n,n\in N}$ une famille dénombrable de stratégies pour J1. Voici comment on construit une stratégie t qui bat toutes les s_n : d'abord on trouve une partie P_1 jouée par s_1 gagnée par J2, et telle qu'à aucun stage de la partie on ne soit passé par une position J2-désespérée. Puis on construit une partie P_2 contre s_2 qui commence comme P_1 tant que s_2 joue comme s_1. Si s_2 se distingue de s_1, on finit la partie en battant s_2 sans passer par une position désespérée. Sinon c'est que $P_1 = P_2$ et ça marche aussi. On continue ainsi jusqu'à avoir fabriqué $P_1, P_2, P_3, \ldots$ qui sont deux à deux consistantes et donc ont toutes été jouées par une même stratégie t pour au moins un t. ⊣

En employant les mêmes idées il vient :

Proposition 35. *Il y a équivalence entre : 1) $R \leq_T I_\kappa$ (ou R est κ-majorée) entraîne R inoffensif; et 2) dans un jeu que l'un des joueurs (mettons J1) ne gagne pas, pour toute famille de κ stratégies pour J1, il existe une stratégie pour J2 qui les bat toutes.*

Démonstration laissée au lecteur

Appliquons la sous-section quand l'ensemble X des coups est au plus dénombrable et supposons HC. Alors $Id_{X^\omega} \equiv_T I_{\omega_1}$ donc le degré I_{ω_1} est offensif, d'après le premier exemple de 3.5. On a vu qu'en revanche tout degré ω-majoré est inoffensif.

Dans des modèles contredisant HC d'une manière appropriée, on peut utiliser la proposition précédente pour obtenir par exemple que le cardinal du continu est $c > \omega_1$, et les degrés κ-majorés sont inoffensifs si et seulement si $\kappa < c$. Et ce n'est qu'un petit exemple : la dépendance du sujet par rapport aux hypothèses de théorie des ensemble est vaste ; les questions et conjectures en conclusion sur les degrés de Tukey se bornent à une partie restreinte des possibilités...

Par quelles opérations la classe des degrés inoffensifs est-elle stable ? Est-ce le cas pour le produit ? Les degrés projectifs sont-ils inoffensifs si on fait l'hypothèse de la détermination projective ? Les réponses à ces questions auraient des applications. Précisons ce qu'on entend par degré projectif.

Définition 36. *Les degrés de Tukey projectifs sont ceux qui sont de la forme suivante : $R := \{(s,t)/M(s,t) \in A \vee \forall z : M(s,z) \notin A\}$ avec A ensemble projectif de réels.*

Conjecture 37. *Si R, S sont inoffensifs alors $R \times S$ est inoffensif.*

Conjecture 38. *Tous les degrés projectifs sont inoffensifs.*

Bien sûr, la deuxième conjecture est plutôt un axiome puisqu'elle entraîne la détermination projective. La question est plutôt de savoir si elle est une conséquence de la détermination projective. Une démonstration d'un tel fait serait la bienvenue entre autre parce qu'elle donnerait dans la foulée une bonne démonstration (sans recours aux dièses) de la détermination des combinaisons booléennes de projectifs. Il nous semble que la conjonction de ces deux conjectures doit motiver des efforts de démonstration.

Chalons [10], dont est tirée la présente section, contient une foule d'autres développements sur les degrés de Tukey : degrés "offensifs", "éternels", "degrés de réalité" ; théorèmes généraux du sujet ; illustration dans les graphes...

4 Degrés ludiques

Reprenons les définitions et remarques de l'Introduction. Soit f une fonction dont les arguments $x^1, \ldots, x^n$ et la valeur $f(x^1, \ldots, x^n)$ sont des κ-uplets (pour un cardinal non nul κ) ; f est dite **locale** si pour chaque $i < \kappa$ la i-ème composante de $f(x^1, \ldots, x^n)$ ne dépend que des i-èmes composantes de $x^1, \ldots, x^n$. Autrement dit – si on note x_i la i-ème composante d'un κ-uplet noté x – il existe une fonction f_i telle que $f(x^1, \ldots, x^n)_i = f_i(x^1_i, \ldots, x^n_i)$. En pratique on aura généralement $n \leq 2$ – sauf pour les "degrés multitables", où n pourra devenir un cardinal infini μ ; le cardinal κ est appelé l'**uplicité** de la fonction et de ses arguments. La réduction **ludique**, c'est la 2-réduction dans le cadre Φ qui prend comme fonctions de réduction toutes les fonctions locales.

Plus précisément, on prend les produits cartésiens de κ ensembles – appelés aussi ensembles d'uplicité κ. Les Φ-relations binaires, ou relations d'uplicité κ sont celles qui ont un domaine de la forme $E \times F$, où E, F sont deux ensembles d'uplicité κ ; on écrit $R \leq_\kappa S$ si R, S sont de telles relations et que R se réduit à S par des fonctions locales (nécessairement d'uplicité κ). Et on note $\equiv_\kappa$ l'équivalence associée à ce préordre. On note $\leq_L, \equiv_L$ la réunion pour tout κ de ces structures et on appelle degrés ludiques les classes d'équivalence modulo $\equiv_L$.

4.1 Opérations de base

1_κ est le degré de la relation toujours vraie d'uplicité κ – autrement dit, 1_κ est le degré de n'importe quelle relation R égale à son domaine, lequel est de la forme $E \times F$ où E, F sont des produits cartésiens de longueur κ. On écrira souvent 1_L sans préciser l'uplicité.

De même ∞_κ – souvent noté ∞_L – est le degré de la relation toujours fausse d'uplicité κ.

Fait 39.

- 1_κ est le degré ludique minimum, appelé aussi degré trivial. Et une relation R d'uplicité κ est de degré trivial ($R \in 1_\kappa$) si et seulement si R admet une fonction de choix locale : $\forall x R(x, f(x))$ où f est locale.
- ∞_κ est le degré ludique maximum. Et $R \in \infty_\kappa$ si et seulement si R est une relation non totale : $\exists x \forall y \neg R(x, y)$.

Notez que toute fonction d'uplicité $\kappa = 1$ est locale ; $\leq_1$ est donc la 2-réduction entre relation binaires, dans le cas où toutes les fonctions de réductions sont permises. On a noté dès la Section 1 que (sous axiome du choix) ce cas est trivial : n'existent que les degrés $1, \infty$. C'est donc le cas $\kappa > 1$ qui nous intéresse désormais. On a vu la motivation du sujet par la Physique, mais l'étude de $\leq_\kappa$ dans son ensemble est loin du Monde physique ; la motivation est alors de prolonger l'étude du pré-ordre de Tukey. Ce prolongement permet notamment de progresser dans l'étude des jeux infinis ; l'étude de $\leq_L$ est ainsi dans le même esprit que l'étude de $\leq_T$: une étude de l'univers mathématique selon une approche unificatrice de son foisonnement – approche que nous qualifions de phénoménologique, pour des raisons déjà invoquées...

Précisons ce que devient l'opération duale R^{-1} pour $R \subset E \times F$, relation binaire d'uplicité κ :

Définition 40.

$$R^{-1} := \{(f, r) \ : \neg R(r, f(r)), \ r \in E, f : E \mapsto F \textbf{ et } f \text{ local}\}$$

Les principales propriétés étendent au préordre ludique celles des deux sections précédentes.

Proposition 41. *R, S étant des relations binaires d'uplicité κ et en écrivant $\leq$ pour $\leq_\kappa$:*

- *$R \leq S$ est un préordre.*
- *$R \leq S \iff S^{-1} \leq R^{-1}$; et $(R^{-1})^{-1} \equiv R$.*

Autrement dit l'opération $R \mapsto R^{-1}$ est partout définie, décroissante et involutive sur les classes de l'équivalence associée au préordre.

Preuve. Les deux premiers points de la proposition ont été prouvés en sous section 2.2 dans tout cadre Φ pour la 2-réduction tel que : 1) les Φ-fonctions sont closes par composition et comprennent les constantes, les projections, les λ-abstractions et les fonctions d'évaluation : $(f, x) \mapsto f(x)$. Et 2) Si $L : {}^E F \Rightarrow F$ est une fonctionnelle entre Φ-ensembles alors le lemme de l'étoile est vérifié par L à l'aide de fonctions de réduction de Φ (c'est-à-dire $\forall y \in E\ L(f_y) = a$ et $f_y(a) = y$, où la fonctionnelle : $y \mapsto f_y$ ainsi que la fonctionnelle : $L \mapsto a$ sont des fonctions de réduction de Φ).

Il s'agit donc ici de vérifier (1+2) quand Φ est le cadre ludique : ensembles et relations d'uplicité κ, et toutes les fonctions locales de cette uplicité comme fonctions de réduction. Appelons i-ième fibre de Φ la collection Φ_i des fonctions f_i pour $f \in \Phi, f : x \mapsto (f_i(x_i))_{i<\kappa}$; il est aussi trivial qu'ennuyeux, et laissé au lecteur, de vérifier que les propriétés (1+2) sont vraies de Φ si elles sont vraies de chacune de ses fibres Φ_i. Ce qui est le cas puisque ces fibres sont toutes égales à V. ⊣

Comme dans le cas de la Section 3, on définit les opérations de somme, de produit et de borne supérieure ou inférieure ; on vérifie qu'elles préservent l'équivalence $\equiv_L$ associée au préordre $\leq_L$ - donc elles opèrent sur les degrés ludiques ; et que la dualité préserve les opérations de somme et produit, qui ont donc une duale. Par exemple $R +^{-1} S := (R^{-1} + S^{-1})^{-1}$. Et la dualité fait permuter borne supérieure et borne inférieure : $(sup_i R_i)^{-1} = inf_i(R_i^{-1})$.

Enonçons ce qu'on entend par la borne supérieure :

Théorème 42. *Soit $R_i, i \in I$ une famille – indexée par un ensemble – de degrés ludiques tous de même uplicité. Il existe alors un degré ludique S qui est le plus petit des majorants de la famille des $R_i, i \in I$. Si les R_i sont des représentants de leur degré alors le S suivant convient : $(x_i)_i S (r_i)_i$ si chaque x_i est de la forme (z_i, p_i), et si dans le cas où tous les p_i sont égaux à un certain p, alors $(z_i)_i R_p (r_i)_i$.*

Preuve. C'est le même genre de raisonnement que pour les degrés de Tukey, en remplaçant les fonctions par des familles de fonctions indexées par $i < \kappa$. Soit donc R_z une famille de relations toutes de même arité. Supposons que toutes les R_z soient $\leq$. Cela signifie qu'il existe une stratégie de réduction f_z, g_z de R_z à S. Chacune de ces f_z est de la forme $(f_z^i)_i$ idem pour pour les g_z. Réduisons maintenant $sup_z R_z$ à S. Soit donc $(x_i)_i$ dans le domaine de $sup_z R_z$; chaque x_i est de la forme (a_i, z_i). À x_i associons $f_{z_i}^i(a_i) \in S_i$. On vient de décrire la i-ième fibre d'une fonction locale qui est peut-être le h d'une réduction h, g. Il reste donc à définir g. À m_i, r_i on associe $g_{m_i}^i(r_i)$. On laisse au lecteur le soin de vérifier que ça marche. Dans l'autre sens c'est évident. ⊣

La borne inférieure existe donc aussi – nous laissons le soin au lecteur d'en donner une d'efinition plus directe que la suivante :

Corollaire 43. $inf_i R_i := (sup_i R_i^{-1})^{-1}$

4.2 Plongement de $\leq_T$ dans les degrés ludiques

Le plongement va dans $\leq_2$:

Définition 44.

- *Soit R une relation binaire représentant un degré de Tukey ; le degré ludique associé à R est celui de la relation d'uplicité 2 suivante, notée L_R : $(x_1, x_2)L_R(r_1, r_2)$ si et seulement si $x_1 R r_2$.*
- *Pour toute permutation σ de κ et tout κ-uplet $x = (x_i)_{i<\kappa}$ on pose $x^\sigma = (x_{\sigma(i)})_{i<\kappa}$; et si T est une relation d'uplicité κ on note T^σ la relation définie par $xT^\sigma r$ si et seulement si $x^\sigma T r^\sigma$.*

Théorème 45.

- *Les degrés Tukey se plongent dans les degrés ludiques au sens suivant : si $R \leq_T S$ alors $L_R \leq_2 L_S$. Et réciproquement.*
- *De plus, $L_{R^{-1}} \equiv_2 ((L_R)^{sym2})^{-1}$, en notant sym2 la transposition (élément non neutre du groupe symétrique à deux éléments).*

La preuve a été donnée pour l'essentiel dans l'introduction (Section 1) ; et la définition de T^{sym2} la formalise exactement.

Remarque. On a hésité à définir le passage d'une relation T d'uplicité κ à une de ces transformées T^σ comme ne devant pas changer le degré ludique de T. Finalement nous n'avons pas fait ce choix, pour des raisons essentiellement techniques de présentation. Cependant, il est clair que la notion de puissance d'un téléphone est invariante par permutation de ses combinés.

4.3 Exemples de degré ludique

Pour illustrer de quelle manière $\leq_L$ prolonge la structure des degrés de Tukey nous décrivons des degrés qui sont 'anti' quelque chose. Par exemple 'anti-théorème de compacité' ou encore 'anti-théorème de récurrence' ou 'anti-théorème de Cantor', etc. L'idée est que plus le théorème x est fort, plus le degré ludique 'anti-théorème x' est faible.

À la section 3 le degré de Tukey *incompact*(E) mesurait la non-compacité d'un espace topologique E ; il est à comparer avec le premier exemple ci-dessous de degré 'anti'.

Définition 46. *anticompact(E) est la relation R d'uplicité 2 suivante : $(F, x)R(U, V)$ est défini si et seulement si F est sous-ensemble fini de E, $x \in E$, U et V sont des ouverts ; alors $(F, x)R(U, V)$ est vrai si et seulement si $F \subseteq U$, $U \neq E$, $x \in V$ et enfin si $[x \in F \Longrightarrow V \subset U]$.*

Proposition 47. *anticompact(E) est trivial si et seulement si E a un recouvrement ouvert sans sous-recouvrement fini.*

Preuve. Supposons que $R := anticompact(E)$ est trivial : R admet une fonction de choix locale – couple f, g tel que $(F, x)R(f(F), g(x))$ pour tout $F \subset E$ fini et tout $x \in E$. Alors la famille $g(x), x \in E$ est un recouvrement ouvert de E ; en vue d'une preuve par l'absurde supposons que E soit compact. Il existe donc F_0 fini tel que $E = \cup_x g(x) = \cup_{x \in F_0} g(x)$; posons $U = f(F_0)$, $V = g(x_0)$ où $x_0 \notin U$ témoigne que $U \neq E$. De $(F_0, x_0)R(f(F_0), g(x_0))$ se déduit $x_0 \in g(x_0) = V \subset U$, donc $x_0 \in U$ contredisant le choix de x_0. Est ainsi démontré par l'absurde que $anticompact(E) = 1_L$ implique E non compact.

Réciproquement supposons que U est un recouvrement ouvert de E sans sous-recouvrement fini ; pour tout $x \in E$ fixons un ouvert V_x de U tel que $x \in V_x$ et pour $F \subset E$ posons $U_F = \cup_{x \in F} V_x$. On vérifie facilement que $(F, x)R(U_F, V_x)$ pour tout (F, x) dans le domaine de R. Donc le couple $F \mapsto U_F, x \mapsto V_x$ est une fonction de choix locale pour R – qui est de degré trivial. ⊣

anticompact(E) ressemble donc à *incompact*(E) ; mais supposons maintenant que E, F soient deux espaces compacts. Alors voici un moyen de comparer leur compacité respective : E est plus compact que F si $anticompact(E) \geq anticompact(F)$.

Définition 48. *$antirec(E, <)$ est le $<_2$-degré de la relation R définie par $(p, q)R(\alpha, \beta)$ si $p, q \in \omega, \alpha, \beta \in E$ et $[p \leq q \Leftrightarrow \beta \leq \alpha]$.*

Notez que cette définition fait sens même quand $(E, <)$ désigne un ordre total quelconque ; mais c'est seulement pour des ordinaux λ que le degré ludique $antirec(\lambda, <)$ est un 'anti-théorème de récurrence' : le théorème de récurrence sur λ voit sa force augmenter avec λ mais $antirec(\lambda, <)$ voit la sienne diminuer. En effet :

Proposition 49.

- *$antirec(E, <)$ est de degré trivial si et seulement si $(E, <)$ admet une suite infinie décroissante.*
- *Pour des ordinaux, $\kappa > \mu$ entraîne $antirec(\kappa) <_2 antirec(\mu)$ ($<_2$ strictement.)*

Preuve du premier point. Supposons $R := antirec(E, <)$ trivial : R admet une fonction de choix locale – couple f, g tel que $\forall p, q \in \omega[p \leq q \iff f(p) \geq g(q)]$. Alors $f(0) > g(1) > f(2) > g(2) > f(3) > \ldots$ est une suite infinie descendante dans $(E, <)$. Réciproquement, si $\alpha_i, i < \omega$ est strictement décroissante dans $(E, <)$, poser $f(i) = \alpha_{2i}, g(i) = \alpha_{2i+1}$ fournit un couple qui est une fonction de choix locale pour R.

Preuve du deuxième point laissée au lecteur. ⊣

L'exemple $antistar(E)$ ci-dessous est l'anti du lemme de l'étoile (d'où le "star"), et aussi du théorème de Cantor qui se trouve derrière ce lemme.

Définition 50. *$antistar(E)$ est le degré de la relation R suivante : pour $f : E \to E$ et $a, y, y' \in E$ on a $(f, a)R(y', y)$ si ($y' \neq a$ ou $f(a) \neq y$)).*

Rappelons que la négation du lemme de l'étoile – énoncé pour une fonctionnelle $L : {}^E E \Longrightarrow E$ - s'écrit : pour tout $a \in E$ on peut trouver $y = g(a) \in E$ tel que $\forall f\ [L(f) \neq a \text{ ou } f(a) \neq y]$. Alors le lemme de l'étoile revient à affirmer que $R := antistar(E)$ n'a pas de fonction de choix locale, autrement dit est non trivial. Car dire qu'un couple (L, g) est une fonction de choix locale pour R, c'est dire que pour tous f, a on a $(f, a)R(L(f), g(a))$, soit : "$L(f) \neq a$ ou sinon $f(a) \neq g(a)$". Or à l'intérieur de ces "...", nous avons exactement la négation du lemme de l'étoile pour la fonctionnelle L.

Mais de plus, les $antistar(E)$ ont une place à part au sein des $\leq_2$-degrés car ils sont co-initiaux dans les degrés non triviaux. Cela va résulter par dualité d'une construction au contraire cofinale dans les degrés $<_2 \infty$:

Proposition 51. *Pour toute relation $S \subset E^2$, $S <_2 \infty \Longrightarrow S \leq_2 R_E$, où $R_E := \{[(x, y), (y, x)]; x, y \in E\}$.*

La collection des R_E, E ensemble quelconque, est cofinale dans les degrés $<_2 \infty$.

Preuve. Puisque $S <_2 \infty$ alors S est une relation totale, autrement dit il existe une fonction de choix (non locale) (f_1, f_2) pour S telle que $\forall x_1, x_2 \in E[(x_1, x_2)S(f_1(x_1, x_2), f_2(x_1, x_2)]$. Posons $h(x_1, x_2) = (x_1, x_2)$ et

$$g((x_1, x_2), (y_1, y_2)) = [f_1(x_1, y_1), f_2(x_2, y_2)];$$

c'est une réduction ludique de S à R_E. En effet elle est locale, et on a les implications $(x_1, x_2)R_E(y_1, y_2) \Longrightarrow y_1 = x_2$ et $y_2 = x_1 \Longrightarrow g = [g_1(x_1, y_1), g_2(x_2, y_2)] = [f_1(x_1, x_2), f_2(x_1, x_2)]$, donc la fonction de réduction g prend tout simplement la valeur de la fonction de choix pour S. De sorte que (h, g) réduit trivialement S à R_E. ⊣

Le tour de passe-passe a été le suivant : on a une fonction de choix pour S non locale $f = (f_1, f_2)$; mais la relation R_E oblige certaines de ses variables à être égales à d'autres de telle manière que f devienne (par la substitution ainsi permise de variables) une fonction de réduction locale de S à R_E.

Les degrés R_E^{-1} fournissent donc une classe co-initiale au-dessus de 1_2 : $(f,g)R_E^{-1}(x,y)$ si $f,g \in^E E, x,y \in E$ et $y \neq f(x)$ ou $x \neq g(y)$. Elle est d'une simplicité satisfaisante, mais on pourrait la remplacer par la classe encore plus simple des $antistar(E)$, en montrant que $antistar(^E E) \leq_2 R_E^{-1}$.

Cette sous-section était dans l'esprit “brefs extraits d'un riche panorama”, comme la section sur les degrés de Tukey. Le reste de la section est consacré à l'ambition principale derrière les degrés ludiques.

4.4 Application à la Mécanique Quantique

Circonscrire de façon mathématique la partie restreinte de $\leq_L$ qui est susceptible d'avoir rapport à la MQ (ou à d'autres parties de la Physique) est la première question qui sera traitée. Nombre de questions posées par la Mécanique Quantique gagnent en clarté, en caractère abordable et scientifique à ce travail de délimitation mathématique. Le plan de ce “tracé de frontière” entre les degrés ludiques qui pourraient avoir un rapport avec la Physique et tous les autres degrés, est le suivant :

• Nous introduisons une première classe de degrés “proches de la Physique” baptisés degrés multitables ; très grossièrement il s'agit des degrés susceptibles d'apparaître dans une expérimentation (réelle ou de pensée) se basant sur une interprétation multi-monde de la Physique (comme celle introduite par Everett, mais en laissant de la place pour tous les progrès et audaces futures). Le lecteur n'a pas besoin de connaître ces interprétations, au contraire la définition même de la classe Multitable est un début d'introduction aux spéculations multi-mondes.

• Nous introduisons une autre classe *a priori* plus grande, dite $\neg TSD$, en partant d'une caractéristique frappante des effets quantiques: la non-transmission de **données** à distance (en un sens qu'il s'agira de préciser, et qui n'interdit pas les **effets** quantiques à action instantanée bien que distante - par exemple les corrélations de spin entre deux particules éloignées).

• Le théorème principal de la section et de tout l'article affirme l'égalité entre la classe Multitable et la classe $\neg TSD$. Comme les deux classes de degrés sont d'apparence et définition différentes, certaines propriétés faciles à voir pour une classe sont ainsi transférées de manière non évidente à l'autre classe. D'un point de vue plus philosophique, le fait que des considérations toutes différentes conduisent à une seule et même classe – disons $\neg TSD$ – parle en faveur du rôle de cette classe pour l'avenir des recherches. (L'article démontre ce théorème en utilisant une forme faible de HCG – Hypothèse du Continu Généralisée - mais A. Khélif a éliminé cette hypothèse et fait la démonstration dans la Théorie des Ensembles usuelle ZFC. C'est un des paradoxes du sujet que cette intrusion de spéculations ensemblistes très postérieures à Cantor, dans un domaine motivé par la Physique.)

• La classe $\neg TSD$ est taillée très large pour les besoins de la Physique ; l'étape suivante est donc de définir des sous-classes de degrés ludiques plus réduites et qui font des pas conceptuels en direction des degrés ludiques directement liés aux connaissances, théories et expérimentations actuelles en MQ. Viendra ainsi la considération des degrés ludiques forçables, déterministes, casino-inoffensifs (une notion inspirée de la notion de degré de Tukey inoffensif), et FMQ ("Fabricable par la Mécanique Quantique").

• Nota Bene - De manière à refléter le monde physique (en l'idéalisant) on peut souhaiter se limiter aux degrés ludiques d'uplicité finie et dont les représentants sont des relations binaires sur des ensembles finis. Mais dans les étapes précédentes, cette limitation n'est pas faite quand elle n'est pas nécessitée par les besoins des démonstrations mathématiques. Car dans ces étapes, il est plus clair de n'introduire une limitation finie que lorsque cela devient utile ou nécessaire pour les besoins d'une preuve.

4.5 Degrés multitables

Définition 52. *Une relation $R \subset E \times F$ d'uplicité 1 est dite* 2-multitable *s'il existe $f^1, f^2 : E^2 \mapsto F$ tels que $\forall x^1, x^2 \in E[x^1 R(f^1(x^1, x^2)$ ou $x^2 R(f^2(x^1, x^2)]$.*

Commentaire.

• On voit le contenu intuitif de cette définition en imaginant deux mondes parallèles : dans un Monde1 se trouve un Casino1 qui propose à ses clients de jouer selon la règle suivante : le Casino1 met $x^1 \in E$ sur une table1 ; le joueur répond par $y \in E$, et gagne si $x^1 Ry$. Dans un Monde2 et son Casino2 il se passe la même chose, à ceci près que l'élément $x^2 \in E$ posé sur la table2 est indépendant de x^1. Imaginons un joueur qui joue dans les deux mondes à la fois, ayant connaissance simultanée de ce qui se passe dans les deux ; dans le Monde1 il joue la stratégie $f^1(x^1, x^2)$, et $f^2(x^1, x^2)$ dans le Monde2. Si de cette manière il est sûr de gagner dans au moins un des Mondes, alors R est 2-multitable...

• Il est facile d'étendre cette définition et son idée dans le cas où au lieu de 2 Mondes il y en a μ - un autre cardinal ≥ 1 : le joueur a μ stratégies, une dans chaque monde, mais chacune est fonction de tout ce qui a été posé sur les μ tables. Quand de surcroit l'uplicité passe de 1 à $\kappa > 1$, il faut veiller à ce qu'être multitable reste une propriété de R qui dépend seulement de son degré ludique. Cela s'obtient en demandant que chacune des μ stratégies f du multijoueur soit une fonction locale de ses μ arguments ; d'où la définition suivante :

Définition 53. *Soit $R \subset E \times F$ une relation d'uplicité quelconque ; R est μ-multitable s'il existe des fonctions locales $f^j : E^\mu \mapsto F, j < \mu$ telles que $\forall X \in E^\mu \exists j < \mu \; X^j R f^j(X)$, où X^j désigne la j-ème coordonnée de X.*

La donnée de ces fonctions $f^j, j < \mu$ est appelé une μ-stratégie gagnante pour R ; vu le caractère local des f^j, κ étant l'uplicité de R cela revient à l'existence de $(f_i^j)_{i<\kappa, j<\mu}$ vérifiant que pour toute famille $((x_i^j)_{i<\kappa})_{j<\mu} \in E^\mu$ il existe $k < \mu$ tel que $(x_i^k)_{i<\kappa} \; R \; [f_i^k((x_i^j)_{j<\mu})]_{i<\kappa}$.

Pour confirmer que le caractère μ-multitable de R ne dépend que de son degré ludique il suffit de vérifier que si $S \leq_L R$, alors de la μ-stratégie $f^j, j < \mu$ gagnante pour R se déduit une μ-stratégie gagnante pour S. Cela montrera également que la classe μ-multitable est une partie initiale de $\leq_L$. Soient donc E', F' tels que $S \subset E' \times F'$ et h, g locales telles que $\forall x, y[R(h(x), y) \Rightarrow S(x, g(x, y))]$. Pour $X = (X^j)_{j<\mu} \in (E')^\mu$ on pose $h^\mu(X) = (h(X^j))_{j<\mu}$; par hypothèse sur R, S il existe $k < \mu$ tel que $R(h^\mu(X))^k, f^k(h^\mu(X))$ et comme $(h^\mu(X))^k = (h(X^k)$ cela entraîne $S[X^k, g(X^k, f^k(h^\mu(X)))]$. Donc $g^j(X) = g[X^j, f^j(h^\mu(X))], j < \mu$ est une μ-stratégie gagnante pour S.

Définition 54. *R est multitable s'il est μ-multitable pour au moins un cardinal μ.*

La classe des degrés ludiques multitables est assez petite : les relations 1-multitables ne sont que celles de degré ludique trivial ; les relations μ-multitables, elles, sont triviales... dans un μ-monde !

Exemples de degrés multitable

Définition 55. *Soient κ, μ deux cardinaux quelconques. On appelle injection virtuelle de κ dans μ notée $inj(\kappa, \mu)$, le degré ludique de la relation R suivante : $(x, y)R(r, s)$ est défini si $x, s \in \kappa$ et $y, r \in \mu$; $(x, y)R(r, s)$ est vrai si de plus $y = r \Rightarrow x = s$.*

Sauf le cas inintéressant où $\mu = 1$ ces degrés sont multitables et ça se démontre facilement.

Proposition 56. *Si $\mu > 1$ alors $inj(\kappa, \mu)$ est multitable.*

Preuve. Il existe un cardinal t assez grand pour que $\mu^t = \kappa^t$. C'est tout ce que contient cette proposition ; car soit ϕ une bijection de l'ensemble de t-uplets κ^t sur l'ensemble de t-uplets μ^t et pour $j < t$ soient ϕ_j, ϕ_j^{-1} les applications : $(X \mapsto \phi(X))_j$ de κ^t vers μ et : $(Y \mapsto \phi^{-1}(Y))_j$ de μ^t vers κ. La famille de paires $f^j = (\phi_j, \phi_j^{-1}), j < t$ est une t-stratégie qui gagne pour la relation $inj(\kappa, \mu)$: elle se paye même le luxe de gagner sur toutes les t tables à la fois ! ⊣

Par ailleurs la classe multitable est close par segment initial, car une 2-réduction locale de R à S permet facilement de déduire une stratégie multitable pour R à partir de n'importe quelle stratégie multitable pour S.

4.6 La classe $\neg TSD$

D'abord le cas d'uplicité 2.

Définition 57. *Soit R un degré ludique binaire. R est TSD de haut en bas si elle satisfait*

$$(1)\ \exists x_2 \forall r_2 \exists x_1 \forall r_1\ \neg R[(x_1, x_2), (r_1, r_2)].$$

On définit de façon analogue TSD de bas en haut : c'est la définition ci-dessus appliquée à la relation $(x_1, x_2) R^\tau (r_1, r_2)$ définie par $(x_2, x_1) R (r_2, r_1)$. Et on dit que R est $\neg TSD$ si R ne transmet d'information ni vers le haut, ni vers le bas.

Proposition 58. *Soit $R \subset E^2 \times F^2$ un degré ludique d'uplicité 2. R est TSD de haut en bas ssi*

$$2)\ \exists (b \in E, f : E \mapsto E, g : F \mapsto E) \forall x \in E, r_1, r_2 \in F\ (f(x), b) R (r_1, r_2) \implies g(r_2) \neq x.$$

Preuve. On suppose (2) ; alors (1) est vérifié en prenant $x_2 = b, x_1 = f(g(r_2))$, car $(f(g(r_2)), b) R (r_1, r_2)$ implique la contradiction $g(r_2) \neq g(r_2)$. Supposons (1) ; soient b, g tels que (2) est vérifié pour tous r_1, r_2 avec $x_2 = b, x_1 = g(r_2) : \neg (f(g(r_2)), b) R (r_1, r_2)$. Alors $(f(x), b) R (r_1, r_2)$ implique $g(r_2) \neq x$, et b, f, g vérifient (2). ⊣

La Proposition ci-dessus donne la raison pour laquelle la définition (1) est baptisée "transmission de haut en bas". Car la condition (2) entraîne que pour tout $x \in E$ un téléphone R-garanti peut transmettre un $y \neq x$: on émet le message $(f(x), b)$ par le téléphone, donc le récepteur obtient (r_1, r_2) tel que $(f(x), b) R (r_1, r_2)$ et transmet $y = g(r_2) \neq x$. On note ici que la plus petite des informations concernant x est de connaître un y tel que $x \neq y$; d'où le $g(r_2) \neq x$ de la définition ci-dessus : c'est le minimum d'information dont on puisse demander la transmission.

On définit maintenant la classe TSD dans le cas général (c'est-à-dire des degrés ludiques d'uplicité finie quelconque). Pour cela on généralise la définition (1) de manière naturelle :

Définition 59. *Soit R un degré ludique d'uplicité n finie. On dit que $R \in TSD$ s'il existe une permutation σ de $\{1, 2.., n\}$, telle que $\exists x_1 \forall r_1 \exists x_2 \forall r_2 \ldots . \exists x_n \forall r_n [(x_{\sigma(i)})_i, (r_{\sigma(i)})_i] \notin R$*

Dans le cas $n = 2$ on retrouve la deuxième définition de TSD, les permutations σ se limitant à l'identité et la transposition τ.

Théorème 60. *Les classes TSD et $\neg TSD$ sont toutes deux closes par produit (fini ou infini) de degrés ludiques (évidemment tous de même uplicité).*

Esquisse de preuve. Si pour tout

$$i, \forall x_1 \exists y_1 \forall x_2 \ldots (x_1, \ldots, x_n) R_i (y_1, \ldots y_n)$$

alors

$$\forall (x_1^i)_i \exists (y_1^i)_i \forall (x_2^i)_i \ldots \forall i \ (x_1^i, \ldots) R_i (y_1^i, \ldots, y_n^i). \qquad \dashv$$

Théorème 61. *Soit R un degré ludique qui est multitable. Alors $R \in \neg TSD$.*

Preuve. Supposons :

$$\exists x_1 \forall r_1 \exists x_2 \forall r_2 \ldots . \exists x_n \forall r_n \neg [(x_1, \ldots, x_n) R (r_1, \ldots r_n)]$$

qui est l'une des manières pour R d'être $\neg TSD$. Autrement dit le joueur J1 a une stratégie gagnante σ dans le jeu où à la i-ème ronde J1 choisit x_i puis J2 r_i, et à la fin J1 gagne si $\neg[(x_1, \ldots, x_n) R (r_1, \ldots r_n)$. Faisons jouer J1 contre J2 sur μ tables distinctes, en laissant J1 appliquer σ sur chacune des tables, et J2 appliquer une stratégie μ-multitable pour le jeu R.

Objection : le premier jeu dure n rondes, alors que la stratégie de J2 concerne un jeu en une seule ronde ; mais cette stratégie est locale d'uplicité n, elle comporte donc n fonctions dont la i-ème donne – sur chacune des μ tables – la réponse de J2 qu'il faut à la i-ème ronde. Sur chaque table, J2 va perdre puisque la stratégie de J1 est gagnante ; ainsi J2 n'a pas de μ-stratégie gagnante et cela montre que R n'est pas μ-multitable, prouvant le théorème sous forme contraposée. $\dashv$

4.7 Multi-tables = $\neg TSD$

L'égalité de ces deux classes résulte du théorème précédent et du suivant :

Théorème 62. *Soit R un degré ludique d'uplicité finie qui n'est pas TSD. Alors R est multitable.*

La classe $\neg TSD$ étant close par produits, ce théorème a une conséquence qui ne se devine vraiment pas, si on ignore l'égalité de Multitable avec $\neg TSD$:

Corollaire 63. *La classe Multitable est stable par produits.*

La preuve du théorème donnée ci-dessous utilisera l'Hypothèse Généralisée du Continu HCG (plus précisément : pour tout ensemble E il existe un cardinal régulier κ tel que $|E|^\kappa = \kappa^+, E^{\kappa^+} = \kappa^{++}$. A. Khélif a éliminé cette hypothèse de la preuve, en utilisant la méthode de Shelah – basée sur le lemme de Ketonen – qui élimine l'hypothèse du continu de la caractérisation due à Keisler des classes élémentaires au moyen d'ultraproduits et ultrapuissances. La preuve s'étend même aux uplicités infinies, moyennant un axiome de théorie des ensembles dont on connaît mal la force : on sait juste qu'il est conséquence de certains axiomes de grand cardinal.)

Proposition 64. *$\neg TSD$ est une partie initiale de $\leq_L$.*

Preuve. Supposons $S \leq_n R \in \neg TSD$: on a une réduction locale (h, g) de S à R, d'autre part

$$1_R)\ \forall x_1 \exists y_1 \forall x_2 \ \dots\ \forall x_n \exists y_n\ (x_1, \dots, x_n) R(y_1, \dots y_n)$$

– et de même avec les relations R^σ. (1_R) s'exprime par l'existence d'une stratégie gagnante τ_R pour le joueur J2 dans le jeu en n rondes utilisé par la preuve du Théorème 61. Nous définissons une stratégie gagnante τ_S pour le joueur J2 dans le même jeu où l'on remplace R par S : ceci montrera (1_S), et en appliquant le même raisonnement à R^σ, S^σ pour toute permutation σ de $\{1, \dots, n\}$ on conclura $S \in \neg TSD$. On note (2_S) la définition :

$$\tau_S(x_1) = g_1(x_1, \tau_R(h_1(x_1)))\ ;\ \tau_S(x_1, x_2) = g_2(x_2, \tau_R(h_1(x_1), h_2(x_2)); \dots$$
$$\tau_S(x_1, \dots, x_n) = g_n(x_n, \tau_R(h_1(x_1), \dots, h_n(x_n)).$$

L'application de τ_R fait que

$$(h_1(x_1), \dots, h_n(x_n)) R[(\tau_R(h_1(x_1)), \dots, \tau_R(h_1(x_1), \dots, h_n(x_n))].$$

Du fait que (h, g) réduit S à R, (2_S) se traduit alors par

$$(x_1, \dots, x_n) S(\tau_S(x_1), \dots, \tau_S(x_1, \dots, x_n)).$$

Autrement dit, τ_S est une stratégie gagnante pour S ce qui montre (1_S) – et montre $S \in \neg TSD$, en faisant jouer tous les S^σ. ⊣

Esquisse de preuve du théorème. Nous le prouvons pour l'uplicité 2 ; la généralisation à l'uplicité finie est de routine. Soit E un ensemble de cardinal régulier κ tel que $2^\kappa = \kappa^+, 2^{\kappa^+} = \kappa^{++}$. Nous allons construire une stratégie multitable universelle pour les relations $R \in \neg TSD$ dont le domaine est E. Cela montre que toutes ces relations sont multitables, démontrant le théorème aussi pour tout $S \in \neg TSD$ dont le domaine est de cardinal $\leq \kappa$. En effet la sous-section 1.5.3 a montré que S a une extension canonique R de même degré ludique et dont le domaine est de cardinal κ; et R multitable entraîne alors S multitable.

On fixe une bijection r de κ^{++} sur ${}^{\kappa^+}E$. Par récurrence transfinie sur $\alpha < \kappa^{++}$ on définit deux applications g_α, puis h_α allant de κ^+ dans E ; ainsi $g_\alpha, h_\alpha, r(\alpha)$ seront tous dans ${}^{\kappa^+}E$. On choisit d'abord g_0 quelconque ; ensuite h_0 est choisi de manière que pour chaque $a \in E$, si $X := g_0^{-1}(a)$ est de cardinal κ^+, l'image de X par h_0 égale E tout entier. Prenons comme hypothèse d'induction (vraie pour $\alpha = 1$) que $g(\beta), h(\beta)$ sont choisis pour tout $\beta < \alpha$. Soient F un ensemble fini d'ordinaux $< \alpha$ et X un sous-ensemble de κ^+ de cardinal κ^+ qui est intersection finie d'images inverses de constantes de E par les applications $g_\beta, \beta \in F, h_\beta, \beta \in F$ ainsi que par l'application $r(\alpha)$. Il est facile de choisir g_α de manière que l'image de X par cette application énumère tout E; de plus pour α fixé, le nombre de ces ensembles F, X est borné par κ^+ car α est de cardinal au plus κ^+. Donc il est possible d'imposer cette exigence à g_α simultanément pour chacun de ces ensembles X. g_α une fois fixé, on choisit ensuite h_α de manière qu'il vérifie les mêmes exigences, augmentées en ajoutant la fonction g_α à toutes celles dont on prend les images inverses.

Soit R n'importe quelle relation $\neg TSD$ de domaine E^2 ; on va constater que :

$$\forall x, y \in {}^{\kappa^+}E \; \exists i < \kappa^+ \; (x(i), y(i)) R (g^*(x)(i), h^*(y))(i),$$

où $g^*(x)$ désigne la fonction $g_{r^{-1}(x)}$. Autrement dit, le couple de fonctions $[(x, i) \mapsto g^*(x)(i), (y, i) \mapsto h^*(y)(i)]$ est une stratégie locale qui réalise R sur κ^+ tables – ou encore, R est κ^+-multitable. Notez que cette stratégie (g^*, h^*) ne dépend que de E, pas de R – c'est son "universalité".

Premier cas : $r^{-1}(x) \leq r^{-1}(y)$ – Puisque R est $\neg TSD$,

$$\exists x_1 \forall y_1 \exists x_2 \forall y_2 \; (x_1, x_2) R (y_1, y_2).$$

Pour toute valeur $a \in E$ prise par la fonction x – et l'une de ces valeurs est prise κ^+ fois – il existe une valeur b de $g^*(x)$ sur $\{i, x(i) = a\}$ prise κ^+ fois telle que pour toute valeur c de y prise κ^+ fois par y sur $\{i : \; x(i) = a \; et \; g^*(x)(i) = b\}$ il existe une valeur d prise par $h^*(y)$ sur $\{i, x(i) = a, g^*(x)(i) = b, y(i) = c\}$ telles que $(a, c) R (b, d)$ – puisque par construction $h(y)$ prend toutes les valeurs possibles sur $\{i, x(i) = a, g(x)^*(i) = b, y(i) = c\}$.

Si $r^{-1}(x) > r^{-1}(y)$ on raisonne de la même manière. $\dashv$

La Proposition 51 a exhibé une famille simple de relations dont les degrés sont cofinaux sous ∞_L ; on peut également donner une famille simple de relations dont les degrés sont cofinaux dans la classe $\neg TSD$.

Proposition 65. *Posons* $(x_0, x_1) S_E (y_0, y_1)$ *ssi* [$y_0 = (x_0, x_1, 0)$ *et* $y_1 = (a, b, 0)$ *ou* $y_0 = (c, d, 1)$ *et* $y_1 = (x_0, x_1, 1)$] - $a, b, c, d, x_0, x_1, y_0, y_1$ *variant sur l'ensemble* E *;* S_E *est* $\neg TSD$ *et* $S \leq_L S_E$ *pour tout* $S \in \neg TSD$ *de domaine* E^2.

Preuve laissée en exercice (en s'inspirant de celle de la Proposition 51).

4.8 Sous-classes remarquables de $\neg TSD$

Comme déjà dit, $\neg TSD$ est taillé grand, il faut découper cette classe plus finement. Une première manière utilise des idées de logicien éloignées de la Physique ; nous ne faisons qu'en dire un mot.

4.8.1 Degrés forçables

Soit R une relation ludique d'arité finie, on dit que R est forçable s'il existe une extension générique de l'univers dans laquelle $R =_L 1$.

Proposition 66. *Être ou ne pas être forçable ne dépend que du degré ludique de* R.

Le fait suivant montre que les degrés forçables sont petits.

Théorème 67. *Si* R *est forçable alors* R *est* $\neg TSD$.

Un fait bien connu sous une autre forme peut s'exprimer avec notre langage.

Proposition 68. *antirec*(κ) *n'est jamais forçable.*

Ceci resulte du fait qu'un forcing ne transforme jamais une relation bien fondée en relation mal fondée.

Corollaire 69. *antirec*(κ) *n'est jamais* $inj(\kappa, \mu)$ *dès que* μ *est infini.*

Les deux sous-classes suivantes de $\neg TSD$ ont été introduites postérieurement à [10], et nous les présentons sans démonstrations.

4.8.2 Degrés casino-inoffensifs

Certains effets quantiques se réalisent chaque fois qu'on en fait l'expérience ; le téléphone GHZ de la Parenthèse heuristique de Section 1 en est un exemple. D'autres consistent à modifier la probabilité d'un événement, de manière qu'elle diffère de la probabilité en l'absence d'effet quantique (cf les inégalités de Bell). Jusqu'ici la classe $\neg TSD$ ne fait aucune place explicite à de tels effets quantiques probabilistes.

Voici une sous-classe de $\neg TSD$ destinée à réparer ce manque. L'idée vient de la notion de degré de Tukey inoffensif : une relation R tellement peu fidéle que si depuis l'avenir un accolyte emploie un téléphone R-garanti pour informer un Joueur vivant dans le présent (dans un jeu infini à information parfaite), et même si l'accolyte sait tout l'avenir de la partie en cours, il ne parvient pas à aider le Joueur... Deux différences ici : 1, il s'agit de jeux finis ; et 2, inoffensif est pris dans le sens : qui ne modifie pas **l'espérance de gain** du Joueur. Cela conduit à la notion suivante:

Définition 70. *Une relation binaire d'uplicité 2* $R \subset (E_1 \times F_1) \times (E_2 \times F_2)$ *-* F_1, F_2 *étant finis – est dite casino-inoffensive ssi il existe* $p : (E_1 \times F_1 \times E_2 \times F_2) \longrightarrow [0,1]$ *avec les propriétés suivantes:*

- *Si* $(x_1, x_2) \in E_1 \times E_2$, *la somme des* $p(x_1, x_2, y_1, y_2)$ *(*(y_1, y_2) *variant dans* $E_1 \times E_2$*) est égale à 1.*
- *Pour un triplet* $(x_1, x_2, y_1) \in E_1 \times F_1 \times E_2$, *la somme des* $p(x_1, x_2, y_1, y)$ *(*y *variant dans* F_2*) ne dépend que* x_1, y_1 *– propriété restant vérifiée si on permute les indices 1 et 2.*
- *Si* $p(x_1, x_2, y_1, y_2) \neq 0$, *on a* $(x_1, x_2, y_1, y_2) \in E$.

Proposition 71. *À équivalence ludique près, la classe Casino-inoffensif est sous-classe propre de* $\neg TSD$.

Cette classe réussit effectivement à contenir (à $\equiv_L$ près) les effets quantiques probabilistes connus. Mais elle contient aussi des relations qui ne doivent rien à la Mécanique Quantique, ce qui n'est pas étonnant, puisque seules les idées probabilistes ont été ajoutées à la notion d'inoffensivité, sans s'inspirer de la MQ. Cela nous amène à la classe suivante, destinée à “coller” à la Théorie Quantique.

4.8.3 Degrés FMQ

Définition 72. *Une relation binaire d'uplicité 2 et de domaine fini est FMQ (Fabricable par la Mécanique Quantique) ssi elle se réduit ludiquement à une relation* $R \subset (E_1 \times F_1) \times (E_2 \times F_2)$ *de la sorte suivante.*

$F_1 = F_2 = \{1, 2, \ldots, k\}$, *$E_1, E_2$ sont l'ensemble des bases orthonormales d'un espace euclidien E de dimension k. On a une forme bilinéaire non nulle A sur E. Soit $(u, i, v, j) \in E_1 \times F_1 \times E_2 \times F_2$, soient u_i, v_i les i-ème vecteurs des bases u, v. Alors $(u, i, v, j) \in R$ si et seulement si $A(u_i, v_j) \neq 0$.*

Cette définition est faite pour pouvoir démontrer que les degrés FMQ sont bien ceux qui ont un représentant correspondant à un effet théorique de la mécanique Quantique – à condition que cet effet se laisse représenter par une relation binaire d'uplicité 2. Le "téléphone GHZ" de la Section 1 montre que ce n'est pas toujours le cas : sa représentation est d'uplicité 3 ; mais la définition de FMQ s'étend de manière évidente à toutes les uplicités finies. Alors FMQ d'uplicité 3 contient le "téléphone GFZ". De tous les effets quantiques connus, ne reste finalement en-dehors de la classe FMQ que "l'inspection non explosive d'une bombe".

Il existe encore d'autres classes fines dans $\neg TSD$; et la réduction ludique a cette souplesse bienvenue faisant que les notions de casino-inoffensif et de FMQ ont des équivalents ludiques constitués par des conditions purement combinatoires [7]. De plus, la caractérisation combinatoire de FMQ possède une interprétation physique multimonde. Mais la présente sous-section n'a contenu qu'un bref aperçu sans démonstrations sur les développements actuels ; pour un exposé systématique voir [7, 11].

4.8.4 Téléphones déterministes

Définition 73. *On dira qu'une relation R i.e. d'uplicité 2 est déterministe si R est le graphe d'une fonction : pour tout $x = (x_1, x_2)$ il existe au plus un $y = (y_1, y_2)$ tel que $R(x, y)$.*

Proposition 74. *Si R est déterministe non triviale alors R est TSD. Et même plus : $I_2 \leq R$, autrement dit un téléphone $R-$garanti permet de transmettre au moins autant d'information qu'un téléphone I_2-garanti.*

La preuve de la proposition est assez simple pour être comprise sur un cas particulier : imaginons que $R \subseteq E^2 \times E^2$ et qu'il y a une fonction (f, g) telle que $R(x, y)$ ssi $y_1 = f_1(x)$ et $y_2 = f_2(x)$. Puisque $R \neq 1$, l'une au moins des deux f_i 'ne dépend pas que de x_i'. Wlog disons que c'est f_2. Soit $a, b, x_2 \in E$ tels que $f_2(a, x_2) \neq f_2(b, x_2)$. L'utilisateur du deuxième combiné peut donc appuyer sur la touche x_2. Avant d'être éloignés l'un de l'autre, les communiquants A,B ont convenu que si A veut envoyer le message '-', il appuiera sur a, pour envoyer le message '+', il appuiera sur b. L'utilisateur du combiné2 sait donc à la réception de $f_2(x_1, x_2)$ si $x_1 =$ '$-$' ou si $x_1 =$ '$+$'. On laisse au lecteur le soin de montrer $I_2 \leq R$ à partir de ce fait. $\dashv$

Remarque heuristique. Si la nature est déterministe alors tous les téléphones qu'elle permet de fabriquer sont déterministes ; et la proposition qui précède en conclut que tous ces téléphones sont $=_L$ 1 ou TSD. D'autre part, la relativité entraîne l'inexistence de téléphones I_2-garantis qui soient instantanés. Comme I_2 est TSD on conclut finalement :

si la nature est déterministe alors tous les téléphones qu'elle permet de fabriquer sont **triviaux**.

Ce raisonnement est exactement celui qui permet d'obtenir une démonstration rigoureuse du non-déterminisme de la théorie quantique. Celui-ci était soupçonné depuis les années 1920–30 (congrès de Copenhague 1927, en particulier) mais non-démontré scientifiquement. Les arguments étaient d'une nature philosophique. Les dates importantes sont celles de 1966, 1982, et les années 1990 (les premiers téléphones $\neg TSD$, plusieurs références dans [4]). Une preuve existe donc à strictement parler seulement depuis les années 90. Un point important est qu'il n'existe aucune démonstration (au sens strict, i.e. non philosophique) connue du non-déterminisme de la MQ sans se servir des hypothèses relativistes. D'où la pertinence de la question traitée ci-dessous.

4.9 Compatibilité de la théorie quantique avec la théorie relativiste

Nous ne faisons que dire un mot des importants et nombreux problèmes soulevés par la consistance d'une théorie physique avec la théorie relativiste. Question : de quelles certitudes dispose-t-on qu'il n'existe pas d'expérience astucieuse (même si on ne voit pas bien laquelle) qui permettrait d'obtenir une contradiction entre la mécanique quantique et la relativité ? ou encore entre une nouvelle théorie et la relativité.

À partir du moment (Bell 1966, expérience d'Alain Aspect 1982), où on a commencé à construire les premiers téléphones ludiques non-triviaux, on est fondé à demander jusqu'à quel point on peut (au moins par la pensée) admettre l'existence de tels téléphones dans un univers relativiste. Notre résultat principal va donner une réponse substantielle. Il est important de comprendre les exigences et difficultés de la question : certes, quel que soit le téléphone T (non TSD), on ne voit pas comment obtenir une contradiction expérimentale entre la relativité et l'existence de T d'une manière canonique c'est-à-dire en utilisant T pour produire la boucle temporelle classique qui permet de démontrer qu'on ne peut informer les gens du présent de ce qui va se produire dans le futur. Mais cela ne veut pas dire *a priori* qu'il n'existe pas une astuce et un montage expérimental très raffiné qui rendrait incompatible l'existence de T et les hypothèses d'Einstein. Un tel montage pourrait *a priori* utiliser un tout autre argument que le caractère $\neq 1$ de T. Voire même pourquoi pas un montage combinant plusieurs téléphones $\neg TSD$. *Et le fait de ne pas trouver de contradiction à une théorie n'implique pas qu'il n'en existe pas.* Nous répondons à la question malgré ces obstacles.

Proposition 75. *Tout montage n'utilisant que des téléphones $\neg TSD$ est compatible avec la relativité.*

Idée intuitive de la preuve.

• soit c un cardinal tel que le produit de tous les téléphones du montage soit $c-$multitable (c existe d'après le Corollaire 63). Il existe donc une stratégie $c-$multitable σ telle que si on l'applique sur c tables, et si sur la i-iéme table le Casino i joue une suite x^i constituée d'une entrée pour chacun des c téléphones, alors pour au moins un $k < c$, la réponse $\sigma_k((x^i)_{i<c})$ fournie par σ sur la k-iéme table respecte la garantie de chacun des c téléphones. Supposons c mondes relativistes avec dans chacun une équipe de physiciens qui effectue et teste le montage. La stratégie $c-$multitable montre que dans au moins l'un de ces mondes pris comme tables, tous les téléphones du montage fonctionnent – donc tout le montage fonctionne dans ce monde relativiste. Les physiciens de ce monde en tirent la conclusion que le montage est compatible avec la Relativité.

• La théorie des ensembles ZF est bien entendu capable de formaliser ces c mondes relativistes et la fantasmagorie multi-monde qui vient de vérifier la consistance relativiste du montage dans l'un des mondes. Car l'existence mathématique, non physique de multi-mondes n'est pas un problème dans ZF ; or l'existence seulement mathématique suffit à la consistance...

5 Conclusion

MQ et déterminisme

Répétons la fin de Section 4 en la résumant : il existe une preuve complète que la MQ est incompatible avec le déterminisme. D'abord la MQ implique l'existence de téléphones ayant un degré ludique non trivial (exemple : GHZ) ; ensuite la Proposition 74 prouve que tout téléphone déterministe non trivial est TSD ; et enfin la relativité pose en **axiome** l'inexistence d'un téléphone TSD (il n'y a pas besoin de raisonnement, c'est l'un principes-clé de la relativité).

Cette preuve combine MQ et Relativité (et sinon on n'en connaît pas) ; mais la Proposition 75 nous autorise cette combinaison.

Pourquoi parler de localité ?

En poussant la preuve qui précède on pourrait montrer : *en MQ il n'y a ni variables cachées locales, ni variables cachées non locales (autrement dit il n'y a pas de variables cachées du tout).* Et ceci reste une conséquence **formellement déductible** de la conjonction de prédictions simples de la MQ, et d'axiomes de la Relativité, le tout ayant une vérification expérimentale.

On n'entre pas ici dans l'exposé systématique de ce fait (en particulier, la notion de variables cachées, locales ou non n'est pas définie). Mais on en signale la conséquence pratique : la discussion philosophique de la MQ est **trop** philosophique, passant à côté de preuves ou réfutations plus scientifiques et formelles.

Les résultats de cet article vont un peu plus loin que cette conséquence, dans la direction suivante : ils développent une manière de mesurer disons *les degrés de magie* de n'importe quelle théorie potentielle, aussi farfelue soit-elle, et ne se cantonnent pas à Relativité + MQ.

Ils permettent surtout de donner corps à un paradigme, que nous pouvons ici appeler *le paradigme téléphonique*, vu l'heuristique que nous avons donnée des degrés ludique (avec des "téléphones" qui n'ont bien sûr rien à voir avec le développement des réseaux téléphoniques mobiles). Le paradigme est utile pour guider l'extension des techniques développées à propos des jeux à information parfaite (généralement par les logiciens et les spécialistes de théorie des ensembles). L'extension met les techniques en question à la portée de tous les autres jeux, à savoir : les jeux à information imparfaite (généralement étudiés par les probabilistes ou les économistes) ; les jeux en équipe, les jeux contre soi-même, les jeux amnésiques, et finalement la grande partie de l'homme "contre" son environnement. Mais cette extension est en cours dans [11].

Le paradoxe EPR (dans ses manifestations formalisées qu'ont été GHZ, etc, à partir des années 1980-1990) est un pied de nez offert par notre environnement à notre équipe de champions physiciens : la Nature gagne alors qu'on peut prouver **qu'il n'y a pas de stratégie gagnante** (Mais comment fait-elle alors pour gagner ?). D'où la conclusion de type "bottage en touche" labelisée par le mot *indéterminisme* ; mais on peut parier qu'il y aura d'autres tournois spectaculaires entre le réel et les physiciens, et seule la formalisation permet d'avoir les idées claires et d'enquêter. À un moment donné, les mots ont une limite dont le flou est trop grand pour affronter les subtilités ludiques qui apparaissent comme **données** par un monde extérieur dont on a fini par s'apercevoir que sa véritable nature nous échappe encore.

Ces défis sur ce qu'est le réel ne peuvent plus être étudiés sans ajouter au langage et aux concepts actuels des philosophes le bagage totalement précis que constitue la théorie des degrés ludiques, dans les développements résumés ici et dans les extensions en cours. En effet on vient de voir que ce bagage nous rend capables de trancher des questions philosophiques sur la MQ qui sont restées incertaines pendant des décennies. . .

Références

[1] Kunen, *Set Theory*, Springer Verlag, 1980.

[2] Kanamori, *Higher infinite*, Springer Verlag, 1994.

[3] Bernard D'espagnat, *Le réel voilé*, Paris, Fayard, 1994.

[4] Roger Penrose, *Les ombres de l'esprit*, traduction de *Shadows of mind*, Paris, InterÉditions, 1995.

[5] Voir chap 12 de [3].

[6] C. Chalons, The Target's theorem, *C. R. Acad. Sci. Paris*, Série I, t. 331, 2000, pp. 1–6.

[7] A. Khélif, *Degrés inoffensifs et degrés FMQ*, 2012, prépublication.

[8] A. C. Elitzur and L. Vaidman, Quantum mechanical interaction-free measurements, *Found. Phys.*, vol. 23, 1993, pp. 987–97.

[9] Daniel M. Greenberger, Michael A. Horne, and Anton Zeilinger, *Going Beyond Bell's Theorem*, 2008, (publication électronique sur le site `http://arxiv.org/abs/0712.0921`).

[10] C. Chalons, *Degrés ludiques*, 1999, prépublication (à trouver avec d'autres travaux connexes sur le site internet : `http://www.logique.jussieu.fr/~chalons/z2009/pouranatole.php`

[11] C. Chalons, *Le paradigme téléphonique*, 2012, prépublication (à trouver sur le même site internet).

[12] A. Kechris, *Classical Descriptive Set Theory*, Springer-Verlag, 1995.

[13] J.W. Tukey, *Convergence and Uniformity in Topology*, Princeton University Press, 1940.

[14] P. Vojtàš, Generalized Galois-Tukey connections between explicit relations on classical objects of real analysis, *Israel Math. Conference Proceedings*, ed. H. Judah, vol. 6, 1991, pp. 619–643.

4

Primes in Models of $I\Delta_0 + \Omega_1$: Density in Henselizations

PAOLA D'AQUINO[1] AND ANGUS MACINTYRE[2]

Abstract: We work in a model $\mathcal{M}$ of $I\Delta_0 + \Omega_1$, with fraction field K. For each prime p of $\mathcal{M}$ we consider the p-adic valuation v_p on K, and prove that K is dense in its Henselization K_p^h. This will be applied in our project of proving quadratic reciprocity law in $\mathcal{M}$.

1 Some personal remarks

Alan's early work, from his Ph.D. thesis [18] and the Paris-Wilkie-Woods paper [16], is surely one of the greatest achievements in the model theory of arithmetic. That subject, and particularly the part where one tries to connect to number theory and complexity theory, remains perhaps the most daunting of all the classical challenges in mathematical logic.

For us, in a collaboration over twenty years, the key notions coming from Alan relate to **prime** and **pigeonhole principle** in fragments strictly below $I\Delta_0+exp$ [13]. In the latter system one generally has little difficulty in adapting classical proofs of classical theorems of arithmetic. All this changes if one descends to $I\Delta_0$ where little has been done. We only know that every element is divisible by a prime and for every prime p there is a p-adic valuation of each element. A version of the Chinese

[1]Department of Mathematics and Physics, Seconda Universitá di Napoli, Viale Lincoln 5, 81100 Caserta, paola.daquino@unina2.it

[2]School of Mathematical Sciences, Queen Mary, University of London, Mile End Road, London E1 4NS, angus@dcs.qmul.ac.uk

Studies in Weak Arithmetics.
Patrick Cégielski, Charalampos Cornaros,
Costas Dimitracopoulos.

Remainder Theorem can be proved in $I\Delta_0$ which allows to define the number of primes dividing an element [6]. In particular one still does not know if one can prove cofinality of the primes. Nor one does know if the Δ_0-Pigeonhole Principle can be proved (the expectation, after Ajtai's work [1], is that it cannot be). But we have Alan's beautiful Theorem that the Δ_0-Pigeonhole Principle (Δ_0-PHP) implies cofinality of the primes, and indeed Bertrand's Postulate. This was the first example of something very congenial to us, namely that only one proof, and that little-known, of a classical result with very many proofs, adapts to a proof in a weak fragment (in this case, the proof is due to Sylvester). Alan knew the classical literature very well indeed, and for us was the natural person to consult as we became involved with our longterm project of proving quadratic reciprocity in $I\Delta_0 + \Omega_1$ see [9, 10]. We could count on his knowledge of the classical literature on primes, as well as his deep insights on hardcore complexity theory.

Beyond number theory, his work on cofinality of primes brought to the forefront weaker versions of Δ_0-PHP, for example that concerning Δ_0-injections from $2n$ to n. This turned out to suffice for cofinality of primes. It, in turn, was shown in [16] to follow from Ω_1, the principle that dyadic lengths are closed under products [7]. This principle, much weaker than *exp*, is naturally connected to polynomial-time computability.

In our joint work on primes we have found that $I\Delta_0 + \Omega_1$ provides a natural setting for developing the classical arithmetic of primes. In the present paper we make a novel combination of these principles with the cell-decomposition theory for Henselian fields begun by Paul Cohen [5] and refined and exploited by Jan Denef [11].

As we entered the last stages of our work on quadratic reciprocity, we took it for granted that we would have comments and improvements from Alan, and our work would not feel complete without his approval. Now we are left with a deep personal and scientific loss.

Alan was a true original. The early work remains inspiring after thirty years, and unsurpassed in its genre. Afterwards he thought long and hard about the heroic problems of complexity theory, and it came naturally to him to share his ideas. One cannot imagine him in priority disputes, or in bragging, or in publishing routine things for career purposes.

We both found him very congenial company, interested in everything, independent, funny, and kind. It was a privilege to work with his ideas and great fun to socialize with him. The ideas will of course survive, but we will greatly miss the meetings.

2 Introduction

In a series of papers [9, 10] we have looked at classical elementary number theory in models of $I\Delta_0 + \Omega_1$, where Ω_1 stands for the axiom $\forall x, y \exists z (x^{[\log_2 y]} = z)$. Let $\sharp$ be a new function symbol, where $\sharp(x, y) = x^{[\log_2 y]}$. If we expand the language of arithmetic by $\sharp$, and we allow induction on bounded formulas in the expanded language $\mathcal{L}(\sharp)$ (the bounds now are terms containing the symbol $\sharp$) we obtain a theory $I\Delta_0^\sharp$ that is well-known to be bi-interpretable with $I\Delta_0 + \Omega_1$. We consider ordered rings $\mathcal{M}$ whose positive parts are models of $I\Delta_0 + \Omega_1$, and let K be the field of fractions of $\mathcal{M}$. It is known (see [8]) that irreducible elements in $\mathcal{M}$ are prime, and indeed maximal. We consider for each prime p the p-adic valuation v_p on $\mathcal{M}$, and then extend it to K. It is known that $v_p : K \to \mathcal{M}$ is a total function with Δ_0-graph (see [6]), and v_p need not be surjective [6]. Indeed, its range is the *Bennett segment* for p, a Σ_1-definable convex subgroup of $(\mathcal{M}, +)$ [13]. The value group Γ_p consists of *lengths in p-adic notation.* Since 2-adic length bounds p-adic length for all p, p-adic length bounds are also allowed in the induction.

Now v_p on K has a unique Henselization K_p^h, algebraic over K, and an immediate extension. K_p^h is not interpretable in $\mathcal{M}$ in general, and in particular if $\mathcal{M}$ is recursively saturated. However in [10], working on the quadratic reciprocity law in $\mathcal{M}$, we found it convenient to use the K_p^h. In [10] we remarked that the Ax-Kochen-Ershov work gives uniformly for all p, standard or not, and all first-order formulas of the language of valued fields, a Δ_0-way of interpreting in $\mathcal{M}$ satisfaction in K_p^h *for elements of* K.

Note that Γ_p is a $\mathbb{Z}$-group [10], but we know little about k_p, the residue field of v_p. For p standard, $k_p = \mathbb{F}_p$, and for p nonstandard k_p has characteristic 0. In [8] we proved that k_p is quasifinite, i.e. has exactly one extension of each standard finite dimension. But we see no reason to think that for nonstandard p, k_p is pseudofinite.

Merely because $\mathcal{M}$ is a model of open induction, K is dense in its real closure, and indeed $\mathcal{M}$ forms an *integer part* for that real closure. It is natural then to ask how K sits inside K_p^h, and in particular whether we avoid the well-known pathology of valued fields not dense in their Henselizations, see [15]. This is a natural issue in the study of quadratic forms, since in the classical theory one does make serious use of approximation theory.

For open induction there is already so much pathology about primes, let alone exponentiation, that we have not looked at possibilities there. But for $I\Delta_0 + \Omega_1$ we have

Theorem 1. *K is dense in K_p^h, for every prime p.*

The proof makes essential use of a major result from the model theory of Henselian fields, namely the Denef-Pas uniform Cell Decomposition [11, 17]. We find it very satisfying that one can exploit in models of arithmetic deep uniformities in the model theory of Henselian fields. Our use of cell-decomposition is deeper than our use of Ax-Kochen-Ershov.

3 The Proof

Fix $\mathcal{M}$, a model of $I\Delta_0+\Omega_1$, with field of fractions K. Fix a prime p in $\mathcal{M}$. Suppose $\alpha \in K_p^h$, $\alpha \notin K$. The element α is algebraic over K, say of degree n (standard), with minimum polynomial $f(x) = x^n + \lambda_1 x^{n-1} + \ldots + \lambda_{n-1}x + \lambda_n$, with $\lambda_1, \ldots, \lambda_n \in K$.

Lemma 1. *There is no greatest $v_p(\beta - \alpha)$, as β runs through K.*

Proof. Fix $\beta \in K$. We need to get ϵ in K so that $v_p(\beta + \epsilon - \alpha) > v_p(\beta - \alpha)$. Now, $\beta \neq \alpha$ since $\alpha \notin K$. Then the above condition gives $v_p\left(\frac{\epsilon}{\beta-\alpha}+1\right) > 0$. So we first require $v_p\left(\frac{\epsilon}{\beta-\alpha}\right) = 0$. Since $v_p(\beta-\alpha)$ is in the value group of K, there is $\tau \in K$ with $v_p(\tau) = v_p(\beta-\alpha)$. Now since the residue field of K_p^h does not extend the residue field of K_p, there is $\mu \in K$ with $v_p(\mu) = 0$ and $v_p\left(\frac{\tau}{\beta-\alpha} - \mu\right) > 0$. Now take $\epsilon = -\tau/\mu$, and the proof is complete. We have used that the immersion $K \to K_p^h$ is immediate (see also [14]). ⊣

Note that this alone has no implications about density [15]. What we want is to start with $\gamma \in \Gamma_p$ and get $\beta_\gamma \in K$ so that $v_p(\beta_\gamma - \alpha) > \gamma$. Note that since the extension $K \to K_p^h$ is immediate we may as well assume $v_p(\alpha) = 0$ (for otherwise replace α by $\alpha^{-v_p(\alpha)}$, and if the latter can be approximated arbitrarily well from K then so can α).

Now we face the key obstruction, that of separating α from all other roots of f in K_p^h, using formulas in the language of fields involving the coefficients $\lambda_1, \ldots, \lambda_n$, and then, in the style of Cohen-Denef-Pas [5, 11, 17] transcribing the meaning of formulas $\psi(\alpha, x_1, \ldots, x_m)$ into that of formulas $\psi^*(\lambda_1, \ldots, \lambda_n, x_1, \ldots, x_m)$. That one can do this is Cell Decomposition Theory, and we refer to [11] and [17] for a full explanation. If one fixes n (the degree of f) and finitely many $\psi_j(y, x_1, \ldots, x_m)$ (as ψ above) then the elimination procedure is uniform for all $p > n$ for some standard n.

We combine these deep insights with some less demanding ones coming from the fact that K_p^h is the Henselization of (K, v_p), and satisfaction in K_p^h for parameters from K is Δ_0-interpretable in $\mathcal{M}$.

Pas worked with a natural many-sorted formalism for valued fields. The sorts are:

- the field sort with $+, \cdot, -, 0, 1$
- the sort $k \cup \{\infty\}$, k the residue field with $+, \cdot, 0, 1$
- the value group $\Gamma \cup \{\infty\}$ with $<, +, -$.

In addition we have a primitive for $v : K \to \Gamma \cup \{\infty\}$, and an angular component $ac : K \to k$, see [17].

We are working with K_p^h as our many-sorted Henselian field. Inside we have the valued field K, with a natural angular component ac_p. k_p, Γ_p, and ac_p are all Δ_0-interpretable in $\mathcal{M}$. Note that we say nothing about interpretability of K_p^h.

The great beauty and utility of what Pas and Denef did is that they show that any (even) many-sorted ψ is equivalent, for residue fields avoiding finitely many finite characteristics, to a formula with no quantified variables of field sort. For such a formula $\psi(\mathbf{w}, \mathbf{z}, \mathbf{s})$, where the $\mathbf{w}$ are of the field sort, $\mathbf{z}$ are of the residue field sort, and $\mathbf{s}$ are of the value group sort, one can define, uniformly in $\mathcal{M}$ and for $p > n$, for some standard n depending only on ψ,

$$\{(\mathbf{x}, ac_p(\mathbf{x}), v_p(\mathbf{x})) : \mathbf{x} = x_1, \dots x_n \in K \textit{ and } K_p^h \models \psi(\mathbf{x}, ac_p(\mathbf{x}), v_p(\mathbf{x}))\}$$

by a Δ_0-formula of $\mathcal{L}(\sharp)$. (Here we cut some corners, pretending the $\mathbf{x}$ are in $\mathcal{M}$, rather than in K. To be correct we should use $y_1, \dots, y_n, w_1, \dots w_n$ from $\mathcal{M}$, and write $x_j = \frac{y_j}{w_j}$.) Fix $\mathcal{M}$, p and $\alpha \in K_p^h$ as before, and $f(x) = x^n + \lambda_1 x^{n-1} + \dots + \lambda_{n-1}x + \lambda_n$, with $\lambda_1, \dots, \lambda_n \in K$. Now we use cell-decomposition for the generic monic polynomial of degree n in the sense of Pas [17] (uniformly for $p \geq p_0 \in \mathbb{N}$) to get finitely many definable (graphs of) functions

$$\rho_1(y_1, \dots, y_n), \dots, \rho_k(y_1, \dots, y_n)$$

so that (uniformly for $p \geq p_0$) the formula $v_p(\rho_j(y_1, \dots, y_n) - w) \geq v_p(t)$ is equivalent in K_p^h to a $\psi_j(\mathbf{y}, \mathbf{w}, \mathbf{t})$ of the form given earlier where the variables $\mathbf{y}, \mathbf{w}, \mathbf{t}$ are all field sorts, and $\alpha = \rho_j(\lambda_1, \dots, \lambda_n)$ for some j.

We use the notations $\psi_j^{(K_p^h)}$ to denote the subset of K_p^h defined by ψ_j, and ψ_j^b to denote the Δ_0-formula defining $\mathcal{M}^2 \cap \psi_j^{(K_p^h)}$, described above. We want to stress that ψ_j^b is a bounded formula which will be used in the induction below. Now suppose α is not arbitrarily closely approximated by elements of K, in the p-adic valuation. Pick $m \in \Gamma_p$, so that $v_p(\alpha - \beta) \leq m$ for $\beta \in K$. By replacing α by $\alpha p^{-v_p(\alpha)}$ (extension is immediate) we can and will assume $v_p(\alpha) = 0$.

Now by Lemma 1 we know there is no best approximation to α

from K. We show by a Δ_0-induction that if we assume that α is not arbitrarily closely approximable from K then α does have a best approximation from K, and we get a contradiction. There is an $r \in \mathcal{M}$ satisfying $v_p(\alpha - r) > 0$, since the residue field does not extend. Also we can choose $0 \leq r < p$ (by Euclidean division). Moreover, by writing (in a Δ_0-way)

$$\alpha - r = c_1 p + c_2 p^2 + \ldots \ldots \quad 0 \leq c_j < p$$

and replacing c_1, if need be, by another residue c_1', and putting $r' = c_0 + c_1' p$ we can choose r' so that $v_p(\alpha - r') = 1$, $0 < r' \leq p + p^2$. We now have the informal basis for a recursion, to be codified in $\mathcal{M}$. If we have $r_j \in \mathcal{M}$, $v_p(\alpha - r_j) = j$, then (uniformly in a Δ_0-way) we can define $r_{j+1} \in \mathcal{M}$, so that $v_p(\alpha - r_{j+1}) = j+1$, and $0 \leq r_{j+1} \leq r_j + p^{j+2}$ (just imitate the above argument). Now consider

$$\left\{ j \leq m \mid \exists r \in \mathcal{M} \left(r \leq \frac{p^{j+3}}{p-1} \wedge v_p(\alpha - r) = v_p(p^j) \right) \right\}.$$

Remember that m is in the Bennett segment for p, and we have an absolute bound $\frac{p^{j+3}}{p-1}$ for r. By the basic remark on Cell Decomposition, the set is Δ_0-definable. Moreover, it has no last element, by the construction just given. So by a standard argument it is all of $[1, m]$, and so we can approximate α to a best approximation.

If p is finite, we change the argument using Denef's Cell Decomposition for fields elementarily equivalent to the standard $\mathbb{Q}_p$, K_p^h is elementarily equivalent to $\mathbb{Q}_p$ by [2, 3, 4, 12]. We have then completed the proof of Theorem 1.

References

[1] Ajtai M., The complexity of the pigeonhole principle, *Combinatorica*, vol. 14 (4), 1994, pp. 417–433.

[2] Ax J. and Kochen S., Diophantine problems over local fields. I. *American Journal of Mathematics*, vol. 87, 1965, pp. 605–630.

[3] Ax J. and Kochen S., Diophantine problems over local fields II, *American Journal of Mathematics*, vol. 87, 1965, pp. 631–648.

[4] Ax J. and Kochen S., Diophantine problems over local fields III. Decidable fields, *Annals of Mathematics*, vol. 83 (2), 1966, pp. 437–456.

[5] Cohen P. J., Decision procedures for real and p-adic fields, *Communications in Pure and Applied Mathematics*, vol. 22, 1969, pp. 131–151.

[6] D'Aquino P., Local behaviour of Chebyshev theorem in models of $I\Delta_0$, *Journal of Symbolic Logic*, vol. 57, 1992, pp. 12–27.

[7] D'Aquino P., Pell equations and exponentiation in fragments of arithmetic, *Annals of Pure and Applied Logic*, vol. 77, 1996, pp. 1–34.

[8] D'Aquino P. and Macintyre A., Non standard finite fields over $I\Delta_0 + \Omega_1$, *Israel Journal of Mathematics*, vol. 117, 2000, pp. 311–333.

[9] D'Aquino P. and Macintyre A., Quadratic forms in models of $I\Delta_0 + \Omega_1$ I, *Annals of Pure and Applied Logic*, vol. 148, 2007, pp. 31–48.

[10] D'Aquino P. and Macintyre A., Quadratic forms in models of $I\Delta_0 + \Omega_1$ part II: local equivalence, *Annals of Pure and Applied Logic*, vol. 162, 2011, pp. 447–456.

[11] Denef J., The rationality of the Poincaré series associated to the *p*-adic points on a variety, *Inventiones Mathematicae*, vol. 77, 1984, pp. 1–23.

[12] Ershov Yu.L., On the elementary theory of maximal normed fields, *Soviet Mathematics. Doklady*, vol. 6, 1965, pp. 1390–1393. (In Russian)

[13] Gaifman H. and Dimitracopoulos C., Fragments of Peano's arithmetic and the MRDP theorem, *Logic and Algorithmic (Zurich, 1980)*, pp. 187–206, Univ. Genève, Geneva, 1982.

[14] Kaplanski I., Maximal fields with valuations, *Duke Mathematical Journal*, vol. 9, 1942, pp. 303–321.

[15] Kuhlmann F.-V., Dense subfields of henselian fields, and integer parts, *Proceedings of the Workshop and Conference on Logic, Algebra, and Arithmetic*, Lecture Notes in Logic n° 26, 2006, pp. 204–226.

[16] Paris J., Wilkie A. and Woods A., Provability of the Pigeonhole Principle and the existence of infinitely many primes, *Journal of Symbolic Logic*, vol. 53 (4), 1988, pp. 1235–1244.

[17] Pas J., Uniform *p*-adic cell decomposition and local zeta-functions, *Journal für die Reine und Angewandte Mathematik*, vol. 399, 1989, pp. 137–172.

[18] Woods A., *Some problems in logic and number theory and their connections*, Ph.D. thesis, Manchester University, 1981. Part II of this volume.

5

A New Proof of Tanaka's Theorem

ALI ENAYAT[1]

Abstract: We present a new proof of a theorem of Kazuyuki Tanaka, which states that every countable nonstandard model of WKL_0 has a non-trivial self-embedding onto a proper initial segment of itself. Moreover, the new proof has an ingredient that yields a novel characterization of models of WKL_0 among countable models of RCA_0.

1 Introduction

Friedman [6, Theorem 4.4] unveiled the striking result that every countable nonstandard model of PA is isomorphic to a proper initial segment of itself. One of the earliest refinements of Friedman's theorem is due to Lessan [10], who showed that a countable model $\mathcal{M}$ of Π_2^{PA} (the Π_2-consequences of PA) is isomorphic to a proper initial segment of itself iff $\mathcal{M}$ is 1-tall and 1-extendable. Here '$\mathcal{M}$ is 1-tall' means that the set of Σ_1-definable elements of $\mathcal{M}$ is not cofinal in $\mathcal{M}$, and '$\mathcal{M}$ is 1-extendable' means that there is an end extension $\mathcal{M}^*$ of $\mathcal{M}$ that satisfies $\mathrm{I}\Delta_0$ and $\mathrm{Th}_{\Sigma_1}(\mathcal{M}) = \mathrm{Th}_{\Sigma_1}(\mathcal{M}^*)$. Dimitracopoulos and Paris [4], in turn, strengthened Lessan's aforementioned result by weakening Π_2^{PA} to $\mathrm{I}\Delta_0+\mathrm{B}\Sigma_1+exp$, and they used this strengthening to show that every countable nonstandard model of $\mathrm{I}\Sigma_1$ is isomorphic to a proper initial segment of itself. This strengthening of Friedman's theorem was independently established, through a different line of reasoning, by Ressayre [12] in the following stronger form indicated in the 'moreover clause'

[1]Department of Philosophy, Linguistics, and Theory of Science, University of Gothenburg, Box 200, 405 30 Gothenburg, Sweden, ali.enayat@gu.se

Studies in Weak Arithmetics.
Patrick Cégielski, Charalampos Cornaros, Costas Dimitracopoulos.

(see [7], Ch. IV, Sec. 2(d) for a detailed exposition of Ressayre's proof).

1.1. Theorem. *Every countable nonstandard model $\mathcal{M}$ of* $\mathrm{I}\Sigma_1$ *is isomorphic to a proper initial segment of itself. Moreover, for any prescribed $a \in M$, there is a proper initial segment I of $\mathcal{M}$, and an isomorphism $\phi_a : \mathcal{M} \to I$ such that $\phi_a(m) = m$ for all $m \leq a$.*[2]

The point of departure for this paper is a theorem of Tanaka [14] that extends Theorem 1.1 to models of the fragment WKL_0 of second order arithmetic. Models of WKL_0 are two-sorted structures of the form $(\mathcal{M}, \mathcal{A})$, where $\mathcal{M} = (M, +, \cdot, <, 0, 1) \models \mathrm{I}\Sigma_1$, and $\mathcal{A}$ is a family of subsets of M such that $(\mathcal{M}, \mathcal{A})$ satisfies the following three conditions:

(1) Induction for Σ^0_1 formulae;
(2) Comprehension for Δ^0_1-formulae; and
(3) Weak König's Lemma: every infinite subtree of the full binary tree has an infinite branch.

It is well known that every countable model $\mathcal{M}$ of $\mathrm{I}\Sigma_1$ can be expanded to a model $(\mathcal{M}, \mathcal{A}) \models \mathrm{WKL}_0$. This important result is due independently to Harrington and Ratajczyk; see [13], Lemma IX.1.8 + Theorem IX.2.1. Therefore Theorem 1.2 below is a strengthening of Theorem 1.1.

1.2. Theorem. (Tanaka) *Every countable nonstandard model $(\mathcal{M}, \mathcal{A})$ of WKL_0 is isomorphic to a proper initial segment I of itself in the sense that there is an isomorphism $\phi : \mathcal{M} \to I$ such that ϕ induces an isomorphism $\widehat{\phi} : (\mathcal{M}, \mathcal{A}) \to (I, \mathcal{A} \restriction I)$. Moreover, given any prescribed $a \in M$, there is some I and ϕ as above such that $\phi_a(m) = m$ for all $m \leq a$.*

In Theorem 1.2, $\mathcal{A} \restriction I := \{A \cap I : A \in \mathcal{A}\}$, and the isomorphism $\widehat{\phi}$ induced by ϕ is defined by: $\widehat{\phi}(m) = \phi(m)$ for $m \in M$ and $\widehat{\phi}(A) = \{\phi(a) : a \in A\}$ for $A \in \mathcal{A}$. Tanaka's proof of Theorem 1.2 in [15] is based on an elaboration of Ressayre's proof of Theorem 1.1, which uses game-theoretic ideas.

Tanaka's motivation for his result, as pointed out in [14, Sec. 3], was the development of non-standard methods within the confines of the frugal system WKL_0. A remarkable application of Tanaka's result appears in the work of Tanaka and Yamazaki [15], where it is used to show that the construction of the Haar measure (over compact groups) can be implemented within WKL_0 via a detour through nonstandard

---

[2] Ressayre also noted that the existence of such a family of embeddings $\{\phi_a : a \in M\}$ characterizes models of $\mathrm{I}\Sigma_1$ among countable models of $\mathrm{I}\Delta_0$.

models. This is in contrast to the previously known constructions of the Haar measure whose implementation can only be accommodated within the stronger fragment ACA_0 of second order arithmetic. As it turns out, the known applications of Tanaka's theorem in the development of nonstandard methods do not need the full force of Theorem 1.2, but rather they rely on the following immediate corollary of the first assertion of Theorem 1.2.

1.3. Corollary. *Every countable nonstandard model $(\mathcal{M}, \mathcal{A})$ of WKL_0 has an extension $(\mathcal{M}^*, \mathcal{A}^*)$ to a model of WKL_0 such that $\mathcal{M}^*$ properly end extends $\mathcal{M}$, and $\mathcal{A} = \mathcal{A}^* \restriction M$.*

In this paper we present a new proof of the first assertion of Tanaka's Theorem (in Section 3). As shown in Theorem 3.6 our work also yields a new characterization of models of WKL_0 among countable models of RCA_0. Before presenting the new proof, we first need to discuss a self-embedding theorem (in Section 2) that plays a key role in our new proof. As indicated in Remark 3.4, our method can be extended to also establish the 'moreover' clause of Tanaka's Theorem. Further applications of the methodology of our proof of Theorem 1.2 will be presented in [5].

2 The Solovay-Paris Self-Embedding Theorem

Paris [11, Theorem 4] showed that every countable recursively saturated model of $I\Delta_0 + B\Sigma_1$ is isomorphic to a proper initial segment of itself, a result that is described by Paris as being 'implicit' in an (unpublished) paper of Solovay.[3] This result of Solovay and Paris can be fine-tuned as in Theorem 2.1 below, following the strategy (with $n = 0$) of [9, Theorem 12.3]. The details of the proof of Theorem 2.1 have been worked out by Cornaros in [3, Corollary 11] and Yokoyama [16, Lemma 1.3].

2.1. Theorem. *Suppose $\mathcal{N}$ is a countable model of $I\Delta_0 + B\Sigma_1$ that is recursively saturated, and there are $a < b$ in $\mathcal{N}$ such that for every Δ_0-formula $\delta(x, y)$ we have:*

$$(*) \qquad \mathcal{N} \models \exists y\, \delta(a, y) \Longrightarrow \mathcal{N} \models \exists y < b\, \delta(a, y).$$

There is an isomorphism $\phi : \mathcal{N} \to I$, where I is an initial segment of $\mathcal{N}$ with $a < I < b$, and $\phi(a) = a$.

2.2. Remark. Condition $(*)$ of Theorem 2.1 can be rephrased as

[3] Solovay's paper (as listed in [11]) is entitled *Cuts in models of Peano*. We have not been able to obtain a copy of this paper.

$$f(a) < b \text{ for all partial } \mathcal{N}\text{-recursive functions } f,$$

with the understanding that a partial function f from N to N is a partial $\mathcal{N}$-recursive function iff the graph of f is definable in $\mathcal{N}$ by a parameter-free Σ_1-formula.

3 Proof of Tanaka's Theorem

The goal of this section is to show that every countable nonstandard model $(\mathcal{M}, \mathcal{A})$ of WKL_0 is isomorphic to a proper initial segment of itself. In Remark 3.4 we will comment on how our method can be adapted so as to also establish the 'moreover clause' of Tanaka's Theorem. Our proof has three stages:

Stage 1: Given a countable nonstandard model $(\mathcal{M}, \mathcal{A})$ of WKL_0, and a prescribed $a \in M$ in this stage we use the 'muscles' of $\mathrm{I}\Sigma_1$ in the form of the strong Σ_1-collection [8, Theorem 2.23, p.68] to locate an element b in $\mathcal{M}$ such that $f(a) < b$ for all partial $\mathcal{M}$-recursive functions f (as defined in Remark 2.2).

Stage 2 Outline: We build an *end extension* $\mathcal{N}$ of M such that the following conditions hold:

(I) $\mathcal{N} \models I\Delta_0 + B\Sigma_1$;
(II) $\mathcal{N}$ is recursively saturated;
(III) $f(a) < b$ for all partial $\mathcal{N}$-recursive functions; and
(IV) $SSy_M(N) = \mathcal{A}$.

Stage 3 Outline: We use Theorem 2.1 to embed $\mathcal{N}$ onto a proper initial segment J of $\mathcal{M}$. By elementary considerations, this will yield a proper cut I of J with $(\mathcal{M}, \mathcal{A}) \cong (I, \mathcal{A} \upharpoonright I)$.

We now proceed to flesh out the above outlines for the second and third stages.

Stage 2 Details

This stage of the proof can be summarized into a theorem in its own right (Theorem 3.2).[4] Before stating the theorem, we need the following definitions.

3.1. Definition. *Suppose $\mathcal{M} \subset_{end} \mathcal{N}$, where $\mathcal{M} \models I\Sigma_1$ and $\mathcal{N} \models I\Delta_0$.*

(a) $\in_{Ack}$ *is Ackermann's membership relation* [8, 1.31, p.38] *based on binary expansions*[5].

[4] We are grateful to Keita Yokoyama for pointing out that Theorem 3.2 is of sufficient interest to be explicitly stated.

[5] Recall: $m \in_{Ack} n$ is defined as "the m-th digit of the binary expansion of n is 1". It is well known that, provably in $I\Delta_0$, if 2^n exists then (a) n can be written

(b) *Suppose* $c \in N$ *and* 2^c *exists in* $\mathcal{N}$. $c_{E,\mathcal{N}}$ *is the '*$\in_{Ack}$*-extension' of* c *in* $\mathcal{N}$, *i.e.,*

$$c_{E,\mathcal{N}} := \{i \in N : \mathcal{N} \models i \in_{Ack} c\}.$$

Note that since exp holds in $\mathcal{M}$, *and* $\mathcal{N}$ *satisfies* $I\Delta_0$, *we can fix some element* $k \in N \backslash M$ *and apply* Δ_0*-overspill to the* Δ_0*-formula*

$$\delta(x) := \exists y < k \ (2^x = y)$$

to be assured of the existence of an element $c \in N \backslash M$ *for which* 2^c *exists in* $\mathcal{N}$.

(c) $SSy_M(\mathcal{N})$ *is the* M*-standard system of* $\mathcal{N}$, *which is defined as the collection of subsets of* M *that are of the form* $M \cap c_{E,\mathcal{N}}$ *for some* $c \in N$ *such that* 2^c *exists in* $\mathcal{N}$.

(d) *Suppose* $b \in M$, *and let* $M_{\leq b} := \{a \in M : \mathcal{M} \models a \leq b\}$. *We say that* $\mathcal{N}$ *is a conservative extension of* $\mathcal{M}$ *with respect to* $\Pi_{1,\leq b}$*-sentences iff for all* Π_1*-formulae* $\pi(x_0, \cdots, x_{n-1})$ *in the language of arithmetic with the displayed free variables, and for all* $a_0, \cdots, a_{n-1}$ *in* $M_{\leq b}$ *we have:*

$$\mathcal{M} \models \pi(a_0, \cdots, a_{n-1}) \text{ iff } \mathcal{N} \models \pi(a_0, \cdots, a_{n-1}).$$

The following theorem supports the existence of the model $\mathcal{N}$ as in the outline of Stage 2. Note that condition (III) of the outline of Stage 2 is implied by the last clause of Theorem 3.2 since if δ is any Δ_0-formula of standard length, and $\overline{a}$ and $\overline{b}$ are constants representing a and b, then the sentence

$$\exists y \ \delta(\overline{a}, y) \rightarrow \exists y < \overline{b} \ \delta(\overline{a}, y)$$

is a $\Pi_{1,\leq b}$-sentence.

3.2. Theorem. *Let* $(\mathcal{M}, \mathcal{A})$ *be a countable model of* WKL_0 *and let* $b \in M$. *Then* $\mathcal{M}$ *has a recursively saturated proper end extension* $\mathcal{N}$ *satisfying* $I\Delta_0 + B\Sigma_1$ *such that* $SSy_M(\mathcal{N}) = \mathcal{A}$, *and* $\mathcal{N}$ *is a conservative extension of* $\mathcal{M}$ *with respect to* $\Pi_{1,\leq b}$*-sentences.*

Proof. Let $True^{\mathcal{M}}_{\Pi_1}$ be the set of 'true' Π_1-sentences with (parameters in M), as computed in $\mathcal{M}$.[6] Fix some nonstandard $n^* \in M$ with $n^* >> b$, e.g., $n^* = supexp(b)$ is more than sufficient[7]. Since $\mathcal{M}$ satisfies $I\Sigma_1$ and $True^{\mathcal{M}}_{\Pi_1}$ has a Π_1-definition within $\mathcal{M}$, there is some element $c \in M$

(uniquely) as a sum of powers of 2 (and hence n has a well-defined binary expansion), and (b) $m \in_{Ack} n$ iff $2 \mid \lfloor m2^{-n} \rfloor$.

[6] It is well-known that $I\Sigma_1$ has enough power to meaningfully define $True^{\mathcal{M}}_{\Pi_1}$ (see, e.g., [8, Theorem 1.75, p.59]). Note that $True^{\mathcal{M}}_{\Pi_1}$ need not be in $\mathcal{A}$.

[7] Here $supexp(b)$ is an exponential stack of length $b + 1$, where the top entry is b, and the rest of the entries consist of 2's.

that codes the fragment of $True^{\mathcal{M}}_{\Pi_1}$ consisting of elements of $True^{\mathcal{M}}_{\Pi_1}$ that are below n^*, i.e.,

$$c_{E,\mathcal{M}} = \{m \in M : m \in True^{\mathcal{M}}_{\Pi_1} \text{ } and \text{ } m < n^*\},$$

We observe that $c_{E,\mathcal{M}}$ contains all $\Pi_{1,\leq b}$-sentences that hold in $\mathcal{M}$. Within $\mathcal{M}$, we define the 'theory' T_0 by:

$$T_0 := I\Delta_0 + B\Sigma_1 + \{m : m \in True_{\Pi_1} \text{ } and \text{ } m < n^*\}.$$

At this point we recall a result of Clote-Hájek-Paris [2] that asserts:

$$(\#) \quad I\Sigma_1 \vdash Con(I\Delta_0 + B\Sigma_1 + True_{\Pi_1}).$$

In light of (#), we have:

$$(*) \quad \mathcal{M} \models Con(T_0).$$

It is clear that T_0 has a Δ_1-definition in $\mathcal{M}$ (note: here we are not claiming anything about the *quantifier complexity of each member of* T_0). Hence by Δ^0_1-comprehension available in WKL_0 we also have:

$$(**) \quad T_0 \in \mathcal{A}$$

We wish to build a certain chain $\langle \mathcal{N}_n : n \in \omega \rangle$ of structures such that the elementary diagram of each $(\mathcal{N}_n, a)_{a \in N_n}$ can be computed by $(\mathcal{M}, \mathcal{A})$ as some $E_n \in \mathcal{A}$. Note that E_n would be replete with sentences of nonstandard length. Enumerate $\mathcal{A}$ as $\langle A_n : n \in \omega \rangle$. Our official requirements for the sequence $\langle \mathcal{N}_n : n \in \omega \rangle$ is that for each $n \in \omega$ the following three conditions are satisfied:

(1) $T_0 \subseteq E_n \in \mathcal{A}$ (in particular: $\mathcal{N}_n$ is recursively saturated).
(2) $\mathcal{M} \subset_{end} \mathcal{N}_n \prec \mathcal{N}_{n+1}$.
(3) $A_n \in SSy_M(\mathcal{N}_{n+1})$.

We now sketch the recursive construction of the desired chain of models. To begin with, we invoke (*), (**), and the completeness theorem for first order logic (that is available in WKL_0, see [13, Theorem IV.3.3]) to get hold of $\mathcal{N}_0$ and E_0 satisfying condition (1). To define $\mathcal{N}_{n+1}$, we assume that we have $\mathcal{N}_n$ satisfying (1). Next, consider the following 'theory' T_{n+1} defined within $(\mathcal{M}, \mathcal{A})$:

$$T_{n+1} := E_n + \left\{\overline{t} \in_{Ack} d : t \in A_n\right\} + \left\{\overline{t} \notin_{Ack} d : t \notin A_n\right\},$$

where d is a new constant symbol, and $\overline{t}$ is the numeral representing t in the ambient model $\mathcal{M}$. Note that T_{n+1} belongs to $\mathcal{A}$ since $\mathcal{A}$ is a Turing ideal and T_{n+1} is Turing reducible to the join of E_n and A_n. It is easy to see that T_{n+1} is consistent in the sense of $(\mathcal{M}, \mathcal{A})$ since $(\mathcal{M}, \mathcal{A})$ can verify that T_{n+1} is finitely interpretable in $\mathcal{N}_n$. Using the completeness theorem within $(\mathcal{M}, \mathcal{A})$ this allows us to get hold of the desired $\mathcal{N}_{n+1}$

and E_{n+1} satisfying (1), (2), and (3). The recursive saturation of $\mathcal{N}_{n+1}$ follows immediately from (1), using a well known over-spill argument. Let

$$\mathcal{N} := \bigcup_{n \in \omega} \mathcal{N}_n.$$

It is evident that $\mathcal{N}$ satisfies the properties listed in Theorem 3.2. $\dashv$

3.3. Remark. The result of Clote-Hájek-Paris that was invoked in the above proof was further extended by Beklemishev, who showed the consistency of $\mathrm{I}\Delta_0 + \mathrm{B}\Sigma_1 + True_{\Pi_2}$ within $\mathrm{I}\Sigma_1$ ($n = 1$ of [1, Theorem 5.1]). Beklemishev's result shows that condition $(*)$ of Stage (2) can be strengthened to assert that $\mathcal{M} \models Con(T_0 + PRA)$, where PRA is the scheme asserting the totality of all primitive recursive functions (recall that the provable recursive functions of $\mathrm{I}\Sigma_1$ are precisely the primitive recursive functions). This has the pleasant consequence that the model $\mathcal{N}$ constructed in Stage 2 can be required to satisfy PRA, and *a fortiori*: exp.

Stage 3 Details

Thanks to properties (I), (II), and (III) of the outline of Stage 2, we can invoke Theorem 2.1 to get hold of a self-embedding ϕ of $\mathcal{N}$ onto a cut J with $a \in J < b$. The image I of M under ϕ is an initial segment of $\mathcal{M}$ and

$$I < J < M.$$

Let $\mathcal{B} := \{\widehat{\phi}(A) : A \in \mathcal{A}\}$. It is clear that ϕ induces an isomorphism

$$\widehat{\phi} : (\mathcal{M}, \mathcal{A}) \to (I, \mathcal{B}).$$

Coupled with the fact that $SSy_M(\mathcal{N}) = \mathcal{A}$, this implies $\mathcal{A} \restriction I = \mathcal{B}$ by elementary considerations. This concludes the proof of (the first clause) of Tanaka's Theorem. $\dashv$

3.4. Remark. The 'moreover' clause of Tanaka's Theorem can be established by the following variations of the above proof: in Stage 1, given a prescribed $a \in M$, let $a^* = supexp(a)$ and use strong Σ_1-collection to find $b \in M$ such that:

(1) For every Δ_0-formula $\delta(x, y)$, and every $i \leq a^*$,

$$\mathcal{M} \models \exists y\ \delta(i, y) \Longrightarrow \mathcal{M} \models \exists y < b\ \delta(i, y).$$

Let I be the smallest cut of $\mathcal{M}$ that contains a and is closed under exponentiation, i.e., $I := \{m \in M : \exists n < \omega \text{ such that } m < a_n\}$, where $\langle a_n : n < \omega \rangle$ is given by $a_0 := a$, and $a_{n+1} = 2^{a_n}$. In light of (1), we have:

(2) For every Δ_0-formula $\delta(x,y)$, and every $i \in I$,

$$\mathcal{M} \models \exists y\ \delta(i,y) \Longrightarrow \mathcal{M} \models \exists y < b\ \delta(i,y).$$

Then in Stage 3, we use (2) in conjunction with the following refinement of Theorem 2.1.

3.5. Theorem [5]. *Suppose $\mathcal{N}$ is a countable recursively saturated model of $I\Delta_0 + B\Sigma_1$, I is a cut of $\mathcal{N}$ that is closed under exponentiation, and for some b in $\mathcal{N}$, the following holds for all Δ_0-formulas $\delta(x,y)$:*

$$\forall i \in I\ \ (\mathcal{N} \models \exists y\ \delta(i,y) \Longrightarrow \mathcal{N} \models \exists y < b\ \delta(i,y)).$$

There is an isomorphism $\phi : \mathcal{N} \to J$, where J is an initial segment of $\mathcal{N}$ with $J < b$, and $\phi(m) = m$ for all $m \in I$.

The proof of Theorem 3.5 is obtained by dovetailing the proof of Theorem 2.1 with an old argument of Hájek and Pudlák in [7, Appendix].

As pointed out by Keita Yokoyama, the results of this paper, coupled with an observation of Tin Lok Wong, yield the following characterization of models of WKL_0 among countable models of RCA_0.

3.6. Theorem. *Let $(\mathcal{M}, \mathcal{A})$ be a countable model of RCA_0. The following are equivalent:*

(1) *$(\mathcal{M}, \mathcal{A})$ is a model of WKL_0.*

(2) *For every $b \in M$ there exists a recursively saturated proper end extension $\mathcal{N}$ of $\mathcal{M}$ such that $\mathcal{N} \models I\Delta_0 + B\Sigma_1 + PRA$, $SSy_M(\mathcal{N}) = \mathcal{A}$, and $\mathcal{N}$ is a conservative extension of $\mathcal{M}$ with respect to $\Pi_{1,\leq b}$-sentences.*

(3) *For every $b \in M$ there is a proper initial segment I of $\mathcal{M}$ such that $(\mathcal{M}, \mathcal{A})$ is isomorphic to $(I, \mathcal{A} \upharpoonright I)$ via an isomorphism that pointwise fixes $M_{\leq b}$.*

Proof. (1) $\Rightarrow$ (2) is justified by Theorem 3.2 and Remark 3.3, and (2) $\Rightarrow$ (3) follows from Theorem 3.5. As pointed out by Tin-Lok Wong, (3) $\Rightarrow$ (1) follows from the same line of argument as in Ressayre's characterization of models of $\mathrm{I}\Sigma_1$ among countable models of $\mathrm{I}\Delta_0$ that is mentioned in footnote 1. ⊣

Acknowledgments. The new proof of Tanaka's theorem was discovered by the author in the midst of inspiring e-mail discussions with Volodya Shavrukov on a range of topics concerning self-embeddings of models of arithmetic. More recently, I have also benefitted from conversations with Keita Yokoyama. I am grateful to both gentlemen for offering insightful suggestions for improving the paper.

References

[1] L.D. Beklemishev, *A proof-theoretic analysis of collection*, **Arch. Math. Log.**, vol. 37, 1998, pp. 275–296.

[2] P. Clote, P. Hájek, and J. Paris, *On some formalized conservation results in arithmetic*, **Arch. Math. Logic**, vol. 30, 1990, pp. 201–218.

[3] C. Cornaros, *Versions of Friedman's theorem for fragments of PA*, manuscript available at: `http://www.softlab.ntua.gr/~nickie/tmp/pls5/cornaros.pdf`.

[4] C. Dimitracopoulos and J. Paris, *A note on a theorem of H. Friedman*, **Z. Math. Logik Grundlag. Math.**, vol. 34, 1988, pp. 13–17.

[5] A. Enayat, *Self-embeddings of models of arithmetic, redux* (in preparation).

[6] H. Friedman, *Countable models of set theories*, **Lecture Notes in Math.** 337, Springer, Berlin, 1973, pp. 539–573.

[7] P. Hájek and P. Pudlák, *Two orderings of the class of all countable models of Peano arithmetic*, **Model theory of algebra and arithmetic**, Lecture Notes in Math, vol. 834, 1980, pp. 174–185.

[8] P. Hájek and P. Pudlák, **Metamathematics of First Order Arithmetic**, Springer, Heidelberg, 1993.

[9] R. Kaye, **Models of Peano Arithmetic**, Oxford Logic Guides, Oxford University Press, Oxford, 1991.

[10] H. Lessan, **Models of Arithmetic**, Doctoral Dissertation, University of Manchester, 1978. Part III of this volume.

[11] J. B. Paris, *Some conservation results for fragments of arithmetic*, in **Lecture Notes in Math.**, vol. 890, Springer, Berlin-New York, 1981, pp. 251–262.

[12] J.-P. Ressayre, *Nonstandard universes with strong embeddings, and their finite approximations.* **Logic and combinatorics** in Contemp. Math., vol. 65, Amer. Math. Soc., Providence, RI, 1987, pp. 333–358.

[13] S. Simpson, **Subsystems of second order arithmetic**, Springer-Verlag, 1999

[14] K. Tanaka, *The self-embedding theorem of WKL_0 and a non-standard method*, **Ann. Pure Appl. Logic**, vol. 84, 1997, pp. 41–49.

[15] K. Tanaka and T. Yamazaki, *A Non-Standard Construction of Haar Measure and Weak König's Lemma*, **J. Symbolic Logic**, vol. 65, 2000, pp. 173–186.

[16] K. Yokoyama, *Various versions of self-embedding theorem*, manuscript (May 2012).

6

Δ_0-definability of the denumerant with one plus three variables

HENRI-ALEX ESBELIN[1]

Abstract: We prove the Δ_0-definability of the graph of the denumerant function d of variables $(n, a, b, c) \in \mathbb{N} \times (\mathbb{N}^*)^3$ defined by $d(n, a, b, c) = Card\{(x, y, z) \in \mathbb{N}^3; ax + by + cz = n\}$. This answers positively a particular case of an open question of Berarducci and Intrigila about the Δ_0-definability of the graph of the denumerant function with an arbitrary fixed number of variables.

1 Introduction

Problem and main results

The *denumerant* function of variables $n \in \mathbb{N}$ and $(a_1, ..., a_k) \in (\mathbb{N}^*)^k$ is defined by

$$d(n, a_1, ..., a_k) = Card\left\{(x_1, ..., x_k) \in \mathbb{N}^k \mid \sum_{i=0}^{i=k} a_i x_i = n\right\}$$

where $Card(E)$ is the number of elements of a finite set E. This function is widely studied in combinatorics. For fixed values of $(a_1, ..., a_k)$, the functions of the variable n defined by $d(n, a_1, ..., a_k)$ have Δ_0-definable graphs as it is proved in [1]; the authors ask whether the graph of the denumerant function d is Δ_0-definable. Building on the work of Woods, we solve positivly a special case of this problem.

[1]Clermont Université, CNRS UMR 6158, LIMOS
Alex Esbelin@univ-bpclermont.fr

Studies in Weak Arithmetics.
Patrick Cégielski, Charalampos Cornaros,
Costas Dimitracopoulos.

In case $k = 2$, a positive answer is an easy consequence of an elementary formula for $d(n, a, b)$ (see details in section 4.):

Proposition 1. *The denumerant function of variables* $(n, a, b) \in \mathbb{N} \times (\mathbb{N}^*)^2$ *defined by* $d(n, a, b) = Card\{(x, y) \in \mathbb{N}^2 \mid ax + by = n\}$ *has a* Δ_0-*definable graph.*

The main result of this paper is the following one:

Theorem 1. *The denumerant function of variables* $(n, a, b, c) \in \mathbb{N} \times (\mathbb{N}^*)^3$ *defined by* $d(n, a, b, c) = Card\{(x, y, z) \in \mathbb{N}^3 \mid ax + by + cz = n\}$ *has a* Δ_0-*definable graph.*

In order to prove the main result, we need the two following lemmas that are of independent interest. Let us first recall notations about the Euclidean algorithm and continued fractions.

Definition 1. *Let a and b be coprime positive integers. The length of the classical Euclidean algorithm, $l(a, b)$, the sequences of its quotients and remainders, $(\alpha_i(a, b))_{0 \le i < l(a,b)}$ and $(r_i(a, b))_{0 \le i \le l(a,b)}$, and the sequences of the numerators and denominators, $(p_i)_{0 \le i < l(a,b)}$ and $(q_i)_{0 \le i < l(a,b)}$, of the convergents of $\frac{a}{b}$ are defined in the following way:*

$$\left\{ \begin{array}{c} r_0(a, b) = a \text{ and } r_1(a, b) = b \\ \forall i \le l(a, b) - 2, \text{ we have } 0 < r_{i+2}(a, b) < r_{i+1}(a, b) \\ r_i(a, b) = \alpha_i(a, b) \times r_{i+1}(a, b) + r_{i+2}(a, b) \\ r_{l(a,b)-1}(a, b) = \alpha_{l(a,b)-1}(a, b) \times r_{l(a,b)}(a, b) \\ p_0(a, b) = a_0 \text{ and } p_1(a, b) = a_0 a_1 + 1 \\ q_0(a, b) = 1 \text{ and } q_1(a, b) = a_1 \\ \alpha_i(a, b) p_{i-1}(a, b) + p_{i-2}(a, b) = p_i(a, b) \\ \alpha_i(a, b) q_{i-1}(a, b) + q_{i-2}(a, b) = q_i(a, b) \end{array} \right\}.$$

Lemma 1. *The relations of variables z, i, a and b defined by $z = l(a, b)$ and $(i < l(a, b)) \wedge (z = \alpha_i(a, b))$ and $(i \le l(a, b)) \wedge (z = r_i(a, b))$ and $(i < l(a, b)) \wedge (z = p_i(a, b))$ and $(i < l(a, b)) \wedge (z = q_i(a, b))$ are Δ_0-definable.*

In the sequel, we often omit the variables (a, b).

Lemma 2. *The relation of variables z, y, a, b, c and d defined by*

$$(b \neq 0) \wedge (d \neq 0) \wedge \left(z = \sum_{i=0}^{i=y} \lfloor \frac{a}{b} i + \frac{c}{d} \rfloor \right)$$

is Δ_0-definable.

Ideas of the proof

The standard way to prove the Δ_0-definablity of the graph of a recursively defined sequence uses the following coding method: the standard code of the sequence $(u_i)_{0\leq i\leq l}$ in base M is $C = \sum_{i=0}^{i=l} u_i M^i$, where $M \geq 1 + Max\{u_i; 0 \leq i \leq l\}$; hence u_i turns out to be the i-th digit of C in base M; it is possible to introduce C in a Δ_0-formula using a quantification, provided that it is bounded by a polynomial of variables. A classical example is $M = 2$ when the u_i are 0 or 1 and $l \leq log_2(x)$ for some variable x. However sequences with some possibly *large* terms (i.e. not polylogarithmically bounded, which is the case of the sequences introduced in definition 1) or with *a lot of small* terms may not be coded in this standard way. In particular, this classical method does not work for polynomially long sums of the first terms of a Δ_0-definable sequence. Although, Alan Woods proved that it is the case for logarithmically long sums of polynomially bounded Δ_0-definable terms (we shall say here *short sums of not too large terms*).

We summarize the relevant material concerning Δ_0-definability in section 2.

The proof of Theorem 1 will be divided in two parts. The first part of the proof provides a formula defining the sum

$$\sum_{i=0}^{i=y} \lfloor \frac{a}{b} i + \frac{c}{d} \rfloor$$

with only *short sums of not too large terms.* This formula uses the sequence of the convergents of $\frac{a}{b}$ and the corresponding Ostrowski numeration of n; the main tools are Klein's theorem about the convex closure of integer points in a cone, and Pick's formula as a way to count the number of integer points in a polygon with integer vertices. Section 3 provides a detailed exposition of this material.

Section 4 contains the proofs of Lemmas 1 and 2 and Theorem 1.

2 Δ_0-definability

Definition 2. *Let us denote by* $\Delta_0^{\mathbb{N}}$ *the class of relations over integers definable by* Δ_0 *formulae. Relations in* $\Delta_0^{\mathbb{N}}$ *are also called* **rudimentary**. *A set* A *is said to be* Δ_0-*definable or* **rudimentary** *whenever the relation* $z \in A$ *is rudimentary.*

This paper is not selfcontained: the reader is supposed to be familliar with the elementary results and classical methods in Δ_0-definability (for an extended overview, see [2]). We just recall here the more special tools. Let us denote $\mathbf{x} = (x_1, ..., x_k)$.

Definition 3. *A function is* **rudimentary** *if it is polynomially bounded (i.e. there exists a polynomial function* ϕ *with positive integer coefficients, such that* $f(\mathbf{x}) \leq \phi(i, \mathbf{x})$*) and its graph is* Δ_0-*definable.*

The following easy proposition is of frequent use:

Proposition 2. *Let R be a Δ_0-definable relation and f a rudimentary function. Then the relation $R(x_1, ..., x_{i-1}, f(x_1, ..., x_n), x_{i+1}, ..., x_n)$ is still Δ_0-definable.*

The following theorem is fundamental in the proof of the main theorem. It is Lemma 4.2 of section *On summing sequences*, chapter *Counting the pigeons - Census functions and the pigeon hole principle* of Woods' Thesis [6].

Theorem 2. *Let $f(i, \mathbf{x})$ and $g(\mathbf{x})$ be two functions, the graphs of which are in $\Delta_0^{\mathbb{N}}$. Suppose that f is polynomially bounded and g is polylogarithmically bounded (i.e. there exists a polynomial function ψ with positive integer coefficients, such that $g(\mathbf{x}) \leq \psi(\lfloor log_2(x_1) \rfloor, ..., \lfloor log_2(x_k) \rfloor)$). Then the relation*

$$z = \sum_{i=0}^{i=g(\mathbf{x})} f(i, \mathbf{x})$$

is Δ_0-definable.

Moreover we need also the following important result of Paris and Wilkie [4]:

Theorem 3. *Let B be a Δ_0-definable subset of $\mathbb{N}$. Suppose that for some $k \in \mathbb{N}$ we have $Card\{x < n \mid x \in B\} \leq log(n)^k$ for all n. Then the function defined by $G(n) = Card\{x < n \mid x \in B\}$ has a Δ_0-definable graph.*

3 Mathematical tools

We first need classical results in continued fractions theory (see for example the classical textbook [3]). In this section, a and b are coprime positive integers.

Proposition 3. *For all n, we have $q_n > 2^{\lfloor \frac{n}{2} \rfloor}$ and $p_{n+1} > 2^{\lfloor \frac{n}{2} \rfloor}$.*

Proposition 4. *We have $b = q_{l-1}$ and $a = p_{l-1}$ and $\frac{p_{2k}}{q_{2k}} < \frac{a}{b} < \frac{p_{2k+1}}{q_{2k+1}}$ for $2k + 1 \leq l - 2$.*

Theorem 4. *If $b \geq 3$, the convergents $\frac{p_i}{q_i}$ of $\frac{a}{b}$ are the fractions of coprime numerators and denominators p and q satisfying the following property:*

$$|v\frac{a}{b} - u| > |q\frac{a}{b} - p|$$

for all $(u, v) \in \mathbb{N} \times \mathbb{N}^*$ such that $0 < v < q$; in the case $b = 2$ and a odd, the convergents are the fractions $\frac{\lfloor \frac{a}{b} \rfloor}{1}$ and $\frac{a}{b}$.

Right: i=2k+1 is odd

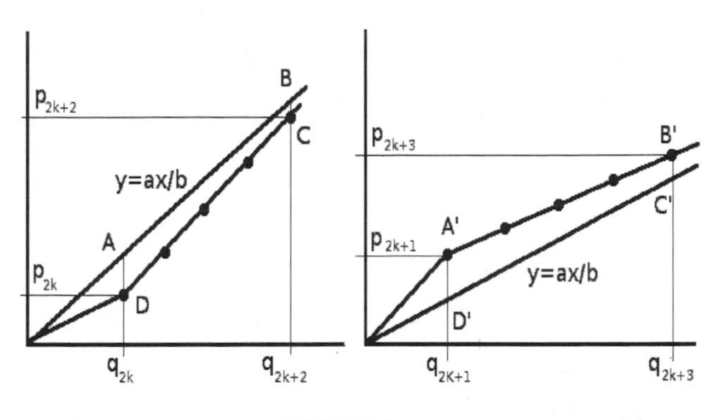

FIGURE 1

Now we need two results from geometry of numbers. The first one is the well known Pick's Formula [5]:

Lemma 3. *The number of lattice points lying in a polygon with lattice points as vertices is equal to its area plus half of the number of the lattice points lying on its boundary plus 1.*

The following result is an old theorem of Klein (see [7]). Let us call *lattice points* the points with integer coordinates.

Lemma 4. *There is no lattice points in the polygon of vertices of coordinates* $(q_i; \frac{a}{b}q_i)$, $(q_{i+2}; \frac{a}{b}q_{i+2})$, $(q_{i+2}; p_{i+2})$ *and* $(q_i; p_i)$ *but the points of coordinates* $(q_i + tq_{i+1}; p_i + tp_{i+1})$ *where* $0 \leq t \leq \alpha_{i+2}$.

Using Lemmas 3 and 4, we intend to get a new formula for the sum $\sum_{j=0}^{j=n} \lfloor \frac{a}{b} j \rfloor$. We first establish formulas for some special sums, namely the case where $j \in [q_i; q_i + tq_{i+1})$. These formulas slightly depend on the parity of i. Let us denote i mod 2 the remainder of the integer i in the Euclidean division by 2.

Lemma 5. *For* $0 \leq t \leq \alpha_{i+2}$ *and* $i + 2 \leq \lambda$, *we have*

$$\sum_{j=q_i}^{j=q_i+tq_{i+1}-1} \lfloor \frac{a}{b} j \rfloor = \frac{t}{2}\left((2p_i + tp_{i+1})q_{i+1} - q_{i+1} - p_{i+1} + 1 - 2(i \bmod 2)\right)$$

Proof. From Klein's Theorem and Pick's Formula,

$$\sum_{j=q_{2k}}^{j=q_{2k}+tq_{2k+1}-1} \lfloor \frac{a}{b} j \rfloor$$

turns to be equal to:

$$\frac{p_{2k} + p_{2k} + tp_{2k+1}}{2} tq_{2k+1} - \frac{p_{2k} + p_{2k} + tp_{2k+1} + t + tq_{2k+1}}{2} + p_{2k} + t$$

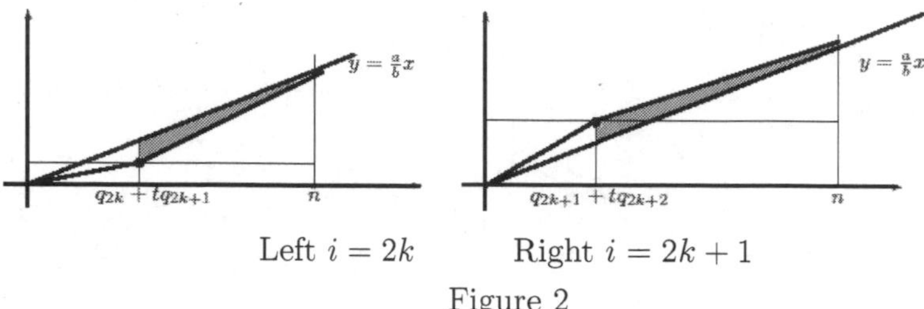

Left $i = 2k$ Right $i = 2k + 1$

Figure 2

hence to $(p_{2k} + \frac{t}{2}p_{2k+1})tq_{2k+1} - \frac{t}{2}(p_{2k+1} + q_{2k+1} - 1)$, which is the announced result for even i. A similar proof shows that:

$$\sum_{j=q_{2k-1}+tq_{2k}}^{j=q_{2k-1}+(t+1)q_{2k}-1} \lfloor \frac{a}{b}j \rfloor = (p_{2k-1} + \frac{t}{2}p_{2k})tq_{2k} - \frac{t}{2}(p_{2k} + q_{2k} + 1) \quad \dashv$$

Lemma 6. *Suppose $q_i + tq_{i+1} + 1 \leq n \leq q_i + (t+1)q_{i+1} - 1$ for $0 \leq t \leq \alpha_{i+2}$; then $\sum_{j=q_i+tq_{i+1}}^{j=n} \lfloor \frac{a}{b}j \rfloor$ is equal to:*

$$\sum_{j=n-q_i-tq_{i+1}}^{j=n-q_i-tq_{i+1}} \lfloor \frac{p_{i+1}}{q_{i+1}}j \rfloor + (n - q_i - tq_{i+1} + 1)(p_i + tp_{i+1} - 1) - (i \bmod 2)$$

Proof. (1) Let us first consider the even case $i = 2k$; according to Klein's theorem, the set of the lattice points satisfying:

$$(q_{2k} + tq_{2k+1} \leq x \leq n) \wedge \left(0 < y \leq \frac{a}{b}x\right)$$

is the same than the set of the lattice points satisfying:

$$(q_{2k} + tq_{2k+1} \leq x \leq n) \wedge$$
$$\left(0 < y \leq p_{2k} + tp_{2k+1} + \frac{p_{2k+1}}{q_{2k+1}}(x - q_{2k} - tq_{2k+1})\right)$$

which is the disjoint union of the rectangle defined by:

$$(q_{2k} + tq_{2k+1} \leq x \leq n) \wedge (0 < y \leq p_{2k} + tp_{2k+1} - 1)$$

and the triangle defined by:

$$(q_{2k} + tq_{2k+1} \leq x \leq n) \wedge$$
$$\left(p_{2k} + tp_{2k+1} \leq y \leq p_{2k} + tp_{2k+1} + \frac{p_{2k+1}}{q_{2k+1}}(x - q_{2k} - tq_{2k+1})\right).$$

But the number of integer points lying inside the first one is $(n - q_{2k} - tq_{2k+1} + 1) \times (p_{2k} + tp_{2k+1} - 1)$. And the second one has the same discrete area than the triangle $(0 \leq x \leq n - q_{2k} - tq_{2k+1}) \wedge \left(0 < y \leq \frac{p_{2k+1}}{q_{2k+1}}x\right)$.

(2) In the odd case $i = 2k + 1$, the proof is similar, just taking care to the point of coordinates $(q_{2k+1} + tq_{2k+2}, p_{2k+1} + tp_{2k+2})$. $\dashv$

Let us introduce now the usual Ostrowski representation of integers (see [8], [9], [10]). With the previous notation, for $1 \leq n \leq b-1$, there exists a unique finite sequence of non negative integers $(\nu_i)_{0 \leq i \leq \mu-1}$ such that $n = \sum_{i=0}^{i=\mu-1} \nu_i q_i$ and $0 \leq \nu_0 \leq \alpha_1$ and $0 \leq \nu_i \leq \alpha_i$ and $\nu_i = \alpha_{i+1}$ implies $\nu_{i-1} = 0$ and $\nu_{\mu-1} \neq 0$. For our purpose, we need a slightly different form:

Lemma 7. *For $1 \leq n \leq b-1$, there exists a finite sequence of integers $(\gamma_i(n,a,b))_{0 \leq i \leq \lambda(a,b)-1}$ such that*

$$n = \sum_{i=0}^{i=\lambda(a,b)-1} \gamma_i(n,a,b) q_i(a,b)$$

and $0 \leq \gamma_i(n,a,b) < \alpha_i$ and $\gamma_i(n,a,b) = 0$ implies $\gamma_{i+1}(n,a,b) \leq 1$ and $\lambda(a,b) \leq l(a,b)$.

Proof. From the canonical representation, do inductively from the digit of maximal weight $\nu_{\lambda-1}$:
if $\gamma_i = 0$ and $\gamma_{i+1} \geq 2$, let $\gamma_i = \nu_i$, and substitute α_i to ν_{i-1} and increase ν_{i-1} by 1.

Notice that if $\nu_{i-1} = \alpha_i$, then ν_{i-2} is equal to 0; hence it is possible to iterate. ⊣

We are able now to give the main formula.

Theorem 5. *Suppose $1 \leq n \leq b-1$. Let us denote $\epsilon_t = t \bmod 2$ and $\epsilon'_t = 2\epsilon_t - 1$ and $\chi_{\{0\}}(x) = 0$ if $x = 0$ and 1 otherwise. Then*

$$\sum_{i=0}^{i=n} \lfloor \frac{a}{b} i \rfloor = \sum_{t=0}^{t=\lambda} \Bigg(\gamma_t \left((p_{t-1} + \frac{2\gamma_t + 1}{2} p_t) q_{t+1} - \frac{1}{2}(p_t + q_t - \epsilon'_t) \right) +$$

$$\sum_{j=0}^{j=\lfloor \frac{t}{2} \rfloor - 1} \frac{\alpha_{2j+1+\epsilon_t}}{2} \left((p_{2j-\epsilon_t} + p_{2j+2-\epsilon_t}) q_{2j+1-\epsilon_t} - p_{2j+\epsilon_t} - q_{2j+\epsilon_t} + \epsilon'_t \right)$$

$$+ (\Gamma_{t-1} - q_{t-1} + 1) \lfloor \frac{p_t}{q_t} (q_{t-1} + \gamma_t q_t) \rfloor - 1 \Bigg) \chi_{\{0\}}(\gamma_{t-1})$$

where

$$n = \sum_{i=0}^{i=\lambda-1} \gamma_i q_i$$

with $0 \leq \gamma_i \leq \alpha_i$ et $\gamma_i = 0$ implies $\gamma_{i+1} \leq 1$, and

$$\Gamma_t = \sum_{i=0}^{i=t} \gamma_i q_i.$$

Lastly, we generalize easily to any positive integer y:

Lemma 8. *Suppose $b \leq y$, then*

$$\sum_{i=0}^{i=y} \lfloor \frac{a}{b} i \rfloor = \sum_{i=0}^{i=y-b\lfloor \frac{y}{b} \rfloor} \lfloor \frac{a}{b} i \rfloor - \lfloor \frac{y}{b} \rfloor^2 \frac{ab}{2} + \frac{1}{2} \lfloor \frac{y}{b} \rfloor (a - b + 1 + 2ay)$$

Proof. Let us first compute $S(k,a,b) = \sum_{i=0}^{i=kb} \lfloor \frac{a}{b} i \rfloor$. We have:

$$S(k,a,b) = \sum_{i=0}^{i=kb} \lfloor \frac{a}{b}(ka - i) \rfloor = \sum_{i=0}^{i=kb} kb + \sum_{i=0}^{i=kb} \lfloor \frac{a}{b}(-i) \rfloor$$

$$S(k,a,b) = (kb+1)ka - \sum_{i=0}^{i=kb} \lceil \frac{a}{b} i \rceil$$

$$S(k,a,b) = (kb+1)ka - \left(\sum_{i=0}^{i=kb} \left(\lfloor \frac{a}{b} i \rfloor + 1 \right) \right) + k + 1$$

Hence $S(k,a,b) = \frac{1}{2}(k^2ab + ka - kb + k)$ which is a particular case of the result. Now in the general case, and denoting $k = \lfloor \frac{y}{b} \rfloor$, we have:

$$\sum_{i=0}^{i=y} \lfloor \frac{a}{b} i \rfloor = \sum_{i=0}^{i=bk} \lfloor \frac{a}{b} i \rfloor + \sum_{i=bk}^{i=y} \lfloor \frac{a}{b} i \rfloor - \lfloor b \frac{a}{b} k \rfloor$$

$$\sum_{i=0}^{i=y} \lfloor \frac{a}{b} i \rfloor = \frac{1}{2}(k^2ab + ka - kb + k) + \sum_{i=0}^{i=y-bk} \lfloor \frac{a}{b}(i + kb) \rfloor - ak$$

$$\sum_{i=0}^{i=y} \lfloor \frac{a}{b} i \rfloor = \frac{1}{2}(k^2ab - ka - kb + k) + ak(1 + y - bk) + \sum_{i=0}^{i=y-bk} \lfloor \frac{a}{b} i \rfloor$$

which is the announced result. ⊣

4 Proof of Δ_0−definability

Proof of Proposition 1

Proposition 1 follows easily from the following proposition:

Proposition 5. *Let a and b be coprime positive integers. Let x_0 and y_0 be integers such that $ax_0 + by_0 = 1$. Then*

$$d(n,a,b) = \lfloor \frac{x_0 n}{b} \rfloor + \lfloor \frac{y_0 n}{a} \rfloor + 1.$$

This elementary result is a variant of *Popoviciu's formula*:

$$d(n,a,b) = \frac{n}{ab} - \left\{ \frac{x_0 n}{b} \right\} - \left\{ \frac{y_0 n}{a} \right\} + 1$$

(see [11] for multiple references).

Proof. First notice that $k \in \mathbb{Z} \cap \left[-\frac{y_0 n}{a}; \frac{x_0 n}{b}\right] \mapsto (x_0 n - kb, y_0 n + ka)$ is one to one on $\{(x,y) \in \mathbb{N}^2; ax + by = n\}$. Now for real numbers α and β we have $\mathbb{Z} \cap [\alpha; \beta] = \mathbb{Z} \cap [\lceil \alpha \rceil; \lfloor \beta \rfloor]$ which have cardinality $\lfloor \beta \rfloor - \lceil \alpha \rceil + 1 = \lfloor \beta \rfloor + \lfloor -\alpha \rfloor + 1$. ⊣

Proof of Lemma 1

Proof. Let us denote P (resp. Q) the set of numerators (resp. denominators) of the convergents of $\frac{a}{b}$. From Theorem 4, these sets turn out to be Δ_0-definable. Notice now that Proposition 3 shows that $Card\{q \leq n; q \in Q\}$ and $Card\{p \leq n; p \in P\}$ are $O(log(n))$. Hence, from Theorem 3, the relations $z = Card\{q \leq n; q \in Q\}$ and $z = Card\{p \leq n; p \in P\}$ are Δ_0-definable.

As a consequence, the length of the Euclidean algorithm have Δ_0-definable graphs because it is $z = Card\{q \leq a; q \in Q\}$.

Moreover, the increasing enumeration $(p_i)_{i<l}$ of the elements of P is definable by: $z = p_i$ iff $(z \in P) \wedge (i = Card\{p \leq z - 1; p \in P\})$. Hence $z = p_i$ is Δ_0-definable. The same for $z = q_i$.

Lastly, let us notice that $\alpha_i = \lfloor \frac{q_i}{q_{i-1}} \rfloor$. Hence $z = \alpha_i$ is Δ_0-definable. ⊣

Proof of Lemma 2

The proof will be divided in two steps: in the first one, we assume $c = 0$; in the second one, we reduce the general case to this particular one. We need the well-known Lamé's Theorem and a technical lemma:

Theorem 6. $l(a,b) = O(log(Max\{a,b\}))$

Lemma 9. *Let two sequences* $(u_i)_{i \geq 0}$ *and* $(v_i)_{i \geq 0}$ *be defined by* $u_0 = 0$ *and* $u_1 = 1$ *and* $u_{i+2} = a_{i+2}u_{i+1} + u_i$ *and* $v_0 = 0$ *and* $v_1 = 1$ *and* $v_{i+2} = b_{i+2}v_{i+1} + v_i$. *Suppose that* $1 \leq b_i \leq 2a_i$. *Then* $v_i \leq 2u_i^3$.

Proof. First step. Let us prove by induction on i that $v_i \leq 2^{i-1}u_i$: we obviously have $v_0 \leq 2^{-1}u_0$ and $v_1 \leq u_1$; the induction step is the following: from $v_{i+2} = b_{i+2}v_{i+1} + v_i \leq 2a_{i+2}2^i u_{i+1} + 2^{i-1}u_i \leq 2^{i+1}(a_{i+2}u_{i+1} + u_i)$ follows $v_{i+2} \leq 2^{i+1}u_{i+2}$.

Second step. Let us prove that $2^{i-1} \leq 2u_i^2$ for $i \geq 1$:
from $u_{i+2} = a_{i+2}u_{i+1} + u_i \geq 2u_i$ follows $u_{2k+2} \geq 2^k$ and $u_{2k+1} \geq 2^k$ hence $u_i \geq 2^{\frac{i-2}{2}}$. ⊣

First part. Suppose $c = 0$. From Lemma 8, there is no loss of generality in assuming that $y < b - 1$. It suffices now to prove that, in the formula of Theorem 5:

- ⋆ the iterated sums are of logarithmic length;
- ⋆ the terms of these sums are polynomially bounded;

$\star$ all the introduced functions are rudimentary.

The first item is a consequence of Lamé's Theorem.

The second item follows clearly from the majoration of all the sums by polynomials of variable a and b.

We turn now to the third item. Let us notice that the non trivial rudimentary functions are those introduced in Definition 1, which have been proved to be rudimentary in Lemma 1 and those introduced in Lemma 7. It remains now to prove that the sequence of the Ostrowski digits of n is rudimentary. We intend to consider it as the sequence of the quotients of the Euclidean algorithm of a convenient couple of integers (a', b'): hence this couple is a code of the sequence. Unfortunately, some γ_i may be null but quotients in the Euclidean algorithm may not.

A way to skip this difficulty is to code the $1+\gamma_i$ in the same way. But now we have to check that the coding integers a' et b' are polynomially bounded. Hence we shall use lemma 9 to get the following Δ_0−formula defining $z = \gamma_i(n, a, b)$:

$$(\exists a')_{\leq 2a^3} (\exists b')_{\leq 2b^3}$$
$$\Big[\Big((\forall j)_{\leq \lambda(a,b)-1} (0 \leq \alpha_j(a', b') - 1 \leq \alpha_j(a, b)) \wedge \Big)$$
$$(\alpha_j(a', b') = 1 \rightarrow \alpha_{j+1}(a', b') \leq 2))$$
$$\wedge \left(n = \sum_{i=0}^{i=\lambda(a,b)-1} (\alpha_j(a', b') - 1) q_i(a, b) \right) \wedge (z = \alpha_i(a', b') - 1) \Big]$$

The proof of lemma 2 in case $c = 0$ is complete.

Second part. We reduce the general case of lemma 2 to the case $c = 0$.

Suppose first that some integers i_0 and j_0 satisfy $\frac{a}{b} i_0 + \frac{c}{d} = j_0$. Then

$$\sum_{i=0}^{i=y} \lfloor \frac{a}{b} i + \frac{c}{d} \rfloor = \sum_{i=0}^{i=y} \lfloor \frac{a}{b}(i - i_0) + j_0 \rfloor = \sum_{i=-i_0}^{i=y-i_0} \lfloor \frac{a}{b} j \rfloor + (1 + y) j_0$$

Now from $\lfloor \frac{a}{b} j \rfloor = -\lceil \frac{a}{b}(-j) \rceil$ we get:

$$\sum_{i=-i_0}^{i=0} \lfloor \frac{a}{b} j \rfloor = -\sum_{i=0}^{i=i_0} \Big(\lfloor \frac{a}{b} j \rfloor + 1 \Big) - Card\{j \in [0; i_0]; b \text{ divides } a \times j\}.$$

But $Card\{j \in [0; i_0]; b \text{ divides } a \times j\} = 1 + \lfloor \frac{i_0}{\frac{b}{gcd(a,b)}} \rfloor$. This completes the proof of the first case.

Suppose now that no integers i_0 and j_0 satisfy $\frac{a}{b}i_0 + \frac{c}{d} = j_0$. Then notice that $\lfloor \frac{a}{b}i + \frac{c}{d} \rfloor = \lfloor \frac{a}{b}i + \frac{gcd(a,b)}{b} \lfloor \frac{bc}{d \times gcd(a,b)} \rfloor \rfloor$. This reduces the general case to the case when $c = 0$.

The proof of lemma 2 is complete. ⊣

Reducing Theorem 1 to Lemma 2

We obviously just need to consider the case $gcd(a,b,c) = 1$. The relation $(ax + by + cz = n) \wedge (x \geq 0) \wedge (y \geq 0) \wedge (z \geq 0)$ is equivalent to

$$\exists t \in \mathbb{Z}\,((ax + by = (t \times gcd(a,b)) \wedge$$
$$(t \times gcd(a,b) + cz = n) \wedge (x \geq 0) \wedge (y \geq 0) \wedge (z \geq 0))$$

Let us introduce positive integers z_0 and t_0 such that $1 = t_0 \times gcd(a,b) - cz_0$. The condition turns out to be equivalent to

$$\exists t \in \mathbb{Z}\ \exists t' \in \mathbb{Z}$$
$$(ax + by = t \times gcd(a,b)) \wedge (z = -nz_0 + t' \times gcd(a,b)) \wedge$$
$$(t = nt_0 - t' \times gcd(a,b)) \wedge (x \geq 0) \wedge (y \geq 0) \wedge (z \geq 0)$$

hence to

$$\exists t' \in \mathbb{Z}\left((x > 0) \wedge (y > 0) \wedge \left(\frac{nz_0}{gcd(a,b)} \leq t \leq \frac{nt_0}{c}\right)\right.$$
$$\left.\wedge\ (ax + by = (nt_0 - t'.c) \times gcd(a,b)) \wedge (z = -nz_0 + t' \times gcd(a,b))\right)$$

This gives

$$d(n;a,b,c) = \sum_{t \geq n\frac{nz_0}{a \wedge b}}^{t \leq \frac{nt_0}{c}} d((a \wedge b)(nt_0 - t'.c)); a, b)$$

Now Theorem 1 follows from Proposition 5. ⊣

Let us notice that other ways of reduction lead to the same result such as the one mentioned in [13].

References

[1] Berarducci, A. and Intrigila, B., Linear reccurence relations are Δ_0 definable, *Logic and Foundations of Mathematics, Selected contributed papers, LMPS '95 (A. Cantini, E. Casari, P. Minari editors)*, vol. 92, 1999, pp. 67–81,Kluwer Academic Publishers, Dordrecht, The Netherlands.

[2] Esbelin, H.-A., and More, M., Rudimentary relations and primitive recursion : a toolbox, *Theor. Comput. Sci.*, vol. 193, 1998, pp. 129–148.

[3] Khinchin, A. Ya., *Continued Fractions*, Dover, 1997.

[4] Paris, J. and Wilkie, A., Counting Δ_0-sets, *Fund. Math.*, vol. 127, 1987, pp. 67–76.

[5] Pick, G., Geometrisches zur Zahlenhre, 1899, *Naturwissenschaft Zeitschrift Lotos*, Prague.

[6] Woods, A., *Some Problems in Logic and Number Theory and their Connections*, Ph.D., University of Manchester, 1981, University of Manchester. Part II of this volume.

[7] Klein, F., Sur une représentation géométrique du développement en fraction continue ordinaire, *Nouv. Ann. Math.*, vol. 3 (15), 1896, pp. 327-331.

[8] Fraenkel, A. S., Systems of numeration, *American Mathematical Monthly*, vol. 92, 1985, pp. 321–331.

[9] Berthé, V., Autour du système de numération d'Ostrowski, *Bull. Belg. Math. Soc.*, vol. 8, 2001, pp. 209–238.

[10] Berthé, V., Multidimensional Euclidean algorithms, numeration and substitutions, *A2 Integers*, vol. 11B, 2011.

[11] Beck, M. and Sinai, R., *Computing the continuous discretely: Integer-point enumeration in polyhedra*, Springer Undergraduate Texts in Mathematics, 2007, 226 pages, Springer.

[12] Reveillès, J.-P., *Géométrie discrète, calcul en nombres entiers et algorithmique*, Thèse d'État, Université Louis Pasteur, Strasbourg, 1991, 226 pages, Université Louis Pasteur, Strasbourg.

[13] D'Alessandro, F. and Intrigila, B. and Varricchio, S., On the structure of the counting function of sparse context-free languages, *Theor. Comput. Sci.*, vol. 3561-2, 2006, pp. 104–117.

7

Techniques in weak analysis for conservation results

António M. Fernandes,[1,2] Fernando Ferreira,[3] Gilda Ferreira[4]

Abstract: We review and describe the main techniques for setting up systems of weak analysis, i.e. formal systems of second-order arithmetic related to subexponential classes of computational complexity. These involve techniques of proof theory (e.g., Herbrand's theorem and the cut-elimination theorem) and model theoretic techniques like forcing. The techniques are illustrated for the particular case of polytime computability. We also include a brief section where we list the known results in weak analysis.

Keywords: Bounded arithmetic. Weak analysis. Conservation results. Cut-elimination. Forcing. Polytime computability.

[1]The authors acknowledge support of FCT-Fundação para a Ciência e a Tecnologia [PEst-OE/MAT/UI0209/2011 and PTDC/MAT/104716/2008]. The third author is also grateful to FCT [grant SFRH/BPD/34527/2006] and Núcleo de Investigação em Matemática (Universidade Lusófona).

[2]Instituto Superior Técnico, Departamento de Matemática, Av. Rovisco Pais, 1049-001 Lisboa, Portugal, `amfernandes@netcabo.pt`

[3]Universidade de Lisboa, Faculdade de Ciências, Departamento de Matemática, Campo Grande, Ed. C6, 1749-016 Lisboa, Portugal, `fjferreira@fc.ul.pt`

[4]Universidade Lusófona de Humanidades e Tecnologias, Departamento de Matemática, Av. do Campo Grande, 376, 1749-024 Lisboa, Portugal, `gildafer@cii.fc.ul.pt`

Studies in Weak Arithmetics.
Patrick Cégielski, Charalampos Cornaros,
Costas Dimitracopoulos.

1 Introduction

Weak analysis can be described as the formalization and development of analysis in very weak systems of second-order arithmetic. These systems, as they are ordinarily understood, are subelementary in the sense that they do not prove the totality of the exponential function. In addition, they are often related with well-known classes of computational complexity, e.g. polytime or polyspace computability. This paper is not, however, concerned with weak analysis as described in the first two lines of this introduction (nevertheless, in Section 8, we briefly review – without proofs – the state of the art in this respect). The aim of this paper is rather to describe and exemplify the fundamental techniques for setting up a system of weak analysis. In the remainder of this introductory section, we give a blueprint for defining theories of weak analysis. We first give a general blueprint and then describe it for the particular case of polytime computability and point to the sections of the paper where the blueprint is executed (for the polytime case).

Systems of analysis must be able to speak about real numbers. Therefore, they are usually framed in a second-order language and each real number is "presented" via a set of natural numbers. It is of course important to be able to define basic real numbers, to show that they are closed under basic operations and, in general, to be able to prove simple facts about the real line. This is achieved essentially by a combination of induction and set-formation. The amount of induction present in a weak theory is intrinsically related with its provably total functions (with appropriate graphs). Hence, if one is given a computational complexity class C and the goal is to set up a theory T whose provably total functions are exactly those of C, then one is immediately constrained with respect to the amount of induction permitted in the system.

The situation is also tight with respect to set formation because we allow set parameters in the induction scheme (therefore, the more sets there are, the more induction is available). There does exist, nevertheless, some leeway with regard to set formation. All the systems of weak analysis considered permit forms of recursive comprehension, i.e., a principle of the form:

$$\forall x\, (\exists y A(x, y) \leftrightarrow \forall z B(x, z)) \rightarrow \exists X \forall x\, (x \in X \leftrightarrow \exists y A(x, y))$$

where $A(x, y)$ and $B(x, z)$ are appropriate formulas (possibly with first and second-order parameters) and X does not occur in $A(x, y)$. Typically, $A(x, y)$ and $B(x, z)$ are sufficiently expressive so that the sets of the form $\{x : \exists y A(x, y)\}$ (resp., $\{x : \forall z B(x, z)\}$) give all the recursively enumerable (resp., co-recursively enumerable) sets in the standard model. The reader who is unfamiliar with the subject can be taken aback by this amount of comprehension. Shouldn't the variables y and z above be restricted in some way so that only sets in C are definable (in the standard model)? No, and this liberty simplifies very much the development of analysis. The *intuitive reason* why the above recursive comprehension is not a strong principle is that, in order to form the set $X = \{x : \exists y A(x, y)\}$, one must first establish the equivalence $\forall x\,(\exists y A(x, y) \leftrightarrow \forall z B(x, z))$. Note that only weak principles are available for doing this. In short, one can form a recursive set *if* one can show that the set is indeed recursive. The *formal reason* that makes possible the inclusion of recursive comprehension relies crucially on the availability (of forms) of bounded collection. Fortunately, it is known that under very general conditions, the addition of bounded collection to a bounded theory of arithmetic results in a conservative extension with respect to Π^0_2-sentences. At this juncture, a second-order theory T^2 is set-up: one that has forms of recursive comprehension and bounded collection.

There is another reason why bounded collection is important. It is because of its relation with weak König's lemma. This lemma guarantees the existence of infinite paths through an infinite binary tree (curiously enough, although the path is a *set* the tree need not be a set – it only needs to be a *class* defined by a bounded formula). Weak König's lemma is a form of compactness and, e.g., is instrumental in analysis to prove the compactness of the closed unit interval (see Section 8 for some facts). The addition of weak König's lemma is first-order conservative over T^2. We use for this a forcing technique originally due to unpublished work of Leo Harrington. The proof of the density of some sets needed for the forcing argument relies crucially on bounded collection. Weak König's lemma can be strengthened to the so-called strict-Π^1_1 reflection. Roughly, this principle differs from weak König's lemma in that the binary trees permitted are constituted not by binary strings but by bounded sets. Observe that, in the absence of the totality of exponentiation, not every bounded set is given by a binary string. Strict-Π^1_1 reflection can be added to T^2 without changing the Π^0_2-consequences and, again, we use a forcing argument. However, in order to show the density of certain sets one must rely on stronger forms of bounded collection (and stronger notions of bounded formula). The authors do not know of any uses of strict-Π^1_1 reflection for the development of analysis in weak systems.

We have described a very general blueprint for setting up a weak system of analysis related with a given class C of computational complexity. In this paper, we illustrate this blueprint in a simple case: when C is the class of polytime computable functions. All the techniques needed to set up systems of weak analysis as described above are already present in this case. Most of the results of this paper have appeared elsewhere in several publications, but in here we conveniently integrate them and discuss their scope and limitations. The detailed blueprint for the polytime case, with appropriate references to the sections of the paper, is the following:

(a) We describe a universal theory (i.e., one that can be axiomatized by universal formulas) PTCA which has function symbols for every polytime computable function. Herbrand's theorem guarantees that the provably total functions (with appropriate graphs) of this theory are the polytime computable functions. PTCA is extended in order to permit induction for formulas that, in the standard model, define the NP sets (the Σ_1^{b}-formulas). This is done using a cut-elimination argument. We will observe in Section 4 that this extra amount of induction is essential to introduce recursive comprehension. As a matter of fact, we actually work with a simplified theory Σ_1^{b}-NIA (this corresponds to the theory T in the general blueprint). These issues are treated in Section 2.

(b) In Section 3, we show that the addition of the bounded collection scheme $\mathsf{B}\Sigma_\infty^{\mathsf{b}}$ to Σ_1^{b}-NIA yields a Π_2^0-conservative extension. We use again an argument based on cut-elimination.

(c) We introduce the second-order theory BTFA (corresponding to T^2 in the general blueprint) enjoying a form of recursive comprehension and show that it is first-order conservative over Σ_1^{b}-$\mathsf{NIA}+\mathsf{B}\Sigma_\infty^{\mathsf{b}}$. This is shown in Section 4 via a simple model theoretic argument.

(d) In the next section, we show that the addition of weak König's lemma to BTFA results in a theory which is a first-order conservative extension of BTFA. We use the forcing method to prove this result. We will point to the need of bounded collection in the forcing argument.

(e) In Section 7, we define the scheme of strict-Π_1^1 reflection and show that adding it to BTFA does not permit proving more Π_2^0-sentences. This is done via a forcing argument similar to the one in (d) above. As discussed, for this forcing argument to be successful, one needs a stronger form of bounded collection. This stronger form is formulated and discussed in the previous Section 6, and an appropriate conservation result is proved. Cut-elimination is used in proving this conservation result.

The paper has a further section where we present in a compact manner the state of the art concerning the formalization of analysis in weak systems. This last section has no proofs but provides appropriate references and mentions some open problems.

2 First-order theories for polytime computability

We start by presenting three first-order theories for polytime computability, namely PTCA, PTCA^+ and Σ^b_1-NIA. The former and the latter systems play, in the context of polytime computability, the same role as (first-order) PRA and Σ^0_1-IND play in the context of primitive recursive computability. The inclusion of the intermediate theory PTCA^+ aims at making more explicit a conservation result concerning induction. We remind that the goal of the present paper does not lie in the introduction of weak theories connected with polytime computability (already described in [11, 12, 13, 3]) but in the survey of conservation techniques, which we try to motivate and present in a simple and clear way.

When working in systems of bounded arithmetic, we must opt between two notations: *unary* or *binary*. Unary notation, the most commonly used in the literature, was the choice of Samuel Buss in his seminal thesis [3], while binary notation was used by the second author in [11]. Due to the possibility of interpreting theories in one notation into theories in the other notation (e.g. [23]), the choice has no intrinsic significance and is basically a question of taste and pragmatics. The authors share the opinion that binary notation is more intuitive and convenient for describing and dealing with weak (subexponential) systems of analysis and this explains why it is adopted in the present paper. Therefore, the language that we use aims at describing the set of finite binary words, denoted by $\{0,1\}^*$ (or $2^{<\omega}$). This is the domain of the intended *standard* model. Let us introduce some notation: ϵ denotes the empty word; for x and y elements in $\{0,1\}^*$, $x\frown y$ represents the *concatenation* of x by y (we usually omit the concatenation symbol $\frown$ and just write xy); $x \subseteq y$ means that x is an *initial subword* of y (x is a string prefix of y); $|x|$ denotes the *length* of the word x; $x_{|_y}$ is the *truncation* of x by the length of y ($x_{|_y}$ is x itself if $|x| \leq |y|$; otherwise, it is the prefix of x with length $|y|$); finally, $x \times y$ is the word x concatenated with itself length of y times (note that $|x \times y| = |x| \cdot |y|$).

Let $\mathcal{L}$ be the first-order language which has three constant symbols ϵ, 0 and 1, two binary function symbols $\frown$ and $\times$ (with the standard interpretations given in the previous paragraph) and two binary relation symbols $=$ and $\subseteq$ (for *equality* and *initial subwordness*, respectively). Let $\mathcal{L}_\mathcal{P}$ be the extension of $\mathcal{L}$ by adding a function symbol for each description (given below) of a polytime computable function.

Definition 1. PTCA (acronym for Polynomial Time Computable Arithmetic) is the first-order theory, in the language $\mathcal{L}_{\mathcal{P}}$, which has the following axioms:

- Basic axioms

  $x\epsilon = x$, $x(y0) = (xy)0$ *and* $x(y1) = (xy)1$;
  $x \times \epsilon = \epsilon$, $x \times y0 = (x \times y)x$ *and* $x \times y1 = (x \times y)x$;
  $x \subseteq \epsilon \leftrightarrow x = \epsilon$, $x \subseteq y0 \leftrightarrow x \subseteq y \vee x = y0$ *and*
  $x \subseteq y1 \leftrightarrow x \subseteq y \vee x = y1$;
  $x0 = y0 \rightarrow x = y$ *and* $x1 = y1 \rightarrow x = y$;
  $x0 \neq y1$, $x0 \neq \epsilon$ *and* $x1 \neq \epsilon$;

- Defining axioms

  (a) *Initial functions*
  $C_0(x) = x0$
  $C_1(x) = x1$
  $P_i^n(x_1, ..., x_n) = x_i$, for $1 \leq i \leq n$
  $Q(x, y) = 1 \leftrightarrow x \subseteq y$; $Q(x, y) = 0 \vee Q(x, y) = 1$

  (b) *Derived functions*
  1. Composition:
  $f(\bar{x}) = g(h_1(\bar{x}), ..., h_k(\bar{x}))$, if f is the description of the composition from g, h_1, ..., h_k
  2. Bounded recursion on notation:
  $f(\bar{x}, \epsilon) = g(\bar{x})$
  $f(\bar{x}, y0) = h_0(\bar{x}, y, f(\bar{x}, y))_{|t(\bar{x}, y)}$
  $f(\bar{x}, y1) = h_1(\bar{x}, y, f(\bar{x}, y))_{|t(\bar{x}, y)}$,
  where t is a term of the language $\mathcal{L}$ and f is the description of the bounded recursion on notation defined from g, h_0, h_1 and t;

- Scheme of induction on notation

$$A(\epsilon) \wedge \forall x\,(A(x) \rightarrow A(x0) \wedge A(x1)) \rightarrow \forall x A(x),$$

  where A is a *polytime decidable matrix*, possibly with parameters. The class of polytime decidable matrices is the smallest class of formulas of $\mathcal{L}_{\mathcal{P}}$ containing the atomic formulas and closed under Boolean operations and quantifications of the form $\forall x\,(x \subseteq t \rightarrow \ldots)$ or $\exists x\,(x \subseteq t \wedge \ldots)$ (usually abbreviated by $\forall x \subseteq t\,(\ldots)$ and $\exists x \subseteq t\,(\ldots)$ respectively).

The above definition gives simultaneously the theory PTCA and the language $\mathcal{L}_{\mathcal{P}}$ in which it is formulated: the *descriptions* of the polytime computable functions are given by (a) and (b) above. The proof that these schemes generate exactly the polytime computable functions can be found in [11]. Note that the polytime decidable matrices define exactly the polytime predicates in the standard model (initial subword quantification can be decided in polytime because it only needs a polytime computable search).

We argue that PTCA is a universal theory. This is not immediate because the induction scheme is not constituted by universal formulas. We rely on a well-known result of Łoś and Tarski that states that it is enough to prove that PTCA is preserved by substructures. Let $\mathcal{M}$ be a model of PTCA and let $\mathcal{N}$ be a substructure of $\mathcal{M}$. As we have pointed, we only need to argue that induction on notation also holds in $\mathcal{N}$. Note that the scheme of induction can be reformulated thus:

$$A(\epsilon) \wedge \forall y \subseteq a(A(y) \rightarrow A(y0) \wedge A(y1)) \rightarrow A(a).$$

Therefore, the result follows if it is shown that polytime decidable matrices are absolute between $\mathcal{M}$ and $\mathcal{N}$. This is a consequence of the following result:

Lemma 1. *For each polytime decidable matrix $A(\bar{z}, x)$ there is a function symbol g in $\mathcal{L}_{\mathcal{P}}$ such that*

$$\mathsf{PTCA} \vdash \exists y \subseteq x A(\bar{z}, y) \rightarrow g(\bar{z}, x) \subseteq x \wedge A(\bar{z}, g(\bar{z}, x)).$$

It is not difficult to define by bounded recursion on notation a function g whose value is the prefix y of x of smallest length such that $A(\bar{z}, y)$, if there is one such value. (The search argument should be clear, but details can be found in [12], page 141-143.) Note that, if we wanted to search along all words with length less than or equal to $|x|$, bounded recursion on notation would not be enough. Such a (exponential) search goes beyond polytime computability.

We need two auxiliary results. The first says that polytime decidable matrices can be expressed in PTCA by means of quantifier-free formulas. The second states that PTCA allows the definition of functions by cases. (These results are easy to prove; see [12].)

Lemma 2. *For each polytime decidable matrix A, there is a function symbol K_A in $\mathcal{L}_{\mathcal{P}}$ such that*

$$\mathsf{PTCA} \vdash (A(\bar{x}) \rightarrow K_A(\bar{x}) = 1) \wedge (\neg A(\bar{x}) \rightarrow K_A(\bar{x}) = 0),$$

where the free variables of A are among $\bar{x}$.

Lemma 3. *Given polytime decidable matrices $A_1(\bar{x}), \ldots, A_n(\bar{x})$ and function symbols $f_1(\bar{x}), \ldots, f_n(\bar{x}), f_{n+1}(\bar{x})$ there is a function symbol $f(\bar{x})$ such that the theory* PTCA *proves*

$$(A_1(\bar{x}) \wedge f(\bar{x}) = f_1(\bar{x})) \vee (\neg A_1(\bar{x}) \wedge A_2(\bar{x}) \wedge f(\bar{x}) = f_2(\bar{x})) \vee \ldots$$
$$\vee(\neg A_1(\bar{x}) \wedge \ldots \wedge \neg A_n(\bar{x}) \wedge f(\bar{x}) = f_{n+1}(\bar{x})).$$

By the previous lemmas and the universal axiomatizability of PTCA, the following is immediate by Herbrand's theorem:

Theorem 1. *If* $\mathsf{PTCA} \vdash \forall \bar{x} \exists y A(\bar{x}, y)$*, where A is a polytime decidable matrix and $\bar{x}$ and y are the only free variables of A, then there is a function symbol f in $\mathcal{L}_{\mathcal{P}}$ such that* $\mathsf{PTCA} \vdash \forall \bar{x}\, A(\bar{x}, f(\bar{x}))$.

We denote by $x \preceq y$ (respectively, $x \equiv y$) the formula $1 \times x \subseteq 1 \times y$ (respectively, $1 \times x = 1 \times y$). In the standard model $x \preceq y$ (respectively, $x \equiv y$) says that the length of x is less than or equal (respectively, equal) to the length of y. Quantifications of the form $\forall x\, (x \preceq t \rightarrow \ldots)$ and $\exists x\, (x \preceq t \wedge \ldots)$ (usually abbreviated by $\forall x \preceq t\, (\ldots)$ and $\exists x \preceq t\, (\ldots)$) are called *bounded quantifications.*

Let us now introduce the auxiliary theory PTCA^+:

Definition 2. PTCA^+ is the first-order theory whose axioms are those of PTCA plus the following scheme of induction on notation:

$$B(\epsilon) \wedge \forall x\, (B(x) \rightarrow B(x0) \wedge B(x1)) \rightarrow \forall x B(x),$$

where $B(x)$ is of the form $\exists y \preceq t\, A(y, x)$, where $A(y, x)$ is quantifier-free (possibly with parameters), and t is a term in which y does not occur.

The formulas of the form B above define exactly the NP-sets in the standard model (see [11]). The increase in induction (from polytime to non-deterministic polytime) does not change the provably total functions (with polytime graphs) of the theory. The result below should be compared with the classic result of Charles Parsons (independently obtained by Gaisi Takeuti and Grigori Mints) that says that the provably total functions of Σ^0_1-IND are the primitive recursive functions. For appropriate references to this classical result and a new proof of it, see [15].

Theorem 2. *If* $\mathsf{PTCA}^+ \vdash \forall \bar{x} \exists y A(\bar{x}, y)$*, where A is a polytime decidable matrix and $\bar{x}$ and y are the only free variables of A, then* $\mathsf{PTCA} \vdash \forall \bar{x} \exists y A(\bar{x}, y)$.

Proof. Let us consider the theory PTCA^+ given by a universal axiomatization $\mathcal{A}$ of PTCA (we take the formulas in $\mathcal{A}$ to be quantifier-free), together with the induction scheme of Definition 2. Now, let us formulate PTCA^+ in Gentzen's sequent calculus by adding to the usual initial sequents and rules for predicate logic with equality, the following initial sequents and rule:

- the initial sequents of the form $\rightarrow A$, with A in $\mathcal{A}$;

- the induction rule

$$\frac{\Gamma, B(x) \to \Delta, B(x0) \qquad \Gamma, B(x) \to \Delta, B(x1)}{\Gamma, B(\epsilon) \to \Delta, B(s)}$$

where $B(x)$ is of the form $\exists w \preceq t\, A(w, x)$, with $A(w, x)$ a quantifier-free formula (possibly with parameters), t is a term in which w does not occur, s is any term, and the variable x does not occur free in Γ or Δ (it is an eigenvariable).

Suppose that $\mathsf{PTCA}^{+} \vdash \forall \bar{x} \exists y A(\bar{x}, y)$, where (w.l.o.g.) A is quantifier-free. Our goal is to prove that $\mathsf{PTCA} \vdash \forall \bar{x} \exists y A(\bar{x}, y)$. Take $\mathcal{P}$ a proof of $\to \exists y A(\bar{x}, y)$ in the sequent calculus described above. By Gentzen's cut elimination theorem adapted to our setting (more precisely partial cut elimination, also called the free-cut elimination theorem) we know that there is a proof $\mathcal{P}'$ of $\to \exists y A(\bar{x}, y)$, in the sequent calculus above, in which every cut formula comes from an initial sequent or from a principal formula in the induction rule. Therefore, by the subformula property of the calculus, every formula in the proof $\mathcal{P}'$ is a subformula of the conclusion or of a cut formula. In both cases, it is a quantifier-free formula or an existential formula.

The result follows if we manage to prove that, for every sequent $\Gamma \to \Delta$ in the proof $\mathcal{P}'$, we have $\mathsf{PTCA} \vdash \wedge\Gamma \to \vee\Delta$ (where $\wedge\Gamma$, respectively $\vee\Delta$, denotes the conjunction, respectively disjunction, of the formulas in Γ, respectively in Δ). The proof is by induction on the number of lines of $\mathcal{P}'$. The only case requiring attention is the induction rule. Suppose that the result holds for the two premisses of the induction rule, i.e. $\mathsf{PTCA} \vdash \wedge\Gamma \wedge B(x) \to \vee\Delta \vee B(xi)$, for $i = 0, 1$. We want to prove that it also holds for the conclusion of the rule, i.e. $\mathsf{PTCA} \vdash \wedge\Gamma \wedge B(\epsilon) \to \vee\Delta \vee B(s)$.

By Theorem 1, it is possible to show that there are function symbols h_0 and h_1 such that PTCA proves:

$$\begin{aligned}(\wedge\Gamma \wedge \neg \vee \Delta) \to \exists \bar{y}_i \forall x \forall w\, (w \preceq t(x) \wedge A(w, x) \to \\ h_i(\bar{y}_i, x, w) \preceq t(xi) \wedge A(h_i(\bar{y}_i, x, w), xi)),\end{aligned}$$

for $i = 0, 1$. Of course, this fact relies crucially on the fact that the formulas in Γ and Δ are either quantifier-free or existential (see [12], pages 144-147, for details).

We now reason in PTCA. Assume that we have $\wedge\Gamma$, $\neg \vee \Delta$ and $B(\epsilon)$. We want to prove $B(s)$. By the property above, we know that there are $\bar{a}_0$ and $\bar{a}_1$ satisfying

$$\begin{aligned}(\wedge\Gamma \wedge \neg \vee \Delta) \to \forall x \forall w\, (w \preceq t(x) \wedge A(w, x) \to \\ h_i(\bar{a}_i, x, w) \preceq t(xi) \wedge A(h_i(\bar{a}_i, x, w), xi)),\end{aligned}$$

for $i = 0, 1$. It is not difficult to define a function f, by bounded recursion on notation, such that: $f(\bar{y}_0, \bar{y}_1, u, \epsilon) = u$ and $f(\bar{y}_0, \bar{y}_1, u, xi) = h_i(\bar{y}_i, x, f(\bar{y}_0, \bar{y}_1, u, x))$, $i = 0, 1$. Since $B(\epsilon)$ holds, take c such that $c \preceq t(\epsilon)$ and $A(c, \epsilon)$. By induction on notation on x it is possible to argue that, for all x, $f(\bar{a}_1, \bar{a}_2, c, x) \preceq t(x)$ and $A(f(\bar{a}_1, \bar{a}_2, c, x), x)$. In particular we have $f(\bar{a}_1, \bar{a}_2, c, s) \preceq t(s)$ and $A(f(\bar{a}_1, \bar{a}_2, c, s), s)$, i.e. $B(s)$. $\dashv$

As it is well-known, it is possible to introduce in the theory Σ^0_1-IND all primitive recursive functions (this is essentially done in the famous incompleteness paper of Gödel) and, therefore, PRA can be considered a subtheory of Σ^0_1-IND. Of course, given the paucity of primitive function symbols in the language of Peano arithmetic, the availability of induction for Σ^0_1-formulas is crucial in this regard (mere bounded quantification in the language of Peano arithmetic obviously does not suffice). A similar phenomenon takes place in our setting. Let us look at the situation.

Let $x \subseteq^* y$ abbreviate the formula $\exists w \subseteq y\, (wx \subseteq y)$. The meaning of this relation is clear: x is a subword (not necessarily an initial subword) of y. The class of *subword quantification formulas*, abbreviated sw.q.-formulas, is the smallest class of formulas of $\mathcal{L}$ containing the atomic formulas and closed under Boolean operations and subword quantifications, i.e. of the form $\forall x\, (x \subseteq^* t \to \ldots)$ or $\exists x\, (x \subseteq^* t \wedge \ldots)$, where t is a term in which x does not occur. As usual, subword quantifications can be abbreviated by $\forall x \subseteq^* t\, (\ldots)$ and $\exists x \subseteq^* t\, (\ldots)$, respectively. The reader should note that the sets of the standard model defined by sw.q.-formulas are polytime computable (but by no means exhaust this class). This is clear because the number of subwords of a word is quadratic in the length of the given word (therefore, the needed searches are polytime computable). The following definition is important:

Definition 3. A Σ^b_1-*formula* is a formula of the form $\exists x\, (x \preceq t(\bar{z}) \wedge A(\bar{z}, x))$, usually written as $\exists x \preceq t(\bar{z})\, A(\bar{z}, x)$, where A is a sw.q.-formula and x does not occur in the term t. Π^b_1-*formulas* are defined dually.

It is possible to show that the Σ^b_1-formulas define exactly the NP-sets in the standard model. (This is somewhat reminiscent of the fact that the Σ^0_1-formulas define the recursively enumerable sets.) The reason for this lies in the fact that sw.q.-formulas are expressive enough to describe computations of Turing machines (see [11] for details).

Definition 4. Σ_1^b-NIA is the first-order theory,[5,6] formulated in the language $\mathcal{L}$, which has the following axioms:

- The *basic axioms* of Definition 1.
- The *scheme of induction on notation*

$$A(\epsilon) \wedge \forall x\,(A(x) \rightarrow A(x0) \wedge A(x1)) \rightarrow \forall x A(x),$$

where A is a Σ_1^b-formula, possibly with other free variables besides x.

In [11], it was shown that the class of polytime decidable matrices is closed under subword quantification (modulo equivalence in PTCA). Hence Σ_1^b-NIA is a subtheory of PTCA^+. On the other hand, it is possible (and relatively easy) to present for each $f(\overline{x}) \in \mathcal{L}_\mathcal{P}$ a Σ_1^b-formula $F_f(\overline{x}, y)$ in the language $\mathcal{L}$ such that Σ_1^b-NIA $\vdash \forall \overline{x}\, \exists^1 y\, F_f(\overline{x}, y)$ and which has the defining properties of f (as given by the axioms in Definition 1). The proof is by induction on the complexity of the description of f (the details can be found in [12]). We can subsume the above discussion by the following result:

Proposition 1. *Every model of* Σ_1^b-NIA *can be extended to a model of* PTCA^+ *by defining the interpretation of each function symbol* $f \in \mathcal{L}_\mathcal{P}$ *via* F_f.

3 Adding bounded collection

In this section we are going to enrich the theory Σ_1^b-NIA with a bounded collection scheme. As already discussed, this principle plays a pivotal role in setting up systems of weak analysis. Once more, we are going to use a cut-elimination technique to prove that the enriched theory is conservative over Σ_1^b-NIA concerning Π_2^0-formulas.

The *bounded formulas* of $\mathcal{L}$ consist of the smallest class of formulas containing the atomic formulas and closed under Boolean connectives and bounded quantifications. The *principle of bounded collection*, denoted by $\mathsf{B}\Sigma_\infty^b$, is the following scheme:

$$\forall x \preceq a \exists y A(x, y) \rightarrow \exists z \forall x \preceq a \exists y \preceq z A(x, y),$$

where A is a bounded formula (possibly with parameters) and z is a new variable.

For technical reasons pertaining to the formulation of an efficient calculus of sequents, in this section (and in Section 6) we consider that the language $\mathcal{L}$ has *primitive* bounded quantifiers. I.e., if A is a formula of $\mathcal{L}$, we introduce the formulas $\forall x \preceq t\, A$ and $\exists x \preceq t\, A$ (with t a term

[5]This theory was originally named Σ_1^b-PIND (see [11, 12]). The current denomination stands for *Notation Induction Axiom* and follows the terminology in [2].

[6]This theory is defined in the spirit of Buss's seminal theory S_2^1 ([4]).

where x does not occur) as new formulas, instead of mere abbreviations of $\forall x\,(x \preceq t \to A)$ and $\exists x\,(x \preceq t \wedge A)$, respectively.

In analogy with the formulation of PTCA^+ given in the proof of Theorem 2, we formulate Σ_1^b-NIA in Gentzen's sequent calculus with the corresponding rule for induction on notation (for Σ_1^b-formulas). We add the following rules for the bounded quantifiers:

$$\frac{\Gamma, A(t) \to \Delta}{\Gamma, t \preceq s, \forall x \preceq sA(x) \to \Delta}\ \forall_{\preceq} : l \qquad \frac{\Gamma, b \preceq t \to \Delta, A(b)}{\Gamma \to \Delta, \forall x \preceq tA(x)}\ \forall_{\preceq} : r$$

$$\frac{\Gamma, b \preceq t, A(b) \to \Delta}{\Gamma, \exists x \preceq tA(x) \to \Delta}\ \exists_{\preceq} : l \qquad \frac{\Gamma \to \Delta, A(t)}{\Gamma, t \preceq s \to \Delta, \exists x \preceq sA(x)}\ \exists_{\preceq} : r$$

where b is an eigenvariable (it is not free in either Γ, Δ or t). Of course, with these rules it is possible to prove the equivalences $\forall x \preceq tA \leftrightarrow \forall x\,(x \preceq t \to A)$ and $\exists x \preceq tA(x) \leftrightarrow \exists x\,(x \preceq t \wedge A)$.

Theorem 3. *The theory Σ_1^b-NIA $+ \mathsf{B}\Sigma_\infty^b$ is Π_2^0- conservative over the theory Σ_1^b-NIA.*

Proof. The theory Σ_1^b-NIA $+ \mathsf{B}\Sigma_\infty^b$ can be formulated in the sequent calculus described above together with the following rule for bounded collection:

$$\frac{\Gamma \to \Delta, \forall x \preceq t \exists y A(x,y)}{\Gamma \to \Delta, \exists z \forall x \preceq t \exists y \preceq zA(x,y)}$$

where A is a bounded formula and b is an eigenvariable.

Suppose that Σ_1^b-NIA $+ \mathsf{B}\Sigma_\infty^b \vdash \forall x \exists y A(x,y)$ with A a bounded formula. Then, there is a proof of $\to \exists y A(x,y)$ in the sequent calculus above. The free-cut elimination theorem ensures that there is a proof $\mathcal{P}$ of $\to \exists y A(x,y)$ without free cuts. As a consequence, all formulas occurring in the sequents $\Gamma \to \Delta$ in $\mathcal{P}$ are Σ_1^0-formulas, i.e. of the form $\exists y B(x,y)$, with B a bounded formula.

Let $\Gamma \to \Delta$ be a sequent in $\mathcal{P}$, where Γ is $\exists x_1 B_1(x_1, x), \ldots, \exists x_n B_n(x_n, x)$ and Δ is $\exists y_1 C_1(y_1, x), \ldots, \exists y_k C_k(y_k, x)$, with $B_1, \ldots, B_n, C_1, \ldots, C_k$ bounded formulas (we admit the absence of existential quantifiers to accommodate bounded formulas in the sequent). It can be proved, by induction on the number of lines of the proof $\mathcal{P}$, that from bounds of the antecedents we can get (in Σ_1^b-NIA) bounds for the consequents in the following way:

$$\Sigma_1^b\text{-NIA} \vdash \forall u \exists v \forall x \preceq u\,(\Gamma^{\preceq u} \to \Delta^{\preceq v}),$$

where $\Gamma^{\preceq u}$ abbreviates $\exists x_1 \preceq uB_1(x_1, x) \wedge \ldots \wedge \exists x_n \preceq uB_n(x_n, x)$ and $\Delta^{\preceq v}$ abbreviates $\exists y_1 \preceq vC_1(y_1, x) \vee \ldots \vee \exists y_k \preceq vC_k(y_k, x)$. The proof of this is not difficult. We do not give a proof because the situation will reappear in a stronger context in Section 6.

If we apply the above result to the last sequent of $\mathcal{P}$, we get

$$\Sigma^b_1\text{-NIA} \vdash \forall u \exists v \forall x \preceq u \exists y \preceq v A(x, y).$$

It follows that Σ^b_1-NIA $\vdash \forall x \exists y A(x, y)$. ⊣

The above result (and proof-technique) is due to Samuel Buss [4]. For model-theoretic proofs see also [4] and [14]. This latter paper has a very general formulation of the above theorem. For a proof using the bounded functional interpretations see [19].

4 A theory of analysis for polytime computability

In this section we define the basic second-order theory for polytime computability. Let $\mathcal{L}_2$ be the second-order language (a two-sorted language) which is the result of adding to $\mathcal{L}$ second-order variables X, Y, Z, ... (intended to vary over subsets of $\{0, 1\}^*$) and a binary relation symbol $\in$ which infixes between a term of $\mathcal{L}$ and a second-order variable. The terms of $\mathcal{L}_2$ are the same as the terms of $\mathcal{L}$ and we have similar definitions of bounded, Σ^b_1, and Π^b_1-formulas: we just have to take into account that in the present setting we have new atomic formulas of the form $t \in X$ (set parameters are permitted).

One needs to be clear about the semantics for second-order theories. The semantics of this paper (as well as of the ordinary studies of subsystems of analysis, weak or strong) is the so-called *Henkin semantics.* This means that a structure for $\mathcal{L}_2$ consists of a first-order structure $\mathcal{M}$ for the first-order part of the language and a subset S of the power set of the domain of $\mathcal{M}$ for the range of the second-order variables. Note that S need not be the full power set of the domain of $\mathcal{M}$. As it is well-known, the semantics with the full power set is not axiomatizable whereas Henkin semantics is essentially a first-order semantics (enjoying a completeness and compactness theorem, for instance).

The theory defined below was introduced by the second author in [13], where some conservation results were proved (like the theorem of this section). The system can play the role of a *feasible* base theory for *reverse mathematics* in analogy with RCA_0 which is commonly chosen as the base system for reverse mathematics [26] in the primitive recursive environment. (At this point, we must refer to the system RCA^*_0 studied in the paper [27] which has some similitudes to weak systems with regard to the treatment of bounded quantifications. It is, however, a system with exponentiation.) In Section 8, we briefly discuss these matters.

Definition 5. BTFA (acronym for *Base Theory for Feasible Analysis*) is the second-order theory, in the language $\mathcal{L}_2$, whose axioms are those of Σ_1^b-NIA + $B\Sigma_\infty^b$ plus the following comprehension scheme:

$$\forall x\,(\exists y A(x,y) \leftrightarrow \forall z B(x,z)) \rightarrow \exists X \forall x\,(x \in X \leftrightarrow \exists y A(x,y))$$

where A is a Σ_1^b-formula and B is a Π_1^b-formula, possibly with first and second-order parameters, and X does not occur in A or B.

In the definition above, we could have required that the formulas A and B be sw.q.-formulas because first-order bounded quantifications can be absorbed into (plain unbounded) first-order quantifications of the same type. We choose the *definition* above for conceptual clarity since the simplified situation is peculiar to polytime computability (nevertheless, we work with the simplified version in the proof of the theorem below). For systems related to stronger complexity classes (e.g., polyspace computability) one must also work with second-order bounded quantifications (see Section 6) and these cannot be absorbed into first-order quantifications.

We will need the following technical lemma. It states that subword quantifications do not lead to formulas outside the Σ_1^b-class. The result is similar to other cases in logic and the proof does not present difficulties (see [12]). Therefore, we omit it.

Lemma 4. *Let $F(x,y)$ be a Σ_1^b-formula, possibly with (first or second-order) parameters. Then the theory* BTFA *proves*

$$\forall x \subseteq^* a\, \exists y \preceq w\, F(x,y) \rightarrow \exists u \preceq w \times a1 \times a1\, \forall x \subseteq^* a\, \exists y \subseteq^* u\, (y \preceq w \wedge F(x,y)).$$

The proof of the next result follows familiar lines. Its idea is the same as when one proves model-theoretically that the theory RCA_0 is first-order conservative over Σ_1^0-IND: one can always enrich a given first-order model of the latter theory by declaring that the second-order domain consists of all its recursive sets (i.e., simultaneously defined by Σ_1^0 and Π_1^0-formulas). The resulting second-order structure is a model of RCA_0. Therefore, if a first-order sentence fails in some model of Σ_1^0-IND it also fails in some model of RCA_0. By the completeness theorem, this shows that RCA_0 is a first-order conservative extension of Σ_1^0-IND. For a reference, see [26]. In the proof below, we focus on the role of bounded collection and on the need for Σ_1^b-induction on notation (as opposed to mere induction on notation for polytime decidable matrices).

Theorem 4. BTFA *is first-order conservative over* Σ_1^b-NIA + $B\Sigma_\infty^b$.

Proof. Let $\mathcal{M}$ be a model of Σ_1^b-NIA + $B\Sigma_\infty^b$. Denote by M the domain of $\mathcal{M}$. Let S be the class of all subsets X of M which can be defined in $\mathcal{M}$ simultaneously by formulas of the form $\exists y A(x,y)$ and

$\forall z B(x,z)$, with $A(x,y)$ and $B(x,z)$ sw.q-formulas (possibly with parameters from M). We show that the second-order structure $(\mathcal{M}, S)$, obtained from $\mathcal{M}$ by adjoining S as the second-order domain, is a model of BTFA. We first argue that, to each sw.q.-formula $C(\overline{x})$ of $\mathcal{L}_2$ with designated free first-order variables $\overline{x}$ (possibly with parameters from M and S), we can associate two formulas $C_\Sigma(\overline{x})$ and $C_\Pi(\overline{x})$ of $\mathcal{L}$ of the form $\exists y A(\overline{x}, y)$ and $\forall z B(\overline{x}, z)$ (respectively), with $A(\overline{x}, y)$ and $B(\overline{x}, y)$ sw.q.-formulas (possibly with parameters from M), such that $\mathcal{M} \models \forall \overline{x}\,(C_\Sigma(\overline{x}) \leftrightarrow C_\Pi(\overline{x}))$ and $(\mathcal{M}, S) \models \forall \overline{x}\,(C(\overline{x}) \leftrightarrow C_\Sigma(\overline{x}))$. It is clear that the existence of these formulas implies that the scheme of comprehension holds in $(\mathcal{M}, S)$. The construction of C_Σ and C_Π is by induction on the complexity of C. For example, in the simple atomic case $C :\equiv x \in X$, we take $C_\Sigma(x)$ and $C_\Pi(x)$ as being, respectively, the formulas $\exists y\, A(x,y)$ and $\forall z\, B(x,z)$, where $A(x,y)$ and $B(x,y)$ are as in the defining formulas of the parameter X of S. The Boolean cases are not difficult to deal and subword quantification only poses a difficulty in the two cases of unmatching types of quantification. So, let us consider the case $C(x,v) :\equiv \forall u \subseteq^* v\, D(x,u)$. The construction of $C_\Sigma(x,v)$ relies on the previous lemma and on $\mathsf{B\Sigma^b_\infty}$. Let us analyze the situation. By induction hypothesis, let $D_\Sigma(x,u)$ be $\exists y\, A(x,u,y)$, with $A(x,u,y)$ a sw.q.-formula. A simple application of $\mathsf{B\Sigma^b_\infty}$ shows the following equivalence: $C(x,v) \leftrightarrow \exists w \forall u \subseteq^* v \exists y \preceq w\, A(x,u,y)$. By Lemma 4, the right-hand side of the above equivalence can be put in the form of an existential quantification followed by a sw.q.-formula. This gives our formula $C_\Sigma(x,v)$.

Let us now argue that induction on notation for Σ^b_1-formulas holds in $(\mathcal{M}, S)$. Suppose that $(\mathcal{M}, S) \models A(\epsilon) \wedge \neg A(a)$, with $a \in M$ and $A(x)$ a Σ^b_1-formula. We want to find a $c \in M$ such that $(\mathcal{M}, S) \models A(c) \wedge \neg A(ci)$ (i=0 or i=1) and $ci \subseteq a$. Let $A(x)$ be $\exists w \preceq t\, B(x,w)$, with $B(x,w)$ a sw.q.-formula. Let $B_\Sigma(x,w) :\equiv \exists y B_1(x,w,y)$ and $B_\Pi(x,w) :\equiv \forall z B_2(x,w,y)$ as above, with $B_1(x,w,y)$ and $B_2(x,w,z)$ sw.q.-formulas satisfying the expected conditions. In particular, we have

$$\mathcal{M} \models \forall x \subseteq a \forall w \preceq t\,(\exists y B_1(x,w,y) \leftrightarrow \forall z B_2(x,w,z)).$$

We claim that there is $b \in M$ such that

$$\mathcal{M} \models \forall x \subseteq a \forall w \preceq t\,(\exists y B_1(x,w,y) \leftrightarrow \exists y \preceq b B_1(x,w,y)).$$

The argument is easy and uses $\mathsf{B\Sigma^b_\infty}$. It follows from what we have that

$$\mathcal{M} \models \forall x \subseteq a \forall w \preceq t\, \exists y, z\,(B_2(x,w,z) \rightarrow B_1(x,w,y)).$$

Hence, by $\mathsf{B}\Sigma^{\mathsf{b}}_{\infty}$, there is b such that

$$\mathcal{M} \models \forall x \subseteq a \forall w \preceq t\, \exists y, z \preceq b\, (B_2(x, w, z) \to B_1(x, w, y)).$$

The reader can argue that this b does the job.

It follows from the claim above that, for x a subword of a, the formula $\exists w \preceq t\, B_\Sigma(x, w)$ is equivalent to a Σ^{b}_1-formula. (Note that a bounded quantifier "$\exists y \preceq b$" intrudes at this point even if we were only considering induction for polytime decidable matrices.) It is now clear that we can get the witness c to the failure of the hypothesis of the (notation) induction principle by applying induction on notation for Σ^{b}_1-formulas in $\mathcal{M}$. (Given the intrusion of the bounded existential quantification discussed above, the extension from induction on notation for polytime decidable matrices to Σ^{b}_1-formulas, given by Theorem 2, is necessary for this part of the argument.)

It remains to see that $\mathsf{B}\Sigma^{\mathsf{b}}_{\infty}$ holds in $(\mathcal{M}, S)$. The proof follows easily from an extension of the construction of C_Σ and C_Π to bounded formulas C (now with *bounded* formulas A and B). The extension uses heavily the $\mathsf{B}\Sigma^{\mathsf{b}}_{\infty}$ scheme available in $\mathcal{M}$. ⊣

5 Adding weak König's lemma

In this section, we are going to add a nonconstructive scheme to BTFA, namely (a version) of *weak König's lemma.* Given A a formula of $\mathcal{L}_2$ and x a (first-order) designated variable, we denote by $Tree(A_x)$ the formula:

$$\forall x \forall y\, (A(x) \wedge y \subseteq x \to A(y)) \wedge \forall u \exists x \equiv u A(x).$$

The second condition is the so-called infinity or limitless condition. Given X a second-order variable, we denote by $Path(X)$ the formula:

$$Tree((x \in X)_x) \wedge \forall x \forall y\, (x \in X \wedge y \in X \to x \subseteq y \vee y \subseteq x).$$

Weak König's lemma for trees defined by bounded formulas, denoted by $\Sigma^{\mathsf{b}}_{\infty}$-$\mathsf{WKL}$, is the following scheme:

$$Tree(A_x) \to \exists X\, (Path(X) \wedge \forall x\, (x \in X \to A(x))),$$

where A is a bounded formula and X is a new variable.

The aim of this section is to prove the following theorem:

Theorem 5. $\mathsf{BTFA} + \Sigma^{\mathsf{b}}_{\infty}$-$\mathsf{WKL}$ *is first-order conservative over* BTFA.

We use the method of forcing to prove this theorem. The argument, in the traditional context of RCA_0, is due to unpublished work of Leo Harrington (recently, [17] presented a proof-theoretic proof of this conservation result in the traditional context). The main lemma is the following:

Lemma 5. *Let* $(\mathcal{M}, S)$ *be a countable model of* BTFA *and let* A *be a bounded formula (possibly with parameters in the domain of* $\mathcal{M}$ *and in* S*) with a designated variable* x *such that* $(\mathcal{M}, S) \models Tree(A_x)$*. There is a subset* G *of the domain of* $\mathcal{M}$ *such that* $(\mathcal{M}, S \cup \{G\})$ *is a model of* $\Sigma_1^b\text{-NIA} + B\Sigma_\infty^b$ *and*

$$(\mathcal{M}, S \cup \{G\}) \models Path(G) \wedge \forall x\, (x \in G \to A(x)).$$

Proof. Let $\mathbb{T}$ be the set of all (limitless) subtrees of A defined by a bounded formula (possibly with parameters in the domain M of $\mathcal{M}$ and in S). The elements of $\mathbb{T}$ are our *forcing conditions.* We order $\mathbb{T}$ by inclusion. If $T, Q \in \mathbb{T}$ we write $T \leqslant Q$ instead of $T \subseteq Q$. Let $\mathbb{G} \subseteq \mathbb{T}$ be a *generic filter* for a sufficiently rich countable class of dense sets.[7] We put $G := \bigcap \mathbb{G}$ and claim that G is indeed a path through the tree given by A (a so-called generic path).

Fact 1. *Let* n *be a tally element of the domain of* $\mathcal{M}$*. The set*

$$\mathbb{D}_n = \{T \in \mathbb{T} : (\mathcal{M}, S) \models \exists^1 x\, (x \equiv n \wedge x \in T)\}$$

is dense.

Proof of fact 1. Let n be a tally sequence and $T \in \mathbb{T}$. We have to show that some subtree of T is a member of $\mathbb{D}_n$. The following cannot happen:

$$(\mathcal{M}, S) \models \forall x \equiv n\, \exists m\, (n \preceq m \wedge \forall z \equiv m\, (x \subseteq z \wedge z \equiv m \to z \notin T)).$$

Otherwise, by bounded collection, there is a tally m such that all elements of T have length smaller than m, contradicting the limitlessness of T. It easily follows that there is x, with $x \equiv n$, such that $\{z \in T : z \subseteq x \vee x \subseteq z\}$ is limitless. This set is obviously a subtree of T and is clearly in $\mathbb{D}_n$. ⊣ (of fact 1)

Given a tally n, since $\mathbb{G}$ is generic, $\mathbb{G} \cap \mathbb{D}_n \neq \emptyset$. Take $T \in \mathbb{G} \cap \mathbb{D}_n$ and let σ_n be the unique element of T of length n. Using the fact that $\mathbb{G}$ is a filter, it is now clear that $\sigma_n \in \bigcap \mathbb{G}$. By the arbitrariness of n, this entails that G is indeed a (limitless) path.

---

[7] A set $\mathbb{D}$ of forcing conditions is called *dense* if, for each forcing condition T, there is a forcing condition Q with $Q \in \mathbb{D}$ and $Q \leqslant T$. The sufficiently rich countable class of dense sets is usually given by the *definable* sets in an extension of the language of set theory containing the *forcing language* (defined later in the proof) and constants for each forcing condition. A *filter* of forcing conditions contains the tree given by A, is closed upwards and, if T and Q are in the filter then there is a forcing condition R in the filter with $R \leqslant T$ and $R \leqslant Q$. A filter is called *generic* if it intersects all the given dense sets. Since the class of dense sets is countable, it is a known fact that for any forcing condition T there is always a generic filter $\mathbb{G}$ such that $T \in \mathbb{G}$.

It is not difficult to argue that Σ_1^b-NIA holds in the generic extension $(\mathcal{M}, S \cup \{G\})$. Actually, this does not depend on the genericity of G. We use the following simple observation (which will be used frequently in the sequel). Let $B(x, p, P, G)$ be a bounded formula with parameters $p \in M$, $P \in S$ (to ease the notation we only consider one parameter of each type) and the generic set G. There is a term $t_B(x, p)$ such that, for any element $a \in M$, we have the following equivalence: For all elements $x \preceq a$ and for all $y \in G$ with $t_B(a, p) \preceq y$,

$$(\mathcal{M}, S \cup \{G\}) \models B(x, p, P, G) \text{ if, and only if, } (\mathcal{M}, S) \models B^*(x, p, P, y),$$

where $B^*(x, p, P, y)$ is the bounded formula obtained from $B(x, p, P, G)$ by replacing every atomic formula of the form $q \in G$ by $q \subseteq y$. The observation is correct because, to decide a bounded formula, one only has to check below a certain point (the formal proof is straightforward, by induction on the complexity of the formula B). Note that the term t_B does not depend on G.

In order to check that Σ_1^b-NIA holds in the generic extension, take $B(x, G)$ (we omit the parameters of the ground model) a Σ_1^b-formula and $a \in M$ such that $B(a, G)$ is false in $(\mathcal{M}, S \cup \{G\})$. Take $y \in G$ with $t_B(a) \preceq y$. By the above claim, $(\mathcal{M}, S) \models \neg B^*(a, y)$. Since Σ_1^b-NIA holds in the ground model, either $(\mathcal{M}, S) \models \neg B^*(\epsilon, y)$ or there is $x \subset a$ such that $(\mathcal{M}, S) \models B^*(x, y) \wedge \neg B^*(xi, y)$, where $i \in \{0, 1\}$ is such that $xi \subseteq a$. It is clear that either counterexample lifts to the generic extension.

It remains to check that $\mathsf{B}\Sigma_\infty^b$ holds in $(\mathcal{M}, S \cup \{G\})$. It is at this stage that we use the *forcing theorem*. Let us briefly review the basic facts of forcing in our setting. The *forcing language* is the extension of the language of $\mathcal{L}_2$ by addition of new constants for the elements of M and of S together with a constant C (for the generic path). We identify the former constants with the elements themselves. The *forcing relation* is a relation involving forcing conditions and sentences of the forcing language. The definition of the forcing relation (at the atomic level) is as follows:

1. $T \Vdash x = y$ iff $x = y$;
2. $T \Vdash x \in X$ iff $x \in X$;
3. $T \Vdash x \in C$ iff $(\mathcal{M}, S) \models \exists m \forall w \equiv m\ (w \in T \Rightarrow x \subseteq w)$;

for x, y in the domain M, $X \in S$ and T a forcing condition.

Fact 2. *If F is an atomic sentence of the forcing language and T is a forcing condition then $T \Vdash F$ if, and only if, $\forall Q \leqslant T\, \exists R \leqslant Q\, (R \Vdash F)$.*

Proof of fact 2. We need only discuss the case of atomic sentences $x \in$

C. The left to right direction is clear. Assume that $T \not\Vdash x \in C$. Then, $Q := \{y \in T : x \not\subseteq y\}$ is a *limitless* subtree of T, i.e., $Q \leqslant T$. Obviously, for no $R \leqslant Q$ one has $R \Vdash x \in C$. ⊣ (of fact 2)

The above fact shows that we are in the presence of a *weak forcing notion.* The definition of the forcing relation to arbitrary sentences is done according to the following recursive scheme:

4. $T \Vdash F \wedge G$ if $T \Vdash F$ and $T \Vdash G$;
5. $T \Vdash \neg F$ if, for all conditions Q with $Q \leqslant P$, $Q \not\Vdash F$;
6. $T \Vdash \forall x\, F(x)$ if, for all elements $p \in M$, $T \Vdash F(p)$;
7. $T \Vdash \forall X\, F(X)$ if $T \Vdash F(C)$ and, for all $P \in S$, $T \Vdash F(P)$.

As it is well-known, all logical axioms are generically forced (i.e., forced by every condition) and the rules of logical inference are preserved by forcing (in the sense that if a condition forces the premisses then it forces the conclusion). We state this fact just by saying that "weak forcing preserves classical logic". In the remainder of the proof of the lemma, we will use the forcing theorem. It says that, for every formula $F(V)$ of the language (possibly with parameters in the ground model) and every forcing condition T, the condition $T \Vdash F(C)$ holds if, and only if, for *every* generic filter $\mathbb{H}$ such that $T \in \mathbb{H}$, $(\mathcal{M}, S \cup \{H\}) \models F(H)$, where $H = \bigcap \mathbb{H}$.

Fact 3. *Let $B(x, X)$ be a bounded formula, possibly with parameters in the ground model, and consider a forcing condition T. The condition $T \Vdash \exists x B(x, C)$ holds if, and only if,*

$$(\mathcal{M}, S) \models \exists m \forall y \equiv m\, (y \in T \rightarrow \exists x \preceq m\, (t_B(x) \preceq m \wedge B^*(x, y))).$$

Proof of fact 3. Assume that the latter condition holds. Let $\mathbb{H}$ be an arbitrary generic filter such that $T \in \mathbb{H}$. By the forcing theorem, we only need to show that the existential statement $\exists x B(x, H)$ holds in $(\mathcal{M}, S \cup \{H\})$. By hypothesis, there is a (tally) element m in M and $x \preceq m$ such that

$$(\mathcal{M}, S) \models t_B(x) \preceq m \wedge B^*(x, y_0),$$

where y_0 is the truncation of the path H at length m. Of course, this x witnesses the existential statement.

Suppose, now, that the latter condition fails. Let

$$Q = \{y \in T : (\mathcal{M}, S) \models \forall x \preceq y\, (t_B(x) \preceq y \rightarrow \neg B^*(x, y))\}.$$

It is clear that Q is a tree (it is a limitless one by the assumption). Hence Q is a forcing condition with $Q \leqslant T$. Take a generic filter $\mathbb{H}$ such that $Q \in \mathbb{H}$ and put $H = \bigcap \mathbb{H}$. We claim that

$$(\mathcal{M}, S \cup \{H\}) \models \forall x \neg B(x, H).$$

Note that, by the forcing theorem, this fact entails $T \not\Vdash \exists x B(x, C)$. To show the claim, consider an arbitrary $x \in M$. Take $y \in H$ with $t_B(x) \preceq y$ and $x \preceq y$. Since $H \subseteq Q$, by definition of Q, $(\mathcal{M}, S) \models \neg B^*(x, y)$. The claim follows. $\dashv$ (of fact 3)

We can now prove that $\mathsf{B\Sigma}^{\mathsf{b}}_{\infty}$ is forced by every condition T and, as a consequence (forcing theorem), that $\mathsf{B\Sigma}^{\mathsf{b}}_{\infty}$ holds in every generic extension. An instance of $\mathsf{B\Sigma}^{\mathsf{b}}_{\infty}$ is a conditional statement. To see that a condition T forces a conditional statement, one must show that if a stronger condition $Q \leqslant T$ forces the antecedent then there is a yet stronger condition $R \leqslant Q$ which forces the consequent. Hence, suppose that $Q \leqslant T$ and $Q \Vdash \forall w \preceq a \exists x\, B(w, x)$, where B is a bounded formula of the forcing language (and a comes from M). Then, for every w with $w \preceq a$, $Q \Vdash \exists x\, B(w, x)$. According to Fact 3, for every w with $w \preceq a$, we have

$$(\mathcal{M}, S) \models \exists m \forall y \equiv m\, (y \in Q \to \exists x \preceq m\, (t_B(x) \preceq m \wedge B^*(w, x, y))).$$

Therefore, $(\mathcal{M}, S) \models \forall w \preceq a \exists m \forall y \equiv m\, (y \in Q \to \exists x \preceq m\, (t_B(x) \preceq m \wedge\ B^*(w, x, y)))$. Since $\mathsf{B\Sigma}^{\mathsf{b}}_{\infty}$ holds in the ground model, there is a (tally) element m_0 in M such that

$$(\mathcal{M}, S) \models \forall w {\preceq} a \forall y {\equiv} m_0\, (y \in Q \to \exists x {\preceq} m_0\, (t_B(x) {\preceq} m_0 \wedge B^*(w, x, y))).$$

By the density of $\mathbb{D}_{m_0}$, take a forcing condition R with $R \leqslant Q$ and $R \in \mathbb{D}_{m_0}$. We get

$$(\mathcal{M}, S) \models \forall w \preceq a \exists x \preceq m_0\, (t_B(x) \preceq m_0 \wedge B^*(w, x, y_0)),$$

where y_0 is the unique element of R of length m_0. It easily follows from the forcing theorem that $R \Vdash \exists m \forall w \preceq a \exists x \preceq m\, B(w, x)$. $\dashv$

The above lemma does not provide a model of BTFA because the recursive comprehension scheme is missing. However, as in the proof of Theorem 4, it is possible to close the given model of $\mathsf{\Sigma}^{\mathsf{b}}_1\text{-}\mathsf{NIA} + \mathsf{B\Sigma}^{\mathsf{b}}_{\infty}$ under this comprehension scheme. In sum, we have accomplished the following: Given a countable model $(\mathcal{M}, S)$ of BTFA and given a bounded formula A (possibly with first and second-order parameters) with designated variable x such that $(\mathcal{M}, S) \models Tree(A_x)$, there is a countable S' with $S \subseteq S' \subseteq \mathcal{P}(M)$ such that $(\mathcal{M}, S')$ is a model of BTFA and

$$(\mathcal{M}, S') \models \exists X\, (Path(X) \wedge \forall x\, (x \in X \to A(x))).$$

There are only countably many trees of $(\mathcal{M}, S)$ defined by bounded formulas (possibly with parameters in the domain M of $\mathcal{M}$ and in S). We can enumerate all these formulas A_1, A_2, ... and get, successively, countable models $(\mathcal{M}, S^1)$, $(\mathcal{M}, S^2)$, ... (with $S^1 \subseteq S^2 \subseteq \ldots$) of BTFA such that, for each positive natural number i, $(\mathcal{M}, S^i)$ has paths through the trees given by $A_1, \ldots A_i$. Let $S^\omega = \bigcup_i S^i$. It is easy to argue that $(\mathcal{M}, S^\omega)$ is still a (countable) model of BTFA and that it has paths through all the trees given by bounded formulas *with parameters in* M *or* S. Of course, there may be other trees: those defined by bounded formulas with second-order parameters in S^ω. However, we can repeat the above operation with ground model $(\mathcal{M}, S^\omega)$ and get a countable model $(\mathcal{M}, S^{\omega+\omega})$ which has paths for all trees defined by bounded formulas with parameters in M or in S^ω. After repeating this process ω-times, we end up with a model $(\mathcal{M}, S^{\omega\cdot\omega})$, where $S^{\omega\cdot\omega} = \bigcup_k S^{\omega\cdot k}$. This is easily seen to be a model of $\mathsf{BTFA} + \Sigma^{\mathsf{b}}_{\infty}\text{-}\mathsf{WKL}$.

We have shown that, given a countable model $(\mathcal{M}, S)$ of BTFA, it is possible to find $S^{\omega\cdot\omega}$ with $S \subseteq S^{\omega\cdot\omega} \subseteq \mathcal{P}(M)$ such that $(\mathcal{M}, S^{\omega\cdot\omega})$ is a model of $\mathsf{BTFA} + \Sigma^{\mathsf{b}}_{\infty}\text{-}\mathsf{WKL}$. Note that the first-order part remains the same. By an easy application of the completeness theorem, it follows that the theory $\mathsf{BTFA} + \Sigma^{\mathsf{b}}_{\infty}\text{-}\mathsf{WKL}$ is first-order conservative over BTFA.

6 On second-order bounded theories of arithmetic

In this section, we make a detour through a setting where the second-order part of the language has a severe restriction: the set variables denote sets which are always bounded by a term of the language. The reason for this diversion is twofold. In the next section, when studying a conservation result concerning a scheme of strict Π^1_1-reflection, we need a stronger form of collection (we need collection for formulas with first and second-order *bounded* quantifiers). More importantly, when working in theories of bounded arithmetic connected with computational complexity classes stronger than polytime computability, e.g. polyspace computability, FCH or $\mathsf{EXPTIME}$, it is convenient to work with bounded second-order variables. For polyspace computability and exponential time computability, see the theories U^1_2 and V^1_2 (respectively) defined by Buss in [4]. The class FCH is related with the *hierarchy of counting functions*, a class defined by Klaus Wagner in [28] based on an iteration procedure that stems from Leslie Valiant's well-known counting class $\sharp P$. The functions of this hierarchy are all polynomial space computable and the theory related to this class, the so-called theory TCA (theory for counting arithmetic), is contained in U^1_2. See [21] and [22].

Let $\mathcal{L}^b_2$ be the second-order language obtained by adding second-order bounded variables to $\mathcal{L}$ and the relation symbol $\in$ which infixes between a term of $\mathcal{L}$ and a second-order bounded variable.[8] The second-

[8] In this definition, $\mathcal{L}$ is the language of Section 3 with primitive bounded first-order quantifiers.

order bounded variables are complexes of the form X^t where X is a (second-order) variable and t is a (first-order) term. As far as the authors are aware, second-order bounded variables were introduced by Buss in his thesis [4]. The intuitive idea is that X^t has only elements x such that $x \preceq t$.

For ease of reading, often we write $\forall X \preceq t\,(...)$ and $\exists X \preceq t\,(...)$ instead of the official notation $\forall X^t\,(...)$ and $\exists X^t\,(...)$ (respectively). These quantifiers are called *bounded* second-order quantifiers.

In the following definition, we introduce an auxiliary theory Aux:

Definition 6. Aux is the theory formulated in the language $\mathcal{L}_2^b$, whose axioms are those of $\Sigma_1^{\mathsf{b}}\text{-NIA} + \mathsf{B}\Sigma_\infty^{\mathsf{b}}$,

$$\forall x \forall X^t (x \in X^t \to x \preceq t)$$

for any term t, and the following comprehension scheme

$$\forall x \preceq t\,(A(x) \leftrightarrow B(x)) \to \exists X^t \forall x \preceq t\,(x \in X^t \leftrightarrow A(x))$$

where A is a Σ_1^{b}-formula, B is a Π_1^{b}-formula, possibly with first and second-order parameters, X^t does not occur in A or B and x does not occur in the term t.

It is clear that the theory Aux can be seen as a subtheory of BTFA in a natural manner. Therefore, its provably total functions (with appropriate graphs) are still the polytime computable functions.

The theory Aux can be formulated in Gentzen's sequent calculus by adding to the corresponding formulation of $\Sigma_1^{\mathsf{b}}\text{-NIA} + \mathsf{B}\Sigma_\infty^{\mathsf{b}}$ presented in the proof of Theorem 3 two new kinds of initial sequents:

$$s \in X^t \to s \preceq t$$

where s and t are terms, and

$$\forall x \preceq t\,(A(x) \leftrightarrow B(x)) \to \exists X^t \forall x \preceq t\,(x \in X^t \leftrightarrow A(x))$$

where A is a Σ_1^{b}-formula, B is a Π_1^{b}-formula, possibly with first and second-order parameters, X^t does not occur in A or B and x does not occur in t. We also add four new rules for the second-order quantifiers:

$$\frac{\Gamma, A(F^t) \to \Delta}{\Gamma, \forall X^t A(X^t) \to \Delta}\ \forall_{2nd}:l \qquad \frac{\Gamma \to \Delta, A(C^t)}{\Gamma \to \Delta, \forall X^t A(X^t)}\ \forall_{2nd}:r$$

$$\frac{\Gamma, A(C^t) \to \Delta}{\Gamma, \exists X^t A(X^t) \to \Delta}\ \exists_{2nd}:l \qquad \frac{\Gamma \to \Delta, A(F^t)}{\Gamma \to \Delta, \exists X^t A(X^t)}\ \exists_{2nd}:r$$

$\frac{\Gamma \to \Delta, A(F^t)}{\Gamma \to \Delta}$, with F^t a second-order variable and C^t a second-order eigenvariable.

We say that a formula is in the class $\Sigma_\infty^{1,\mathsf{b}}$ if it belongs to the smallest class of formulas containing the atomic formulas and closed under

Boolean operations, bounded first-order quantifications and second-order (bounded) quantifications. These are also known as the bounded formulas in the language of $\mathcal{L}_2^b$. The principle of bounded collection for the language $\mathcal{L}_2^b$, denoted by $\mathsf{B}^1\Sigma_\infty^{1,\mathsf{b}}$, is the following scheme:

$$\forall X \preceq t \exists y A(X^t, y) \to \exists z \forall X \preceq t \exists y \preceq z A(X^t, y),$$

where A is a $\Sigma_\infty^{1,\mathsf{b}}$-formula (possibly with other first and second-order free variables). It is easy to argue that this form of bounded collection implies the principle $\mathsf{B}\Sigma_\infty^{\mathsf{b}}$.

Proposition 2. $\mathsf{Aux} + \mathsf{B}^1\Sigma_\infty^{1,\mathsf{b}} \vdash \mathsf{B}\Sigma_\infty^{\mathsf{b}}$.

Proof. Suppose that $\forall x \preceq t \exists y A(x, y)$, with A a bounded formula. We want to prove that $\exists z \forall x \preceq t \exists y \preceq z A(x, y)$. Let $B(y, X^t)$ be the $\Sigma_\infty^{1,\mathsf{b}}$-formula, expressing that when X^t is empty or has more than one element then $y = \epsilon$, and when X^t has a unique element, say x, then y satisfies $A(x, y)$. Formally, $B(y, X^t)$ is

$$(y = \epsilon \wedge \forall x \preceq t\ x \notin X^t) \vee (y = \epsilon \wedge \exists x, x' \preceq t\,(x \neq x' \wedge x, x' \in X^t)) \vee$$
$$\exists x \preceq t\,(A(x, y) \wedge x \in X^t \wedge \forall x' \preceq t\,(x \neq x' \to x' \notin X^t)).$$

By definition of $B(y, X^t)$, we have $\forall X^t \exists y B(y, X^t)$. Applying $\mathsf{B}^1\Sigma_\infty^{1,\mathsf{b}}$, we conclude that $\exists z \forall X^t \exists y \preceq z B(y, X^t)$. Fix such a z. But then $\forall x \preceq t \exists y \preceq z A(x, y)$. Just think of the y which corresponds to the singleton set $X^t :\equiv \{x\}$. $\dashv$

Theorem 6. *The theory* $\mathsf{Aux} + \mathsf{B}^1\Sigma_\infty^{1,\mathsf{b}}$ *is conservative over the theory* Aux *with respect to formulas of the form* $\forall x \exists y A(x, y)$*, with* A *a* $\Sigma_\infty^{1,\mathsf{b}}$*-formula.*

Proof. The theory $\mathsf{Aux} + \mathsf{B}^1\Sigma_\infty^{1,\mathsf{b}}$ can be formulated in Gentzen's sequent calculus (as above), making use of the following new inference rule, we call $\mathsf{B}^1\Sigma_\infty^{1,\mathsf{b}}$-*rule*:

$$\frac{\Gamma \to \Delta, \exists y A(y, C^t)}{\Gamma \to \Delta, \exists z \forall X \preceq t \exists y \preceq z A(y, X^t)}$$

where A is a $\Sigma_\infty^{1,\mathsf{b}}$-formula (possibly with other free variables), C^t is a second-order eigenvariable, and y does not occur in the term t.

Suppose that $\mathsf{Aux} + \mathsf{B}^1\Sigma_\infty^{1,\mathsf{b}} \vdash \forall x \exists y A(x, y)$, with A a $\Sigma_\infty^{1,\mathsf{b}}$-formula. Then, there is a proof of $\to \exists y A(x, y)$ in the sequent calculus above. Applying the free-cut elimination theorem, we know that there is a proof $\mathcal{P}$ of $\to \exists y A(x, y)$ without free cuts. Therefore all formulas occurring in the sequents $\Gamma \to \Delta$ in $\mathcal{P}$ are of the form $\exists y B(y, x, X^p)$, with B a $\Sigma_\infty^{1,\mathsf{b}}$-formula. As usual, the existential quantifier can be absent.

Let $\Gamma \to \Delta$ be a sequent in $\mathcal{P}$, where Γ is

$$\exists x_1 B_1(x_1, x, X^{p(x)}), \ldots, \exists x_n B_n(x_n, x, X^{p(x)})$$

and Δ is $\exists y_1 C_1(y_1, x, X^{p(x)}), \ldots, \exists y_k C_k(y_k, x, X^{p(x)})$, where $B_1, \ldots, B_n, C_1, \ldots, C_k$ are $\Sigma_\infty^{1,\mathsf{b}}$-formulas. To ease reading, we show only a first-order free variable x and a second-order bounded variable $X^{p(x)}$, instead of tuples of such. Also, we have the same term $p(x)$ for every formula (this can be assumed without loss of generality). We are also only displaying the variable x in the term $p(x)$.

Let us prove, by induction on the number of lines of $\mathcal{P}$, that from bounds of the antecedents we can get (in Aux) bounds for the consequents in the following way:

$$\mathsf{Aux} \vdash \forall u \exists v \forall x \preceq u \forall X^{p(x)} (\Gamma^{\preceq u} \to \Delta^{\preceq v}),$$

where $\Gamma^{\preceq u}$ abbreviates $\exists x_1 \preceq u B_1(x_1, x, X^{p(x)}) \wedge \ldots \wedge \exists x_n \preceq u B_n(x_n, x, X^{p(x)})$ and $\Delta^{\preceq v}$ abbreviates $\exists y_1 \preceq v C_1(y_1, x, X^{p(x)}) \vee \ldots \vee \exists y_k \preceq v\, C_k(y_k, x, X^{p(x)})$. Note that applying this result to the last sequent of $\mathcal{P}$, we conclude that Aux proves $\forall u \exists v \forall x \preceq u \exists y \preceq v A(x, y)$. Thus $\mathsf{Aux} \vdash \forall x \exists y A(x, y)$.

The proof above is going to be illustrated by considering the cut-rule, the $\exists : r$-rule, the $\forall_\preceq : r$-rule and the $\mathsf{B}^1\Sigma_\infty^{1,\mathsf{b}}$-rule. The other cases are trivial (including the initial sequents and the rules for the second-order bounded quantifiers) or similar to the ones presented; the rules for $\forall : e$ and $\forall : d$ do not occur in $\mathcal{P}$ and there is no need to analyze the rule for $\mathsf{B}\Sigma_\infty^{\mathsf{b}}$ because we have shown that this principle is derivable from $\mathsf{B}^1\Sigma_\infty^{1,\mathsf{b}}$.

Cut-rule:

$$\frac{\Gamma \to \Delta, A \qquad A, \Gamma \to \Delta}{\Gamma \to \Delta}$$

If A is a $\Sigma_\infty^{1,\mathsf{b}}$-formula the result is immediate taking $v :\equiv v_1 \hat{} v_2$, where v_1 and v_2 the bounds that exist by induction hypothesis.

Suppose that A is of the form $\exists z D(z, x, X^{p(x)})$ with D a $\Sigma_\infty^{1,\mathsf{b}}$-formula. By induction hypothesis we have:

1) $\mathsf{Aux} \vdash \forall u \exists v \forall x \preceq u \forall X^{p(x)} (\Gamma^{\preceq u} \to \Delta^{\preceq v} \vee \exists z \preceq v D(z, x, X^{p(x)}))$

2) $\mathsf{Aux} \vdash \forall u \exists v \forall x \preceq u \forall X^{p(x)} (\exists z \preceq u D(z, x, X^{p(x)}) \wedge \Gamma^{\preceq u} \to \Delta^{\preceq v})$.

We want to prove that $\mathsf{Aux} \vdash \forall u \exists v \forall x \preceq u \forall X^{p(x)} (\Gamma^{\preceq u} \to \Delta^{\preceq v})$.

Fix u. By 1) there is v_1 such that $\forall x \preceq u \forall X^{p(x)} (\Gamma^{\preceq u} \to \Delta^{\preceq v_1} \vee \exists z \preceq v_1 D(z, x, X^{p(x)}))$. Assume that $u \preceq v_1$ (if not just replace v_1 by $v_1 \hat{} u$ that the assertion above remains valid). By 2), there is v_2 such that $\forall x \preceq v_1 \forall X^{p(x)} (\exists z \preceq v_1 D(z, x, X^{p(x)}) \wedge \Gamma^{\preceq v_1} \to \Delta^{\preceq v_2})$. We get $\forall x \preceq u \forall X^{p(x)} (\exists z \preceq v_1 D(z, x, X^{p(x)}) \wedge \Gamma^{\preceq u} \to \Delta^{\preceq v_2})$, using the fact that if $u \preceq u'$ then $\Gamma^{\preceq u} \to \Gamma^{\preceq u'}$. Let $v :\equiv v_1 \hat{} v_2$. We conclude that $\forall x \preceq u \forall X^{p(x)} (\Gamma^{\preceq u} \to \Delta^{\preceq v})$.

$\exists$:r-rule:

$$\frac{\Gamma \to \Delta, A(t)}{\Gamma \to \Delta, \exists x A(x)}$$

Fix u. By induction hypothesis, there exists v_1 so that

$$\forall x \preceq u \forall X^{p(x)} (\Gamma^{\preceq u} \to \Delta^{\preceq v_1} \vee A(t)).$$

Note that by the construction of $\mathcal{P}$, A is a $\Sigma^{1,\mathsf{b}}_{\infty}$-formula. We want to prove that there is v such that $\forall x \preceq u \forall X^{p(x)} (\Gamma^{\preceq u} \to \Delta^{\preceq v} \vee \exists x \preceq v A(x))$. Just take v as being $v_1 \hat{\ } t(u)$.

$\forall_{\preceq}$:r-rule:

$$\frac{a \preceq t, \Gamma \to \Delta, A(a)}{\Gamma \to \Delta, \forall y \preceq t A(y)}$$

with a an eigenvariable and A a $\Sigma^{1,\mathsf{b}}_{\infty}$-formula.

We want to prove that $\mathsf{Aux} \vdash \forall u \exists v \forall x \preceq u \forall X^{p(x)} (\Gamma^{\preceq u} \to \Delta^{\preceq v} \vee \forall y \preceq t A(y))$. Fix u. Let $u' :\equiv t(u) \hat{\ } u$. By induction hypothesis, there is v' such that

$$\forall x \preceq u' \forall X^{p(x)} \forall y \preceq u' (y \preceq t \wedge \Gamma^{\preceq u'} \to \Delta^{\preceq v'} \vee A(y)).$$

Thus, $\forall x \preceq u \forall X^{p(x)} \forall y \preceq u' (y \preceq t \wedge \Gamma^{\preceq u} \to \Delta^{\preceq v'} \vee A(y))$. Just let v be v'.

$\mathsf{B}^1\Sigma^{1,\mathsf{b}}_{\infty}$-rule:

$$\frac{\Gamma \to \Delta, \exists y A(y, C^t)}{\Gamma \to \Delta, \exists z \forall Y \preceq t \exists y \preceq z A(y, Y^t)}$$

By induction hypothesis we know that

$$\mathsf{Aux} \vdash \forall u \exists v \forall x \preceq u \forall X^{p(x)} \forall Y^{t(x)} (\Gamma^{\preceq u} \to \Delta^{\preceq v} \vee \exists y \preceq v A(y, Y^{t(x)})).$$

To prove, as we want, that

$$\mathsf{Aux} \vdash \forall u \exists v \forall x \preceq u \forall X^{p(x)} (\Gamma^{\preceq u} \to \Delta^{\preceq v} \vee \exists z \preceq v \forall Y^{t(x)} \exists y \preceq z A(y, Y^{t(x)})),$$

consider an arbitrary u, take v as the element that exists by induction hypothesis, and let $z :\equiv v$. ⊣

The complete proof, in a slightly different context, is given in [20].

Theorem 7. *The theory* $\mathsf{BTFA} + \mathsf{B}^1\Sigma^{1,\mathsf{b}}_{\infty}$ *is conservative over the theory* $\mathsf{Aux} + \mathsf{B}^1\Sigma^{1,\mathsf{b}}_{\infty}$ *with respect to sentences of the form* $\forall x \exists y A(x, y)$, *where* $A \in \Sigma^{1,\mathsf{b}}_{\infty}$.

Proof. The proof uses a strategy similar to the one applied in Theorem 4 for dealing with the comprehension scheme. Let $\mathcal{M}$ be a model of $\mathsf{Aux} + \mathsf{B}^1\Sigma^{1,\mathsf{b}}_{\infty}$ with first-order domain M and second-order domain S_b. Let S be the class of all subsets $X \subseteq M$ which can be defined in $\mathcal{M}$ simultaneously by formulas of the form $\exists y A(x, y)$ and $\forall y B(x, y)$, with

A a Σ_1^b-formula and B a Π_1^b-formula (possibly with first and second-order parameters from $\mathcal{M}$). Using $\mathsf{B}\Sigma_\infty^b$, it can be proved that $S_b = \{X^a : X \in S \wedge a \in M\}$ (where $X^a = \{x \in X : x \preceq a\}$). It can also be argued that the second-order structure $\mathcal{M}^*$, obtained from $\mathcal{M}$ by keeping the same first-order domain and replacing the second-order domain S_b by S, is a model of $\mathsf{BTFA} + \mathsf{B}^1\Sigma_\infty^{1,b}$. The argument uses the following fact: if $F(x, X)$ is a $\Sigma_\infty^{1,b}$-formula, there is a term $t(x)$ such that, for $c \in M$ and $C \in S$, we have $\mathcal{M}^* \models F(c, C)$ if and only if $\mathcal{M} \models F(c, C^b)$, whenever $t(c) \preceq b$.[9] $\dashv$

7 Adding strict Π_1^1-reflection

In this section, we go back to the (unbounded) second-order setting and add to BTFA a principle that, in the presence of the totality of the exponential function, is equivalent to weak König's lemma. The principle in question, called *strict Π_1^1-reflection*, is defined by the following scheme:

$$\forall X \exists x A(x, X) \rightarrow \exists u \forall X \exists x \preceq u A(x, X),$$

with A a $\Sigma_\infty^{1,b}$-formula (possibly with parameters) in which u does not occur. It is sometimes convenient to present the strict Π_1^1-reflection scheme in its contrapositive form:

$$\forall u \exists X \forall x \preceq u\, A(x, X) \rightarrow \exists X \forall x\, A(x, X),$$

with A in the conditions above.

The original version of strict Π_1^1-reflection was introduced by Jon Barwise in the context of admissible set theory (see chapter VIII of [1]). This is perhaps a good place to remark that fragments of arithmetic (weak arithmetic) have similarities with admissibility (the preface of [24] calls attention to these similarities). Strict Π_1^1-reflection was first considered in the weak context by Andrea Cantini in [5]. In this paper, Cantini asks whether strict Π_1^1-reflection is equivalent to weak König's lemma (Cantini shows that the former implies the latter).[10] We conjecture that it is not. The following observation is (essentially) due to Cantini [5]:

Proposition 1. *The theory* $\mathsf{BTFA} + \text{strict } \Pi_1^1\text{-reflection}$ *proves* $\mathsf{B}^1\Sigma_\infty^{1,b}$.

[9] For the details (in a slightly different context, but easily adapted to the present case), see [20], pp 65-70.

[10] Our formulation of strict Π_1^1-reflection is incomparable with Cantini's formulation. On the one hand, it admits a matrix of $\Sigma_\infty^{1,b}$-formulas, instead of just Σ_∞^b-formulas. On the other hand, the form of Cantini's formulation is more general. With his formulation of strict Π_1^1-reflection, Cantini shows that this principle is a Π_2^0-conservative extension of BTFA. For a detailed analysis of a generalization of Cantini's reflection scheme see [8].

Proof. Let us reason in $\mathsf{BTFA}+\mathsf{strict}\ \Pi^1_1\text{-}\mathsf{reflection}$. Suppose that $\forall X \preceq t\exists y A(X,y)$, with A a $\Sigma^{1,\mathsf{b}}_\infty$-formula (possibly with parameters). Of course, $\forall X\exists y\tilde{A}(X,y)$, where $\tilde{A}(X,y)$ is the $\Sigma^{1,\mathsf{b}}_\infty$-formula $A(X^t,y)$. By strict Π^1_1-reflection, $\exists z\forall X\exists y \preceq z\tilde{A}(X,y)$. It is easy to check that such a z satisfies $\forall X \preceq t\exists y \preceq zA(X,y)$. ⊣

So, $\mathsf{B}^1\Sigma^{1,\mathsf{b}}_\infty$ is a consequence of $\mathsf{BTFA}+\mathsf{strict}\ \Pi^1_1\text{-}\mathsf{reflection}$. The next theorem says that, in a sense, the formulas of $\mathsf{B}^1\Sigma^{1,\mathsf{b}}_\infty$ are the only non-trivial consequences of *strict* Π^1_1*-reflection* which do not have second-order unbounded quantifiers.

Theorem 8. *The theory* $\mathsf{BTFA}+\mathsf{strict}\ \Pi^1_1\text{-}\mathsf{reflection}$ *is conservative over* $\mathsf{BTFA}+\mathsf{B}^1\Sigma^{1,\mathsf{b}}_\infty$ *with respect to formulas without second-order unbounded quantifiers.*

Proof sketch. Let $(\mathcal{M},S)$ be a countable model of $\mathsf{BTFA}+\mathsf{B}^1\Sigma^{1,\mathsf{b}}_\infty$ and $A(x,X)$ be a $\Sigma^{1,\mathsf{b}}_\infty$-formula (possibly with first and second-order parameters) such that

$$(\mathcal{M},S) \models \forall u\exists X\forall x \preceq u\, A(x,X).$$

We are going to show that there is a subset G of the domain M of $\mathcal{M}$ such that the structure $(\mathcal{M},S\cup\{G\})$ is a model of $\Sigma^{\mathsf{b}}_1\text{-}\mathsf{NIA}+\mathsf{B}^1\Sigma^{1,\mathsf{b}}_\infty$ in which $\forall x\, A(x,G)$ holds. It is possible to close $S\cup\{G\}$ under our form of recursive comprehension by the argument of the proof of Theorem 7. Let us denote this closure structure by $(\mathcal{M},S^1)$: this structure is again a model of $\mathsf{BTFA}+\mathsf{B}^1\Sigma^{1,\mathsf{b}}_\infty$ and, moreover, has the same bounded sets as $(\mathcal{M},S)$. (I.e., the sets of the form X^a with a an element of M coincide for X ranging in S or in S^1). Now, using a chain argument as in the last part of the proof of Theorem 5, there is $S^{\omega\cdot\omega}$, with $S^1 \subseteq S^{\omega\cdot\omega} \subseteq \mathcal{P}(M)$ such that the structure $(\mathcal{M},S^{\omega\cdot\omega})$ is a model of $\mathsf{BTFA}+\mathsf{B}^1\Sigma^{1,\mathsf{b}}_\infty+\mathsf{strict}\ \Pi^1_1\text{-}\mathsf{reflection}$ with the same bounded sets as the initial structure. This fact, of course, entails our result.

The adjoining of G mimics the forcing construction of the proof of Theorem 5. However, we now use a different notion of tree. In Theorem 5, the forcing conditions are (limitless) trees defined by bounded formulas. The elements of these trees are, of course, the 0-1 strings of the model $\mathcal{M}$ (i.e., the elements of M). Roughly, in our case, the bounded sets of S play the role of the 0-1 strings. One should intuitively view a bounded set as a very long 0-1 string (the "characteristic function" of the bounded set) along the linear order $\leq_l$ defined by

$$x \leq_l y :\Leftrightarrow (x \preceq y \wedge x \not\equiv y) \vee (x \equiv y \wedge \exists z \subseteq x(z0 \subseteq x \wedge z1 \subseteq y)) \vee (x = y).$$

Note that $\leq_l$ is defined first according to length and, within the same length, lexicographically. (In the presence of the totality of the exponential function such "long strings" can be obtained from the usual ones.)

After this interlude, take the given $\Sigma^{1,\mathsf{b}}_{\infty}$-formula $A(x, X)$. In analogy to similar properties in other contexts, it is easy to show that there is a term $t_A(x)$ such that, for all a and $x \preceq a$, the equivalence between $A(x, X)$ and $A(x, X^b)$ holds in second-order structures of Σ^{b}_1-NIA for b such that $t_A(a) \preceq b$. For simplicity, we write t instead of t_A and we assume, without loss of generality, that $x \preceq t(x)$.

We can now describe our forcing conditions. These are sets T of the form

$$\{(a, D^{t(a)}) : a \in M, D \in S \text{ and } (\mathcal{M}, S) \models B(a, D^{t(a)})\},$$

where $B(x, X)$ is a $\Sigma^{1,\mathsf{b}}_{\infty}$-formula (possibly with first and second-order parameters), such that:

1. $(\mathcal{M}, S) \models \forall x \exists X B(x, X^{t(x)})$;
2. $(\mathcal{M}, S) \models \forall X \forall x\, (B(x, X^{t(x)}) \to \forall y \leq_l x\, B(y, X^{t(y)}))$;
3. $(\mathcal{M}, S) \models \forall x \forall X\, (B(x, X^{t(x)}) \to \forall u \preceq x\, A(u, X^{t(x)}))$.

Intuitively, the first two conditions say that the formula B defines some kind of limitless tree. The third condition just says that T is a subtree of the (limitless, by assumption!) tree associated with the formula A.

Given conditions T and Q, we say that $Q \leqslant T$ if $Q \subseteq T$. We can now take a generic filter $\mathbb{G}$ of forcing conditions and define:

$$G := \{x \in M : \forall T \in \mathbb{G}\, \exists a \in M, D \in S$$
$$(x \in D \wedge x \preceq t(a) \wedge (\mathcal{M}, S) \models (a, D^{t(a)}) \in T)\}.$$

By the genericity of $\mathbb{G}$, it can be shown that, for all $x \in M$, $(\mathcal{M}, S) \models \forall u \preceq x\, A(u, G^{t(x)})$. We use the density of the sets $\mathbb{D}_x$ whose members are the forcing conditions containing a single member of the form $(x, X^{t(x)})$. Of course, the proof of the density property of these sets uses the scheme of $\mathsf{B}^1\Sigma^{1,\mathsf{b}}_{\infty}$.

As before, the forcing language has constants for each element of M and of S and an extra constant C (standing for the generic set G). We define the forcing relation for the relevant atomic sentences, i.e., for sentences of the form $s \in C$. We say that $T \Vdash s \in C$ if

$$(\mathcal{M}, S) \models \exists x \forall X^{t(x)}((x, X^{t(x)}) \in T \to s \in X^{t(x)}).$$

It is easy to verify that the above is a good weak forcing notion. The forcing relation is extended to all sentences of the forcing language in the standard way. Note that $T \Vdash s \in C$ is a $\exists\Sigma^{1,\mathsf{b}}_{\infty}$-formula.

It can be shown that the structure $(\mathcal{M}, S \cup \{G\})$ is a model of Σ^{b}_1-$\mathsf{NIA} + \mathsf{B}^1\Sigma^{1,\mathsf{b}}_{\infty}$ in which $\forall x\, A(x, G)$ holds. The reader can check the details in [8] (there is also related information in [6] and [21]). $\dashv$

8 Weak analysis digest

In [25], Wilfried Sieg posed the following question: to find a mathematically significant subsystem of analysis whose class of provably recursive functions consists only of the computationally feasible ones. In the last part of his Ph.D. dissertation [11], the second author made a first attempt at investigating mathematics in a weak setting. He considered some basic theorems of analysis in the Cantor space setting. In fact, the discussion of these theorems (e.g., the Heine/Borel covering theorem) is very natural in the Cantor space setting (the case of the closed unit interval poses some minor technical difficulties). Pursuing this line of research, in the mid-nineties the theory BTFA of Section 4 was defined and a few years later, in collaboration with the first author, the paper "Groundwork for weak analysis" [9] was written. As the title suggests, the paper aims at giving the groundwork for developing analysis in a weak setting – in our case in BTFA. The paper defines the real number system and the notion of a real-valued continuous function of a real variable. It proves the intermediate value theorem and, as a consequence, it is shown that the reals form a real closed ordered field. In the last section of the paper, it is shown that BTFA is interpretable in Robinson's theory of arithmetic Q and, as a corollary, one obtains that Tarski's theory of real closed ordered fields is interpretable in Q (a result independently due to Harvey Friedman). It is an open question whether $\mathsf{BTFA} + \Sigma^{\mathsf{b}}_{\infty}\text{-}\mathsf{WKL}$ is interpretable in Q. We conjecture that it is. The same question can be posed for the principle of strict Π^1_1-reflection.

The paper [10] investigates the role of weak König's lemma over BTFA. As discussed in Section 5, the adjunction of $\Sigma^{\mathsf{b}}_{\infty}\text{-}\mathsf{WKL}$ results in a theory that is a first-order conservative extension of BTFA. In [10] it is shown that, over BTFA, the Heine/Borel covering theorem for the closed unit interval is equivalent to weak König's lemma for trees defined by Π^{b}_1-formulas. The Heine/Cantor theorem (which says that every real-valued continuous function on the closed unit interval is uniformly continuous) was also investigated. It was shown that the Heine/Cantor theorem implies weak König's lemma for *set* trees and is implied by weak König's lemma for trees defined by Π^{b}_1-formulas. The authors were not able to close the gap.

In reverse mathematics over RCA_0, it is well known that weak König's lemma is sufficient to guarantee that real valued continuous functions defined on the closed unit interval have a supremum and attain it (see [26]). However, over BTFA, weak König's lemma does not seem to be sufficient to guarantee the existence of the supremum because this fact implies the so-called "slow" principle of induction for Σ_1^b-predicates, where "slow" induction is the following form of induction:

$$A(\epsilon) \wedge \forall x\,(A(x) \to A(S(x))) \to \forall x A(x).$$

Here, the "successor function" S is defined thus: $S(\epsilon) = 0, S(x0) = x1$ and $S(x1) = S(x)0$. The successor of x is the *next* element after x in the ordering $\leq_l$ of the previous section. In the unary framework of Buss, the above scheme of "slow" induction corresponds to the ordinary "+1" scheme of induction for Σ_1^b-formulas (the mark of Buss' theory T_2^1 of [3]). The paper [10] shows that, over $\mathsf{BTFA} + \Sigma_\infty^b$-$\mathsf{WKL}$, the scheme of slow induction for Σ_1^b-formulas is equivalent to the existence of the maximum for real-valued continuous functions defined on the closed unit interval. In [29], Takeshi Yamazaki considers the stronger principle ($\star$): given a real-valued continuous function F defined in $[0,1] \times [0,1]$, there exists a real-valued continuous function G defined on the closed unit interval such that, for all $x \in [0,1]$, $G(x) = \sup_{0 \leq y \leq x} F(x,y)$. Yamazaki, however, uses a different notion of continuous function. Our notion is an adaptation of the usual notion given in [26] to our weak setting. Yamazaki's notion defines a continuous function by approximations of piecewise linear functions with a *modulus of approximation* (see [29] for details). This is a stronger notion than our notion and, in particular, a continuous function in the sense of Yamazaki is automatically uniformly continuous. With this form of continuity, Yamazaki shows that ($\star$) is equivalent to comprehension for bounded formulas.

On a different direction, the first author considers in [7] a *variation* of the base theory BTFA and shows (by a forcing argument) that a version of Baire's category theorem is first-order conservative over the base theory considered.

The Ph.D. dissertation of the third author [21], as well as [18], studies Riemann integration for real-valued continuous functions defined in closed bounded intervals. It is shown that in the theory TCA^2, a theory related to the computational "counting class" FCH (see the beginning of Section 6), it is possible to define and develop Riemann integration, up to the fundamental theorem of calculus, for continuous functions with a modulus of uniform continuity. It does not seem possible to develop Riemann integration over a weaker base (for instance, over BTFA). The reason is the following: the existence of the Riemann integral implies the possibility of counting the number of elements of a bounded polytime decidable set. This is shown in [16]. Note that the possibility of this counting goes beyond polytime computability (unless certain classes of computational complexity collapse). Nevertheless, it may be possible to define Riemann integration in BTFA using Yamazaki's definition of continuous function, or a related definition (e.g., where the approximating functions are polynomials). It would be nice to see if this is possible, specially if the class of continuous functions considered is sufficiently robust (e.g., contains many analytic functions). Of course, this restricted class of continuous functions does not seem to coincide with the wider class (the one based on the standard definition in [26]). We believe that this is related to the eventual failure of Weierstrass' approximation theorem in weak settings. Indeed, we conjecture that this theorem is equivalent to the totality of the exponential function over BTFA.

References

[1] J. Barwise. *Admissible sets and structures.* Perspectives in Mathematical Logic. Springer-Verlag, 1975.

[2] W. Buchholz and W. Sieg. A note on polynomial time computable arithmetic. *Contemporary Mathematics*, vol. 106, 1990, pp. 51–55.

[3] S. R. Buss. *Bounded arithmetic.* PhD thesis, Princeton University, Princeton, New Jersey, 1985. A revision of this thesis was published by Bibliopolis (Naples) in 1986.

[4] S. R. Buss. A conservation result concerning bounded theories and the collection scheme. *Proceedings of the American Mathematical Society*, vol. 100, 1987, pp. 109–116.

[5] A. Cantini. Asymmetric interpretations for bounded theories. *Mathematical Logic Quarterly*, vol. 42(1), 1996, pp. 270–288.

[6] A. M. Fernandes. *Investigações em sistemas de análise exequível* (in Portuguese). PhD thesis, Universidade de Lisboa, 2001.

[7] A. M. Fernandes. The Baire category theorem over a feasible base theory. In Stephen Simpson, editor, *Reverse Mathematics 2001*, *Lecture Notes in Logic* 21, pp. 164–174. A K Peters, Massachusetts, 2005.

[8] A. M. Fernandes. Strict Π^1_1-reflection in bounded arithmetic. *Archive for Mathematical Logic*, vol. 49(1), 2010, pp. 17–34.

[9] A. M. Fernandes and F. Ferreira. Groundwork for weak analysis. *The Journal of Symbolic Logic*, vol. 67(2), 2002, pp. 557–578.

[10] A. M. Fernandes and F. Ferreira. Basic applications of weak König's lemma in feasible analysis. In Stephen Simpson, editor,

Reverse Mathematics 2001, *Lecture Notes in Logic* 21, pp. 175–188. A K Peters, Massachusetts, 2005.

[11] F. Ferreira. *Polynomial time computable arithmetic and conservative extensions.* PhD thesis, Pennsylvania State University, USA, 1988.

[12] F. Ferreira. Polynomial time computable arithmetic. *Contemporary Mathematics*, vol. 106, 1990, pp. 137–156.

[13] F. Ferreira. A feasible theory for analysis. *The Journal of Symbolic Logic*, vol. 59(3), 1994, 1001–1011.

[14] F. Ferreira. A note on a result of Buss concerning bounded theories and the collection scheme. *Portugaliae Mathematica*, vol. 52, 1995, pp. 331–336.

[15] F. Ferreira. A simple proof of Parson's theorem. *Notre Dame Journal of Formal Logic*, vol. 46, 2005, pp. 83–91.

[16] F. Ferreira and G. Ferreira. Counting as integration in feasible analysis. *Mathematical Logic Quaterly*, vol. 52(3), 2006, pp. 315 – 320.

[17] F. Ferreira and G. Ferreira. Harrington's conservation theorem redone. *Archive for Mathematical Logic*, vol. 47(2), 2008, pp. 91–100.

[18] F. Ferreira and G. Ferreira. The Riemann integral in weak systems of analysis. *Journal of Universal Computer Science*, vol. 14(6), 2008, pp. 908–937.

[19] F. Ferreira and P. Oliva. Bounded functional interpretation and feasible analysis. *Annals of Pure and Applied Logic*, vol. 145, 2005, pp. 115–129.

[20] G. Ferreira. *Aritmética Computável em Espaço Polinomial* (in Portuguese). Master thesis, Universidade de Lisboa, 2001.

[21] G. Ferreira. *Sistemas de Análise Fraca para a Integração* (in Portuguese). PhD thesis, Universidade de Lisboa, 2006.

[22] G. Ferreira. The counting hierarchy in binary notation. *Portugaliae Mathematica*, vol. 66, 2009, pp. 81–94.

[23] G. Ferreira and I. Oitavem. An interpretation of S_2^1 in Σ_1^b-NIA. *Portugaliae Mathematica*, vol. 63(4), 2006, pp. 427–450.

[24] G. Sacks. *Higher Recursion Theory.* Perspectives in Mathematical Logic. Springer-Verlag, 1990.

[25] W. Sieg. Hilbert's program sixty years later. *The Journal of Symbolic Logic*, vol. 53, 1988, pp. 338–348.

[26] S. G. Simpson. *Subsystems of Second Order Arithmetic*. Perspectives in Mathematical Logic. Springer, Berlin, 1999.

[27] S. G. Simpson and R. Smith. Factorization of polynomials and Σ^0_1 induction. *Annals of Pure and Applied Logic*, vol. 31, 1986, pp. 289–306.

[28] K. W. Wagner. Some observations on the connection between counting and recursion. *Theoretical Computer Science*, vol. 47(2), 1986, pp. 131–147.

[29] T. Yamazaki. Reverse mathematics and basic feasible systems of 0-1 strings. In Stephen Simpson, editor, *Reverse Mathematics 2001*, *Lecture Notes in Logic* 21, pp. 394–401. A K Peters, Massachusetts, 2005.

8

On the Notion of *Proving its own Consistency*

Doukas Kapantaïs[1]

Abstract: Almost since the date of its publication, there is the widespread opinion that Gödel's Second Incompleteness Theorem has excluded once and for all that Peano Arithmetic (or any interesting formal Arithmetic) can establish its own consistency by its own means. I argue that, although this is true, it does not come as a consequence of the Incompleteness Theorem itself. More precisely, I argue that no formal Arithmetic (no formal Theory in general) can under any circumstances establish that it is consistent, without some piece of evidence independent of this Arithmetic itself. If this is exact, it also implies the following: Assume some interestingly strong formal Arithmetic, which is free from the limitations of Gödel's Second Incompleteness Theorem (e.g. Willard [2001]). Assume also that some formula of the same Arithmetic expressing that it is consistent is the end-formula of some formal proof of it. This *per se* says nothing about the semantical question; i.e. is this Arithmetic consistent or not?

Keywords: Second Incompleteness Theorem, syntax vs. semantics, Hilbert's program, finitist attitude, formal Theories, informative content.

[1]Research Center for Greek Philosophy, Academy of Athens, Anagnostopoulou 14, 10673, Athens
dkapa@academyofathens.gr

Studies in Weak Arithmetics.
Patrick Cégielski, Charalampos Cornaros, Costas Dimitracopoulos.

0 Formal theories as sets of formulas

One way for one to conceive of a formal Theory T is to think of it as a subset of the well-formed formulas of some formal language L. This subset takes the name "the provable formulas of T" and is the extension of the predicate "T-provable". Another way to do the same is by a set of axioms and/or deductive rules operating on the same set of well-formed formulas. Instead of directly focusing on T's provability predicate, one now focuses on its axioms and deductive rules. As a matter of fact, one is now identifying the extension of "T-provable" with the set of formulas that can be inferred from T's axioms *via* the deductive rules. Things are usually done in the latter way, and for good reasons too. It is impossible for one to go through the infinite number of provable formulas every interesting Theory processes. On the other hand, one can easily go through the finite number of its axioms/axiom schemata. In substance, these two methods to conceive of a formal Theory are not antagonistic. Axioms and deductive rules can be seen as operators that, given a formal language, generate some predicate: the provability predicate of the Theory they are axioms/deductive rules of. This is why they have been put forward in the first place. In order to generate the set of provable formulas of the Theory they define[1]. Which suggests that, in principle, a formal Theory can be adequately identified with the extension of its provability predicate.

A consequence of the above is that there is no limit as for how "stupid" a formal (logical) Theory can become. If I set myself the task to create a formal Theory having but a single provable formula, I can easily succeed in doing so. It would be an unexciting Theory, one not based on propositional and lower predicate calculus, and not p.r. adequate either. What of it though? Not being any of these things would not imply not being a "formal Theory". Some (formal) Theories have to be more "stupid" than others, and stupidity is no formal notion.

Luckily, this is only in principle. Formal Theories in general, and formal Arithmetics specifically, are never put forward without some considerable amount of overpowering "intended interpretation" behind them. This safeguards them from the general kind of ridicule I have just described, although this ridicule is perfectly allowable by what "formal Arithmetic" and/or "formal Theory" means. So, luckily, formal Arithmetics are usually built upon standard propositional and first order predicate calculus and to begin with they prove anything that can be proved by the latter. The Peano axioms are also built into these Arith-

[1] As Gödel puts it: "*A formal system can simply be defined to be any mechanical procedure for producing formulas, called provable formulas.*" [18]: 369.

metics, either directly or as theorems. So, actual formal Arithmetics never look (very) stupid. Unless...

Unless, of course, they are contradictory and prove anything whatsoever. Hence, the importance of establishing that they are not. Otherwise, and despite all the intellectual labor that has been spent in order to create them, they would appear, at the end, just as silly as anyone for whom everything is provable.

1 The difference between (i) a proof about numbers, (ii) a proof about Theories and (iii) a proof about Theories that is also a proof about numbers

In a sense, no formal Theory has any reliable claim to make about anything whatsoever, on the sole basis of it being a formal Theory. There are formal Theories of all sorts, among which unsound ones too. So, to say that formal Theory T proves φ, and to understand thereby that whether φ obtains or not has been decided on the sole basis of (i) T having proved φ and (ii) T being a formal Theory, does not make any sense. You need to know (or at least assume) that T is sound. In other words, you need to have faith in T, either because you know it to be sound, or because you assume that much.

So, in general, if some formal Arithmetic A proves φ, φ obtains, on the assumption that A is sound. Therefore, if this Arithmetic is PA:

(1) If PA is sound, then: if PA proves φ, φ is the case.

For example. Because I have overwhelming evidence that PA is sound, I will believe any formal decision PA will eventually make with respect to Goldbach's Conjecture. PA's soundness is something I put my money on, and so I will be putting my money on Goldbach's Conjecture, if PA proves it. There is no irrational element in this way of thinking, especially when φ is a formula expressing some property of the universe of numbers. One would perhaps object at this point that any formula of any formal Arithmetic expresses (*via* its intended interpretation) some property of the canonical universe of natural numbers[2]. True as this observation may be (Arithmetic is about numbers after all), it misses some crucial distinction. For, although all properties of numbers are just that, i.e. numeric properties, quite a few of them are (because of some acceptable codification)[3] *theoretic* properties (of

[2]With respect to formulas, I will be using "expressing" for properties that might or might not obtain, and "captures" or "represents" for properties that do obtain.

[3]By "acceptable", I mean a codification satisfying the conditions as put forward in [15].

numbers)[4]. Consequently, quite a few of them obtain for numbers if and only if some Metamathematical properties obtain for Arithmetics. To be precise, the distinction I wish to establish is between: (i) numeric properties that imply properties of some Arithmetic, and (ii) numeric properties having no such implications. I will call the former "**theoretic** properties of numbers" and the latter "**plain** properties of numbers". More specifically, I will say that property F of number n is "theoretic" if and only if, according to some acceptable codification of some formal language L, n-being-F expresses some property of some formal Arithmetic[5].

The next question is why this is a crucial distinction that needs to be stressed. Remember the elementary *rationale* behind my betting at the beginning. I was inclined to put my money on anything PA proves, because I was assuming that PA is sound. I was not, however, inclined to make any move, before PA decides on the matter. This, with respect to plain properties, is as rational as rational can be. It is rational, because I assume that PA is sound and because the soundness of PA does not imply any of the plain properties, unless trivially, i.e. unless they obtain. But, then again, how am I supposed to know that (i.e. that they obtain), if I am currently relying on PA for illuminating me on the matter? So, in case φ expresses some plain property of numbers, I am never allowed to argue *from* the soundness of PA to the fact that φ obtains/doesn't obtain. And, therefore, since I am assuming PA's soundness and I also know nothing about whether φ obtains or not, my strategy should be to wait for PA to formally decide on φ, and then put my money on what PA has decided. This means that, since I recognize (1) to be rational, I also recognize (2) to be rational:

(2) If I believe that PA is sound and I am ignorant about φ, and φ expresses some plain property, I should first wait for PA to decide on φ, and then put my money on what PA has decided.

Which says that my belief in PA's soundness, and this belief alone, does not suffice in order to make me decide on how to bet with respect to φ.

---

[4]Here I wish to establish a distinction between (i) a property of numbers (numeric property), (ii) a property of a Theory: a Metamathematical property, and (iii) the numeric counterpart of this latter: a "theoretic property", as I will call it. See *infra*.

[5]Notice that, primarily, it is formal languages that are codified by numerals, not formal Arithmetics. Formal Arithmetics are codified only derivatively and on the basis of the codification of the formal language. Because of that, codified formulas of some Arithmetic might be expressing properties of some other Arithmetic too. E.g. a number codifying a formula of Peano Arithmetic might be saying something about Q as well.

So, suppose that PA has "made up its mind" about φ, and I bet accordingly. Is my money safe? Well, it depends. If my initial assumption is correct, and PA is sound, they are. If not, I might lose it all. Which implies that:

(3) If I believe that PA is sound and I am ignorant about φ, and φ expresses some plain property, and PA decides on φ, and I bet on whatever PA has decided, I will win some money, in case PA is sound.

Now, for the distinction. Whereas plain properties are properties of numbers and not of PA, some properties of PA are also properties of numbers, namely, theoretic properties of numbers. This means that if φ expresses some theoretic property, and since soundness is a property of formal systems, it is not to be excluded that PA's soundness and φ are interconnected; especially if φ happens to be *about* PA. More specifically, it is not to be excluded that PA's soundness implies φ. If this is so, it would be quite silly of me to wait for PA's verdict on φ, before making my move on φ. For consider it this way. In case φ expresses some theoretic property, which is implied by PA's soundness, (3) will still hold. So:

(4) If I believe that PA is sound and I am ignorant about φ, and φ expresses some theoretic property, which is implied by PA's soundness, and PA decides on φ, and I bet on whatever PA has decided, I will win some money, in case PA is sound.

Again, to the question "Is my money safe?" the answer is that it is, only in case my initial assumption concerning the soundness of PA is correct. Otherwise, it depends. This is what the last clause in both (3) and (4) stands for. But, on the other hand, the equivalent of (2) for theoretic properties obtains no longer. For, in case I believe in PA's soundness, and in case PA's soundness implies φ, I should put my money on φ, as soon as I realize the implication. Because, obviously:

(5) If PA is sound and PA's soundness implies φ, and I put my money on φ, I will win some money, *in case PA is sound.*

Which suggests that, in case I believe that PA is sound, and I know that PA's soundness implies φ, the (only) rational thing for me to do is to put my money on φ right away. I have no reason to wait for PA to formally decide on the matter and then put my money "on whatever PA has decided", because again my money will be safe only in case PA is sound. So:

(6) If I believe that PA is sound and I am ignorant about φ, and I

know that PA's soundness implies φ, I should put my money on φ right away.

Waiting PA's verdict on the matter would not make our chances any better. It would all depend, still, on PA's soundness. So:

(7) If I believe that PA is sound and I know that PA implies φ, PA proving φ will not make φ any more probable than it now is.

(6) is OK! At least, it remains OK for as long as the following principle is recognised as rational:

(**PR**) If I believe that X, and I know that X implies Y, I must believe that Y as well.

In order to recapitulate. If I believe that PA is sound and I know that PA implies φ, there is nothing for me to gain out of waiting for PA to eventually prove φ and then put my money on φ. For PA's having proved φ will not ameliorate my chances. It is still the case that φ obtains, on the assumption that PA is sound. So, if I say that I believe in PA's soundness, and I am determined to put some money on the matter, I should put my money on φ with no further ado. And, in any event, I should never wait for PA to formally decide in favour of φ, in order just to "make sure".

If I wait, it is either because I have not decided as yet on whether I should put any money at all, or because I cannot see through the *rationale* behind (6).

2 Why to trust/not to trust a numeric proof, and why to trust/not to trust a theoretic proof

Forget all about betting. The moral behind §1 is this: In case φ is not implied in any known way by T's soundness, and on the assumption that T is sound, T proving φ can be informative with respect to φ in the following two respects: (i) we come to know that T formally proves φ, and (ii) we come to know that φ obtains, on the assumption that T is sound. On the other hand, in case it is already known that T's soundness implies φ, and on the assumption that T proves φ, T proving φ is informative only with respect to (i); (ii) we knew already.

In case PA proves the Goldbach conjecture, we come to know something about PA's provability predicate *and* something about the semantics of the Conjecture itself. For, although not unconditionally, i.e. not in case PA is not sound, we come to know that the Conjecture obtains, *in case* PA is sound. Now, depending on how much the scientific community trusts PA, this piece of information might be appreciated as a major breakthrough. (Have no doubt that it will in case one day PA proves the Conjecture.) On the other hand, suppose there is some alternative formulation of PA's consistency, which does not imply any contradiction in any known way. Call this formulation $Con'(PA)$, as opposed to the canonical formulations, which we will note by $Con(PA)$[6]. In case PA proves $Con'(PA)$, we come to know something about the syntax of PA, and nothing more; i.e. we come to know nothing about the semantics of $Con'(PA)$, and so neither do we come to know anything about whether PA is consistent or not, either conditionally or unconditionally. (That PA is consistent, *in case PA is sound*, we knew already.) Finally, suppose there is some equipotent (or quasi equipotent) Arithmetic to PA (call it PA'), such that $Con(PA')$ can be proved in it without implying any contradiction in any known way[7]. Again, and by PA' proving $Con(PA')$, we come to know nothing about the semantics of $Con(PA')$ for exactly the same reasons as before.

$Con(PA')$ and $Con'(PA)$ express properties of PA' and PA respectively, about which we already know that they are implied by PA''s and PA's soundness; they both fall under (6).

And so the natural question: Why should anyone ever take pains to make PA formally prove an alternative formulation of its own consistency, or find some other formal Arithmetic, that proves some canonical formulation of its own consistency?

3 Formal proofs and factive (formal) proofs.

While trying to answer the questions at the end of §2, let me begin by some terminological clarifications.

I will call a proof of φ "**formal**" if there is a formal Theory T, such that a sequence of well-formed formulas of T, which is a proof of T, ends with φ[8].

I will call a formal proof of φ "**factive**" if this proof, by being the kind of proof it is, implies that what φ expresses under its intended interpretation obtains[9].

[6]By "canonical formulations of PA's consistency", I have in mind all formulations of PA's consistency that *via* the 2nd Incompleteness Theorem can be deductively connected to the proof of a formal contradiction.

[7]$Con(PA')$ being the counterpart of $Con(PA)$ within PA'. I.e. some canonical formulation of PA''s consistency, which, as assumed, no longer implies any contradiction in any known way.

[8]For Hilbert systems. For the rest, *mutatis mutandis*.

[9]Notice that according to these definitions a real proof is not necessarily a factive proof, for according to the same definitions all factive proofs are formal proofs, whereas not all real proofs are formal proofs. Besides, not all factive proofs are real proofs either.

I will say that a proof of φ "**establishes**" φ if this proof is factive.

"**Having evidence**" for φ will mean that claiming φ is justified by the way things have been shown to be.

Obviously, and since unsound Theories exist, "formal proof" does not imply "factive" proof. Moreover, even if it does (e.g. assume T being sound under its intended interpretation) the fact that it does might be concealed from us, and so it might be concealed from us that the formal proof really "establishes" what its end-formula says. Which means, on the occasion, that we might not be in position to draw the inference from "s is a proof of T" to "what the end-formula of s says, under its intended interpretation, obtains"[10], i.e. we might not be in position to realize that s, by being the kind of proof it is (on the occasion, by being a proof of the *sound* Theory T), implies that its end-formula speaks the truth.

Now, this is far from being a desperate situation. For there are plenty of cases, where we can recognize proofs of formal Theories as sound proofs.

First, as we saw, we might already know the Theory to be sound.

Another occasion, the most unexciting one, is when, instead of knowing the Theory to be sound, we (already) know the end-formula to be true. Take any explosive system you care to mention and pickup any proof of that system, which proves something you already know. You know that what this formal proof proves obtains, but not, of course, on the basis of being a formal proof of an explosive system.

A third possibility is when a formal Theory proves something that you ignore, but about which you know that, *if* it is proved by this Theory, it will be shown to be the case. This is because it just might happen that we are ignorant as to whether the Theory is sound, being also ignorant as to whether some specific formula is true, but have evidence that, in case this particular Theory proves this particular formula, it proves a true formula. For example: we do not know that PA is inconsistent, and we do not know that PA proves that it is inconsistent either. However, we do have evidence that, in case PA proves $\neg Con(PA)$, it will have proved a true formula[11]. Notice, that the way by which PA proves $\neg Con(PA)$, if it ever does, might not be formalizing a valid piece of argument.

---

[10] Henceforth, I will occasionally be dropping the clause "under its intended interpretation".

[11] Notice here that the Theory that emerges in case one adds that PA is not consistent to PA is consistent (and in case PA is consistent, it is unsound). But this Theory is not PA. PA, in case it proves that it is inconsistent, is both inconsistent and unsound.

A fourth possibility, where we might recognize a proof of a Theory T as factive, although we do not know T to be sound, is when we realize that this particular proof of T formalizes a pattern of argument we know to be valid and the proof begins with premises we know to be true. Example: choose any proof of PA, in which no unbound quantification is involved. If you believe that this part of PA formalizes some part of the "finitary way of thinking", and you have blind confidence in this way of thinking, you cannot but endorse what PA proves on the occasion.

So, and in order to sum up. A formal proof is factive if and only if one can infer, based on some particular characteristic of the proof, that what it (formally) proves obtains. Now, there are four things a formal proof might be, such that the above can be safely inferred. It can be either:

(A) A formal proof of a sound formal Theory.

(B) A formal proof of a formal Theory having a true formula as its end-formula.

(C) A formal proof of a formal Theory ending with a specific formula, about which we know that it will be revealed to be true, if proved by the Theory.

(D) A formal proof of a formal Theory, which formalizes a valid pattern of argument and begins with premises we know to be true[12].

In (A), we focus on the Theory, in (B), on the conclusion of some proof of the Theory, in (C) on both the Theory and some specific proof of the Theory, in (D), on the general form of some proof of the Theory. Besides: In (A), (C) and (D), the proof might turn out to be a genuine discovery. In (B), this is impossible. Finally: In (A) and (D), the proof must be sound. In (B) and (C), it might not even be valid[13].

[12]The list aims at being complete. I have no *a priori* argument that this one is. Any new member that does not already fit in is welcomed! Now, since my general argument in §4 and §5 rests upon the list being complete, the overall argument would need to be checked, once any new member comes to be added.

[13]What must be sound in (B) is the argument leading to the conclusion that the end-formula of this particular proof of T says something that obtains, and in (C) the argument leading to the conclusion that if T proves this formula, T proves a true formula. These arguments are not the same with the arguments these T-proofs formalize. It is only in (D) that the arguments T-proofs formalize are always real proofs. Notice that one might be believing what the end-formula of a formal proof says, according to its intended interpretation, without this proof being a real proof, under the same interpretation. Of course, there is a real proof at the background here too, namely, some valid argument establishing that the end formula of the formal proof in question is true.

However, and despite the above discrepancies, all four cases allude to four different "kinds of thing" a proof might be, and all four of them provide sufficient evidence that the proof doesn't lie. Hence, the crucial question: Can a formal Theory T establish by itself any among (A) to (D) to obtain with respect to some proof of it? If it can, this would imply that T is in position to provide by itself the piece of evidence we need for reaching the conclusion that the proof is factive.

More specifically. Suppose that PA arrives at an alternative formulation of its own consistency, which, *prima facie*, cannot be validly transformed within PA to any formal contradiction, or suppose that some alternative formal Arithmetic $PA^{'}$ has an unproblematic canonical proof of its own consistency. Can PA (or $PA^{'}$) provide by itself some evidence that one of the above characteristics is a characteristic of that proof, and, thereby, allow the conclusion that PA ($PA^{'}$) is consistent?

4 Asking the right question to the wrong person

In this section, I will argue that neither PA, nor any other formal Arithmetic can prove (factive sense) that they are consistent.

Suppose that researcher R is asking (wishes to know) whether formal Arithmetic T can prove its own consistency. Since formal Arithmetics are what they are (i.e. formal Theories) it is clear that the above situation can be described as follows:

(8) R wishes to know whether there is a sequence of well-formed for mulas, written in the formal language of T, such that it is a proof of T, and ends with a formula expressing that T is consistent.

The obvious reaction to this is to further ask why, on the occasion, is R wishing to know such a thing. The safe reply is that, on the occasion, R's curiosity must be syntactical, at least; i.e. R wishes to know whether there is any allowed transformation of the axioms of T into $Con(T)$, or into some equivalent formula. Can, on the occasion, R's curiosity be any more ambitious than that? The answer, again, is that it might. For R might be interested not only in whether there is one such transformation of T's axioms, but also in whether the formula these axioms have been transformed into is true. So R might be wishing to know:

(9) Whether there is a sequence of well-formed formulas, written in the formal language of T, such that it is a proof of T, and ends with a formula expressing that T is consistent, and what this formula says is true. (I.e. [...] and ends with a formula *capturing* that T is consistent).

Now, suppose that (8) has been answered by the affirmative, i.e. suppose that we have found some such sequence of T's well-formed formulas. To begin with, this sequence is a proof. Trivially so. It is a formal proof of some formal system, and any formal proof of any formal system is a (formal) proof. The crucial question can, then, be formulated as follows.

(10) Can this hypothetical proof of T's consistency be a factive proof?

Which says: can this hypothetical proof of T's consistency imply, *just by being the kind of proof it is*, that what its end-formula says obtains? Now, this proof is quite a few things. In order for us to safely conclude that it is a factive proof, we need to have evidence that at least one of these "things" implies that what the proof says obtains.

According to the conclusions in §3, this formal proof cannot be factive, unless either (A) T is sound, or (B) the end-formula of the proof is true, or (C) it is a formal proof ending with a true formula, *if* proved by T, or (D) it is a formal proof formalizing a sound argument. So, on the assumption that T has a formal proof ending with a formula saying that it is consistent, what we are looking for in order to recognize it as factive is some evidence establishing one of the above. Which straightforwardly implies that:

(11) T can establish by itself that it is consistent only in case, beside the hypothetical formal proof ending with a formula saying that it is consistent, T can also establish by itself one of (A) to (D) above.

This is impossible. For let us consider them one by one.

(A) Suppose that T has provided by itself some evidence that it is a sound Theory. How could this evidence be looking like? Again, T, by being a formal system, has a single quite monotonous way of providing "evidence-provided-by-itself". It does so (can do so) only by its formal proofs. So, the assumption that T has provided by itself some such evidence is equivalent with the assumption that we have discovered a T-proof ending with a formula saying that much; i.e. saying that T is sound. But, again, the previous counterexample is still effective. How can one know that T is telling the truth here, since any contradictory (and therefore unsound) formal system necessarily provides such "evidence" too? In order to be trustworthy/reliable, any evidence that T has proved its soundness not *qua* an unsound system must come from a source other than T.

(B) It is quite impossible for T to provide by itself any supplementary evidence that its hypothetical proof of a formula saying that it is consistent terminates with a formula that is true. If T has the aforementioned proof, T has done the best it could already. For how else can T (how can any formal Theory) "argue" that a certain formula of its formal language is true, unless by proving the formula. So, we are, at present, either asking from T to repeat the same formal proof over and over again, or asking from T to produce a proof of some formula implying what the first proof "proves" together with a proof that it really implies it, which is absurd, for all these proofs, *qua* proofs of T, will be as reliable as the first one is.

(C) Imagine that there is some formula φ, expressing that T is consistent, and that we have evidence that, *if* T proves φ, T will prove something true. Again, just like in (A), even if we manage to find, within T, a formal proof of the above conditional, it won't be because of that that we will believe in φ. Again, even if T were inconsistent, it would have proofs for both these formulas (i.e. both for φ and for $Prov_T(\varphi) \rightarrow \varphi$).

(D) This is perhaps the trickier case. For one could (*prima facie* at least) think that in case T arrives at some formal proof of its own consistency, which no one could possibly doubt, then T would have arrived at some factive proof of its own consistency *by itself.* And indeed, in a sense this would have been the case since "proved by itself" can also have the meaning of "written in its own formal language and following its own rules of inference". This observation misses some crucial parameter in the enterprise of "being after a factive proof of T's consistency" though. For, when we are asking T to establish (i.e. to prove in the factive sense) that it is consistent, we are not merely asking from T to provide some reliable proof of its own consistency; we are further asking from T to provide the evidence that this proof is reliable. In that sense, T cannot possibly establish that it is consistent, unless, besides the (possibly reliable) proof that it is consistent, provides also some reliable (formal) proof that this proof is reliable. And, then, we would be needing of a further formal proof saying that the latter proof is reliable, for no formal proof has ever being rescued by a bad argument saying how reliable the former is. And so *ad infinitum.* Again, just like in (A) to (C), the crucial question is: And how can T argue that any of these formal arguments are reliable? And once more the same monotonous answer follows: By the only way T, *qua* a formal system, knows how to argue, i.e. by its formal proofs. The typical counterexample being again any contradictory (explosive) system.

The following is a less abstract example on the same topic.

Assume, for the sake of the argument, that one day some finitary proof of PA's consistency is found so that it can, as any finitary proof can[14], be formalized within PA. Why do we believe in what this proof says? In other words, the question is: What has turned this proof of PA into a factive proof? And the obvious answer is that this "thing" cannot be any (other) formal proof of PA. So not any formal proof of PA saying that the first proof belongs to some kind of proofs that never lie. The explanation of our inclination to believe in this or that particular proof of PA is that we (being in position to contemplate PA in some non formal way) scrutinize the proof and see that it belongs to some specific category of proofs that we trust in general[15]. So, the reason cannot be that we trust these proofs because PA invites us to do so. Hence, what has turned this proof into factive cannot be PA *per se*. The crucial parameter in our hypothetical scenario is that this proof is finitary and that finitary proofs are reliable, not that PA says that they are.

(Here an eclectic affinity is certainly Lewis Carroll on Modus Ponens. We can never reach a formal explanation of why it is that we can detach B out of (i) $A \to B$ and (ii) A. For, in trying to do so, we have to introduce into the system a formal rule saying that we can always detach ψ out of any φ and any ψ, such that $\varphi \to \psi$ and φ. So now, we are formally entitled to detach B out of (i) $A \to B$, (ii) A, and, in addition, (iii) the above rule, when formalized into the system. And again, why we can always do so, begs for some further formal explanation. And so, we are now in need of yet another formal rule allowing us to always do so out of (i)-(iii), and so *ad infinitum*. The affinity between our hypothetical scenario and Carroll on MP is that both examples pinpoint some general characteristic of formal arguments. Formal arguments are based on some syntactical (i.e. mechanical) rules allowing one to manipulate formal tokens in some predetermined ways. Now, the explanation as for why one is logically justified in manipulating these formal tokens in these ways cannot be making part of the formal argument itself. For if it did make part of the formal argument itself, it would also have to make part of the tokens to be manipulated, and so we would be needing of yet another formal explanation for establishing why these tokens can be thus manipulated, and so *ad infinitum*.

[14] Or so believes the great majority of logicians.

[15] For some relative "anti-formalistic" arguments see [36] and [33].

Therefore, in our hypothetical scenario, any formal proof of why some proofs are to be trusted, would be in need of some further formal proof of why it itself needs to be trusted. *Mutatis mutandis*, and for Modus Ponens, any formal explanation of why we can detach B out of some tokens becomes itself a token out of which B is to be detached.)[16]

If we are after some convincing argument that some formal Theory T is consistent, this argument cannot be provided by T itself. In reality, we are after an argument following some reliable pattern of reasoning. Now, if this argument can be formalized within T, so much the better for the expressive strength of T. It would never be though that we trust this argument, because it is a formal argument *of* T, as it would neither be that we trust this argument, because T has, in addition, a formal proof "saying" that this argument is to be trusted. Even in that case, we would rather need to check back the reliability of this proof by making sure that it formalizes reliable patterns of reasoning, the reliability of which has no formal foundation of its own.

All in all, and as we were saying back in §0, T is a formal system, and as such T is a *mechanical* procedure for producing formulas, called provable formulas. Meaning that nothing excludes (*logically*) that up to now we have discovered only the "good news", while none of the "bad news" (i.e. the contradictory ones) have been produced as yet. If, however, T is inconsistent some "bad news" will eventually reach us.

Let us now go back to (8) and adjust it to Peano Arithmentic *sub specie* of what we were saying in §1. We know that if there is a sequence of formulas which is a proof of PA and ends with some canonical formulation of PA's consistency, PA is inconsistent. Researcher R is asking to know whether there is some alternative formulation of PA's consistency, $Con'(PA)$, or some equally reassuring formula (e.g. an alternative formulation of PA's Reflection), such that they can be proved in PA without revealing the system to be inconsistent. R also asks to know whether there is some Arithmetic, PA', such that the canonical formulations of its own consistency or some equally reassuring formulas can be proved by PA', and PA' is consistent indeed.

The interesting part in this hypothetical endeavor concerns R's motivation (see (9) above). As we have already noticed, R's curiosity must be syntactical at least. The very nature of the questions R raises is cer-

[16]This does not say that Lewis Carroll excludes in general that there is any account as for why Modus Ponens is sound, e.g. some specific, paradigmatic instance of it, or an explanation of what Modus Ponens *is/stands for*. What Lewis Carroll presents as a paradox is that, once any such explanation is formalized, the system needs of yet another formalized explanation in order to give a formal account of why the system with the "added explanation" is sound, and so *ad infinitum*.

tainly such. "Is there any allowed manipulation of the axioms of PA such that it transforms them into $Con'(PA)$?" "Is there any alternative Arithmetic PA', such that one can transform its axioms into $Con(PA')$ without revealing them to be contradictory?" "Is there any sequence of numbers falling under a specific well-defined category of numerical sequences, called PROOFS[17], and ending with $\ulcorner Con(PA) \urcorner$?" So, through that angle, R's curiosity is combinatorial in nature. Now, how difficult this and similar questions might be, becomes evident as soon as one realizes that number theorists have yet to answer them, and that they will never do so algorithmically, in case the Church-Turing thesis obtains and PA is consistent.

Assume the first goal of researcher's R quest having being accomplished. Meaning: assume that either R has discovered $Con^{'}(PA)$ or that R has created $PA^{'}$. Seeing the extreme combinatorial difficulty of the relevant questions about $Con(PA)$, it would come as no surprise if the combinatorial difficulty of the new questions were equally elevated. Which suggests that any eventual answer to these latter could be proved to be a milestone in combinatorics. Could it be proved to be anything more?

I answer this question in the next section.

5 Semantical import in a strong and in a weak sense

Assume that formal Theory T has proved up to now no two contradictory formulas, and that its formal language disposes of some formula $Con(T)$ expressing the consistency of T in such a way that no contradiction has yet to be derived from $Con(T)$ by the deductive rules of T. Now, in case we discover some T-proof of $Con(T)$, should we believe in what $Con(T)$ says? The answer, as we have seen, is straightforward: not because the proof of $Con(T)$ is a proof of T, or a formal proof in general.

And we now ask: What further evidence do we need to have in order to believe in what this eventual proof says? We need to know that one of (A) to (D) obtains. Next question in line: Can T convince us by itself that one of (A) to (D) obtains? Again, this is impossible for the exact same reasons that no hypothetical formal proof of $Con(T)$ by T is reliable *per se*.

Let us examine what happens, in case we know already one of (A) to (D) to obtain. The question now becomes: Under these circumstances, will any eventual formal proof of $Con(T)$ by T have any semantical

[17]Following [44], I will be writing arithmetized Metamathematical properties in capital letters.

import?

This is an ambiguous question deserving a more detailed presentation.

Semantics, being what it is, i.e. the discipline occupying itself with the laws governing the relations between world and language, is certainly interested in any proof of any system that establishes any previously unknown truth. If system S proves φ, and we come to know that φ is the case because system S has formally proved φ (i.e. we did not know φ before that, and we come to know φ partly because S has proved it), it goes without saying that this formal proof of S is of some semantical interest. But semantics is also interested, it ought to be, in us having discovered that some system or other is capable of proving some true formula, and this remains so even if the formula was known to be true already. For consider it this way: the discovery of the fact that system S is capable of proving the truth expressed by φ enriches our knowledge on how world and language relate to one another. We learn something about the number of *truths* a certain system can prove, not only about the *formulas* it can transform its axioms into; and, so, we thereby enrich our knowledge on how this system relates to the world, not only on its transformation rules and on how they relate to its axioms.

Let us, then, rigidly distinguish between the first and the second situation above. We will say that "T proving φ has semantical import in the **strong sense**", when by coming to know that T has a formal proof ending with φ, we come to know that φ is the case. We will say that "T proving φ has semantical import in the **weak sense**", when, by coming to know that T has a formal proof ending with φ, we come to know that T can prove some fact about the world we knew already.

It is obvious that, in case we have at our disposal some T-independent piece of evidence of the kind we need to have in order to establish any among (A) to (D), the additional discovery of some formal proof of T ending with $Con(T)$ will certainly have some semantical import in the weak sense. T, no matter how it proves it, proves some truth about the world. This is straightforward enough. If we have this formal proof at hand, and we also have one of (A) to (D) at hand, we can be sure that T is consistent and that T can formally prove that it is consistent. Notice here that the "no matter how" qualification above is not stylistic. It means that the particular way in which T has reached this truth is of no importance. The formal proof might even be formalizing some non valid piece of reasoning. Nonetheless, it is still semantically important (minimally perhaps but still important) that this particular true formula belongs to the set of T's *provable* formulas[18].

So, if we know any among (A) to (D) to obtain, and we discover some proof of T ending with $Con(T)$, this has to have some weak semantical import.

Can the same discovery be of some strong semantical import as well?

This has to be excluded as for (A) and (B).

We begin with (A). In case we know T to be sound, and we do not believe there are true contradictions, we also know that T is consistent.

With respect to (B) the point is self-evident.

On the other hand, and as for (C) and (D), the discovery of some formal proof of T ending with $Con(T)$ might have semantical import in the strong sense as well; i.e. we might come to know that T is consistent after (and partly because of) this formal proof of T ending with a formula saying that T is consistent.

We begin with (C). What (C) says is that if ever we come across some proof of T ending with $Con(T)$, what the proof asserts will be the case. Therefore, if we already know (C), as soon as we discover a formal proof of T ending with $Con(T)$, we *eo ipso* discover that T is consistent.

With respect to (D) this will also be the case. For it is possible that we come to know that T is consistent at the same time (and partly because of) we have discovered some formal proof of T ending with $Con(T)$, we have checked this proof step by step, and realized that it formalizes some general pattern of reasoning that we trust. Therefore, here too, the discovery of this formal proof would have semantical import in the strong sense as well.

However, one must notice -again- that in none of these last two cases must one think that T has proved (factive sense) its own consistency by itself, for the T-independent piece of evidence is still indispensable. As for (C), it is the T-independent piece of evidence that, *if* we ever discover some proof of T saying that it is consistent, what the proof claims is true. As for (D), it is the equally T-independent piece of evidence coming down to the fact that some particular formal proof of T formalizes a reliable pattern of reasoning.

In the final section, I will go briefly through some historical aspects of the controversy surrounding Gödel's Second Incompleteness Theorem

[18] Especially with respect to Peano Arithmetic, such an eventuality will have far-reaching semantical consequences. For, in case PA is consistent, PA is also Π_1 sound, and if some Arithmetic is provably Π_1 sound, this Arithmetic realizes, (according to some scholars at least), Hilbert's dream. See [21, 16, 39]. Notice, however, that -again- one believes that PA's consistency implies its Π_1 soundness not because of some formal proof of PA.

and its philosophical significance.

6 Some historical feedback

Most of the time things are accounted as follows.

STORY: Gödel, in 1931, delivered two fatal blows to Hilbert's Program. Hilbert wanted Arithmetic to be in position to prove all true formulas about numbers, as he also wanted Arithmetic to be in position to prove its own consistency. By the First Theorem of Incompleteness, Gödel succeeded in showing that any sufficiently strong Arithmetic is incomplete, if consistent. By the Second Theorem of Incompleteness, Gödel succeeded in showing that any sufficiently strong Arithmetic cannot prove its own consistency, if consistent it is. Therefore, Gödel showed that neither Arithmetic is complete, if consistent, nor can it prove that it is consistent, under the same assumption.

Literally speaking, STORY speaks the truth. As for the implications of the First Theorem, there is no doubt about it. The first part of Gödel's paper excludes the possibility of there being any p.r. adequate consistent Arithmetic that can prove all true arithmetical formulas. And as for the Second Theorem, STORY speaks the truth as well, if literarily understood. For the Second Theorem excludes the possibility that any Arithmetic belonging to a great variety (almost all-encompassing) of formal Arithmetics[19] can formally prove its own consistency, if consistent it is. However, much often, and with respect to the Second Theorem, STORY gets loaded with a much stronger meaning. STORY is understood as implying that, by the time we became aware that PA, if consistent, cannot formally prove any among the canonical formulations of its own consistency, we have abandoned any hope that one day PA will eventually *establish* its own consistency by its own means. And this is wrong, or at least misleading. For while it is true to say that by the discovery of the theorem we did come to realize that PA cannot prove any among the canonical formulations of its own consistency, if consistent it is, it is not the case that it is *by this discovery* that we came to know that PA cannot establish its own consistency by its own means. This was something that we did know or -to put it better- something we *ought* to have known/realized already, and by the mere observation that PA is what it is, i.e. a formal system, and so PA cannot possibly establish its own consistency by *itself*. So, STORY gets misleading as soon as "to prove" in statements like, "Gödel's Second Incompleteness Theorem has made us aware that

[19] See [16], as in [44].

Arithmetic cannot prove its own consistency, if consistent it is", is understood in the strong, factive sense.

From one confusion, several follow. And so in our case.

From the initial misconception, i.e. that PA could, in principle, have established its own consistency by its own means, if only it wasn't for Gödel's Second Incompleteness Theorem, many theorists came to look down on any proof of PA's consistency that was coming from another (stronger than PA) source. They were considering it as "not quite the thing"; the "thing" being some proof of PA's consistency that has a witness within the universe of natural numbers; therefore an arithmetical proof, therefore, a proof performable "by PA itself". As a consequence, the *Dialectica* proof by Gödel, Gentzen's proof using transfinite induction, several attempts coming from modal and intuitionistic logic, even the proof ZF can provide, were all, and despite the consensus that they are sound, welcomed not with the greatest of enthusiasms. The reason being none other than the above.

And here comes the third misconception, a historical one this time. Because of some reason that it is easy for one to conjecture, wrong nonetheless, the effort to force Peano Arithmetic into proving its own consistency has been confused with Hilbert trying to produce an "absolute" proof of the same consistency. So people were now thinking that by "absolute proof" Hilbert had in mind some proof performable by PA itself. The same impression was reinforced by the fact that Hilbert himself, when referring to this "absolute" proof, was describing it as the opposite of the relative proof of Geometry's consistency he had delivered already. I.e. Hilbert has been suggesting that, unlike what happened in the case of Geometry, where its consistency has been made dependent on the consistency of something else (i.e. on the consistency of Analysis), a similar enterprise was not welcomed in the case of Arithmetic. So people were thinking that what Hilbert had in mind was something along the following lines: "What happened in the case of Geometry was that the consistency of a certain system S has been reduced to the consistency of another system S'; what we must try to do for Arithmetic, where this in not welcomed, is to prove its consistency within Arithmetic itself". Never once -at least as far as I know- has Hilbert expressed himself that way, or even close. And for good reasons too. For what Hilbert had in mind, while "being after an 'absolute' proof of Arithmetic's consistency", was not a proof performable by Arithmetic itself, but a proof so transparently valid that no one would dare putting a trial on, i.e. a finitary proof. Now, since PA disposes of finitary as well as non finitary elements, it is clear that Hilbert could not possibly have simply been after a proof of PA's consistency within

PA, since theoretically this proof could reveal itself to be non finitary, and a non finitary proof of PA's consistency would simply not do for Hilbert. On the other hand, the reason why Hilbert thought that no relative proof of PA's consistency "would do" is the utter simplicity of Arithmetic. Arithmetic is the bare essential of Mathematics, and so its consistency cannot be made dependent on the consistency of any other (i.e. less elementary) part of Mathematics without circularity. The latter would have to contain the former, and so the consistency of the latter could not possibly be prior to the consistency of the former. And so the general command was: Find some finitary proof of Arithmetic's consistency; no proof that would make it dependent on some non finitary part of Mathematics would do. Evidence for that comes both from Hilbert's general analysis of the situation in [20, 21, 22, 23][20] but also from the *rationale* behind the several (failed) attempts by Ackermann and Hilbert to prove the consistency of Arithmetic by the use of the ϵ-function[21]. The general idea behind these attempts was to find some non finitary proof (by employing of the ε-function) that would be provably reducible to a finitary one. It was not to find a proof that would be performable within PA itself; back then it was not clear that there can be no finitary proofs outside of PA. Now, today's general consensus that any finitary proof is a proof performable within PA (i.e. that finitary $\subseteq PA$), came much later[22]. Hilbert, at the time, could not possibly have been excluding there being finitary proofs outside PA. The best evidence that Hilbert was not identifying these two commands comes from Gödel in a much quoted passage of the legendary paper:

I wish to note expressly that Theorem XI (and the corresponding results for M and A) do not contradict Hilbert's formalistic viewpoint. For this viewpoint presupposes only the existence of a consistency proof in which nothing but finitary means of proof is used, and it is conceivable that there exist finitary proofs that cannot be expressed in the formalism of P[23].

From which it follows that the main goal of the Program was not to produce a proof-performable-by-PA-itself, but a *finitary* proof. Now, where this proof was supposed to be found is a subsidiary concern. At

[20]See especially how Hilbert defines Proof Theory and Metamathematics (i.e. the niche within which this proof was supposed to be performed), and see also [24], on the same topic.

[21]See [22].

[22]Gödel came to explicitly acknowledge this no sooner than in [17], (see also the note added in [44]:616).

[23][44]:615.

least, this is how Gödel was thinking about the matter back then, and we have no reason to doubt his exegesis on Hilbert with respect to that point.

That this was the core of Hilbert's Program makes good sense for another reason too: Exegetical charity towards formalism. Formal Mathematics was initially conceived as ordinary Mathematics that "has been formalized". So formal (or "Axiomatic" as Hilbert would have put it), Mathematics is no more powerful a tool than ordinary Mathematics[24]. Or -to put it better- it is, but not *in principle*; i.e. not in the sense that it can prove more things than ordinary Mathematics can. (Think of the latter as performed by an ideal mathematician.) Which is to say that formal Mathematics, together with its informal Metamathematics[25], do help us in proving things more easily and more securely than when we do so in ordinary informal Mathematics. From this point of view, formal Mathematics renders us more self-confident and more reliable at once. Moreover -and this is an achievement apart- formalization has transformed some humanly impossible tasks into humanly possible ones. (Try to imagine *Principia Mathematica* in ordinary English.) Despite all that, the fact remains that there is nothing one can prove formally that one cannot, in principle, prove informally as well. So, a formal proof of formal Mathematics, according to which Mathematics is consistent would not be any more reliable than its informal counterpart. Which means that the possibility that the latter can be formalized, or even arithmetized, would not make the former any more secure, except that it would have minimized the possibility of our having made a mistake out of carelessness. In substance, formal Mathematics can argue no more convincingly in favor of its own consistency than some informal argument can. And, in any event, no such formal argument (i.e. formal proof) would have vindicated the informal conviction mathematicians have that, when they prove things, they follow some non contradictory way of thinking, and the pre-formal conviction people have that the laws governing human thought are consistent. These convictions and these potential formal proofs either stand together or fall together[26].

I will now go briefly through the specific attempts that have been made in order to find interestingly strong Arithmetics that "prove"

---

[24]It is no extension of it, in any sense.

[25]Which is to be distinguished from informal ordinary Mathematics. (Notice here that most of the time, when we speak of "formal proofs", we have in mind their informal Metamathematical counterparts together with the conviction that they really are "their counterparts", i.e. that the proof can be performed formally as well. See [2].)

[26]See also the Appendix on that.

their own consistency[27]. It is important to notice here that the debate over these Arithmetics is not only formal but foundational as well. I.e. people who have invented them are not only concerned with the formal aspects of the consistency proofs these Arithmetics dispose, but also (not to say mainly), with the question of whether or not these consistency predicates really capture consistency. (See (i) and (ii) bellow.)

Following Feferman (see [19]:282-83), one may classify these attempts in three main groups. They are departures from PA with respect to: either (i) their provability predicate, or (ii) their set of axioms, or (iii) their set of deductive rules. Examples of (i) are Arithmetics following an intuition by Kreisel [26, 28] and changing the classical provability predicate into Rosser-provability. An example of (ii) comes from [10] and is based on an ingenious technique to -so to say- rectify PA's axioms in case they are inconsistent, or leave them intact, in case they are not. An example of (iii) comes from [43], and a system with cut-elimination. Since Feferman's classification, another possibility came to be added: (iv) Willard [46] has succeeded in proving relative consistency by imposing restrictions on the code numbers of proofs.

Let us consider these attempts one by one.

(i) Rosser-provability systems do prove (in an alarmingly easy way) that they are consistent; they miss the mark for the following reason. Although, in case the system is consistent in the traditional sense, Rosser-provability is equivalent with standard provability, it is not equivalent with standard provability, in case the system is not consistent in the traditional sense. A consequence of this is that all systems (no matter whether consistent or not in the traditional sense), are Rosser-consistent[28]. Meaning that any, e.g., traditionally explosive system is Rosser-consistent! Hardly, the thing one is after, if after some non contradictory system. In some more detail, what happens is as follows. These systems take a system to be consistent, if no two proofs in the traditional sense contradict one another *and* the one is proved in a stage earlier than the other. Now, clearly, every contradictory system in the traditional sense is Rosser-consistent, for the simple reason that out of any two contradictory formulas it proves in the traditional sense, the one has been proved earlier than the other. Moreover, any system that is order-adequate can capture this fact about

[27] In a forthcoming paper, I will present them in some more detail.

[28] This accounts for the "easiness" with which Rosser-consistency can be proved for and within these systems.

itself, and so it can even prove its Rosser-consistency. Feferman [10] is right in calling this solution "intensionally" unsatisfactory.

(ii) Feferman's ingenious device provides one with some finitary method to either test that PA is consistent and stay content with that (mission accomplished), or, and in case it is not, rectify it into some other system that is provably consistent (mission accomplished again). But, although it *is* the case that this method is finitary and so can be formalized within PA itself, it requires an infinite number of stages in order to reach any of the above results. I.e. we will never make sure after any finite number of stages that PA is consistent, if consistent it is, nor, if it is not, will we ever manage to rectify it into some consistent Arithmetic, such that we know about it that it is consistent. So, our old acquaintance, researcher R, who is now pictured as asking to know whether PA is consistent or not, or, in case it is not, wishes to know the shape of its rectified alternative, will never live to satisfy her curiosity. For, if PA is consistent, R will never live to know it, and if it is not, although R will eventually find that out, she will never witness the ultimate shape of PA's rectified alternative[29]. Hardly the thing Hilbert was interested in. (Besides, it must be added that, even if R had an infinite amount of moments at her disposal, and did eventually come to contemplate this formal proof, the reason why she would have believed in what the proof says would not have been a formal one. It would -again- be some independent piece of non formal evidence.)

(iii) Takeuti's system, by adopting cut-elimination, does manage to formally prove that it does not prove $0 = 1$. However, the problem raised in §4 is still pertinent. Do we believe in this proof, and, if yes, why? I.e. if we do, is it because this particular formal system says so, or for some other non formal reason? The same answer as within the parentheses in (ii) above must be provided here as well. Moreover, -my conjecture is- the remote explanatory reason that this system manages to formally prove its own consistency is that it can formally prove about itself that it is a cut-free system. (I have no space to elaborate this conjecture here.) Another cut-free system that can formally prove a good amount of Σ Reflection is to be found in [27]. (The same, *mutatis mutandis*, questions have to be raised with respect to this system as well.)

[29]The reason being that the test will be going on *ad infinitum* for all rectified alternatives of PA, and when some rectified alterative will be reached, such that it is not contradictory, R will never make sure that this is the one.

(iv) Willard's self-justifying Arithmetics manage to prove their relative consistency. They are systems such that they can formally prove that no proof of them contradicts some already given proof of them, if the code number of the latter is of some "reasonable size". This number is no constant but a function of the number of the proved formula. It is evident that Willard's self-justifying Arithmetics constitute a first class case study for the theoretical thought experiment this paper tries to put forward. They are systems of sufficiently (in an appropriate sense of sufficiently) strong Arithmetics[30] that not only are they free from the limitations of Gödel's Second Incompleteness Theorem, but also manage to prove their consistency in the above mentioned way. So, let us raise the central question addressed throughout this paper: What do these formal proofs imply as for the semantical question? Do these proofs prove true statements? Willard's systems are consistent if and only if Peano Arithmetic is (see Theorems 3.4., 4.3. in [46]). Despite the technical *tour de force* these systems certainly are, with respect to the substantial question we are left exactly where we were. It all depends on the consistency of Peano Arithmetic itself. If we assume that PA is consistent, we are happy to see that there are consistent systems that can formally prove that they are consistent. If we do not, we are, once more, exactly where we were at the beginning as for the epistemic parameter of the problem. They both could turn out to be inconsistent. (I don't believe this for a moment, but, and in line with what the main argument of my paper has been trying to show, the formalism *per se* can provide me with no extra argument.)[31]

Appendix

Throughout the paper, I have systematically refrained from elaborating more on the -so to say- "polemical" side of my arguments. This was done for the following reasons. First, because exegetical charity obliges one to exhaust all means, before proclaiming that any author "is wrong" or "misleading". Now, the everyday language, which is precisely the language Metamathematical essays are written in, disposes

[30] Willard has argued (private communication) that his Arithmetics are, in some respects, even stronger than PA, while in others are not.

[31] Many thanks to Michael Detlefsen for having read a draft of this paper and clarifying things to me on his work on the Second Theorem, to Dimitris Portides for his remarks and comments and, especially, to Dan Willard, who has read several versions of it, and kindly helped me to stay away from several pitfalls. Special thanks to the participants of JAF31 and to two anonymous referees for some valuable comments.

of no notational distinction between "to prove" in the factive and "to prove" in the formal sense. This has the following consequence. Quotations from the authors "under attack" would need to be exceedingly lengthy for the reader to make sure that, when the author proclaims that, e.g., "there is an interest in finding a Theory that proves its own consistency", 'to prove' is meant in a foundational rather than a formal sense. Hence, the need for very long quotations, word by word commentaries and careful reconstructions of overall arguments. This is hardly feasible within a single paper. I have tried to stay focused on the heuristic side of the argument, instead[32].

Nonetheless, I will still present an instance of the misfortunes that a not-so-clear distinction between the factive and the formal sense of "to prove" can lead one to.

It is often claimed that systems that can (formally) prove their own consistency would be foundationally useful in so far as they could provide us with valuable insights on human reasoning[33]. In general, these discussions are to be found in a literature attempting to connect Gödel's Second Theorem to the Mind and Machines problem[34]. The argument runs (roughly) as follows: Conventional formal Arithmetics are incapable of coming forward with any explanation as for how we came to

[32]Let it be noted however that several thinkers have expressed similar, if not identical, worries. Raymond Smullyan, in [42]: 84, last paragraph, assimilates the claim that no proof of PA's consistency can ever be found, if PA is consistent, with the claim that any proof of PA's consistency by PA itself would mean that PA is consistent. He claims that (i) both these things are equally "foolish", that (ii) the foolishness of the former rests upon the foolishness of the latter, and that (iii) such misunderstandings do "unfortunately" occur. Georg Kreisel, in his usual sparkly tone, says in [28]: "As Gödel himself stressed, back in 1931 on p. 197 of (4), his second theorem is irrelevant to any sensible consistency problem. In any case, if $ConF$ is in doubt, why should it be proved in F (and not in an incomparable system)? [...] He [Gödel] knew only too well the publicity value of this catch word ["consistency"], which 'contrary to his own view of the matter' had made his second incompleteness theorem more spectacular than the first'." Jeremy Avigad, [1], quotes Gödel and understands the quotation as claiming that "[T]his, I take it, explains his disdain for mathematicians and philosophers who expect too much from syntactic methods. They are the ones who expect 'to derive something from nothing' while avoiding 'the appearance of philosophy'" (the 'ones' referred to by Gödel is "Hilbert's school").

[33]E.g. [46], 542: "(*) How do Human Beings manage to muster the physical energy and psychological desire to think (and prove theorems) when the many generalizations of Gödel's Incompleteness Theorem assert that no reasonable conventional axiom system can confidently assume its own consistency? The partial virtues of $IS^{\lambda}(A)$ and $ISREF(A)$ axiom systems is that their Tangibility Reflection Principles will offer at least a partial answer to the Paradoxical Question (*)" [$IS^{\lambda}(A)$ and $ISREF(A)$ are systems that prove their own consistency.] For the general frame of this line of approach to the Theorem see esp. [30] and [34].

[34]Visser, [45]:161, has identified this line of approach with one of the main motivations for building consistency within a system.

the conclusion that the rules of thought are, in principle, consistent. For, no matter which is the mechanism human minds use in order to think, they certainly entertain the belief that, in principle, they are consistent in their thinking. Now, if the way we, 'reasoning human beings', arrive at withholding beliefs is by arguments, conventional formal Arithmetics are no possible models for our minds. For conventional Arithmetics suffer from the limitations of Gödel's Second Theorem, and can come with no "argument" to the effect that they are consistent, if consistent they are. On the other hand, less canonical formal Arithmetics that do prove their own consistency are, at least with respect to this parameter, no alien to what we do when we think.

Assume, *per impossibile*, that formal systems could, in principle, help us in finding explanations as for the reasons why we believe we are consistent. Obviously, what is hoped for here is that the way some formal system proves its own consistency could throw some light on the way we arrive at the conclusion that we are consistent. So, assume the presence of a Machine that implements the algorithms (*via* programs) that correspond to the recursive functions constituting the system under consideration. Now, imagine that one day the Machine prints out the formula corresponding to the "I am consistent" statement. If the system can unproblematically prove its own consistency, then there is no problem with the Machine printing this formula either. But the question is: And what this can possibly have in common with our belief in the consistency of our thinking? The only way to connect these two is to imagine that we are a Machine that prints out some similar sentence, but this is hardly enough. For this, up to now, has nothing to do with any *factive* proof. When, in this setting, we claim that we are consistent, we do not claim that a specific output came out of our minds, but that a specific output came out of our minds and that this specific output describes, when interpreted, something that is true, and that, on the occasion, this thing is that we are consistent; hence, our ability "to muster the physical energy and psychological desire to think" etc. The question, therefore, is: On which basis do we become convinced that what this output says, when interpreted, is true?, which is another way of asking: Why do we believe in the interpreted formula? If our analysis in the paper is exact, the evidence has to be provided from some other, exterior, source, i.e. a source different from the implemented algorithm. Which also means both (i) that the fact that some Arithmetic proves its own consistency is no sufficient evidence for what this proof "proves", and (ii) that the fact that we are convinced that we are consistent cannot be paralleled with the way we mean that the above Arithmetic "proves its own consistency". In the first, it is the formal sense of "to prove", while in the latter, it is the factive[35].

References

[1] J. Avigad, Gödel and the Metamathematical Tradition, in Kurt Gödel, Solomon Feferman, Charles Parsons & Stephen G. Simpson (eds.), *Kurt Gödel: Essays for His Centennial. Association for Symbolic Logic*, 2010, pp. 45–60.

[2] J. Azzouni, The Derivation-Indicator View in Mathematical Practice, *Philosophia Mathematica*, vol. 3, 2004, pp. 81–105.

[3] E. J. Brouwer, [1927], Intuitionistic Reflections on Formalism, in [44].

[4] C. Cellucci, Review of G. Kreisel G. Takeuti [1974], *The Journal of Symbolic Logic*, vol. 50 (1), 1985, pp. 244–46.

[5] M. Davis, *The Undecidable. Basic Papers on Undecidable Propositions, Unsolvable problems and Computable Functions*, Hewlett, NY: Raven Press, 1965. Reed. Dover, 1993.

[6] M. Detlefsen, On Interpreting Gödel's Second Theorem, *Journal of Philosophical Logic*, vol. 8, 1979, pp. 297–313, 1979; also in [38] with a Postscript.

[7] M. Detlefsen, *Hilbert's Program*, Dordrecht: Reidel, 1986.

[8] M. Detlefsen, What does Gödel's Second Incompleteness Theorem Say?, *Philosophia Mathematica*, vol. 9, 2001, pp. 37–71.

[9] H. B. Enderton, *A Mathematical Introduction to Logic*, NY: Academic Press, 1972.

[10] S. Feferman, Arithmetization of metamathematics in a general setting, *Fundamenta Mathematicae*, vol. 49, 1960, pp. 35–92.

[11] S. Feferman, Hilbert's Program Relativized: Proof-Theoretical and Foundational Reductions, *Journal of Symbolic Logic*, vol. 53 (2),

[35] In Willard's terms ([46]: 542, 573), our point excludes that such systems can constitute even "partial answers" to Willard's question (*). By "partial answer", Willard has in mind "answers to logical paradoxes" in general, as he thinks that "there can obviously never exist any completely satisfactory answer to a logical paradox". This obscures further the matter, for it raises the questions of (i) what is "a partial solution to a logical paradox", of (ii) why one should not try making logical paradoxes go away by altering the system, in which they occur (logical paradoxes appear within systems), and (iii) why (*) is a paradox in the first place, if "to prove" is meant in the formal sense with respect to formal systems that "prove their own consistency", and in the factive, with respect to human beings that "prove to themselves" that same thing?

1988, pp. 364–84.

[12] S. Feferman, Reflecting on Incompleteness, *Journal of Symbolic Logic*, vol. 56, 1991, pp. 1–49.

[13] S. Feferman, Are There Absolutely Unsolvable Problems? Gödel's Dichotomy, *Philosophia Mathematica*, vol. 14, 2006, pp. 134–152.

[14] Friedman, Harvey and Michael Sheard, An Axiomatic Approach to Self-Referential Truth, *Annals of Pure and Applied Logic*, vol. 33, 1987, pp. 1–21.

[15] K. Gödel, On Formally Undecidable Propositions of Principia Mathematica and Related Systems, 1931, in [44].

[16] K. Gödel, On Completeness and Consistency, 1931, in [44].

[17] K. Gödel, On Undecidable Propositions of Formal Mathematical Systems, in [5].

[18] K. Gödel, *Collected Works*, vol. 1, Oxford University Press, 1986.

[19] K. Gödel, *Collected Works*, vol. 2, Oxford University Press, 1990.

[20] D. Hilbert, On the Foundations of Logic and Arithmetic, 1904, in [44].

[21] D. Hilbert, On the Infinite, 1925, as in [44].

[22] D. Hilbert, The Foundations of Mathematics, 1927, in [44].

[23] D. Hilbert and P. Bernays, *Grundlagen der Mathematik I*, Berlin: Springer, 1934.

[24] S. Kleene, *Introduction to Metamathematics*, North-Holland, 1952.

[25] S. Kleene, *Mathematical Logic*, NY: John Wiley, 1967.

[26] G. Kreisel, Mathematical Logic, in *Lectures on modern mathematics*, edited by Th. L. Saaty, New York, Wiley, vol. 3, 1965, pp. 95–195.

[27] G. Kreisel and. Takeuti, Formally self-referential propositions for cut free analysis and related systems, *Dissertationes Mathematicae*, 118, 1974.

[28] G. Kreisel, Kurt Gödel. 28 April 1906–14 January 1978, *Biographical Memoirs of the Fellows of the Royal Society*, vol. 26, 1980, pp. 148–224.

[29] G. Kreisel, Hilbert's Program, in *Philosophy of Mathematics*, edited by P. Benacerraf and H. Putnam, Cambridge University Press, 1983.

[30] J. R. Lucas, Minds, Machines and Gödel, *Philosophy*, vol. 36, 1961, pp. 112–27.

[31] E. Mendelson, *Introduction to Mathematical Logic*, CRC Press, 2010.

[32] E. Nagel and J. Newman, *Gödel's proof*, NY University Press, 1958.

[33] A. Pelc, Why Do We Believe Theorems?, *Philosophia Mathematica* (III), vol. 15, 2009, pp. 291–320.

[34] R. Penrose, *The Emperor's New Mind*, Oxford University Press, 1989.

[35] R. Penrose, *Shadows of the Mind*, Oxford University Press, 1994.

[36] Y. Rav, A Critique of a Formalist-Mechanist Version of the Justification of Arguments in Mathematicians' Proof Practices, *Philosophia Mathematica* (III), vol. 17, 2007, pp. 84–94.

[37] M. D. Resnik, On the Philosophical Significance of Consistency Proofs, *Journal of Philosophical Logic*, vol.3 (1-2), 1974, pp. 133–47; reproduced in [38].

[38] S. G. Shanker, *Gödel's Theorem in Focus*, Routledge, 1988.

[39] P. Smith, *An Introduction to Gödel's Theorems*, Cambridge University Press, 2007.

[40] C. Smoryński, The Incompleteness Theorems, in *Handbook of Philosophical Logic*, edited by J. Barwise, Amsterdam: North Holland, 1977, pp. 821–65.

[41] R. Smullyan, *Gödel's Incompleteness Theorems*, Oxford University Press, 1992.

[42] R. Smullyan, Gödel's Incompleteness Theorems, in *The Blackwell Guide to Philosophical Logic*, edited by Lou Goble, Oxford, Blackwell, 2001, pp. 72–89.

[43] G. Takeuti, On the Fundamental Conjecture of GLC, *Journal of Mathematical Society of Japan*, vol. 7, 1955, pp. 394–408.

[44] J. van Heijenoort, *From Frege to Gödel. A Source Book in Mathematical Logic 1879-1931*, Harvard University Press, 1977.

[45] A. Visser, Peano's Smart Children: A Provability Logical Study of Systems with Built-in Consistency, *Notre Dame Journal of Symbolic Logic*, vol. 30 (2), 1989, pp. 161–96.

[46] D. E. Willard, Self-Verifying Systems, the Incompleteness Theorem and Related Reflection Principles, *Journal of Symbolic Logic*, vol. 66, 2001, pp. 536–96.

[47] D. E. Willard, An Exploration of the Partial Respects in which an Axiom System Recognizing Solely Addition as a Total Function Can Verify Its Own Consistency, *Journal of Symbolic Logic*, vol. 70, 2005, pp. 1171–209.

[48] D. E. Willard, Passive induction and a solution to a Paris-Wilkie question, *Annals of Pure and Applied Logic*, vol. 146, 2007, pp. 124-49.

[49] D. E. Willard, Some Specially Formulated Axiomizations for $I\Sigma_0$ Manage to Evade the Herbrandized Version of the Second Incompleteness Theorem, *Information and Computation* 207: 1078-93, 2009.

[50] D. E. Willard, On the Results of a 14-Year Effort to Generalize Gödel's Second Incompleteness Theorem and Explore Its Partial Exceptions, *Collegium Logicum*, IX, 2007, pp. 81–86.

9

Adding standardness to nonstandard arithmetic

RICHARD KAYE,[1] ROMAN KOSSAK,[2] TIN LOK WONG[3,4]

Abstract: We discuss structures (M, ω) where M is a nonstandard model of PA and ω is a unary predicate for the standard cut. For a given theory T of a cut in a model of PA, we give a Henkin–Orey style characterization of theories of models (M, ω), where $(M, \omega) \models T$. In the rest of the paper, we study sets of reals that are representable in theories of models (M, ω) and sets of reals which are definable in (M, ω). In particular, we give a generalization of the easier direction of the theorem of Kanovei characterizing countable Scott sets that are standard systems of models (M, ω).

Keywords: standard cut, Peano arithmetic, nonstandard models, ω-logic, standard systems

1 Introduction

The content of this paper concerns nonstandard models of arithmetic M when an additional unary relation named N is added to represent the standard cut ω. Throughout, the larger model M will be called the *ground model* and in this paper will always be a model of first-order Peano Arithmetic, PA.

[1]University of Birmingham, R.W.Kaye@bham.ac.uk
[2]City University of New York, rkossak@nylogic.org
[3]Ghent University, wtl@cage.ugent.be
[4]Financial support from the John Templeton Foundation under the *Philosophical Frontiers in Reverse Mathematics* project is gratefully acknowledged.

Studies in Weak Arithmetics.
Patrick Cégielski, Charalampos Cornaros,
Costas Dimitracopoulos.

Such structures (M, ω) arise naturally in the attempt to learn more about the arithmetic of $\mathbb{N}$ (i.e. the *standard model of arithmetic* with domain ω) by nonstandard means: one initially takes an ultrapower M of $\mathbb{N}$ and studies its first-order properties. This M has a copy of $\mathbb{N}$ as an initial segment, and so M is much more than a first-order structure, and has in particular the (non-definable) predicate $\mathsf{N}(x)$ expressing the standardness or finiteness of x. More generally, the same is true when M is an arbitrary nonstandard model of weaker theories than $\mathrm{Th}(\mathbb{N})$. In this paper we consider arguably the most natural of these, PA.

Working with the structure (M, ω), one's attention is naturally drawn to the cut given by the predicate N and the additional structure on it induced from M. With a suitably general idea of 'coding', structure induced by M corresponds to subsets and relations on N coded by elements of M. Thus one important motivation for studying models (M, ω) is that they interpret a number of ω-models of second-order arithmetic, $(\omega, \mathcal{X})$ where the sets $A \in \mathcal{X}$ are coded in M in some sense. An important example is the standard system $\mathcal{X} = \mathrm{SSy}(M)$ of the ground model. Another consists of the sets represented in its theory $\mathrm{Th}(M)$. These constructions generalise to the expanded structure (M, ω) and we can look at the set $\mathcal{X}$ of sets coded or represented in (M, ω). This will be a major topic for discussion in this paper, starting at Section 3.

Parallel to the literature on the theory of structures such as $(\omega, \mathcal{X})$ is a sequence of work starting with Paris and Kirby [13] on cuts: their programme is to consider the ways in which an initial segment such as our N embeds in its end extension M, and in particular (motivated by the idea that N is analogous to a limit ordinal λ in set theory) to consider the combinatorial and Ramsey properties of N in M, such as strength—analogous perhaps to measurability of cardinals. In the particular case of the standard cut ω or $\mathbb{N}$ surprisingly little is known: ω is regular in M and may be, but need not be, strong.

This paper reviews a number of known results on such structures (M, ω) in a way that is as self-contained as possible, and presents some additional results and observations on these structures, as well as some alternative proofs of these known theorems. This is intended to facilitate further study of models of this form, and more generally models of the form (M, I) where I is an initial segment of M. Further work in these directions has already been started by Kaye and Wong and we give some indications of forthcoming results here.

The notation and terminology we use is standard in the literature on models of PA [8, 14]. In particular, the *language of arithmetic* $\mathcal{L}_{\mathrm{A}}$ is the language with non-logical symbols $+, \cdot, <, 0, 1$. The standard model of PA with domain ω is denoted $\mathbb{N}$ and its complete theory denoted TA.

As already mentioned, we shall add a unary predicate N to $\mathcal{L}_{\mathrm{A}}$ representing an initial segment and denote the expanded language by ${\mathcal{L}_{\mathrm{A}}}^{\mathrm{cut}}$. All models of ${\mathcal{L}_{\mathrm{A}}}^{\mathrm{cut}}$ in this paper will implicitly satisfy PA together with

$$\mathsf{N}(1) \wedge \forall x\ (\mathsf{N}(x) \to \mathsf{N}(x+x) \wedge \mathsf{N}(x^2) \wedge \forall y < x\ \mathsf{N}(y)) \wedge \exists x\ \neg\mathsf{N}(x).$$

This base theory will be denoted $\mathrm{PA}^{\mathrm{cut}}$. In the sequel, we shall often use the predicate N in a more natural way, writing $x \in \mathsf{N}$ for $\mathsf{N}(x)$, and using the obvious quantifiers $\forall x{\in}\mathsf{N}$ and $\exists x{\in}\mathsf{N}$. When N is interpreted by the standard cut ω in a nonstandard model $M \models \mathrm{PA}$, we obtain the expansion $(M, \omega) \models \mathrm{PA}^{\mathrm{cut}}$.

2 The theory

This section contains a short investigation into the theory of models of the form (M, ω). More precisely, we attempt to determine the relevant extensions of $\mathrm{PA}^{\mathrm{cut}}$.

1. Given $M \models \mathrm{PA}$ what is $\mathrm{Th}(M, \omega)$?
2. What is $\mathrm{Th}(\mathrm{PA}, \omega) = \bigcap\{\mathrm{Th}(M, \omega) : M \models \mathrm{PA}\}$?
3. What is $\mathrm{Th}(\mathrm{TA}, \omega) = \bigcap\{\mathrm{Th}(M, \omega) : M \succ \mathbb{N}\}$?

Note that all such theories include the full induction scheme for all ${\mathcal{L}_{\mathrm{A}}}^{\mathrm{cut}}$ formulas over the predicate N, as well as (of course) full induction for $\mathcal{L}_{\mathrm{A}}$ over the ground model.

The Completeness Theorem for ω-logic gives plenty of useful information here. This is of course a more general result than is needed for our context. We outline the main results here with sketch proofs for the set-up we are considering. For detailed expositions in more general settings, see Grzegorczyk–Mostowski–Ryll-Nardzewski [3], Keisler [10], Leblanc–Roeper–Thau–Weaver [16] and Isaacson [6].

A *theory* is to be considered as just a set of axioms, i.e. not necessarily complete nor even necessarily deductively closed. As is common, we carelessly use the + operation to mean $\cup$ and identify a single axiom σ with the theory $\{\sigma\}$. We say two theories are equal when their deductive closures are equal. (So, for us, two equivalent axiomatisations of the same theory are equal.) Our logic is first-order throughout.

Given an ${\mathcal{L}_{\mathrm{A}}}^{\mathrm{cut}}$ theory T we define

$$\Omega_1(T) = T + \{\forall x{\in}\mathsf{N}\ \theta(x) : \theta \in {\mathcal{L}_{\mathrm{A}}}^{\mathrm{cut}},\ T \vdash \theta(\underline{k}) \text{ for all } k \in \omega\},$$

where $\underline{k}$ is the canonical $\mathcal{L}_{\mathrm{A}}$ notation for the natural number k. Then by recursion on ordinals we set $\Omega_0(T) = T$, $\Omega_{\alpha+1}(T) = \Omega_1(\Omega_\alpha(T))$ and

$\Omega_\lambda(T) = \bigcup\{\Omega_\alpha(T) : \alpha < \lambda\}$ for limit λ. We write $\Omega(T)$ for

$$\bigcup_{\alpha \in \mathrm{On}} \Omega_\alpha(T + \mathrm{PA}^{\mathrm{cut}}),$$

noting that a full union over all ordinals is of course not needed: a union over countable ordinals or actually even less suffices [1].

By an easy induction on ordinals, the equation $\Omega_\alpha(\Omega_\beta(T)) = \Omega_{\beta+\alpha}(T)$ can be verified.

The Ω_1 operation encodes the familiar ω-rule, and the following is obvious.

Proposition 2.1. *If $M \models \mathrm{PA}$ is nonstandard and $(M, \omega) \models T$ for some ${\mathcal{L}_\mathrm{A}}^{\mathrm{cut}}$ theory T, then $(M, \omega) \models \Omega(T)$.*

The converse to this uses the following lemma.

Lemma 2.2. *Given an ${\mathcal{L}_\mathrm{A}}^{\mathrm{cut}}$ theory T and a single ${\mathcal{L}_\mathrm{A}}^{\mathrm{cut}}$ sentence σ,*

$$\Omega_\alpha(T + \sigma) = \Omega_\alpha(T) + \sigma$$

for all $\alpha \in \mathrm{On}$.

Proof. The $\supseteq$ direction is clear, and the other direction is proved by a simple induction on ordinals. ⊣

Note that the lemma fails for a set of sentences Σ replacing our σ. This is associated with the failure of compactness for ω-logic—see Proposition 2.4.

Theorem 2.3. (Henkin [5], Orey [17]) *For an ${\mathcal{L}_\mathrm{A}}^{\mathrm{cut}}$ theory T,*

$$\Omega(T) = \bigcap\{\mathrm{Th}(M, \omega) : (M, \omega) \models T\}.$$

Proof. One direction has already been noted. For the converse, suppose that $\Omega(T) \nvdash \sigma$. We shall show that σ is false in some $(M, \omega) \models T$.

¿From the assumption we have that $\Omega(T) + \neg\sigma$ is consistent, and hence by the lemma $\Omega(T + \neg\sigma)$ is consistent. We now find a complete theory $S = \bigcup_{i \in \omega} T_i$ (complete in first-order logic) extending $T_0 = T + \neg\sigma$ such that $\Omega(T_i)$ is consistent for each i and such that if $S \vdash \exists x{\in}\mathsf{N}\ \theta(x)$ then $S \vdash \theta(\underline{k})$ for some $k \in \omega$. Given a finite extension T_i of T_0 such that $\Omega(T_i)$ is consistent, and considering $\theta(x)$ at this stage, if at all possible we let $T_{i+1} = T_i + \theta(\underline{k})$ for some k. This is not possible when all such $\Omega(T_i + \theta(\underline{k}))$ are inconsistent, and by the lemma this holds just in case $\Omega(T_i) \vdash \neg\theta(\underline{k})$ for all k. In such a case there is then $\alpha \in \mathrm{On}$ such that $\Omega_\alpha(T_i) \vdash \neg\theta(\underline{k})$ for all k, hence $\Omega_{\alpha+1}(T_i) \vdash \forall x{\in}\mathsf{N}\ \neg\theta(x)$. In this case let $T_{i+1} = T_i + \forall x{\in}\mathsf{N}\ \neg\theta(x)$. Note too that if $\forall x{\in}\mathsf{N}\ \neg\theta(x)$ is added to the T_is then $S\nvdash\exists x{\in}\mathsf{N}\ \theta(x)$ as by construction it is consistent in first-order logic.

By applying the omitting types theorem to the complete theory S just constructed and the type $p_\omega(x) = x \in \mathsf{N} + \{x \neq \underline{k} : k \in \omega\}$ we obtain a model $(M, N) \models S$ omitting $p_\omega(x)$ and hence $N = \omega$. But then $(M, \omega) \models T + \neg\sigma$, which is as required. ⊣

Compactness does not hold for ω-logic. It is not hard to give examples of theories in $\mathcal{L}_{\mathrm{A}}{}^{\mathrm{cut}}$ with an additional constant symbol. A more elaborate example in pure $\mathcal{L}_{\mathrm{A}}{}^{\mathrm{cut}}$ is given in the next proposition. The third author would like to thank Yue Yang for some helpful discussions related to the proof.

Proposition 2.4. *Let σ_0 be an $\mathcal{L}_{\mathrm{A}}{}^{\mathrm{cut}}$ sentence saying*

> *'there is a subset of N coded in the ground model that is Σ_1- but not Π_1-definable in N',*

and σ_{k+1} be an $\mathcal{L}_{\mathrm{A}}{}^{\mathrm{cut}}$ sentence saying

> *'if the kth Σ_1-definable subset of N is not Π_1-definable, then it is not coded in the ground model',*

for each $k \in \omega$. Then each of the theories $T_k = \mathrm{PA}^{\mathrm{cut}} + \sigma_0 + \cdots + \sigma_k$ has an ω-model (M_k, ω) but $T_\omega = \mathrm{PA}^{\mathrm{cut}} + \{\sigma_k : k \in \omega\}$ does not.

Proof. It is easy to see that T_ω has no ω-model. So we only need to show for any particular $k \in \omega$, the theory T_k has an ω-model. We will prove this using the omitting types theorem directly. It can also be proved using Fact 3.4 and Theorem VIII.2.24 in Simpson [20].

Fix $k \in \omega$. Let $S_0, S_1, \ldots, S_{k-1}$ be the first k recursively enumerable sets. Set $S = \{(i, n) : i < k \text{ and } n \in S_i\}$. Assuming k is large enough, we may suppose S is not recursive. Pick a recursively enumerable nonrecursive set X such that S is not recursive in X. (This is possible by a theorem of Muchnik from 1956. See Exercise 10-11 in Rogers [18].) Our aim is to produce a nonstandard model $M \models \mathrm{PA}$ in which X is coded but S is not.

Define PA_X^+ to be the $\mathcal{L}_{\mathrm{A}}(c)$ theory

$$\mathrm{PA} + \{\underline{n} \in c : n \in X\} + \{\underline{n} \notin c : n \in \omega \setminus X\}.$$

This is consistent and recursive in X. Consider the type

$$p(v) = \{\underline{n} \in v : n \in S\} \cup \{\underline{n} \notin v : n \in \omega \setminus S\},$$

which is clearly finitely consistent with PA_X^+. If it has no support over PA_X^+, then we are done by the omitting types theorem. Suppose $p(v)$ has a support in $\mathcal{L}_{\mathrm{A}}(c)$, say $\theta(c, v)$, over PA_X^+. Then $\underline{n} \in S$ is equivalent in $\mathbb{N}$ to the $\Sigma_1(X)$-formula 'there is a proof of $\forall v\ (\theta(c,v) \to \underline{n} \in v)$ from PA^+_X'

as well as the $\Pi_1(X)$-formula

'there is no proof of $\forall v\ (\theta(c,v) \to \underline{n} \notin v)$ from PA^+_X'.

Hence S is $\Delta_1(X)$. This contradicts the choice of X. ⊣

We feel that cuts N of some $K \models \mathrm{PA}$ are of particular interest when they satisfy the first order properties of ω, in the language $\mathrm{PA}^{\mathrm{cut}}$.

Definition 2.5. *If (K, N) is an ${\mathcal{L}_{\mathrm{A}}}^{\mathrm{cut}}$ structure that is elementarily equivalent to some $(M, \omega) \models \mathrm{PA}^{\mathrm{cut}}$, then we shall say that the cut N of K is* pseudo-standard.

Corollary 2.6. *There is no ${\mathcal{L}_{\mathrm{A}}}^{\mathrm{cut}}$ theory T such that, for all ${\mathcal{L}_{\mathrm{A}}}^{\mathrm{cut}}$ structures $(M, N) \models \mathrm{PA}^{\mathrm{cut}}$, we have $(M, N) \models T$ if and only if N is a pseudo-standard cut of M.*

Proof. If T were such a theory, then it must contradict T_ω in Proposition 2.4, and so by the Compactness Theorem, it contradicts T_k for some $k \in \omega$. This violates our choice of T. ⊣

The Henkin–Orey theorem answers two of our questions.

Corollary 2.7. (a) $\mathrm{Th}(\mathrm{PA}, \omega) = \bigcap\{\mathrm{Th}(M, \omega) : M \models \mathrm{PA}\} = \Omega(\mathrm{PA}^{\mathrm{cut}})$.

(b) $\mathrm{Th}(\mathrm{TA}, \omega) = \bigcap\{\mathrm{Th}(M, \omega) : M \succ \mathbb{N}\} = \Omega(\mathrm{TA} + \mathrm{PA}^{\mathrm{cut}})$.

In contrast, we do not get a complete answer to the remaining question.

Corollary 2.8. *For a nonstandard $M \models \mathrm{PA}$,*

$$\mathrm{Th}(M, \omega) \supset \bigcap\{\mathrm{Th}(K, \omega) : M \equiv K\} = \Omega(\mathrm{PA}^{\mathrm{cut}} + \mathrm{Th}(M)).$$

Proof. This proof uses the notion of *strength* of ω, something that will be discussed in more detail in the next section—see in particular Fact 3.6. For now it suffices that for any $\mathcal{L}_{\mathrm{A}}$ completion T of PA there are models K_1, K_2 of T such that ω is strong in K_1 but not in K_2 and that the strength of ω is a first-order property of (K_i, ω). This shows that $\Omega(\mathrm{PA}^{\mathrm{cut}} + T)$ is not complete. ⊣

With reference to Corollary 2.7(a), it is perhaps worth pointing out that there is still something mysterious about $\Omega(\mathrm{PA}^{\mathrm{cut}})$, in that its most visible consequence is the scheme of full induction on ${\mathcal{L}_{\mathrm{A}}}^{\mathrm{cut}}$ formulas restricted to N, but this scheme is recursive so cannot axiomatise $\Omega(\mathrm{PA}^{\mathrm{cut}})$ which decides all ${\mathcal{L}_{\mathrm{A}}}^{\mathrm{cut}}$ sentences where quantifiers are restricted to N.

The theories $\Omega(\mathrm{PA}^{\mathrm{cut}}+T)$ and $\Omega_1(\mathrm{PA}^{\mathrm{cut}}+T)$ where T is an $\mathcal{L}_\mathrm{A}$ completion of PA are also of interest in their own rights. A sample (easy) proposition is as follows. The notation used here is standard: for a syntactic object X, X^K represents its realisation or interpretation in a structure K.

Proposition 2.9. *Suppose $(K, N) \models \Omega_1(\mathrm{PA}^{\mathrm{cut}} + \mathrm{Th}(M))$ for some $M \models \mathrm{PA}$. Then for each closed Skolem term t of $\mathcal{L}_\mathrm{A}$ with $t^K \in N$ we have $t^K \in \omega$.*

Proof. If t is such a term and $t^K \in N$ then $\mathrm{Th}(M) \vdash t = \underline{k}$ for some $k \in \omega$. For if not, as $\mathrm{Th}(M)$ is complete $\mathrm{Th}(M) \vdash \neg t = \underline{k}$ for all k hence $\Omega_1(\mathrm{PA}^{\mathrm{cut}} + \mathrm{Th}(M)) \vdash \forall x{\in}\mathsf{N}\ \neg t = x$. $\dashv$

In particular, each nonstandard $\mathcal{L}_\mathrm{A}$ definable element of a model $M \models \mathrm{PA}$ must be greater than all cuts $N \subseteq_\mathrm{e} M$ such that $(M, N) \models \Omega_1(\mathrm{PA}^{\mathrm{cut}}+\mathrm{Th}(M))$. It follows that there are recursively saturated models of PA in which the standard cut is the only pseudo-standard cut.

We end this section with a simple characterisation of pseudo-standard cuts in terms of definable sets and definable elements. Notice the 'parameter-free' condition in (a) below is clearly necessary, but that in (b) is not.

Proposition 2.10. *Let $(K, N) \models \mathrm{PA}^{\mathrm{cut}}$. Then N is pseudo-standard in K if and only if both of the following hold.*

(a) *If $a \in N$ is a parameter-free ${\mathcal{L}_\mathrm{A}}^{\mathrm{cut}}$ definable element of (K, N), then $a \in \omega$.*

(b) *If $I \subseteq_\mathrm{e} N$ that is parameter-free ${\mathcal{L}_\mathrm{A}}^{\mathrm{cut}}$ definable in (K, N), then $I = N$.*

Proof. For the 'only if' direction, pick $M \models \mathrm{PA}$ such that $(M, \omega) \equiv (K, N)$. The proof of (a) is similar to that of Proposition 2.9. For (b), simply note that all $I \subseteq_\mathrm{e} \omega$ definable in (M, ω) must be ω itself. This transfers to (K, N) by elementary equivalence.

For the 'if' direction, suppose (a) and (b) hold. We show that $\Omega(\mathrm{Th}(K, N))$ is consistent, which suffices by the omitting types theorem. Suppose $\phi(x)$ is an ${\mathcal{L}_\mathrm{A}}^{\mathrm{cut}}$ formula such that $\phi(\underline{k}) \in \mathrm{Th}(K, N)$ for each $k \in \omega$. Replacing $\phi(x)$ with $\forall x'{\leq}x\ \phi(x')$ if necessary, we may assume $\phi^{(K,N)}$ is closed downwards. For convenience, we denote this downward-closed set by X. If $X \supseteq N$, then we are done. So suppose $X \subset N$. If X has a biggest element, then $\max X \in \omega$ by (a), which is a contradiction. If X has no biggest element, then $X = N$ by (b), which is also a contradiction. $\dashv$

3 Coded reals

A *real* is a set of natural numbers $A \subseteq \omega$. Reals are *coded* in nonstandard models of arithmetic, and the set of reals that are coded forms an important isomorphism invariant of the model under study. There are two related notions of 'coded' that are studied.

Definition 3.1. *The* standard system, $\mathrm{SSy}(M)$, *of a nonstandard model* $M \models \mathrm{PA}$ *is the set of reals* $A \subseteq \omega$ *such that there is* $a \in M$ *and a formula* $\theta(x, y)$ *in two free variables such that*

$$A = \{n \in \omega : M \models \theta(n, a)\}.$$

One can show, using Gödel's β-function perhaps, that the formula θ here can be taken to be the same formula for all reals, i.e. that for a careful choice of θ the only quantification needed in the definition above is over the parameter $a \in M$. Thus it is common to say 'a codes A'.

The parameter-free version of the standard system is as follows.

Definition 3.2. *The* external Scott set, $\mathrm{Rep}(M)$, *of a nonstandard model* $M \models \mathrm{PA}$ *is the set of reals* $A \subseteq \omega$ *such that there is a formula* $\theta(x)$ *in one free variable such that*

$$A = \{n \in \omega : M \models \theta(n)\}.$$

Obviously $\mathrm{Rep}(M) \subseteq \mathrm{SSy}(M)$. Unlike $\mathrm{SSy}(M)$, the set $\mathrm{Rep}(M)$ only depends on the first-order theory of M so it may or may not be the case that these sets of reals are the same. Models M for which $\mathrm{Rep}(M) = \mathrm{SSy}(M)$ are called *thrifty* in the literature. Clearly, if P_T is the prime model of some (false) completion T of PA, then P_T is thrifty. There are many other thrifty models of PA. For instance, Hájek [4] showed that for every nonstandard $M \models \mathrm{PA}$ and every $n \in \omega$, there are thrifty models $I, K \models \mathrm{PA}$ such that $I \subseteq_{\mathrm{e}} M \subseteq_{\mathrm{e}} K$ and $\mathrm{Th}_{\Sigma_n}(I) = \mathrm{Th}_{\Sigma_n}(M) = \mathrm{Th}_{\Sigma_n}(K)$.

We now repeat the definitions above for a model of $\mathrm{PA}^{\mathrm{cut}}$. In the following, the cut N will usually be the standard cut ω.

Definition 3.3. *The* standard system, $\mathrm{SSy}(M, N)$, *of a model* (M, N) *of* $\mathrm{PA}^{\mathrm{cut}}$ *is the set of reals* $A \subseteq \omega$ *such that there is* $a \in M$ *and a formula* $\theta(x, y)$ *of* ${\mathcal{L}_{\mathrm{A}}}^{\mathrm{cut}}$ *in two free variables such that*

$$A = \{n \in \omega : (M, N) \models \theta(n, a)\}.$$

The external Scott set, $\mathrm{Rep}(M, N)$, *of* (M, N) *is the set of reals* $A \subseteq \omega$ *such that there is a formula* $\theta(x)$ *of* ${\mathcal{L}_{\mathrm{A}}}^{\mathrm{cut}}$ *in one free variable such that*

$$A = \{n \in \omega : (M, N) \models \theta(n)\}.$$

Thus for (M, N) a model of $\mathrm{PA}^{\mathrm{cut}}$ we have that $\mathrm{Rep}(M)$ is contained in both $\mathrm{Rep}(M, N)$ and $\mathrm{SSy}(M)$, and that these two are contained in

SSy(M, N). Examples can be given (using the tools outlined in this section and in Section 4) for which these are all different. On the other hand if the pair (M, N) is recursively saturated then SSy(M, N) = SSy(M).

Note that SSy(M) can be defined using a single formula θ (such as one based on Gödel's β-function) but the same is not obviously true of SSy(M, N), and in fact this is not true in general.[5]

Scott [19] proved that a countable set of reals is realised as SSy(M) or Rep(M) for some $M \models$ PA if and only if it is what is now called a *Scott set*, i.e., a boolean algebra closed under relative recursion and Weak König's Lemma. In Friedman's terminology [20], this is stated as follows.

Fact 3.4 (Scott). (a) *Both* $(\omega, \mathrm{SSy}(M))$ *and* $(\omega, \mathrm{Rep}(M))$ *satisfy* WKL_0 *when* $M \models$ PA *is nonstandard.*

(b) *Let* $(\omega, \mathcal{X}) \models \mathrm{WKL}_0$ *be countable. Then for all recursive consistent extensions* $T \supseteq$ PA, *there exist nonstandard models* $M \models T$ *such that* $\mathrm{SSy}(M) = \mathrm{Rep}(M) = \mathcal{X}$.

The proof of part (a) is somewhat different for Rep and for SSy: the argument that Rep(M) is a Scott set (Fact 3.4) is by writing down appropriate formulas, mimicking the proof of König's Lemma, and the proof of (b) is by a clever construction using infinitely many diagonalisations. The same argument as in (a) applies, showing Rep(M, ω) and SSy(M, ω) are also Scott sets. The analogue of (b) for Rep(M, ω) and SSy(M, ω) appears to be much harder: we shall investigate other possible closure conditions on Rep(M, ω) and SSy(M, ω) below. Since Rep(M) only depends on the theory T of M, we shall also write it as Rep(T). One may ask, in general, what the relationship between a (false) completion $T \supseteq$ PA and the second-order structure $(\omega, \mathrm{Rep}(T))$ is. Scott's theorem tells us that no nontrivial recursive proper extension Γ of WKL_0 can be captured by a recursive extension S of PA, in the sense that

$$(\omega, \mathrm{Rep}(T)) \models \Gamma \Leftrightarrow T \supseteq S$$

for all false completions T of PA. The following simple proposition further suggests that the answer to our question is probably not simple.

Proposition 3.5. *Let* Γ *be an extension of* WKL_0 *in second-order*

[5] In forthcoming work, Kaye shows that if there a single formula $\theta(n, x)$ of $\mathcal{L}_{\mathrm{A}}{}^{\mathrm{cut}}$ such that each $A \in \mathrm{SSy}(M, \omega)$ is defined by $\theta(n, a)$ for some $a \in M$, then $(\omega, \mathrm{SSy}(M, \omega))$ satisfies the full comprehension scheme of second order arithmetic, and that this does not happen in general.

arithmetic such that

$$\{\mathcal{X} \subseteq \mathscr{P}(\omega) : \mathcal{X} \text{ is countable and } (\omega, \mathcal{X}) \models \Gamma\}$$

is nonempty but does not exhaust all countable Scott sets.

(a) *For all consistent recursive* $S \supseteq \mathrm{PA}$, *there are false completions* T_1, T_2 *of* S *such that* $(\omega, \mathrm{Rep}(T_1)) \models \Gamma$ *and* $(\omega, \mathrm{Rep}(T_2)) \not\models \Gamma$.

(b) *No recursive* $S \supseteq \mathrm{PA}$ *satisfies*

$$(\omega, \mathrm{Rep}(T)) \models \Gamma \Leftrightarrow T \supseteq S$$

for all false completions T *of* PA.

Proof. Since (b) follows from (a), we will concentrate on (a). Let S be a consistent recursive extension of PA. Find countable Scott sets $\mathcal{X}_1, \mathcal{X}_2$ such that $(\omega, \mathcal{X}_1) \models \Gamma$ and $(\omega, \mathcal{X}_2) \not\models \Gamma$. By Fact 3.4, there are completions $T_1, T_2 \supseteq S + \neg \mathrm{Con}(S)$ such that $\mathcal{X}_1 = \mathrm{Rep}(T_1)$ and $\mathcal{X}_2 = \mathrm{Rep}(T_2)$. ⊣

A great deal more can be said about the standard system $\mathrm{SSy}(M)$ in the language $\mathcal{L}_{\mathrm{A}}{}^{\mathrm{cut}}$, and indeed the second-order structure for arithmetic $(\omega, \mathrm{SSy}(M))$ can be interpreted in each (M, ω). In particular, Keisler [11, 12] found, for each of the Big Five theories in reverse mathematics, a natural $\mathcal{L}_{\mathrm{A}}{}^{\mathrm{cut}}$ theory that corresponds to it.

Because $\mathrm{SSy}(M)$ can be defined using a single formula θ important closure conditions on $\mathrm{SSy}(M)$ can be expressed in $\mathcal{L}_{\mathrm{A}}{}^{\mathrm{cut}}$ as axioms for (M, ω). Perhaps the most important of these is closure under jump. The following fact takes this much further and is essentially due to Kirby and Paris [13].

Fact 3.6. *For nonstandard* $M \models \mathrm{PA}$ *the following are equivalent:*

(a) $\mathrm{SSy}(M)$ *is closed under jump;*

(b) $(\omega, \mathrm{SSy}(M)) \models \mathrm{ACA}_0$;

(c) ω *is* strong *in* M *i.e. for each* $a \in M$, *there is a nonstandard* b *such that for all* $n \in \omega$, $(a)_n > \omega$ *iff* $(a)_n > b$;

(d) *For all* $\theta(x, y)$ *of* $\mathcal{L}_{\mathrm{A}}$ *there is a canonical choice of* $\phi(x, y)$ *in* $\mathcal{L}_{\mathrm{A}}$ *such that,*

$$(M, \omega) \models \forall a\ \exists b\ \forall n{\in}\mathsf{N}\ (\phi(b, n) \leftrightarrow \exists m{\in}\mathsf{N}\ \theta(a, \langle n, m\rangle)).$$

There are models of arithmetic (indeed, elementarily equivalent ones) in which ω is strong in one but not the other—see, for example, Chapter 13 of Kaye [8]. This was how we proved $\mathrm{Th}(M, \omega) \neq \Omega(\mathrm{PA}^{\mathrm{cut}} + \mathrm{Th}(M))$ for every $M \models \mathrm{PA}$ in Corollary 2.8. Given a (false) completion $T \supseteq \mathrm{PA}$, the most obvious completion of $\Omega(\mathrm{PA}^{\mathrm{cut}} + T)$ is $\mathrm{Th}(P_T, \omega)$, where P_T denotes the prime model of T. Here, ω is strong in P_T if and only if $\mathrm{Rep}(T)$ is closed under jump, and by Fact 3.4, for any countable Scott set $\mathcal{X}$ (including ones closed and not closed under jump) there is a completion $T \supseteq \mathrm{PA} + \neg\,\mathrm{Con}(\mathrm{PA})$ such that $\mathcal{X} = \mathrm{Rep}(T)$.

For $\mathrm{SSy}(M, \omega)$ the situation appears to be more complicated since we know of no analogue to the β-function that gives a canonical way of encoding reals $A \in \mathrm{SSy}(M, \omega)$. Some limited progress can be made by assuming the model M is 'full', i.e. that $\mathrm{SSy}(M, \omega) = \mathrm{SSy}(M)$, and then use the usual Gödel β-function to code reals. Such models do exist, and will be discussed below.

Another useful simplification that can be made is in assuming some saturation properties of M. Countable recursively saturated models M of PA are characterised exactly by their complete theory and their standard system, and the only restriction on pairs $\mathrm{Th}(M)$, $\mathrm{SSy}(M)$ that can occur is that $\mathrm{Th}(M) \in \mathrm{SSy}(M)$ relative to some suitable Gödel numbering [8]. It is convenient (but arguably not always correct) to think of recursively saturated models as being 'more saturated' the larger their standard system. Perhaps the most useful notion of this type is that of *arithmetical saturation* which for PA means that the model M is recursively saturated and $\mathrm{SSy}(M)$ is closed under jump.

It is also clear that for any nonstandard model $M \models \mathrm{PA}$ both $\mathrm{Rep}(M, \omega)$ and $\mathrm{SSy}(M, \omega)$ are closed under jump. Thus models M with $\mathrm{Rep}(M) \neq \mathrm{Rep}(M, \omega)$ and $\mathrm{SSy}(M) \neq \mathrm{SSy}(M, \omega)$ are easy to find. The following beautiful result, due to Kanovei, takes this observation further and settles the question of which (countable) Scott sets occur as $\mathrm{Rep}(M, \omega)$ where M is a nonstandard model of true arithmetic.

Fact 3.7. (Kanovei [7])

(a) *For all* $M \succ \mathbb{N}$, $\mathrm{Rep}(M, \omega)$ *is closed under jump and contains* $\mathbf{0}^{(\omega)}$, *the set of Gödel numbers of sentences in* TA.

(b) *Suppose* $\mathcal{X}$ *is a countable Scott set that is closed under jump and which contains* $\mathbf{0}^{(\omega)}$. *Then there is an elementary extension* M *of* $\mathbb{N}$ *such that* $\mathrm{Rep}(M, \omega) = \mathcal{X}$.

We will provide a different argument for Fact 3.7(a) in Section 4 below. The best known proof of part (b) of this result is currently rather long and technical.

Since $\mathrm{Rep}(M, \omega)$ depends only on $\mathrm{Th}(M, \omega)$, Fact 3.7 immediately tells us how many $\mathcal{L}_{\mathrm{A}}{}^{\mathrm{cut}}$ theories of standard cuts there are.

Corollary 3.8. *There are continuum many* $\mathrm{Th}(M, \omega)$, *where* $M \models \mathrm{TA}$.

¿From Fact 3.7, we might expect that some notion of 'super saturation', meaning here that the model $M \models \mathrm{PA}$ is recursively saturated and SSy(M) contains $\mathbf{0}^{(\omega)}$ and is closed under jump, would give interesting consequences. We are unaware of any such results though.

We have already considered the possibility that SSy(M,ω) and SSy(M) may be the same. Another possible meaning of 'super saturation' is recursive saturation plus the condition SSy(M,ω) = SSy(M).

Definition 3.9. *Let $M \models$ PA be nonstandard. We say M is* full *iff* SSy(M,ω) = SSy(M) *and* fully saturated *if it is full and recursively saturated.*

Notice that the property of fullness of M can be expressed in $\mathcal{L}_\mathrm{A}{}^{\mathrm{cut}}$, and it is straightforward to check that fullness of M implies $(\omega, \mathrm{SSy}(M)) \models \mathrm{CA}_0$, the full second-order comprehension axiom scheme. Kaye (to appear in a forthcoming paper) has recently shown a converse to the assertion here: that if $M \models$ PA is nonstandard and $(\omega, \mathrm{SSy}(M)) \models \mathrm{CA}_0$ then SSy(M,ω) = SSy(M). It follows that if M is full and $N \models$ PA is an end extension of M, then N is full as well, and this in turn implies that SSy(M,ω) = SSy(N,ω). The fact that the standard system is preserved by end extensions is a very useful property of models of PA. We suspect the corresponding result does not hold for SSy(M,ω), but at the moment it is an open problem.

Question 3.10. Let M be a nonstandard models of PA.

(a) Is SSy(M,ω) preserved under end extensions of M?

(b) Is SSy(M,ω) preserved under elementary end extensions of M?

If M is countable and recursively saturated and N is a countable recursively saturated elementary end extension of M, then $(M,\omega) \cong (N,\omega)$, hence (b) has positive answer for such extensions.

Fullness is related to work by Engström and Kaye [2] who introduce the notion of *transplendent* models, which are essentially like resplendent models but (under an appropriate consistency condition) allow one to expand a model M to a larger language without changing the domain, satisfying some recursive or coded theory, and simultaneously omit a recursive or coded type. Countable transplendent models exist, but necessarily have very rich standard systems. A 'nice' characterisation of the standard systems of transplendent models of PA is not known but it is known that if $M \models$ PA is transplendent and $\mathcal{X} = \mathrm{SSy}(M)$ then $(\omega, \mathcal{X})$ is an elementarily submodel of $(\omega, \mathscr{P}(\omega))$.

The following is a sample application of the theory of transplendency.

Proposition 3.11. *Suppose M is a countable transplendent model of* PA. *Then M is full, i.e.* SSy(M,ω) = SSy(M).

Proof. Given a real $A \in \mathrm{SSy}(M,\omega)$ defined by $A = \{n \in \omega : (M,\omega) \models$

$\theta(n, b, \omega)\}$ we consider the theory in $\mathcal{L}_{\mathrm{A}}{}^{\mathrm{cut}}$ together with additional constants a, b stating that the subset of N coded by a (using the β-function) is exactly the set of $n \in \mathsf{N}$ such that $(M, \mathsf{N}) \models \theta(n, b, \mathsf{N})$. We also omit the type $p_\omega(x) = \{x \in \mathsf{N}\} + \{x \neq \underline{k} : k \in \omega\}$, so that N represents ω. By transplendency there is an expansion of (M, b) satisfying all this, which suffices, so $A \in \mathrm{SSy}(M)$. $\dashv$

It follows that whenever $\mathcal{X}$ is a countable Scott set which is the standard system of a transplendent model there is a countable $M \models \mathrm{PA}$ with $\mathrm{SSy}(M, \omega) = \mathcal{X}$. This result has been improved recently by Kaye (forthcoming) who pointed out that any countable Scott set $\mathcal{X}$ such that $(\omega, \mathcal{X})$ satisfies the full comprehension scheme is the standard system of a full model of PA.

These results on transplendent models and full models at least provide some conditions on a Scott set $\mathcal{X}$ which allow us to realise $\mathcal{X}$ as $\mathrm{SSy}(M, \omega)$ for some $M \models \mathrm{PA}$, but are rather unsatisfactory as they rely on very strong hypotheses. Work is currently underway on reducing these hypotheses, and we would make the following conjecture.

Conjecture 3.12. *Let $\mathcal{X}$ be a countable Scott set.*

(a) *$\mathcal{X}$ is realised as $\mathrm{SSy}(M, \omega)$ for some $M \models \mathrm{PA}$ if and only if $\mathcal{X}$ is closed under jump.*

(b) *$\mathcal{X}$ is realised as $\mathrm{SSy}(M, \omega)$ for some model of true arithmetic $M \succ \omega$ if and only if $\mathcal{X}$ contains $\mathbf{0}^{(\omega)}$ and is closed under jump.*

The necessity of the condition concerning $\mathbf{0}^{(\omega)}$ in part (b) will be shown in the following section. It would follow from this, for example, that not all nonstandard models of PA are full. We hope that similar additional information on the Scott sets $\mathrm{Rep}(M, \omega)$ might be obtained.

4 Truth definitions

Kanovei's observation, Fact 3.7(a), that $\mathbf{0}^{(\omega)}$ is in $\mathrm{Rep}(M, \omega)$ for any nonstandard model of true arithmetic M says that there is a truth definition for ω definable in the pair (M, ω). This is of particular interest because of the connections between truth definitions and recursive saturation which we would like to explore further. But before we do that we present in more detail an argument why it is true. The following argument is somewhat different from Kanovei's but useful as it generalises in a number of interesting directions.

Let $M \models \mathrm{PA}$ be nonstandard and $K \subseteq M$ a substructure. We shall attempt to give a definition of truth in K. (One could consider, more generally, the case when K is an interpreted structure over M, but we see no obvious application in our context here.)

Definition 4.1. *An element $c \in M$ is a* certificate *(of truth of certain formulas in K) if it is an M-finite set of triples $\langle \phi, a, t \rangle$ where ϕ is a formula (in the sense of M) of the language $\mathcal{L}_A$, a is an element of M (thought of as coding a sequence $(a_0, a_1, \ldots)$) and t is 0 or 1, such that the following are all satisfied for all ϕ, ψ and a.*

1. $\langle \phi, a, 0 \rangle \in c \Rightarrow \langle \phi, a, 1 \rangle \notin c$
2. $\langle \phi, a, 1 \rangle \in c \Rightarrow \langle \phi, a, 0 \rangle \notin c$
3. $\langle \neg\phi, a, 1 \rangle \in c \Rightarrow \langle \phi, a, 0 \rangle \in c$
4. $\langle \neg\phi, a, 0 \rangle \in c \Rightarrow \langle \phi, a, 1 \rangle \in c$
5. $\langle \phi \wedge \psi, a, 1 \rangle \in c \Rightarrow \langle \phi, a, 1 \rangle \in c$ *and* $\langle \psi, a, 1 \rangle \in c$
6. $\langle \phi \wedge \psi, a, 0 \rangle \in c \Rightarrow \langle \phi, a, 0 \rangle \in c$ *or* $\langle \psi, a, 0 \rangle \in c$
7. $\langle \forall x_i\ \phi, a, 1 \rangle \in c \Rightarrow \forall b \in K\ \langle \phi, a[b/i], 1 \rangle \in c$
8. $\langle \forall x_i\ \phi, a, 0 \rangle \in c \Rightarrow \exists b \in K\ \langle \phi, a[b/i], 0 \rangle \in c$
9. $\langle \phi, a, 1 \rangle \in c \Rightarrow K \models \phi[a]$, *when ϕ is atomic and $a_i \in K$ for all i such that x_i is free in ϕ*
10. $\langle \phi, a, 0 \rangle \in c \Rightarrow K \not\models \phi[a]$, *when ϕ is atomic and $a_i \in K$ for all i such that x_i is free in ϕ*

Here, $a[b/i]$ is the code for the sequence obtained from a by replacing a_i by b, the free variables in our language are x_i for various i, and $\phi[a]$ is the statement obtained by replacing in ϕ each free variable x_i with a_i.

When c is a certificate, we write $\models_c \phi[a]$ for $\langle \phi, a, 1 \rangle \in c$ and $\not\models_c \phi[a]$ for $\langle \phi, a, 0 \rangle \in c$. The reader can add clauses for $\vee$ and $\exists$ if he/she wishes, following the above model.

Definition 4.2. *The certificate c* decides *$\phi[a]$ if $\models_c \phi[a]$ or $\not\models_c \phi[a]$.*

Definition 4.3. *The certificate c is K*-complete *if given a, ϕ, if c decides $\phi[a]$ then c decides $\phi[b]$ for all b such that $b_i \in K$ for all i such that x_i is free in ϕ.*

The following is a straightforward induction (in the meta-theory) on the complexity of ϕ.

Lemma 4.4. *Suppose M, K are as above, c_1, c_2 are K-complete certificates, ϕ is a standard formula, $a_i \in K$ for each x_i free in ϕ, and both c_1, c_2 decide $\phi[a]$. Then*

$$\models_{c_1} \phi[a] \;\Leftrightarrow\; \models_{c_2} \phi[a].$$

Definition 4.5. *Given M, K as above, we define S to be the parameter-free definable set of pairs $\langle \phi, a \rangle$ (parameter-free definable in (M, K, ω)) such that $\phi \in \omega$, for all i with x_i free in ϕ we have $a_i \in K$, and*

$$\forall c\ (c \text{ is a } K\text{-complete certificate deciding } \phi[a] \text{ implies } \models_c \phi[a]).$$

We write $\models_S \phi[a]$ for $\langle \phi, a \rangle \in S$.

As we shall show shortly, under certain circumstances the set S so defined is a definition of full truth for K in (M, K, ω) and S is also defined by

$$\exists c\ (c \text{ is a } K\text{-complete certificate deciding } \phi[a] \text{ and } \models_c \phi[a]).$$

By the previous lemma, the main thing to check is of course that for all standard ϕ and all appropriate a there is a K-complete certificate deciding $\phi[a]$. This works in the following two cases.

Proposition 4.6. *Suppose $K \prec M$ and is not cofinal in M. Then S is a definition of truth for K.*

Proof. That K-complete certificates exist for each $\phi[a]$ is proved by an induction (in the meta-theory) on ϕ. Note that the definition of certificate is not in $\mathcal{L}_A$ but uses K as a predicate, so an internal induction is not available. The only tricky step is the case for the quantifier, as the other cases can be dealt with by taking the union of certificates, and/or the union of a certificate and a set obtained from the separation principle for M-finite sets, provable in PA. (It is this separation principle that requires K to be not cofinal in M. See Kaye–Wong [9] for the details.)

Given a K-complete certificate c deciding $\phi[a]$ for all appropriate a, we need to observe that the set A of appropriate a such that $\exists b \in K\ \langle \phi, a[b/i], 0\rangle \in c$ is coded in M, for then

$$\{\langle \forall x_i\ \phi, a, 1\rangle : a \notin A\} \cup \{\langle \forall x_i\ \phi, a, 0\rangle : a \in A\} \cup c$$

would be coded by a certificate c' deciding $\forall x_i\ \phi$. But this follows from $K \prec M$ since for any a with $a_j \in K$ for x_j free in $\forall x_i\ \phi$ we have

$$\exists b \in K\ \neg\phi[a[b/i]] \Leftrightarrow \exists b \in M\ \neg\phi[a[b/i]] \Leftrightarrow \exists b \in M\ \langle \phi, [a[b/i]], 1\rangle \in c$$

by elementarity, and the last statement is definable in $\mathcal{L}_A$ so the set of such $a[b/i]$ is coded. ⊣

If a set parameter $A \subseteq \omega$ or a parameter $a \in M$ coding such a real $A \in \mathrm{SSy}(M, \omega)$ were present, we would require a corresponding predicate $A^* \subseteq M$ such that $(\omega, A) \prec (M, A^*)$ for the above proof to work to answer Question 4.10.

The previous result proves Kanovei's observation about models of true arithmetic. For models of false arithmetic we have the following.

Proposition 4.7. *Suppose $M \models \mathrm{PA}$, ω is strong in M and $K = \omega$. Then S is a definition of truth for ω.*

Proof. We deal with the quantifier step in a different way, using strength of ω instead of elementarity. Given an ω-complete certificate c

deciding $\phi[a]$ for all appropriate a, we need the set A of appropriate a such that $\exists b \in \omega\ \langle \phi, a[b/i], 0\rangle \in c$ to be coded in M. This can be shown by using the strength of ω on the coded function $f\colon \omega^2 \to M$ defined by

$$f(\phi, a) = \begin{cases} (\min b)(\langle \phi, a[b/i], 0\rangle \in c), & \text{if it exists;} \\ d, & \text{otherwise,} \end{cases}$$

where d is some fixed nonstandard number. $\dashv$

Note that any countable arithmetically saturated model M satisfies the hypothesis of this theorem.

Corollary 4.8. *Let $M \models$ PA be a nonstandard model, and suppose either that $M \models$ TA or ω is strong in M. Let (K, N) be a model of $\mathrm{Th}(M, \omega)$ where N is nonstandard. Then N has a full inductive satisfaction class. In particular, N is recursively saturated.*

Proof. By Proposition 4.7, the set defined by the definition of S in (K, N) is a full inductive satisfaction class for N. Since N is nonstandard, it must be recursively saturated. $\dashv$

In general, we cannot expect this N to be any more than recursively saturated. For example, given $T = \mathrm{Th}(M, \omega)$ there is a countable Scott set $\mathcal{X}$ containing T and not closed under jump. (Indeed there are continuum-many such $\mathcal{X}$.) Let K be a countable recursively saturated model of $\mathrm{Th}(M)$ such that $\mathrm{SSy}(K) = \mathcal{X}$. Since K is resplendent, there is N such that $(K, N) \models T$ and (K, N) is recursively saturated. In particular, N is recursively saturated with $\mathrm{SSy}(K) = \mathrm{SSy}(N) = \mathcal{X}$, and as $\mathcal{X}$ is not closed under jump N is not arithmetically saturated.

Using a theorem of Smith's [21, Theorem 3.14], one can find models of PA that contain a nonstandard pseudo-standard cut, but are not recursively saturated themselves.

More generally, Proposition 4.6 gives other variants of Corollary 4.8.

Corollary 4.9. *Suppose $M \models$ PA is nonstandard, and $K \prec M$ is not cofinal. Then for any $(M', K', N) \models \mathrm{Th}(M, K, \omega)$ with N nonstandard the substructure K' is recursively saturated.*

We are unable to show that $\mathrm{SSy}(M, \omega)$ is closed under the ωth jump without any additional hypothesis. It is tempting to conjecture from Kanovei's theorem and Proposition 4.7 that this is at least true when ω is strong or when M satisfies true arithmetic.

Question 4.10. Suppose $M \models$ PA in which ω is strong, or $M \models$ TA. Is it the case that $\mathrm{SSy}(M, \omega)$ is closed under the ωth jump $\mathbf{a} \mapsto \mathbf{a}^{(\omega)}$?

Kotlarski [15] defined the *recursively saturated part* of a model to be its largest elementary cut that is recursively saturated. In connection with Corollary 4.8 we can ask:

Question 4.11. Can a pseudo-standard cut in a model of TA be its recursively saturated part?

5 Conclusions

It is perhaps surprising that the model theory of structures of the form (M, ω) is so difficult compared to that of M. It seems that this may reflect some intrinsic difficulty, or else model theoretic tools are not available (or are not known by the present authors).

There are many questions one can ask about the model theory of pairs (M, N), where M is a countable model of PA and N is a pseudo-standard cut. Using the omitting types theorem, Tin Lok Wong managed to prove several results on existence of elementary end extensions and cofinal extensions of such structures. The fact that N is pseudo-standard, rather than just a cut in M, plays a crucial role in the proofs. It is interesting that none of the results seems to generalise to the uncountable case. One area we have not addressed in this short paper is definability in models (M, ω). Smith [21] makes a start in this direction, and this suggests that expressive power (for such models when the predicate N is interpreted as ω) is not much stronger. But there is still much work to do in these directions, in particular in unravelling the distinctions between what is definable by a single first-order formula and what is definable by a simple infinitary formula, such as an infinitary conjunction of first-order formulas.

References

[1] Jon Barwise. *Admissible Sets and Structures.* Perspectives in Mathematical Logic. Springer-Verlag, Berlin, 1975.

[2] Fredrik Engström and Richard Kaye. Transplendent models: expansions omitting a type. *Notre Dame Journal of Formal Logic*, 53(3):413–428, 2012.

[3] Andrzej Grzegorczyk, Andrzej Mostowski and Czesław Ryll-Nardzewski. The classical and the ω-complete arithmetic. *The Journal of Symbolic Logic*, 23(2):188–206, June 1958.

[4] Petr Hájek. Completion closed algebras and models of Peano arithmetic. *Commentationes Mathematicae Universitatis Carolinae*, 22(3):585–594, 1981.

[5] Leon Henkin. A generalization of the concept of ω-completeness, *The Journal of Symbolic Logic*, 22(1):1–14, March 1957.

[6] Daniel Isaacson. Necessary and Sufficient Conditions for Undecidability of the Gödel sentence and its truth. In David DeVidi, Michael Hallet and Peter Clark, editors, *Logic, Mathematics, Philosophy: Vintage Enthusiasms*, volume 75 of *The Western Ontario Series in Philosophy of Science*, pages 135–152, Dordrecht, 2011. Springer.

[7] Vladimir Kanovei. On external Scott algebras in nonstandard models of Peano arithmetic. *The Journal of Symbolic Logic*, 61(2): 586–607, 1996.

[9] Richard Kaye. *Models of Peano Arithmetic.* Oxford University Press, Oxford, 1991.

[9] Richard Kaye and Tin Lok Wong. On interpretations of arithmetic and set theory. *Notre Dame Journal of Formal Logic*, 48(4):497–510, October 2007.

[10] H. Jerome Keisler. Some model theoretic results for ω-logic. *Israel Journal of Mathematics*, 4(1):249–261, March 1966.

[11] H. Jerome Keisler. Nonstandard arithmetic and reverse mathematics. *The Bulletin of Symbolic Logic*, 12(1):100–125, March 2006.

[12] H. Jerome Keisler. Nonstandard arithmetic and recursive comprehension. *Annals of Pure and Applied Logic*, 161:1047–1062, 2010.

[13] Laurence A.S. Kirby and Jeff B. Paris. Initial segments of models of Peano's axioms. In Alistair Lachlan and Marian Srebrny and Andrzej Zarach, editors, *Set Theory and Hierarchy Theory V*, volume 619 of *Lecture Notes in Mathematics*, pages 211–226, Berlin, 1977, Springer.

[14] Roman Kossak and James H. Schmerl. *The Structure of Models of Peano Arithmetic*, volume 50 of *Oxford Logic Guides.* The Clarendon Press, Oxford University Press, Oxford, 2006. Oxford Science Publications.

[15] Henryk Kotlarski. The recursively saturated part of models of Peano arithmetic. *Zeitschrift für Mathematische Logik und Grundlagen der Mathematik*, 32(4):365–370, 1986.

[16] Hugues Leblanc, Peter Roeper, Michael Thau and George Weaver. Henkin's Completeness Proof: Forty Years Later. *Notre Dame Journal of Formal Logic*, 32(2):212–232, Spring 1991.

[17] Steven Orey. On ω-consistency and related properties. *The Journal of Symbolic Logic*, 21(3):246–252, September 1956.

[18] Hartley Rogers, Jr. *Theory of Recursive Functions and Effective Computability.* McGraw-Hill Series in Higher Mathematics. McGraw-Hill Book Company, New York, 1967.

[19] Dana Scott. Algebras of sets binumerable in complete extensions of arithmetic. In *Recursive Function Theory*, volume V of *Proceedings of Symposia in Pure Mathematics*, pages 117–121, Providence, R.I., 1962. American Mathematical Society.

[20] Stephen G. Simpson. *Subsystems of Second Order Arithmetic.* Perspectives in Mathematical Logic. Springer, Berlin, 1999.

[21] Stuart T. Smith. Extendible sets in Peano arithmetic. *Transactions of the American Mathematical Society*, 316(1): 337–367, November 1989.

10

Randomness, pseudorandomness and models of arithmetic

PAVEL PUDLÁK[1]

Dedicated to Alan Woods

Abstract: Pseudorandomness plays an important role in number theory, complexity theory and cryptography. Our aim is to use models of arithmetic to explain pseudorandomness by randomness. To this end we construct a set of models $\mathcal{M}$, a common element ι of these models and a probability distribution on $\mathcal{M}$, such that for every pseudorandom sequence s, the probability that $s(\iota) = 1$ holds true in a random model from $\mathcal{M}$ is equal to $1/2$.

1 Introduction

A pseudorandom sequence is an infinite sequence of -1s and 1s computable in nondeterministic polynomial time that is not correlated with any polynomial time computable function. Such sequences can also be viewed as sets in $\mathbf{NP} \cap \mathbf{coNP}$; thus we can also talk about pseudorandom sets. Intuitively, a pseudorandom set splits every set in $\mathbf{P}$ into two sets of equal density. There are some natural and important candidates for pseudorandom sequences in number theory such as the Liouville function (closely related to the Möbius function).

Our main result is a construction of a set of models $\mathcal{M}$, a common element ι of these models and a probability distribution on $\mathcal{M}$, such

[1]Mathematical Institute, Academy of Sciences of Czech Republic, Prague, pudlak@math.cas.cz

Studies in Weak Arithmetics.
Patrick Cégielski, Charalampos Cornaros,
Costas Dimitracopoulos.

that for every pseudorandom sequence s, the probability that $s(\iota) = 1$ holds true in a random model from $\mathcal{M}$ is equal to 1/2. Thus pseudorandomness of sequences manifests itself in $\mathcal{M}$ as genuine randomness. Admittedly, this result is weak, because it concerns only one common element of the models. We present it as a proof of the concept that results of this kind are possible. We suggest some ways of extending this result in Section 4.

We prove our result by using restricted ultrapowers, which are ultrapowers in which the sets of the ultrafilter and the functions are elements of suitable complexity classes. The history of restricted ultrapowers goes back to Skolem (see [5]). We will start with a model M_0, a core of our construction, constructed from the set of all polynomial time computable functions reduced by a suitable ultrafilter on the complexity class **P**. By extending the class of functions and the ultrafilter in various ways, we obtain a set of models $\mathcal{M}$ extending the core model. Remarkably, there is a natural way of defining a probability measure on $\mathcal{M}$.

In the last section we present some philosophical speculations about the nature of pseudorandomness.

2 Preliminaries

Random sequences

We will study sequences $s : \mathbb{N} \to \{-1, 1\}$. Let $\mathcal{S}$ denote the set of all such sequences with the uniform distribution. Intuitively, we have $Pr[s(n) = 1] = Pr[s(n) = -1] = \frac{1}{2}$ and these events are independent for different numbers. Formally, there is a Lebesgue measure m on $\mathcal{S}$ that is uniquely determined by

$$m(\{s \in \mathcal{S};\ s(1) = a_1 \wedge \cdots \wedge s(n) = a_n\}) = 2^{-n}$$

for all n and all strings a of $-1, 1$s. We say that a *random sequence satisfies* P, if *the probability that a randomly chosen sequence satisfies* P *is 1.* If a random sequence satisfies $P_1, P_2, \ldots$, then it also satisfies $\bigwedge_i P_i$.

The basic fact about random sequences is the Law of Large Numbers

$$Pr\left[\lim_{n\to\infty} \sum_{i=0}^{n-1} s(i)/n = 0\right] = 1.$$

This theorem, however, does not provide information about the rate of convergence. Much more precise theorems have been proven, in particular Khinchin's Law of Iterated Logarithm

$$Pr\left[\limsup_{n\to\infty}\sum_{i=0}^{n-1} s(i)/\sqrt{2n\ln\ln n}=1\right]=1,$$

which, in particular, implies that for every $\epsilon > 0$,

$$Pr\left[\lim_{n\to\infty}\sum_{i=0}^{n-1} s(i)/n^{\frac{1}{2}+\epsilon}=0\right]=1. \qquad (eRH)$$

Algorithmic randomness and pseudorandomness

When we say "a random sequence" it is just a *façon de parler.* As explained above, formally, we can only talk properties satisfied with probability 1. However, one can express our intuition about a random sequence using computability theory.

The basic idea used to formalize the concept that a *specific* sequence s "looks like a random sequence" is that s must satisfy many natural properties that random sequences satisfy with probability 1. E.g., we certainly want the property used in the Law of Large Numbers. We cannot require that s satisfies all properties that have probability 1; we can only choose some. A natural approach is to use computable properties, properties that can be algorithmically decided. Therefore it is more natural to talk about satisfying *tests* instead of properties.

The study of such concepts has long history and many researchers contributed to it, including Kolmogorov, Chaitin, Levin, Schnorr and Martin-Löf. This research area is called *algorithmic randomness.* We mention one of the concepts that is studied there, so that we can compare it with the concept that we will introduce.

A *martingale* is a function $F : \{-1,1\}^* \to \mathbb{R}^+$ such that

$$F(a_1,\dots,a_n)=\frac{1}{2}(F(a_1,\dots,a_n,-1)+F(a_1,\dots,a_n,+1)).$$

The following definition is due to Schnorr [8].

Definition 1. *A sequence* $s : \mathbb{N} \to \{-1,1\}$ *is* random, *if for every computable martingale*

$$\limsup_{n\to\infty} F(s(0),s(1),\dots,s(n-1))<\infty.$$

Weaker concepts defined using restricted classes of computable martingales have also been considered. Say that a sequence s is **P**-*random* if the condition above is satisfied for every martingale F computable in polynomial time.

Another concept, which is relevant to this paper, is the concept of *pseudorandom number generator*, which will be abbreviated by PRG. A

PRG is an algorithm to produce a long string of numbers, usually just 0s and 1s, from a short random string, called the *seed.* So in this case we do not have only one infinite sequence, but a small set of finite strings with a probability distribution. This concept plays an important role in the theories that study computational complexity and cryptography.

There is one important difference in how the computational resources are bounded in the two mentioned approaches. When testing a sequence s for **P**-randomness, the testing algorithm receives the whole string $(s(0), s(1), \ldots, s(n-1))$ as an input and can use time polynomial in n. In contrast, algorithms testing pseudorandomness get n represented in binary and can use time polynomial in the length of n, i.e., they run in time polynomial in $\log n$.

We have mentioned the concept of algorithmic randomness only to put our definition of a pseudorandom sequence into the context of other definitions of a similar kind. The concept of pseudorandom generators, however, is directly connected with our definition. Namely, the existence of pseudorandom generators implies the existence of pseudorandom sequences in our sense. We will not prove this implication; we will only state it for a specific related conjecture and sketch a proof of it (see Proposition 2.5).

Pseudorandom sequences

Our concept is closer to the theory of pseudorandomness than to algorithmic randomness. That is why we use the word *pseudorandom.* The reader should, however, be cautioned that there are concepts with similar names that differ significantly from ours.

Definition 2. *A sequence* $s : \mathbb{N} \to \{-1, 1\}$ *will be called* pseudorandom *if*

1. *s is computable in nondeterministic polynomial time,*
2. *for every polynomial time computable function $f : \mathbb{N} \to \{-1, 1\}$,*

$$\lim_{n\to\infty} \sum_{i=0}^{n-1} f(i)s(i)/n = 0. \qquad (e0)$$

Condition 1. means that although we may be not able to compute the value $s(n)$ in polynomial time, if somebody gives us a "witness" we are able to check the correct value of $s(n)$ in polynomial time. (There should always be witnesses for either the value 1 or the value -1, but never for both.) We can identify s with the set $\{n;\ s(n) = 1\} \in \mathbf{NP} \cap \mathbf{coNP}$, so we can also talk about *pseudorandom sets.*

Condition 2. means that s is little correlated with any sequence computable in polynomial time. One can consider various modifications of this condition. For instance, one can allow stronger tests, say, functions computable in appropriately defined subexponential time. One can also impose stronger bounds on correlation. If, for example, we required that

$$\lim_{n\to\infty}\sum_{i=0}^{n-1} f(i)s(i)/n^{\alpha}=0,$$

for some $1/2 < \alpha < 1$, then the correlation would be exponentially small (on finite initial segments; recall that the input size is $\log n$).

Another motivation for the definition above is the *Möbius Randomness Principle* proposed by Peter Sarnak (see [4]). According to this principle the Möbius function μ is not correlated to any "low-complexity" function $F : \mathbb{N} \to [-1, 1]$ in the sense that

$$\lim_{n\to\infty}\sum_{i=1}^{n} F(i)\mu(i)/n=0.$$

This is like saying that μ is pseudorandom, except for a few minor differences. First, μ takes on not only the values $-1, 1$, but also 0. This does not seem very important for studying the concept of pseudorandomness (see Proposition 2.3 below). Second, it is not specified what "low complexity" means. This leaves open the possibility of studying various specific versions of the conjecture. Third, the tests are functions F whose range is in the whole interval $[-1, 1]$. We will show below that this is also an irrelevant difference.

Proposition 2.1. *Suppose that $s : \mathbb{N} \to \{-1, 1\}$ is pseudorandom. Let $F : \mathbb{N} \to [-1, 1]$ be a polynomial time computable function whose values are binary rationals. Then*

$$\lim_{n\to\infty}\sum_{i=1}^{n} F(i)s(i)/n=0.$$

Proof. Let s and F be given. One can easily show that s is pseudorandom also with respect to polynomial time computable functions $f : \mathbb{N} \to \{-1, 0, 1\}$. (Hint: write $f = \frac{1}{2}f_+ + \frac{1}{2}f_-$, where $f_+(n) = 1$ if $f(n) = 1$, otherwise $f_+(n) = -1$, and $f_-(n) = -1$ if $f(n) = -1$, otherwise $f_-(n) = 1$.)

Represent F as a weighted sum of such functions,

$$F(n)=\sum_{j=0}^{\infty} 2^{-j} f_j(n).$$

Then

$$\lim_{n\to\infty}\sum_{i=1}^{n} F(i)s(i)/n = \lim_{n\to\infty}\sum_{i=1}^{n}\sum_{j=0}^{\infty} 2^{-j} f_j(n)s(i)/n$$

$$= \sum_{j=0}^{\infty} 2^{-j} \lim_{n\to\infty}\sum_{i=1}^{n} f_j(n)s(i)/n = 0$$

because the infinite sum converges absolutely. ⊣

Say that a set of numbers A has *positive density,* if

$$\liminf_{n\to\infty} |A \cap [0, n-1]|/n > 0.$$

Corollary 2.2. *If X is a pseudorandom set, then neither X nor its complement contains a set $A \in \mathbf{P}$ of positive density.*

A possible way of stating the Möbius Randomness Principle is to say that μ is pseudorandom in the sense of our definition (extended to sequences of -1, 0 and 1). A closely related function is the Liouville function λ. It is defined by $\lambda(n) = (-1)^k$, where k is the number of prime factors of n counted with their multiplicity.

Proposition 2.3. *μ is pseudorandom if and only if λ is.*

Proof. 1. Suppose μ is pseudorandom. Let a polynomial time computable function $f : \mathbb{N} \to \{-1, 1\}$ and $\epsilon > 0$ be given. Let n_0 be such that

$$\sum_{k>n_0} k^{-2} < \epsilon.$$

We have

$$| \lim_{n\to\infty}\sum_{i=1}^{n} f(i)\lambda(i)/n| \leq$$

$$\sum_{1\leq k\leq n_0} | \lim_{n\to\infty} f(k^2 i)\lambda(k^2 i)/n |+$$

$$\sum_{1\leq k\leq n_0} | \lim_{n\to\infty} \sum_{1\leq k^2 i\leq n} f(k^2 i)\mu(i)/n | + \epsilon = \epsilon.$$

2. Now suppose that μ is not pseudorandom. Let $f : \mathbb{N} \to \{\pm 1\}$ be a polynomial time computable function and $\epsilon > 0$ such that

$$\lim_{n\to\infty}\sum_{i=1}^{n} f(i)\mu(i)/n = \epsilon.$$

Let n_0 be as above. Furthermore, we can suppose that

$$\lim_{n\to\infty}\sum_{i=1}^{n} f(i)\lambda(i)/n = 0,$$

because otherwise we would be done. In a similar fashion as above, decompose the $\sum_{i=1}^{n} f(i)\lambda(i)/n$ into three terms

1. the sum over square free numbers i,
2. the sum over numbers i divisible by k^2 for some $k \leq n_0$, and
3. the sum over the remaining numbers i.

By our assumptions, the limit of the first sum is ϵ, the sum of the limits of the sums 2. and 3. is $-\epsilon$, the sum 3. is $> -\epsilon$. Hence, if we define

$$g(n) = \begin{cases} -f(n) & \textit{if } n \textit{ is divisible by } k^2 \textit{ for some } k \leq n_0, \\ f(n) & \textit{otherwise}, \end{cases}$$

we obtain $\lim_{n\to\infty} \sum_{i=1}^{n} g(i)\lambda(i)/n > 0$. ⊣

Some special cases of the Möbius randomness principle have been proven. The first one was the Prime Number Theorem, which is the case of $f \equiv 1$. (The question whether the bound on correlation can be improved to the form (eRH) is the Riemann Hypothesis.) Recently B. Green proved the principle for AC^0, see [4]. Let us say that a sequence is AC^0-pseudorandom if it satisfies Definition 2 with the condition 2. weakened to AC^0-computable. Then one can state Green's result as follows.

Theorem 2.4. *([4]) The Möbius function is AC^0-pseudorandom.*

The Liouville function is also AC^0-pseudorandom.

It will be very difficult to prove that some sequence is pseudorandom, because the existence of pseudorandom sequences implies $\mathbf{P} \neq \mathbf{NP}$. For specific functions, it may be even harder. If the Möbius function is pseudorandom, then integers cannot be factored in polynomial time.

In the opposite direction, we know that hardness of factoring implies the existence of pseudorandom sequences (we are not able to prove that it implies the pseudorandomness of the Möbius function, though). This is because

1. there are constructions of permutations that are *one-way functions* provided that factoring is hard,
2. there is a construction of a *hard-core predicate* from any one-way permutation, and
3. hard-core predicates are very closely related to pseudorandom sequences.

We will now explain this connection in more detail, but for the sake of brevity, we will skip the definition of a one-way function. Let $1 \leq k_1 < k_2 < \ldots$ be a sequence of integers that grow at most polynomially, and let $F_j : \{0,1\}^{k_j} \to \{0,1\}^{k_j}$, $j = 1,2,\ldots$, be a sequence of permutations. Suppose that these numbers and functions are uniformly computable in polynomial time. We say that functions $B_j : \{0,1\}^{k_j} \to \{0,1\}$, $j = 1,2,\ldots$, are hard-core predicates for the functions F_j, if B_j are uniformly computable in polynomial time, and for every function $g(x)$ computable by a randomized polynomial time algorithm

$$Pr[g(F_j(x)) = B_j(x)] = \frac{1}{2} \pm \frac{1}{k_j^{\omega(1)}}, \tag{1}$$

where the probability is taken over uniformly distributed $x \in \{0,1\}^{k_j}$ and random bits of the algorithm for g; further, $\omega(1)$ is the standard notation for functions going to infinity. In plain words, this means that $B_j(x)$ can be predicted from $F_j(x)$ only with negligible probability (which, in particular, implies that it is difficult to invert F_j).

The concept that is closely related to pseudorandom sequences (as defined in this paper) is the sequence $B_j(F_j^{-1}(y))$, $j = 1,2,\ldots$. To get a pseudorandom sequence s, we only need to connect the bits $B_j(F_j^{-1}(y))$ into one infinite sequence:

$$s(n) = (-1)^{B_j(F_j^{-1}(n - \sum_{i<j} 2^{k_i}))},$$

where $\sum_{i<j} 2^{k_i} \leq n < \sum_{i \leq j} 2^{k_i}$ and where we are identifying $\{0,1\}^{k_j}$ with $0,1,,\ldots,2^{k_j-1}-1$.

In order to show that $s(n)$ is pseudorandom, we have to address only one small complication. While in (2) of the definition of pseudorandomness we consider all initial segments, in (3) of the definition of the hard-core predicate we only consider correlation over the entire interval $[0, 2^{k_j} - 1]$ (but we have better convergence). We need to show that the correlation of $B_j(F_j^{-1}(y))$ with polynomial time functions is also low on any initial segments $[0,a]$ of $[0, 2^{k_j} - 1]$.

Suppose g has positive correlation with $B_j(F_j^{-1}(y))$ on $[0,a]$. If we knew a, we could define g' that has positive correlation with $B_j(F_j^{-1}(y))$ on the entire interval by putting $g'(x) = g(x)$ on $[0,a]$ and $g'(x) = 1 - g(x)$ on the rest. Since we cannot assume that we know a, we have to do something slightly more complicated. Note that we are actually assuming that there are infinitely many indices j for which g has positive correlation with $B_j(F_j^{-1}(y))$ on some interval $[0, a_j]$. The ratios $a_j/2^{k_j}$ have some limit point α, $0 < \alpha < 1$. Take a rational number β close to α (or α itself if it is rational). Then use $\lceil \beta 2^{k_j} \rceil$ as switching points.

This finishes a sketch of the proof of the following proposition.

Proposition 2.5. *If there exists no probabilistic polynomial time algorithm for factoring integers, then there exists a pseudorandom sequence.*

Remarks. 1. The construction actually gives a concrete sequence, but since its definition is rather complicated, we do not present it here.

2. We have not fully used the assumption about factoring; one can show pseudorandomness of the constructed sequence in a little stronger sense.

3. For the concepts and results used above, see [3].

Theories

We need a theory in which it is possible to formalize polynomial time computations. A natural theory in which this is possible is Cook's PV, [1, 6]. This theory has function symbols for all polynomial time computable function. The function symbols correspond to algorithms based on recursion on notation. Our result is quite general and, as such, holds for the stronger theory $PV^{\mathbb{N}}$ defined below.

Definition 3.

1. *$PV^{\mathbb{N}}$ is the theory axiomatized by all true universal sentences in the language of PV.*

2. *$AR^{\mathbb{N}}$ is the theory consisting of all true sentences in the language of PV.*

The theory $PV^{\mathbb{N}}$ is a conservative extension of the theory of all true Π^0_1 arithmetical sentences plus the axiom $\forall x \exists y \; y = x^{\lceil \log(x+1) \rceil}$ (this axiom guarantees that the provably total functions grow sufficiently fast and thus enable us to define polynomial time computations). The theory $AR^{\mathbb{N}}$ is essentially *True Arithmetic,* except that we use the richer language of PV.

We focus on the complexity class $\mathbf{P}$, as this is the most interesting case, but in fact the same result can be proven for concepts based on other classes. An interesting case is the class $\mathbf{AC^0}$ because of the result of Green mentioned above. The theory corresponding to this class is V^0, see [2].

A pseudorandom sequence s is, by definition, computable by a nondeterministic Turing machine. Therefore there exists a PV formula $\sigma(x)$ that defines $s(x) = 1$. Since in $AR^{\mathbb{N}}$ we have all true sentences in the language of PV, it does not matter which formula we use. In our theorem below we will only need that the sequence is arithmetically definable. This is rather weakness than strength of our result.

Random models

Our aim is to represent pseudorandomness in a different way. The basic idea is to study this concept using a set of nonstandard models equipped with a probability distribution. Let σ be a first order formula in the language of PV defining a sequence $s \in \mathcal{S}$, and suppose that we have a set of models $\mathcal{M}$ with a probability distribution ν. Let ϕ be a first order formula. Then we say that s *satisfies the property* ϕ *with probability* p if the probability that in a random model from $\mathcal{M}$ the sequence defined by σ satisfies ϕ is p.

If we want to use formulas ϕ with parameters, i.e., free variables for elements of models, we need to impose some structure on $\mathcal{M}$. In this paper we will only consider the following structure. There is one distinguished model M_0 such all other models are its extensions. This enables us to speak about properties parameterized by elements of M_0.

In this paper we say that a model N is an *extension* of a model M, if M is a substructure of N.

In general the structure defined on $\mathcal{M}$ can be more complicated. We can use various frames, like in the Kripke semantics. If N is one of the "alternative worlds" of M, then N should be an extension of M (not necessarily proper).

An alternative approach is to use a Boolean valued model M with a boolean algebra $\mathcal{B}$ equipped with a probability measure, an approach studied in [7]. This is, however, not fundamentally different from the approach sketched above. Having such a model, we can construct a set of models by taking all ultrafilters on $\mathcal{B}$. Vice versa, having $\mathcal{M}$ and ν as above, we can take the Boolean algebra of measurable subsets of $\mathcal{M}$ and define a measure on this algebra in a natural way.

Ultrapowers

We assume that the reader is familiar with the standard ultrapower construction. In this construction, one uses an ultrafilter defined on the Boolean algebra of *all* subsets of some set X and *all* functions defined on X. Here we will use ultrapowers with $X = \mathbb{N}$ with the Boolean algebra containing only some subsets of $\mathbb{N}$ and use only some functions defined on $\mathbb{N}$. Łoś's theorem can easily be generalized to this setting. We state this generalization without a proof.

Let L be a language interpreted in $\mathbb{N}$, let $\mathcal{B}$ be a Boolean algebra of subsets of $\mathbb{N}$, let F be some a of functions $f : \mathbb{N} \to \mathbb{N}$, and let $\mathcal{U}$ be an ultrafilter on $\mathcal{B}$. In order to be able to define the ultrapower $F/\mathcal{U}$, we need the following conditions to be satisfied:

(a) F contains constant functions for every constant of L and is closed under functions and operations of L.

(b) For every atomic formula $\alpha(x_1, \ldots, x_k)$ of L and every $f_1, \ldots, f_k \in F$,

$$\{n;\ \alpha(f_1(n), \ldots, f_k(n))\} \in \mathcal{B}.$$

In all interesting cases, one can represent every number $n \in \mathbb{N}$ by a closed term, thus F contains all constant functions.

Lemma 2.6. *Let Φ be a subset of formulas of L. Suppose (a) and (b) are satisfied and, moreover, the following two conditions are satisfied*

(c) Φ is closed under subformulas, and

(d) for every formula $\psi(x_1, \ldots, x_k)$ such that either $\exists y\ \psi(x_1, \ldots, x_k) \in \Phi$, or $\forall y\ \psi(x_1, \ldots, x_k) \in \Phi$, for every set $X \in \mathcal{B}$ and all functions $f_1, \ldots, f_k \in F$, if

$$\forall n \in X \exists m\ \psi(m, f_1(n), \ldots, f_k(n)).$$

Then

$$\exists g \in F \forall n \in \mathbb{N}\ \psi(g(n), f_1(n), \ldots, f_k(n)).$$

Then for every formula $\phi \in \Phi$ and all $f_1, \ldots, f_k \in F$

$$F/\mathcal{U} \models \phi([f_1], \ldots, [f_k]) \quad \Leftrightarrow \quad \{n;\ \phi(f_1(n), \ldots, f_k(n)\} \in \mathcal{B}.$$

Hence if F contains all constants, then $F/\mathcal{U}$ is a Φ-elementary extension of $\mathbb{N}$.

Note that if Φ contains only quantifier free formulas, then we do not need condition (d). Thus we get the following corollary by applying the lemma to all quantifier-free formulas of L.

Corollary 2.7. *Suppose conditions (a) and (b) are satisfied. Then every true universal sentence of L is satisfied in $f/\mathcal{U}$.*

We will construct extensions of an ultrapower model by extending the set of functions, the Boolean algebra and the ultrafilter.

Corollary 2.8. *Suppose $F_1 \subseteq F_2$, $\mathcal{B}_1 \subseteq \mathcal{B}_2$, $\mathcal{U}_1 \subseteq \mathcal{U}_2$ and $\Phi_1 \subseteq \Phi_2$. Then $F_2/\mathcal{U}_2$ is a Φ_1-elementary extension of $F_1/\mathcal{U}_1$.*

3 The result

Theorem 3.1. *There exists a model M_0 of $PV^{\mathbb{N}}$, an element $\iota \in M_0$, a set $\mathcal{M}$ of models of $AR^{\mathbb{N}}$ and a probability measure ν on a sigma algebra $\mathcal{B}$ of subsets of $\mathcal{M}$ such that*

1. *models of $\mathcal{M}$ are extensions of M_0,*
2. *for every PV formula $\phi(x_1, \ldots, x_k)$ and every string of elements $a_1, \ldots, a_k \in M_0$, the set $\{M \in \mathcal{M};\ M \models \phi(a_1, \ldots, a_k)\}$ is in $\mathcal{B}$,*

3. *for every pseudorandom sequence s,*

$$Pr_\nu[M \models s(\iota) = 1] = \frac{1}{2}.$$

Recall that $s(\iota) = 1$ is definable by $\sigma(\iota)$, where σ is a PV formula, hence $Pr_\nu[M \models s(\iota) = 1] = \nu(\{M;\ M \models \sigma(\iota)\})$ is well-defined.

Proof. Let $K \subseteq \mathbb{N}$ be an infinite set. We define *density of sets of numbers with respect to K*, or *K-density*, to be the partial function defined by

$$dens_K X = \lim_{k \in K, k \to \infty} \frac{|X \cap [0, k-1]|}{k}, \tag{2}$$

i.e., $dens_K X$ is undefined when the limit does not exist.

The proofs of the following easy facts are left to the reader.

1. $dens_K$ is finitely additive.
2. If $dens_K X = dens_K Y = 1$, then $dens_K X \cap Y = 1$.
3. If C is a countable set of sets of numbers, then there exists an infinite K such that $dens_K X$ is defined (i.e., the limit (2) exists) for every $X \in C$.

Let **AR** denote arithmetically definable sets of natural numbers (which is the same as sets first-order definable in the language of PV). Let K be an infinite set of numbers such that $dens_K X$ is defined for every $X \in \mathbf{AR}$.

The following fact follows easily from Proposition 2.1.

4. Let $X \in \mathbf{P}$ and Z be a pseudorandom set. Then $dens_K X \cap Z = \frac{1}{2} dens_K X$.
   (Define $F(n) = 1$ for $n \in X$ and 0 otherwise.)

Let $\mathcal{F}_0$ be the filter in **P** consisting of all sets of K-density 1. Let $\mathcal{U}_0$ be an ultrafilter extending $\mathcal{F}_0$. Hence all sets in $\mathcal{U}_0$ have positive density. Let **FP** denote the set of functions computable in polynomial time. We define M_0 to be the ultrapower constructed by taking **FP** modulo $\mathcal{U}_0$,

$$M_0 = \mathbf{FP}/\mathcal{U}_0,$$

with the PV function symbols interpreted in the natural way. The fact that in M_0 all true universal PV sentences are satisfied is an immediate consequence of Łoś's theorem (see Corollary 2.7). The distinguished element $\iota \in M_0$ is defined to be the element of $\mathbf{FP}/\mathcal{U}_0$ represented by the identity function id on $\mathbb{N}$, in symbols $\iota = [id]_{\mathcal{U}_0}$.

Let $u_0 = \{U_1 \supset U_2 \supset \dots\}$ be a cofinal chain in $\mathcal{U}_0$. We define the *density of a set with respect to* u_0, or u_0*-density*, to be the partial function defined by[2]

$$dens_{u_0} X = \lim_{n\to\infty} \frac{dens_K X \cap U_n}{dens_K U_n}. \tag{3}$$

Again, one can easily prove that we can pick u_0 so that u_0-density is defined for every $X \in \mathbf{AR}$. The following facts are immediate corollaries of 1. and 4.:

5. $dens_{u_0}$ is finitely additive.
6. $dens_{u_0} Z = \frac{1}{2}$ for every pseudorandom set Z.

The set of models that extend M_0 will be defined using the following set of ultrafilters on the boolean algebra $\mathbf{AR}$.

$$\Omega = \{\mathcal{U};\ \mathcal{U}\ \textit{ultrafilter on}\ \mathbf{AR},\ \textit{and}\ \forall V \in \mathcal{U}\ dens_{u_0} V > 0\}.$$

Lemma 3.2. *$\mathcal{U}_0 \subseteq \mathcal{U}$ for every $\mathcal{U} \in \Omega$.*

Proof. Suppose that $\mathcal{U}$ does not contain $\mathcal{U}_0$. Since u_0 is cofinal in $\mathcal{U}_0$, $U_i \notin \mathcal{U}$ for some $U_i \in u_0$. Hence $\mathbb{N} \setminus U_i \in \mathcal{U}$. But $dens_{u_0} \mathbb{N} \setminus U_i = 0$. So $\mathcal{U}$ is not in Ω. ⊣

The following is also easy.

7. If $\mathcal{F} \subseteq \mathbf{AR}$ is a filter such that $dens_{u_0} U > 0$ for all $U \in \mathcal{F}$, then $\mathcal{F}$ can be extended to an ultrafilter belonging to Ω. In particular, for every $X \in \mathbf{AR}$ such that $dens_{u_0} X > 0$, there exists a $\mathcal{U} \in \Omega$ such that $X \in \mathcal{U}$.

Let **FAR** denote arithmetically definable functions. Define

$$\mathcal{M} = \{M;\ M = \mathbf{FAR}/\mathcal{U},\ \mathcal{U} \in \Omega\}.$$

Note that for $\mathcal{U}_1 \neq \mathcal{U}_2$, the ultrapower models are different (they may be isomorphic, though). Since $\mathbf{FP} \subseteq \mathbf{FAR}$ and $\mathcal{U}_0 \subseteq \mathcal{U}$, for every $\mathcal{U} \in \Omega$, we have:

8. Every $M \in \mathcal{M}$ is an extension of M_0.

The fact that these models are models of $AR^{\mathbb{N}}$ is a consequence of Łoś's theorem (see Lemma 2.6).

For $X \in \mathbf{AR}$, let

$$\Omega[X] = \{\mathcal{U};\ X \in \mathcal{U}\},$$

and put

$$\mathcal{A}_0 = \{\Omega[X];\ X \in \mathbf{AR}\}$$

9. $\mathcal{A}_0$ is a Boolean algebra.

[2]More precisely, we should also use the index K, but there is no danger of confusion, since K is fixed for the rest of the proof.

Indeed, $X \mapsto \Omega[X]$ is a homomorphism of the Boolean algebra **AR** onto $\mathcal{A}_0$.

Lemma 3.3. *$\Omega[X] = \Omega[Y]$ if and only if $dens_{u_0} X \triangle Y = 0$ (where $\triangle$ denotes the symmetric difference).*

Proof. Let C be the kernel of the homomorphism $X \mapsto \Omega[X]$. The lemma is equivalent to the statement that $C = \{X;\ dens_{u_0} X = 0\}$. If $dens_{u_0} X = 0$, then $\Omega[X] = \emptyset$ by definition. If $dens_{u_0} X > 0$, then $\Omega[X] \neq \emptyset$ by 7. ⊣

This lemma enables us to define an additive measure ν_0 on $\mathcal{A}_0$ by putting

$$\nu_0(\Omega[X]) = dens_{u_0} X.$$

In particular, $\nu_0(\Omega) = 1$.

Lemma 3.3. *If $A_1, A_2, \dots \in \mathcal{A}_0$ are pairwise disjoint and $\bigcup_n A_n \in \mathcal{A}_0$, then $\nu_0(A_n) = 0$ for all n except for a finite number of them.*

Proof. To prove the claim, suppose the contrary. Let $X, X_1, X_2, \dots \in$ **AR** be such that $A_n = \Omega[X_n]$, for $n = 1, 2, \dots$, and $\bigcup_n A_n = \Omega[X]$. We observe that $dens_{u_0} X_i \cap X_j = 0$ for $i \neq j$, because $A_i \cap A_j = \emptyset$.

Let $Y_n = X \setminus \bigcup_{i=1}^{n-1} X_i$. We will show that $dens_{u_0} Y_n > 0$ for all n. Suppose that for some n, $dens_{u_0} Y_n = 0$. Let $m \geq n$ such that $dens_{u_0} X_m > 0$. Since $dens_{u_0} X_m \cap \bigcup_{i=1}^{n-1} X_i = 0$, we have $dens_{u_0} X_m \cap X = 0$. This implies that $\Omega[X_m] \cap \Omega[X] = \emptyset$. But this is impossible, because $\Omega[X_m] \neq \emptyset$. Thus $dens_{u_0} Y_n > 0$ for all n.

Extend the filter $\{Y_n;\ n = 1, 2, \dots\}$ to an ultrafilter $\mathcal{U} \in \Omega$. Clearly $\mathcal{U} \in \Omega[X]$, but for no n, $\mathcal{U} \in \Omega[X_n]$. ⊣

An immediate corollary is:

10. ν_0 is σ-additive.

According to a basic theorem about extensions of measures, we can extend the σ-additive probability measure ν_0 defined on the Boolean algebra $\mathcal{A}_0$ to a σ-additive probability measure ν_1 defined on a σ-algebra $\mathcal{A}_1$. Using the bijection $\mathcal{U} \mapsto \mathbf{FAR}/\mathcal{U}$ we translate the measure ν_1 defined on a σ-algebra $\mathcal{A}_1$ to a measure ν defined on a σ-algebra $\mathcal{B}$ of subsets of $\mathcal{M}$.

We will now prove the second condition of the theorem. Let $\phi(x_1, \dots, x_k)$ be a *PV* formula, $a_1, \dots, a_k \in M_0$, let $\mathcal{U} \in \Omega$ and let $M = \mathbf{FAR}/\mathcal{U}$. Further, let $f_1, \dots, f_k \in \mathbf{FAR}$ be the functions representing $a_1, \dots, a_k$ (in symbols, $a_i = [f_i]_{\mathcal{U}_0}$).

According to Łoś's theorem, $M \models \phi(a_1, \ldots, a_k)$ if and only if

$$\{n;\ \mathbb{N} \models \phi(f_1(n), \ldots, f_k(n))\} \in \mathcal{U}.$$

Hence the set of ultrafilters for which the models satisfy $\phi(a_1, \ldots, a_k)$ has the form $\Omega[X]$, for $X \in \mathbf{AR}$. Therefore

$$\{M;\ M \models \phi(a_1, \ldots, a_k)\} \in \mathcal{B}.$$

It remains to prove the third condition of the theorem. Let s be a pseudorandom sequence and let $\sigma(x)$ be a formula defining $s(x) = 1$. By Łoś's theorem, the set of models satisfying $\sigma(\iota)$ corresponds to the set of ultrafilters such that $\{n;\ \mathbb{N} \models \sigma(n)\} \in \mathcal{U}$ (recall that $\iota = [id]_{\mathcal{U}_0}$). Thus we have

$$\begin{aligned} \nu(\{M;\ M \models \sigma(\iota)\}) &= \nu_1(\{\mathcal{U};\ \{n;\ \mathbb{N} \models \sigma(n)\} \in \mathcal{U}\}) \\ &= dens_{\mathcal{U}_0}\{n;\ \mathbb{N} \models \sigma(n)\} \\ &= \tfrac{1}{2}, \end{aligned}$$

by 6. Thus the theorem is proved. ⊣

4 Remarks

1. We will generalize the concept of pseudorandom sequences and sets to cover sequences in which 1 occurs with frequency $p \neq \frac{1}{2}$.

Definition 4. *Let p be a real number, $0 < p < 1$. We will say that a sequence $s : \mathbb{N} \to \{-1, 1\}$ is p-biased pseudorandom, if s is computable in nondeterministic polynomial time and*

$$\lim_{n \to \infty} \sum_{i=0}^{n-1} f(i)((s(i) + 1)/2 - p)/n = 0,$$

for every polynomial time computable function $f : \mathbb{N} \to \{-1, 1\}$.

A set $Z \subseteq \mathbb{N}$ will be called p-biased pseudorandom if it is the set of arguments for which a p-biased pseudorandom sequence is 1.

Intuitively, a set $Z \subseteq \mathbb{N}$ is p-biased pseudorandom if its density on every set X, $X \in \mathbf{P}$, is p. This can be expressed formally using the notation of the proof of the main theorem by

$$dens_K X \cap Z = p \cdot dens_K X$$

for every set $X \in \mathbf{P}$ and for every K for which $dens_K X$ is defined.

Several propositions proved above generalize to p-biased pseudorandom sequences and sets. In particular, we would like to draw the reader's attention to Corollary 1 and Theorem 3.1. The condition 3. of Theorem 3.1 holds true for all real numbers p, $0 < p < 1$ and all p-biased pseudorandom sequences simultaneously.

2. The main weakness of Theorem 3.1 is that condition 3. is stated only for one element, only for ι. One can show that in the constructed system $\mathcal{M}$, condition 3. holds for several other elements of M_0; in particular, it is true for all elements of the form $a\iota + b$ for $a \in \mathbb{N}$ and $b \in \mathbb{Z}$. This can easily be proved using the following lemma.

Lemma 4.1. *Let s be a pseudorandom sequence. Let $g \in$ **FP** be increasing and invertible in polynomial time on an infinite interval $[n_0, \infty)$. Furthermore, suppose that $Rng(g)$, the range of g, has positive density. Then the sequence s' defined by $s'(x) = s(g(x))$ is also pseudorandom.*

Proof. Let s and g be given and suppose s' is not pseudorandom. Let f' be a function that witnesses that s' is not pseudorandom. We define a function f that witnesses that s is not pseudorandom.

$$f(n) = \begin{cases} f'(g^{-1}(n)) & \text{if } n \geq n_0 \text{ and } n \in Rng(g), \\ 0 & \text{otherwise.} \end{cases} \qquad \dashv$$

We certainly cannot expect condition 3. to hold for all numbers of M_0. For small numbers n, $s(n)$ is defined in M_0, because s is computable in exponential time. If $\alpha \in M_0$ is larger than all numbers $c\iota$, for $c \in \mathbb{N}$, then $\alpha = [g]_{\mathcal{U}_0}$ for some g that grows more than linearly. The range of such a g has density 0, hence we cannot deduce anything about it. For example, $\lambda(n^2) = 1$ for all n, whence $M_0 \models \lambda(\iota^2) = 1$.

3. We also cannot expect stronger properties of random sequences to hold in the system of models of Theorem 3.1 unless we assume more about the sequences. For example, $s(\iota)$ and $s(\iota + 1)$ do not have to be independent in $\mathcal{M}$, because we do not assume any kind of independence for pairs $s(n)$ and $s(n+1)$, $n \in \mathbb{N}$. A more specific example (assuming that λ is pseudorandom) is the fact that $\lambda(2n) = -\lambda(n)$.

5 Philosophical speculations

Some cosmologists believe that when the universe emerged from a singularity some physical properties of it were decided randomly. Others even believe that there is a *multiverse* consisting of many different universes, one of which is our universe. In contrast to this, philosophers have never doubted that the basic mathematical structures, namely the natural and real numbers, are unique. These structures are unique in the sense that they must be same in all conceivable physical worlds.

There are good reasons to believe that the natural numbers are absolute in the sense that there are no possible alternatives to them. The only structures that satisfy the basic arithmetical laws and are different from the natural numbers are nonstandard models. Though it has been proposed that the actual natural numbers have the structure of a nonstandard model, e.g., in [9], most philosophers do not accept such a possibility. The problem is that a nonstandard model contains the standard model as an initial part, and so we should identify the natural numbers with this initial part. Thus viable alternatives should use the same numbers with different arithmetical operations. Since we can prove that the operations of addition and multiplication are uniquely determined by the basic axioms (the axioms of Robinson Arithmetic), it is inconsistent to assume that on the set of standard numbers different kinds of addition and multiplication are possible.

However, what is inconsistent in our world may be consistent in a different one, and vice versa. Consider a pair of (necessarily nonstandard) models of Peano Arithmetic M and N that have the same elements, the same addition, but different multiplication. (Such pairs can be easily constructed using recursively saturated models.) In a world in which M is the standard natural numbers, it is inconsistent to assume that anything like N exists. Yet, it does.

We may secretly ponder over such scenarios, but there is a strong reason not to talk about the possibility of different arithmetics openly. If a concept is inconsistent, we cannot talk about it and there cannot be any theory around it. Therefore, any conjecture of this kind would be neither provable nor disprovable and, as such, should be discarded as meaningless.

However, some phenomena can be studied even if they are not directly observable—because they have side effects. The presence of these effects is a proof of the phenomenon. A side effect of the origin of our integers in a random process could be the randomness present in the structure of integers. It is not genuine randomness, because the integers are a single structure. What we rather observe are some properties that are satisfied by truly random objects. Therefore we call this *pseudorandomness*. Number theorists are familiar with this; they use assumptions about random behavior in heuristic arguments when they are not able to prove theorems rigorously and some conjectures are also justified in this way (including the Riemann Hypothesis).

Acknowledgment

I would like to thank Emil Jeřábek and Jan Krajíček for their useful comments and suggestions.

This research was partially supported by grants IAA100190902 of GA AV ČR and GAP202/12/G061 of GA ČR.

References

[1] S.A. Cook, Feasibly constructive proofs and the propositional calculus, *Proc. seventh annual ACM symposium on Theory of computing*, ACM New York, 1975, pp. 83–97.

[2] S.A. Cook, P. Nguyen, *Logical Foundations of Proof Complexity*, ASL series Perspectives in Logic, Cambridge Univ. Press, 2010.

[3] O. Goldreich, *Foundation of Cryptography: Basic Tools*, Cambridge Univ. Press, 2001.

[4] B. Green, On (not) computing the Möbius functions using bounded depth circuits, *Combinatorics, Probability and Computing*, vol. 21(6), 2012, pp. 942–951.

[5] S. Kochen, S. Kripke, Non-standard models of Peano arithmetic, *Logic and arithmetic, int. Symp., Zürich 1980*, Enseign. Math., II. Sér. 28, 1982, pp. 211–231.

[6] J. Krajíček, *Bounded arithmetic, propositional logic, and complexity theory*, Encyclopedia of Mathematics and Its Applications, Vol. 60, Cambridge University Press, Cambridge - New York - Melbourne, 1995.

[7] J. Krajíček, *Forcing with random variables and proof complexity*, London Mathematical Society Lecture Note Series, No.382, Cambridge University Press, 2011.

[8] C.P. Schnorr, Zufälligkeit und Wahrscheinilichkeit. Eine algorithmische Begründung der Wahrscheinilichkeitstheorie. *LNM* 218, Springer, 1971.

[9] P. Vopěnka, *Mathematics in the Alternative Set Theory*, Teubner, Leipzig, 1979.

11

The asymptotic behaviour of the number of trees in certain classes

JAN-CHRISTOPH SCHLAGE-PUCHTA[1]

Abstract: When counting combinatorial structures, asymptotic questions are often encoded into the singularity of least absolute value of the corresponding generating function. In this article we describe the rate of convergence of this singularity, as the structures run over a sequence of sets of trees, which in the limit exhaust the set of all trees. This question is motivated by applications in proof theory, more precisely, phase transitions in unprovability results.

1 Introduction and results

Consider an interesting class $\mathcal{C}$ of finite rooted trees. A natural question to ask is how many trees with a given number of nodes are contained in this class. Of course the answer depends not only on the class $\mathcal{C}$, but also on the question which trees are to be regarded as equal; for example we can count trees up to isomorphism, up to cyclic permutation of the daughter trees of each node, or we can completely fix the ordering of the daughter trees. Further, if we consider r-ary trees we may or may not allow for empty daughter trees.

In view of these possibilities it appears reasonable to restrict oneself to problems stemming from real applications. However, it turns out that in many important cases one can give a nice description of the generating series $T_{\mathcal{C}}(x)$ of the number of trees in $\mathcal{C}$, which can be used to determine the asymptotic behaviour of the number of trees by com-

[1] jan-christoph.schlage-puchta@uni-rostock.de

Studies in Weak Arithmetics.
Patrick Cégielski, Charalampos Cornaros,
Costas Dimitracopoulos.

plex analytic means. If $T_{\mathcal{C}}(x)$ satisfies an algebraic equation, then the location and the order of the singularity of this equation of least absolute value determines the asymptotic behaviour with great precision. Hence, the determination of this singularity is one of the first things to do.

In this article we consider the question as to how the singularity behaves as $\mathcal{C}$ varies.

Our first result deals with the number of r-ary trees for varying r. We denote by the degree of a node the number of daughter nodes, that is, unless the node is the root, one less than the graph theoretic degree.

Theorem 1. *Let $t_k(n)$ be the number of trees with n nodes and degree at most k, $t(n)$ be the of all trees with n nodes. Put $T_k(x) = \sum_{n\geq 1} t_k(n)x^n$, and $T(x) = \sum_{n\geq 1} t(n)x^n$. Then T_k has a real branch point of order 2 at ρ_k, and no other singularity of absolute value $\leq \rho_k+\delta$ for some positive constant δ. We have $\rho + e^{-c_1 k} < \rho_k < \rho + e^{-c_2 k}$ for some constants $c_1 > c_2 > 0$, where ρ is the radius of convergence of $T(x)$.*

One can get the values $c_1 = 2.17$, $c_2 = 0.893$ by just keeping track of the constants in our proof, however, we refrain from giving the details, since doing so involves quite some work, no new ideas, and the numerical value of c_1, c_2 is of dubious interest.

The interesting part here is the dependence of ρ_k on k, the other statements are well known (confer [2], Section IV.7), and are restated here for the convenience of the reader as well as for the definition of ρ_k.

Next we fix r, and consider the set of r-ary planar trees, where each node either has precisely r daughter trees, or is a leaf. Let $s(n)$ be the number of all such trees of height n, and $s_h(n)$ be the number of such trees, such that in each path in the tree the left-most branch is taken at most h times. In particular, $s_0(n)$ is the number of $(r-1)$-ary trees, and $s_1(2) = r$. Then we have the following.

Theorem 2. *Define $s_h(n)$ as before, and set $S_h(x) = \sum_{n\geq 0} s_h(n)x^n$, and $S(x) = \sum_{n\geq 0} s(n)x^n$. Then S and S_h have a real branch point of order 2 at ρ and ρ_h, respectively, and no other singularity of absolute value $\leq \rho_h+\delta$ for some positive constant δ. Then we have $\rho_h - \rho \asymp h^{-2}$.*

Here $f \asymp g$ means that $f = \mathcal{O}(g)$ and $g = \mathcal{O}(f)$; similarly we shall later use $f \ll g$ to denote $f = \mathcal{O}(g)$. For large h we have the explicit bounds

$$\frac{1}{5(C_1+1)(C_2+1)h^2} < \rho_h - \rho < \frac{56}{C_1 C_2 h^2}$$

where

$$C_1 = \left(\frac{r}{r-1}\right)^{r-2}\binom{r}{2}, \qquad C_2 = \left(\frac{r}{r-1}\right)^{r}.$$

Neither the upper nor the lower bound is close to optimal, however, we do not really care about numerical values.

These questions are motivated by problems in proof theory. Bovykin and Weiermann [1] had given analytic statements which are unprovable in Peano arithmetic. More precisely, they gave a parametric version of Kruskal's theorem and showed that as a certain parameter varies there is a sharp phase transition from provable to unprovable (within the frame work of Peano arithmetic), provided that a weak version of Theorem 1 holds true. Weiermann (personal communication) showed that in the situation of Theorem 1 and 2 we have that $\rho_k \to \rho$ holds true, and noted that if one had explicit bounds for the speed of this convergence, one could sharpen these results. In this context Woods [18] showed that the speed of convergence in Theorem 2 is subexponential, however, his proof was indirect and did not give an explicit subexponential bound.

Questions concerning the enumeration of certain types of trees arise repeatedly in the study of phase transitions as introduced by Weiermann [3]. If one works with proof systems different from PA, the question of enumerating trees is replaced by the question of determining for a fixed ordinal α the number of ordinals $\beta < \alpha$, which satisfy $N(\beta) \leq n$, where N is a suitable complexity measure, e.g. the length of the Cantor normal form of β. In general the generating series of these counting functions do not satisfy any reasonable functional equation, which makes the related problems much harder. However, the case $\alpha = \omega^{\omega^{\cdots^{\omega}}}$ was successfully treated by Weiermann [4].

One advantage of reducing problems to the enumeration of trees is that this area of analytic combinatorics is well developed. The specific question on the convergence speed of the singularities ρ_h might look artificial to a combinatorist, still we can use the machinery described in [2], in particular chapter II and VI.

2 Proof of Theorem 1

It is well known that the series $T(x)$ is the unique solution of the functional equation

$$T(x) = x \exp\left(\sum_{\nu\geq 1} \frac{T(x^\nu)}{\nu}\right)$$

satisfying $T(0) = 0$ and $T'(0) = 1$. Our first result shows that T_k can be defined as a solution of an approximation of this equation. Define the polynomial $P_k(X_1, \ldots, X_k)$ as follows. Expand $\exp\left(\sum_{\nu\geq 1} \frac{X_\nu}{\nu}\right)$ into a power series in the infinitely many variables $X_1, \ldots$, and let P_k be the part consisting only of monomials $X_1^{a_1} \ldots X_m^{a_m}$ satisfying $\sum j a_j \leq k$. The following is essentially [2, Prop. I.6], note that while in the statement of that proposition only the existence of some polynomial is shown, the precise form is contained in the proof.

Lemma 2.1. *The function T_k is the unique solution of the functional equation $T_k(x) = xP_k(T_k(x), T_k(x^2), \ldots, T_k(x^k))$.*

Lemma 2.2. *The least positive singularity of T_k is given by the system of equations*

$$\begin{aligned} T_k(\rho_k) &= \rho_k P_k(T_k(\rho_k), \ldots, T_k(\rho_k^k)) \\ 1 &= \rho_k \frac{\partial}{\partial x_1} P_k(T_k(\rho_k), \ldots, T_k(\rho_k^k)). \end{aligned}$$

Proof. This follows from the implicit function theorem, see [2, Chapter VII.4]. ⊣

Since $P_k(x_1, \ldots, x_k)$ consists of the monomials of weight $\leq k$ in $\exp(\sum_i \frac{x_i}{i})$, and the partial derivative of $\exp(x_1 + f(x_2, \ldots, x_k))$ with respect to x_1 is again $\exp(x_1 + f(x_2, \ldots, x_k))$, we see that

$$\frac{\partial}{\partial x_1} P_k(x_1, \ldots, x_k) = P_{k-1}(x_1, \ldots, x_k).$$

Hence, we obtain that

$$T_k(\rho_k) = 1 + \rho_k\Big(P_k(T_k(\rho_k), \ldots, T_k(\rho_k^k)) - P_{k-1}(T_k(\rho_k), \ldots, T_k(\rho_k^k))\Big). \tag{1}$$

We first prove the upper bound. We set $Q_k = P_k - P_{k-1}$. We have $\rho_k \leq \rho_2 = 0.4202$, $\rho = 0.338$ (Confer [2, p. 455]), in particular, $\rho_k^\ell < \rho$ hold true for all $k, \ell \geq 2$. Clearly, all coefficients in the Taylor series of T_k are at most as large as the corresponding coefficient of T, hence, for $x \in [0, \rho_k]$ we have

$$\begin{aligned} &T_k(x) \leq 1 + \rho_k Q_k(T_k(x), T(\rho_2^2) \ldots, T(\rho_2^k)) \\ &\leq 1 + \rho_k \exp\left(T_k(x) + \sum_{i=2}^{\infty} \frac{T(\rho_2^i)}{i}\right) - \rho_k P_{k-1}(T_k(x), T(\rho_2^2) \ldots, T(\rho_2^k)). \end{aligned}$$

We claim that for k sufficiently large this implies $T_k(x) \leq 2$ for all $x \leq \rho_k$. Assume to the contrary that the equation $T_k(x) = 2$ has a solution $x_0 \leq \rho_k$. Then we have

$$2 \leq 1 + \rho_k \exp\left(2 + \sum_{i=2}^{\infty} \frac{T(\rho_2^i)}{i}\right) - \rho_k P_{k-1}(2, T(\rho_2^2) \ldots, T(\rho_2^k)),$$

on the other hand we have

$$\lim_{k\to\infty} P_{k-1}(x_1,\dots,x_{k-1}) = \exp\Big(\sum_{i=1}^{\infty}\frac{x_i}{i}\Big)$$

for all non-negative sequences (x_i) for which the series on the right hand side converges, hence, for k sufficiently large we have

$$\left|\exp\Big(2+\sum_{i=2}^{\infty}\frac{T(\rho_2^i)}{i}\Big) - P_{k-1}(2, T(\rho_2^2)\dots, T(\rho_2^k))\right| < 1,$$

and we obtain $2 < 1+\rho_k$, contradicting the fact that $\rho_k \leq \rho_2 = 0.4202$. Hence, for k sufficiently large we have $T_k(\rho_k) \leq 2$.

Let $a_1,\dots,a_k$ be non-negative integers satisfying $a_1+2a_2+\dots,+ka_k = k$. The coefficient of $x_1^{a_1}\cdots x_k^{a_k}$ in Q_k is

$$\frac{1}{(\sum_{i=1}^k a_i)!}\binom{\sum_{i=1}^k a_i}{a_1,\dots,a_k}\prod_{i=1}^k\frac{1}{i^{a_i}} = \frac{1}{\prod_{i=1}^k a_i! i^{a_i}}.$$

Since the coefficients of T are non-negative, we have $T(\rho^\ell) < \rho^{\ell-2}T(\rho^2)$ for all $\rho \geq 2$, since on the other hand $T(\rho^2) > \rho^2$, we obtain $T(\rho^\ell) \leq T(\rho^2)^{\ell/2}$ for all $\ell \geq 2$. The contribution of the monomial corresponding to the tuple $(a_1,\dots,a_k)$ is therefore

$$\frac{2^{a_1}}{a_1!}\prod_{i=2}^k\frac{T(\rho^2)^{\frac{ia_i}{2}}}{a_i! i^{a_i}} \leq \frac{2^{a_1}T(\rho^2)^{\frac{k-a_1}{2}}}{a_1!} = T(\rho^2)^{k/2}\frac{2^{a_1}T(\rho^2)^{\frac{-a_1}{2}}}{a_1!} \ll T(\rho^2)^{\frac{k}{2}}.$$

Monomials in Q_k correspond to partitions of k, hence, the number of such monomials is $< e^{c\sqrt{k}}$, and we obtain that

$$Q_k(T_k(\rho_k), T(\rho^2),\dots,T(\rho^k)) \ll e^{-ck}$$

for some positive c, that is, $T_k(\rho_k) = 1+\mathcal{O}(e^{-ck})$. From Lemma 2.2 we now deduce

$$\rho_k^{-1} = \big(1+\mathcal{O}(e^{-ck})\big)P_k(T_k(\rho_k),\dots,T_k(\rho_k^k)). \tag{2}$$

The polynomial P_k consists of a subset of the series $\exp(\sum\frac{x_i}{i})$, hence, in the domain $x_1 \leq 2$, $x_\ell \leq cT(\rho^2)^{\ell/2}$ the partial derivative of P with respect to any of the x_i is bounded independent of k. Hence,

$$\begin{aligned}&P_k(T_k(\rho_k),\dots,T_k(\rho_k^k)) - P_k(1, T(\rho_k^2),\dots,T(\rho_k^k))\\ &\quad\ll |1-T_k(\rho_k)| + \sum_{i=2}^{\infty}|T(\rho_k^i) - T_k(\rho_k^i)|\end{aligned}$$

The series T and T_k coincide in the first k coefficients, hence, for real positive x the difference $|T(x)-T_k(x)|$ can be bounded by the difference of $T(x)$ and the partial sum consisting of the terms of degree $\leq k$.

Hence, $|T(x) - T_k(x)| \ll (x/\rho)^k$ uniformly for $x \leq \rho$, and we obtain

$$\begin{aligned} P_k(T_k(\rho_k), \dots, T_k(\rho_k^k)) - P_k(1, T(\rho_k^2), \dots, T(\rho_k^k)) \\ \ll e^{-ck} + \sum_{i=2}^{\infty} \left(\tfrac{\rho_k^i}{\rho}\right)^k \\ = e^{-ck} + \tfrac{\rho_k^{2k}}{\rho^k} \cdot \tfrac{1}{1-\rho_k} \ll e^{-ck} + \left(\tfrac{\rho_k^2}{\rho}\right)^k \leq e^{-ck} + \left(\tfrac{\rho_2^2}{\rho}\right)^k \ll e^{-ck}, \end{aligned}$$

since $\frac{\rho_2^2}{\rho} = 0.5225\ldots < 1$. Note that in order to get the last bound we might have to decrease the value of c. Since ρ_k is bounded against 0 and ∞, we can plug this estimate into (2) to obtain

$$\rho_k^{-1} = \big(1 + \mathcal{O}(e^{-ck})\big) P_k(1, T(\rho_k^2), \dots, T(\rho_k^k)).$$

Since $\rho_k \geq \rho$, replacing ρ_k by ρ decreases the right-hand side, and we obtain

$$\rho_k^{-1} \geq \big(1 + \mathcal{O}(e^{-ck})\big) P_k(1, T(\rho^2), \dots, T(\rho^k)).$$

On the other hand we have

$$\rho^{-1} = \lim_{k\to\infty} P_k(1, T(\rho^2), \dots, T(\rho^k)).$$

Define σ_k as

$$\sigma_k^{-1} = P_k(1, T(\rho^2), \dots, T(\rho^k)).$$

Then $\sigma_k^{-1} - \sigma_{k-1}^{-1} = Q_k(1, T(\rho^2), \dots, T(\rho^k)) \ll e^{-ck}$, hence,

$$|\sigma_k - \rho| = |\sigma_k - \lim_{\ell\to\infty} \sigma_\ell| \ll \sum_{\ell \geq k} e^{-c\ell} \ll e^{-ck},$$

that is, $\rho_k^{-1} \geq \big(1 + \mathcal{O}(e^{-ck})\big)\rho^{-1}$, and we obtain $\rho_k \leq \rho + \mathcal{O}(e^{-ck})$. By choosing c_2 somewhat smaller than c the upper bound follows.

For the lower bound note that (1) implies $T_k(\rho_k) > 1$. On the other hand there exists a tree with $k+2$ vertices which is counted by T, but not by T_k. Hence, $T(x) \geq T_k(x) + x^{k+2}$ holds true for all $x \in [0, \rho]$, in particular $T_k(\rho) \leq 1 - \rho^{k+2}$. This implies $\rho_k \neq \rho$. Differentiating the functional equation yields for $x \in [0, \rho_k)$

$$\begin{aligned} T_k'(x) &= P_k(T_k(x), \dots, T_k(x^k)) \\ &\quad + x \sum_{j=1}^{k} j x^{j-1} T_k'(x^j) \Big(\tfrac{\partial}{\partial x_j} P_k\Big)(T_k(x), \dots, T_k(x^k)) \\ &= \tfrac{T_k(x)}{x} + x P_{k-1}(T_k(x), \dots, T_k(x^{k-1})) \\ &\quad + x \sum_{j=2}^{k} j x^{j-1} T_k'(x^j) \Big(\tfrac{\partial}{\partial x_j} P_k\Big)(T_k(x), \dots, T_k(x^k)) \\ &\leq \tfrac{T_k(x)}{2} + x T_k + k T'(\rho_2) \sum_{j=2}^{k} \Big(\tfrac{\partial}{\partial x_j} P_k\Big)(T_k(x), \dots, T_k(x^k)). \end{aligned}$$

Since P_k is a polynomial of weight k, differentiating $P_k(x_1, \dots, x_k)$ with respect to x_i increases the value by $\frac{k}{x_i}$ at most, and we obtain

$$\begin{aligned}\left(\frac{\partial}{\partial x_j}P_k\right)(T_k(x),\dots,T_k(x^k)) &\leq \frac{k}{T_k(x^j)}P_k((T_k(x),\dots,T_k(x^k))\\ &\leq \frac{jT_k(x)}{x^{j+1}} \leq \frac{2k}{x^{k+1}},\end{aligned}$$

and therefore $\frac{T_k'(x) \ll k^3}{x^{k+1}}$. Since $T(\rho_k) - T(\rho) > \rho^{k+2}$, we deduce

$$\rho_k - \rho \gg \frac{\rho^{k+2}\rho_k^{k+1}}{k^3} \gg e^{-ck},$$

and the claimed lower bound follows as well.

3 Proof of Theorem 2

We begin by recalling some properties of S_h.

Proposition 3.
(1) *We have* $S_0(z) = 1$, $S_{h+1}(z) = 1 + zS_h(z)S_{h+1}(z)^{r-1}$.
(2) *For* $h \geq 2$ *we have* $S_h(\rho_h) = \frac{r-1}{r-2}$.
(3) *We have* $S(\rho) = \frac{r}{r-1}$.
(4) *We have* $\rho = \frac{(r-1)^{r-1}}{r^r}$.

Proof. The recursive relation follows from the definition of the h-th family. The second and third claim follows by singularity analysis from this. The last statement was proven by Weiermann. Note that here the determination of the value of ρ is difficult; once the value of ρ is known, it can be verified by a tedious yet trivial calculation. ⊣

For a real number $x \in [0, \rho_m]$ we put $\delta_m = \frac{r}{r-1} - S_m(x)$. We will study the sequence (δ_m) for fixed x slightly larger than ρ.

Lemma 3.1. *For every* $\epsilon > 0$ *there exists an* m_0 *and a* $\delta > 0$, *such that* $S_{m_0}(x) \in [\frac{r}{r-1} - \epsilon, \frac{r}{r-1}]$ *for all* $x \in [\rho, \rho + \delta]$.

Proof. We have $S_m(\rho) \to S(\rho) = \frac{r}{r-1}$, hence we can choose m_0 such that $S_{m_0}(\rho) \geq \frac{r}{r-1} - \epsilon$. Since S_{m_0} is a power series with non-negative coefficients, it is non-decreasing, and we obtain $S_{m_0}(x) \geq \frac{r}{r-1} - \epsilon$ for all $x \in [\rho, \rho_m]$. On the other hand there exist trees which are counted by S, and not by S_{m_0}, hence $S_{m_0}(\rho) < S(\rho)$, and since both S_{m_0} and S are continuous, this inequality holds true in some neighbourhood of ρ. Hence if δ is sufficiently small, then $S_{m_0}(x) < \frac{r}{r-1}$ for $x < \rho + \delta$. ⊣

Lemma 3.2. *If* $x \in [\rho, \rho_{m+1}]$ *and* $\delta_m > er(x - \rho)$, *then*

$$\delta_{m+1} = \delta_m - C_1\delta_m^2 - C_2(x - \rho) + \mathcal{O}(\delta_m^3 + \delta_m(x - \rho))$$

with

$$C_1 = \left(\frac{r}{r-1}\right)^{r-2}\binom{r}{2}, \qquad C_2 = \left(\frac{r}{r-1}\right)^r.$$

Proof. Inserting the definition of δ_m into the recurrence relation for T_m we obtain

$$\begin{aligned} \tfrac{r}{r-1} - \delta_{m+1} \quad = \quad & 1 + \rho\left(\tfrac{r}{r-1} - \delta_m\right)\left(\tfrac{r}{r-1} - \delta_{m+1}\right)^{r-1} \\ & +(x-\rho)\left(\tfrac{r}{r-1} - \delta_m\right)\left(\tfrac{r}{r-1} - \delta_{m+1}\right)^{r-1} \end{aligned} \tag{3}$$

We claim that given $\delta_m \in [e(x-\rho), 1]$ this equation has a unique solution $\delta_{m+1} \in [0, \delta_m]$. View the difference of the right and the left hand side as a function of δ_{m+1}. If we insert $\delta_{m+1} = 0$, we get

$$\begin{aligned} & \tfrac{r}{r-1} - 1 - \rho\left(\tfrac{r}{r-1} - \delta_m\right)\left(\tfrac{r}{r-1}\right)^{r-1} - (x-\rho)\left(\tfrac{r}{r-1} - \delta_m\right)\left(\tfrac{r}{r-1}\right)^{r-1} \\ & \quad > \tfrac{\delta_m}{r} - (x-\rho)\left(\tfrac{r}{r-1} - \delta_m\right)\left(\tfrac{r}{r-1}\right)^{r-1} \geq \tfrac{\delta_m}{r} - \tfrac{ec}{h^2} \geq 0, \end{aligned}$$

while for $\delta_{m+1} = \delta_m$ we have

$$\begin{aligned} & \tfrac{r}{r-1} - \delta_m - 1 - \rho\left(\tfrac{r}{r-1} - \delta_m\right)^r - (x-\rho)\left(\tfrac{r}{r-1} - \delta_m\right)^r \\ & < \underbrace{\frac{r}{r-1} - 1 - \rho\left(\frac{r}{r-1}\right)^r}_{=0} - \delta_m - (x-\rho)\left(\tfrac{r}{r-1} - \delta_m\right)^r < 0. \end{aligned}$$

The derivative of this function with respect to δ_{m+1} is

$$\begin{aligned} & -1 + (r-1)\rho\left(\tfrac{r}{r-1} - \delta_m\right)\left(\tfrac{r}{r-1} - \delta_{m+1}\right)^{r-2} \\ & +(r-1)(x-\rho)\left(\tfrac{r}{r-1} - \delta_m\right)\left(\tfrac{r}{r-1} - \delta_{m+1}\right)^{r-2} \\ & < -1 + (r-1)\rho\left(\tfrac{r}{r-1}\right)^{r-1} + e(r-1)(x-\rho) \\ & \quad = -\tfrac{1}{r} + e(r-1)(x-\rho) < 0, \end{aligned}$$

and we conclude that $\delta_{m+1} \in [0, \delta_m]$ is uniquely determined by (3).

Inserting the value for ρ into (3), expanding the second term on the right, and bounding the last term on the right rather crudely, we obtain

$$\begin{aligned} \delta_{m+1} = \quad & \delta_m - \left(\tfrac{r}{r-1}\right)^{r-2}\tbinom{r-1}{2}\delta_{m+1}^2 - \left(\tfrac{r}{r-1}\right)^{r-2}(r-1)\delta_{m+1}\delta_m \\ & -(x-\rho)\left(\tfrac{r}{r-1}\right)^r + \mathcal{O}(\delta_m^3 + \delta_m(x-\rho)), \end{aligned} \tag{4}$$

which leads to

$$\delta_{m+1} \geq \delta_m - C_1\delta_m^2 - C_2(x-\rho),$$

provided that $\delta_m \in [er(x-\rho), \epsilon]$, where $\epsilon > 0$ is a constant depending only on r. In particular $\delta_m - \delta_{m+1} = \mathcal{O}(\delta_m^2 + (x-\rho))$, hence for $\delta_m \in [er(x-\rho), \epsilon]$ we can replace δ_{m+1} on the right by δ_m without increasing the error term, and our claim follows.

If $\delta_m > \epsilon$, the stated error term does not tend to 0 as $m \to \infty$, that is, we can replace the error term by $\mathcal{O}(1)$, and the statement follows from the trivial bounds $\delta_m \le \frac{1}{r-1}, \delta_{m+1} > -\frac{1}{r-2}$. ⊣

We now obtain the lower bound for ρ_h. In fact, we show that if we fix $x = \rho + \frac{c}{h^2}$ with c sufficiently small, then $\delta_h > 0$, that is, $S_h(t) < \frac{r}{r-1}$ for $t < x$, and therefore $x < \rho_h$. Choose $\epsilon > 0$ in such a way that the constant implied by the error term in Lemma 3.2 is $< \epsilon^{-1}$. Then we obtain for $\delta_m \in [\frac{ec}{h^2}, \epsilon]$ the recursive bound

$$\delta_{m+1} > \delta_m - (C_1+1)\delta_m^2 - \frac{c(C_2+1)}{h^2}.$$

Choose m_0 in such a way that $\delta_{m_0} < \min(\epsilon, \frac{1}{2(C_1+1)})$, which is possible in view of Lemma 3.1. Put $m_1 = \lceil \delta_{m_0}^{-1} \rceil$. We now prove by induction that

$$\delta_m \ge \frac{1}{2(C_1+1)(m-m_0+1)} - \frac{cm(C_2+1)}{h^2}.$$

For $m = m_0$ this inequality is obviously true. If we assume that it holds for m, then

$$\delta_{m+1} > \frac{1}{2(C_1+1)(m-m_0+1)} - \frac{1}{4(C_1+1)(m-m_0+1)^2} - \frac{c(m+1)(C_2+1)}{h^2}$$

since the right hand side is increasing as a function of δ_m on $[0, \frac{1}{2(C_1+1)}]$. Our claim now follows from the relation

$$\frac{1}{x} - \frac{1}{2x^2} \ge \frac{1}{x} - \frac{1}{x(x+1)} = \frac{1}{x+1}$$

valid for all $x > 1$. We conclude

$$\delta_h > \frac{1}{2(C_1+1)(h-m_0+1)} - \frac{c(C_2+1)}{h} > \frac{1}{4(C_1+1)h} > \frac{erc}{h^2},$$

provided that h is sufficiently large and $c < \frac{1}{5(C_1+1)(C_2+1)}$. Hence we obtain $\rho_h > \rho + \frac{1}{5(C_1+1)(C_2+2)h^2}$.

For the upper bound we show that if c is a sufficiently large constant, then for $x_h = \rho + \frac{c}{h^2}$ the sequence δ_m becomes large and negative quite fast, contradicting the bound $\delta_h \ge -\frac{1}{r-2}$. Assume that $x_h < \rho_h$. Then $T_{m+1}(x_h)$ is given the unique solution of the recursive relation within $[1, \infty)$, that is, we can use (3) without worrying about the existence of a solution. In particular (4) holds true for all δ_m with $m \le h$, since the restriction $\delta_m > e(x-\rho)$ was only used to prove the existence of the solution of the recursion.

First note that the same computation used for the proof of the lower bound gives $\delta_m < \frac{2}{C_1(m-m_0+m_1)}$, provided that m is sufficiently large. Put $m_2 = \lceil h/3 \rceil$. Then for h sufficiently large we conclude $\delta_{m_2} < \frac{7}{C_1 h}$.

Next we use (4) in the weaker form

$$\delta_{m+1} < \delta_m - (C_2 - 1)(x - \rho),$$

which is valid for $|\delta_m| < \epsilon$. Put $m_3 = \lceil 2h/3 \rceil$. Then

$$\delta_{m_3} < \delta_{m_2} - \frac{hC_2}{4}(x - \rho),$$

hence, if $x - \rho > \frac{56}{C_1 C_2 h^2}$, then $\delta_{m_3} < -\frac{7}{C_1 h}$.

If we expand the right hand side of (3) using the binomial theorem, and assume that δ_m is negative, then all terms are positive. Hence deleting terms of total degree ≥ 3 in δ_m and δ_{m+1} gives a value for δ_{m+1} which is too large, that is, for δ_m negative the error term in (4) is negative. Moreover, $\delta_{m+1} < \delta_m$, and we obtain the inequality

$$\delta_{m+1} < \delta_m - C_1\delta_m^2 - C_2(x - \rho) < \delta_m - C_1\delta_m^2$$

valid for all m such that $\delta_m < 0$ and $m < h$ and all x with $\rho < x < \rho_h$. We now define m_i starting with $i = 4$ by $m_i = m_{i-1} + \lceil 1/(C_1\delta_{m_{i-1}}) \rceil$, this sequence is defined as long as the sequence δ_m is defined, that is, for all i such that $m_i \leq h$. Let $i_{\max}$ be the largest index for which this sequence is defined. Then

$$\delta_{m_{i+1}} < \delta_{m_i} - (m_{i+1} - m_i)C_1\delta_{m_i}^2 \leq \delta_{m_i} - \frac{C_1\delta_{m_i}^2}{C_1\delta_{m_i}} = 2\delta_{m_i}.$$

Since $\delta_h > \frac{r}{r-1} - \frac{r-1}{r-2} = \frac{1}{(r-1)(r-2)} < 1$, we obtain $\frac{7 \cdot 2^{i_{\max}}}{C_1 h} < 1$, and therefore $i_{\max} \leq \log h$. On the other hand we have

$$m_{i_{\max}} \leq m_3 + \sum_{i=1}^{i_{\max}-1} \frac{1}{C_1\delta_{m_i}} + 1 \leq \frac{2h}{3} + \frac{2}{C_1\delta_{m_3}} + i_{\max} \leq \frac{20h}{21} + i_{\max}.$$

For h sufficiently large this implies $i_{\max} \geq \frac{h}{22}$, and the two bounds for $i_{\max}$ contradict each other for h sufficiently large. Hence we obtain that the two assumptions $x < \rho_h$ and $x - \rho > \frac{56}{C_1 C_2 h^2}$ contradict each other, and we conclude $\rho_h \leq \rho + \frac{56}{C_1 C_2 h^2}$ for h sufficiently large. Hence our theorem follows.

4 Conclusion and outlook

We have considered a filtrations $\mathcal{T}_1 \subset \mathcal{T}_2 \subset \dots$ of the set of trees. We determined the speed of convergence of the dominating singularity of the generating functions of these subsets to the dominating singularity for the set of all trees, and similarly for a certain set of planar trees. In the first case the speed of convergence was exponential, while in the second it was only quadratic. An intuitive explanation for this phenomenon is that in the first case we have that $T_i(\rho_i) \to T(\rho)$, while in the second we have $S_i(\rho_i) = \frac{r-1}{r-2}$, which is bounded away from $S(\rho) = \frac{r}{r-1}$. However, we are not yet able to formulate and prove a strict statement underlying this intuition.

The research in this article is motivated by its application in the theory of logical phase transitions, in fact, an unpublished version of Theorem 1 has already been applied by Bovykin and Weiermann [1].

This leads to further research in two quite different directions. The first is motivated by the analytic behaviour of the generating functions. Suppose F is a function in two variables, analytic in a neighbourhood of the origin. Assume there exists a unique sequence of functions (f_i) satisfying $f_0 = 1$, $F(f_{i+1}, f_i) = 0$. Under what conditions on F does there exist a unique function f satisfying $F(f, f) = 0$, and how does the sequence (f_i) approximate f? Similarly if F_i is a series of functions in one variable, approximating a function F in a suitable way, are there reasonable conditions such that the solutions f_i of the functional equations $F_i(f_i) = 0$ approximate the solution f of $F(f) = 0$?

The second direction of research is motivated by the proof theoretic applications. If one wants to generalize the phase transitions obtained in [1] to other systems of arithmetic, one has to solve counting problems involving ordinal numbers below a certain ordinal and with bounded complexity. This line of research appears to be more demanding then the first, as in general we cannot expect that the counting functions are described by simple functional equations. Therefore the well-developed machinery of analytic combinatorics cannot be applied anymore.

References

[1] A. Bovykin, A. Weiermann, Unprovability, phase transitions and the Riemann zeta-function, *New directions in value-distribution theory of zeta and L-functions*, Shaker Verlag, Aachen, 2009, pp. 19–37.

[2] P. Flajolet, R. Selfridge, *Analytic Combinatorics*, Cambridge University Press, Cambridge, 2009.

[3] A. Weiermann, An application of graphical enumeration to PA, *J. Symbolic Logic*, vol. 68, 2003, pp. 5–16.

[4] A. Weiermann, Analytic combinatorics for a certain well-ordered class of iterated exponential terms, *International Conference on Analysis of Algorithms DMTCS proc. AD*, Vol. 409, 2005, pp. 409–416.

[18] A. Woods, unpublished manuscript

12

What is Sequentiality?

ALBERT VISSER[1,2]

Abstract: In this paper we give an informally semi-rigorous explanation of the notion of *sequentiality*. We argue that the classical definition, due to Pudlák, is slightly too narrow. We propose a wider notion *polysequentiality* as the notion that precisely captures the intuitions behind *sequentiality*. The paper provides an example of a theory that is polysequential but not sequential.

Keywords: Interpretations, Interpretability

1 Introduction

Sequential theories are, as a first approximation, theories with sufficient coding machinery for the construction of partial satisfaction predicates of a certain sort. Specifically, we want to be able to develop satisfaction for classes of formulas with complexity below n for a complexity measure like *depth of quantifier alternations*. Our aim entails that we need to be able to code sequences of all objects of our theory.

We have a reasonably robust explication of the intended class of theories. Yet, we will argue that the usual definition is slightly too narrow to capture the intuitive motivations behind the notion of *sequentiality*. In essence the point is that the current definition contains the unmoti-

[1]I thank Ali Enayat for his helpful comments. I am grateful to three referees for their insightful remarks and for spotting many small mistakes. I thank Emil Jeřábek for his permission to reproduce an argument that shows the non-polysequentiality of Q.

[2]Department of Philosophy, Utrecht University, Janskerkhof 13A, 3512BL Utrecht, The Netherlands, albert.visser@phil.uu.nl

Studies in Weak Arithmetics.
Patrick Cégielski, Charalampos Cornaros,
Costas Dimitracopoulos.

vated constraint that the definition of the sequences in the given theory must be one-dimensional. We propose a wider notion, *polysequentiality* of which we claim that it captures the intuitions precisely. In fact, *polysequentiality* is simply *sequentiality* without the unmotivated constraint.

1.1 What is a Sequential Theory? *Sequentiality* is an explication of a specific idea of *theory with coding.* There are other legitimate notions of *theory with coding*, but they are not the ones we want to explicate here. Can we give an informally rigorous determination of the idea of coding behind sequentiality? We think that such a rigorous determination is not to be had, at least not quite. The notion does have a lot of intuitive content, but not all choices follow unavoidably from that content. We can only give, say, an informally semi-rigorous determination that will have some gaps in the explanation. Such an explanation cannot quite stand on its own and should be supplemented with other considerations. The considerations are in this case the successful employment of *sequentiality* in theorems and the robustness of sequentiality with respect to variations on its definition. We will return to these points later. We proceed with an informally semi-rigorous determination of *sequentiality.*

As a first step, towards the determination of the intended notion of *theory with coding*, we must ask: coding for what? We can have different needs for coding and we should expect different answers, depending on those needs, to the question *what is a theory with coding*? A first approximation to the question to which sequentiality is (supposed to be) the answer is to say: *we want coding in order to provide* partial satisfaction predicates *and as a consequence*, restricted consistency statements. Even this answer is not quite satisfactory: in theories of pairing we also have partial satisfaction predicates. See e.g. [28] and [40].[3] However, these satisfaction predicates are, in a sense, more finitistic than the ones provided by sequentiality. So, we could add to our specification: *these satisfaction predicates concern classes of formulas with decent verifiable closure properties. We want our satisfaction predicates to work for classes of formulas corresponding to a good complexity measure of formulas, like depth of quantifier alternations.*

What do we need to specify partial satisfaction predicates? Clearly, the very formulation of such predicates involves (i) the possibility to code formulas, so we will need a modicum of the theory of syntax. Moreover, this formulation involves (ii) the possibility to define finite

[3] Such theories also embody a perfectly legitimate idea of a theory of coding. It is just not the one we want to explicate here.

functions or finite sequences. Fortunately, if we have (ii) we have (i) automatically. Thus, the demand that an elementary theory of sequences is present follows directly from our desire to define partial satisfaction. The demand for decent closure properties makes it plausible that such sequences are closed under the operations of projection and of pushing an element on top of a sequence considered as stack. To have projections we need numbers as inputs of the projection, hence we need a modicum of number theory.

A further desideratum is this: *we want partial satisfaction for the full language and also restricted consistency for the theories under consideration themselves.* This means that we want to have sequences involving *all* objects of the original theory. It is easy to give examples that fail this condition. E.g., in Robinson's arithmetic Q we do have sequences for a definable cut in the Q-numbers, but not for the full universe of Q-numbers themselves (See [35], [12] and Example 4.9.).

Do we need more? The good news is that the meager means of having a theory of finite sequences are sufficient. We can start with a bare minimum and build everything more we need from that. For example, we do not need addition and multiplication for our numbers since we can define these operations using the sequences to code the recursive mechanisms.

Finally, it seems reasonable to allow parameters in the definition of our sequences. So, in the general definition, we allow parameters.

1.2 Robustness. Well, but couldn't we also have started with a theory of finite functions? Here the prediction is that all such alternative formulations are equivalent to the one involving sequences. Our notion is substantially *robust* w.r.t. alternative formulations and this robustness is an argument for the correctness of the choices made. We will say a bit more about robustness in Section 3.

1.3. Memorable Results. Here are some memorable results concerning sequential theories.

I. Sequential theories are essentially undecidable (since they interpret Q).

II. Pudlák in his [22] proves a variant of Dedekind's theorem that all models of second-order arithmetic are isomorphic. In Pudlák's variant any pair of number systems definable in a sequential theory U has a pair of definably isomorphic definable cuts. Here a definable number system may be explicated as an interpretation of S^1_2 in U.

III. We can define partial satisfaction relations for formula classes of complexity $\leq n$ in sequential theories. Here we can take as our notion of complexity e.g. depth of quantifier alternations.

IV. Computationally enumerable sequential theories U have a normal form $\mho(U)$ modulo mutual local interpretability.[4] The theory $\mho(U)$ is given by $\mho(U) = \mathsf{S}^1_2 + \{\mathtt{con}_n(U) \mid n \in \omega\}$.[5] Here $\mathtt{con}_n$ is consistency of the axioms of U with Gödel numbers smaller than n, where we allow only formulas in the proofs of complexity $\leq n$. Our complexity measure is depth of quantifier alternations.

We have U is mutually locally interpretable with V iff $\mho(U) = \mho(V)$. An alternative normal form is the theory

$$\Pi(U) := \mathsf{S}^1_2 + \{P \in \Pi^0_1 \mid U \text{ interprets } (\mathsf{S}^1_2 + P)\}.$$

See e.g., [39]. The normal form theorem is related to both the Orey-Hájek characterization and the Friedman characterization.[6]

V. Every finitely axiomatized sequential theory A interprets $\mathsf{S}^1_2 + \mathtt{con}^\star(A)$, where $\mathtt{con}^\star$ can stand for cut-free consistency, tableaux consistency, restricted consistency or Herbrand consistency.

VI. For any finitely axiomatized sequential theory U, there is an interpretation of S^1_2 in U that is Σ^0_1-sound. See e.g. [25], [15], [31], [33].[7]

This fact implies, using methods due to Per Lindström, that if a finitely axiomatized sequential theory A interprets a theory U, then it also faithfully interprets U. This means that there is an interpretation K of U in A such that for all U-sentences B, $A \vdash B^K$ iff $U \vdash B$. This result is due to Harvey Friedman. See [25].

VII. There is a sequential theory U with p-time decidable axiom set such that the predicate logic of U is complete Π^0_2. The theory was first given in [15]. The result is proved in [33].

VIII. A theory U *model-interprets* V if all models of U contain an internal model of V. It is easily seen that model-interpretability is between interpretability and local interpretability. One can show that there are sequential theories U and V such that U model-interprets V, but U does not interpret V, and that there are

[4] Local interpretability is explained in Subsection 2.6.

[5] We pronounce $\mho$ to rhyme with *Joe*.

[6] It is unknown whether there is anything like a normal form for sequential theories modulo mutual global interpretability. An even more modest question is the following: is every sequential theory U mutually interpretable with an extension of S^1_2 in the same language?

[7] It is an open question whether this insight generalizes to wider classes like all finitely axiomatized theories that interpret S^1_2.

sequential theories U and V, such that U locally interprets V, but U does not model-interpret V.

1.4. Sequentiality versus Polysequentiality. We will explain in the paper that the usual definition of sequentiality contains an unmotivated restriction. Moreover, sequential theories are not closed under a fundamental relation of sameness of theories, to wit bi-interpretability. Here bi-interpretability (see Subsubsection 2.3.2) is a very good notion of sameness of theories. Polysequentiality, the generalization of sequentiality that we propose, does not have these disadvantages.

We will show that a theory is polysequential if and only if it is bi-interpretable with a sequential theory.

All the salient results about sequentiality that we know of, including the results I-VI of the list in Subsection 1.3, are preserved under the generalization to polysequentiality. More precisely one can easily see that the results I-VI are preserved under bi-interpretation, so they can be transferred from sequential to polysequential theories. See Example 2.2 in Subsubsection 2.3.2 for an illustrative result. The observed preservation of salient results leads us to formulate the following hypothesis.

Hypothesis: *All good global properties of sequential theories also hold for polysequential theories.*

Admittedly, our Hypothesis contains the problematic word 'good'. Let's say that what is *good* is open for negotiation.

1.5. Contents of the Paper. In Section 2, we provide some basic definitions and facts needed to understand the rest of the paper. We discuss the current definition of *sequentiality* in Section 3. We briefly present the state of the art in Subsection 3.2, and, in Subsection 3.3, we give reasons why the state of the art is not quite satisfactory.

In Section 4, the definition of *polysequentiality* is presented, plus several characterizations. For example, it is shown that a one-sorted theory is polysequential iff it is bi-interpretable with a sequential theory. It follows that polysequential theories are closed under bi-interpretability. In Section 5, it is proved that, on the one hand, there are sequential theories that essentially involve parameters to witness their sequentiality, and, on the other hand, that parameters can always be eliminated to witness polysequentiality. So, in a sense, polysequentiality is a parameter-free notion. Finally, in Section 6, we give a proven example of a polysequential theory that is not sequential. The fact that it is not all that easy to find such an example, is, perhaps, an indication that the usual notion is only *slightly* wrong. The polysequential theories we meet in practice are all sequential. On the other hand, the example is not 'unnatural', which could mean that the practice we are looking at is, until now, rather restricted.

2 Basic Notions

In this section, we formulate the basic notions employed in the paper.

2.1. Theories. The primary focus in this paper is on one-sorted theories of first order predicate logic. Specifically, we analyze sequentiality for the case of one-sorted theories. However, many-sorted theories will occur as an auxiliary. They could be eliminated but the approach is less natural if we do so. We only consider theories with a finite number of sorts. We use $\mathbf{a}$, $\mathbf{b}$, ... for sorts. We take identity to be a logical constant. In the many-sorted case, we have identity for each sort.

Our official signatures are relational, however, via the term-unwinding algorithm, we can also accommodate signatures with functions.

Our signatures need not be finite and we need no constraint on the complexity of the axiom set.

2.2. Interpretations: the One-sorted Case. We describe the notion of an m-dimensional interpretation for a one-sorted language. In Subsection 2.4 we will indicate how to adapt the definition to the many-sorted case.

An interpretation $K : U \to V$ is given by the theories U and V and a translation τ from the language of U to the language of V. The translation is given by a domain formula $\delta(\vec{x})$, where $\vec{x}$ is a sequence of m variables, and a mapping from the predicates of U to formulas of V, where an n-ary predicate P is mapped to a formula $A(\vec{x}_0, \ldots, \vec{x}_{n-1})$, where the $\vec{x}_j$ are appropriately chosen, pairwise disjoint, sequences of m variables. We lift the translation to the full language in the obvious way making it commute with the propositional connectives and quantifiers, where we relativize the translated quantifiers to the domain δ. We demand that V proves all the translations of sentences of U.

We can compose interpretations in the predictable way. Note that the composition of an n-dimensional interpretation with an m-dimensional interpretation is $m \cdot n$-dimensional.

An m-dimensional interpretation is *m-identity preserving*, if it translates identity to pointwise-identity of the corresponding elements of the sequences assigned to the relata. I.o.w., $x = y$ translates to $\bigwedge_{i<m} x_i = y_i$. An m-dimensional interpretation is *m-unrelativized*, if its domain consists of all the sequences of length m of the interpreting theory. A m-dimensional interpretation is *m-direct* if it is unrelativized and pre-

serves identity. Note that all these properties are preserved by composition. We use *identity preserving*, *unrelativized* and *direct*, for: 1-identity preserving, 1-unrelativized and 1-direct.

Each interpretation $K : U \to V$ gives us an inner model construction that builds a model $\widetilde{K}(\mathcal{M})$ of U out of a model $\mathcal{M}$ of V. Note that $\widetilde{(\cdot)}$ behaves contravariantly.

2.3. Sameness of Interpretations. If we want to use interpretations to analyze e.g. sameness of theories, we need to be able to say when two interpretations are 'equal'. Strict identity of interpretations is too fine grained: it is too much dependent of arbitrary choices like which bound variables to use. We specify a first notion of equality between interpretations: two interpretations are *equal* when the *target theory thinks they are*. Modulo this identification, the operations identity and composition give rise to a category $\mathtt{INT}_0$, where the theories are objects and the interpretations arrows.[8]

For each sufficiently good notion of sameness of interpretations there is an associated category of theories and interpretations: the category of interpretations modulo that notion of sameness. Any such a category gives us a notion of isomorphism of theories which can function as a notion of sameness.

We present a list of notions of sameness. For all notions, it is easily seen that sameness is preserved by composition. (There are many more more notions, but they are irrelevant for this paper.)

2.3.1. Equality. The interpretations $K, K' : U \to V$ are equal when V 'thinks' K and K' are identical. By the Completeness Theorem, this is equivalent to saying that, for all V-models $\mathcal{M}$, $\widetilde{K}(\mathcal{M}) = \widetilde{K'}(\mathcal{M})$. This notion gives rise to the category $\mathtt{INT}_0$.

Isomorphism in $\mathtt{INT}_0$ is *synonymy* or *definitional equivalence*. This notion was introduced by K.L. de Bouvère in 1965. See [4] and [5]. There is an anticipation in [23]. It is the strictest notion of sameness of theories except simple sameness of theorem sets.

2.3.2. i-Isomorphism. An i-isomorphism between interpretations K, $M : U \to V$ is given by a V-formula F. We demand that V verifies that "F is an isomorphism between K and M", or, equivalently, that, for each model $\mathcal{M}$ of V, the function $F^{\mathcal{M}}$ is an isomorphism between $\widetilde{K}(\mathcal{M})$ and $\widetilde{M}(\mathcal{M})$. Two interpretations $K, K' : U \to V$, are *i-isomorphic* if there is an i-isomorphism between K and K'. Wilfrid

[8]For many reasons, the choice for the reverse direction of the arrows would be more natural. (See e.g. [34].) However, our present choice coheres with the extensive tradition in degrees of interpretability. So, we opted to adhere to the present choice.

Hodges calls this notion: *homotopy*. See [11], p. 222.

The notion of i-isomorphism can be characterized in a third way. We need only demand that, for every V-model $\mathcal{M}$, there is a V-formula F, which defines in $\mathcal{M}$ an isomorphism between $\widetilde{K}(\mathcal{M})$ and $\widetilde{K'}(\mathcal{M})$. By a simple compactness argument one may show that, for finite signatures, the third definition defines the same notion as the first two.

Clearly, if K, K' are equal, they will be i-isomorphic.

It is intuitively clear that as soon as the interpreting theory has an internal pairing operation, then we can reduce any interpretation K in U to an interpretation $K^\star$ with the dimension of the pairing definition. We formulate a theorem that illustrates how bi-interpretations can be used to give a sharper statement of this intuitive insight.

First we need to be precise about what we mean when we say 'a theory has internal pairing'. The theory of unordered pairing $\mathtt{UP}$ has as non-logical relation symbol the binary predicate $\in$.[9] It is axiomatized as follows.

$\mathtt{UP1}$ $\vdash \forall x, y\, \exists z\, \forall u\, (u \in z \leftrightarrow (u = x \vee u = y))$.

Let's say that U is a *weak m-pair theory* if there is an m-direct interpretation of $\mathtt{UP}$ in U. We have the following theorem.

Theorem 2.1. *Consider any theory U. Then, U is a weak m-pair theory iff, for every n-dimensional interpretation $K : W \to U$, there is an m-dimensional interpretation $K^\star : W \to U$ that is i-isomorphic to K.*

In Appendix B, we give the proof of the theorem.

The notion of i-isomorphism gives rise to a category of interpretations modulo i-isomorphism. We call this category $\mathtt{INT}_1$. Isomorphism in $\mathtt{INT}_1$ is *bi-interpretability*. The notion of *bi-interpretability* was first introduced by G. Ahlbrandt and M. Ziegler in 1986. See [1]. See also [11].

Bi-interpretability preserves many important properties of theories. The class of these properties is very diverse. For example, parameter-free bi-interpretability preserves the automorphism group of a model.[10] (In case we have parameters, we need a slightly more involved statement.) Bi-interpretability preserves κ-categoricity of theories (for infinite κ). It preserves decidability and it preserves finite axiomatizability of theories. Below we show that one of the properties of the list of Subsection 1.3 is preserved.

[9] This theory is the theory $\mathrm{PAIR}_{\mathrm{uno}}$ of [35].

[10] Bi-interpretability does not preserve the *action* of the automorphism group on the domain of the model. That is preserved by definitional equivalence.

Example 2.2. We illustrate the preservation of Pudlák's result II of our list of salient results in Subsection 1.3 under bi-interpretability. We simplify the argument a bit by only considering interpretations without parameters.

Suppose we have theories U and V that are bi-interpretable, say via $K : U \to V$, $M : V \to U$, $F : (M \circ K) \to \mathtt{id}_U$ and $G : (K \circ M) \to \mathtt{id}_V$. Suppose further that V has *the Pudlák property*: for every $N_0 : \mathrm{S}^2_1 \to V$, $N_1 : \mathrm{S}^2_1 \to V$, there is a H such that "H is an isomorphism between V-definable cuts J_0 and J_1 of N_0 resp. N_1". Consider any $Q_i : \mathrm{S}^2_1 \to U$, for $i = 0, 1$. Then $N_i := K \circ Q_i : \mathrm{S}^2_1 \to V$. Let H be the promised isomorphism between V-definable cuts J_i of the N_i. We assume that the domain of H is precisely J_0 and that the range of H is precisely J_1. Then $M \circ H := H^M$ is an U-definable isomorphism between U-definable cuts $M \circ J_i$ of the $M \circ N_i$. We note that F gives us an isomorphism between $M \circ N_i = M \circ K \circ Q_i$ and Q_i. Thus $F^{-1} \circ H \circ F$ will give us the desired isomorphism between definable cuts of the Q_i.

We note that the argument does not use the existence of G. So we proved more than we set out to do: the Pudlák property is preserved to retracts in the category $\mathtt{INT}_1$. This is a common pattern. In fact each of I-IV of the list in Subsection 1.3 is preserved to a retract. As we will see polysequentiality is also preserved to retracts in $\mathtt{INT}_1$. This follows from Propositions 4.11 and 4.12.

2.3.3. Isomorphism. Our third notion of sameness of the basic list is that K and K' are the same if, for all models $\mathcal{M}$ of V, the internal models $\widetilde{K}(\mathcal{M})$ and $\widetilde{K'}(\mathcal{M})$ are isomorphic. We will simply say that K and K' are isomorphic. Clearly, i-isomorphism implies isomorphism. We call the associated category $\mathtt{INT}_2$. Isomorphism in $\mathtt{INT}_2$ is *iso-congruence.*

2.3.4. Elementary Equivalence. The fourth notion is to say that two interpretations are the same if, for each $\mathcal{M}$, the internal models $\widetilde{K}(\mathcal{M})$ and $\widetilde{K'}(\mathcal{M})$ are elementary equivalent. We will say that K and K' are elementary equivalent. By the Completeness Theorem, we easily see that this notion can be alternatively defined by saying that K is the same as K' iff, for all U-sentences A, we have $V \vdash A^K \leftrightarrow A^{K'}$. It is easy to see that isomorphism implies elementary equivalence. We call the associated category $\mathtt{INT}_3$. Isomorphism in $\mathtt{INT}_3$ is *elementary congruence* or *sentential congruence.*

2.3.5. Identity of Source and Target. Finally, we have the option of abstracting away from the specific identity of interpretations completely, simply counting any two interpretations $K, K' : U \to V$ the same. The associated category $\mathtt{INT}_4$ is simply the preorder of the degrees of interpretability. Isomorphism in this preorder is *mutual interpretability.*

2.4. The Many-sorted Case. Interpretability can be extended to interpretability between many-sorted theories. The *profile* of an interpretation is a mapping from the sorts of the interpreted theory to sequences of the sorts of the interpreting theory. We write $(\mathbf{a}_0 : \vec{\mathbf{b}}_0) \ldots (\mathbf{a}_{l-1} : \vec{\mathbf{b}}_{l-1})$ to present a profile. (The order of the sorts is chosen arbitrarily.) We abbreviate a sequence of n $\mathbf{a}$'s to $\mathbf{a}^n$. The default sort of one-sorted theories is $\mathbf{o}$, the sort of objects.

We specify a domain $\delta^{\mathbf{a}}(x_0^{\mathbf{b}_0}, \ldots, x_{k-1}^{\mathbf{b}_{k-1}})$, for each sort $\mathbf{a}$ of the interpreted theory, where $\vec{\mathbf{b}}$ is the sequence associated to $\mathbf{a}$ by the profile and where $\vec{\mathbf{b}}$ has length k. We write $\vec{x}^{\vec{\mathbf{b}}}$ for $x_0^{\mathbf{b}_0}, \ldots x_{k-1}^{\mathbf{b}_{k-1}}$.

With each predicate P of the interpreted theory of type $\mathbf{a}_0, \ldots, \mathbf{a}_{n-1}$, we associate a formula $A(\vec{x}_0^{\vec{\mathbf{b}}_0}, \ldots, \vec{x}_{n-1}^{\vec{\mathbf{b}}_{n-1}})$, where $(\mathbf{a}_i : \vec{\mathbf{b}}_i)$ is in the profile. We demand that the target theory verifies, for $i < n$, the formula $A(\vec{x}_0^{\vec{\mathbf{b}}_0}, \ldots, \vec{x}_{n-1}^{\vec{\mathbf{b}}_{n-1}}) \to \delta^{\mathbf{a}_i}(\vec{x}_i^{\vec{\mathbf{b}}_i})$.

Our translation commutes with the propositional connectives. The translation of a quantifier $\forall x^{\mathbf{a}}$ is $\forall \vec{x}^{\vec{\mathbf{b}}}\,(\delta^{\mathbf{a}}(\vec{x}^{\vec{\mathbf{b}}}) \to \ldots)$. Similarly, for the existential quantifier.

An interpretation is $(\mathbf{a}{:}\vec{\mathbf{b}})$-direct if $(\mathbf{a} : \vec{\mathbf{b}})$ is in the profile and if identity translates to pointwise identity and if the translated domain consists of all sequences of sort $\vec{\mathbf{b}}$.

Remark 2.3. Every many-sorted theory U has a one-sorted flattening $U^\flat$, where the sorts are simulated by predicates. The (poly)sequentiality of U could be defined as the (poly)sequentiality of $U^\flat$.

We refrain from exploring this perspective, since we feel that it only becomes meaningful against the background of an explanation in which sense U and $U^\flat$ are the same. To do this we would need to develop the theory of *piecewise interpretations* a bit further. Moreover, if we have piecewise interpretations available the present definitions would undoubtedly lift without detour over flattening to the many-sorted case. ⊣

2.5. Parameters. We can extend our notion of interpretation to *interpretation with parameters* as follows. Say our interpretation is $K : U \to V$. In the target theory, we have a parameter domain $\alpha(\vec{z})$, which is provably non-empty. The definition of interpretation remains the same but for the fact that the parameters $\vec{z}$ may occur in the domain formula and in the translations of the predicate symbols. We have to take appropriate measures to avoid variable clashes. It is clear, but somewhat laborious to work out, that this can be done in a systematic way. Our condition for K to be an interpretation becomes:

$$U \vdash A \;\Rightarrow\; V \vdash \forall \vec{z}\,(\alpha(\vec{z}) \rightarrow A^{K,\vec{z}}).$$

We note that an interpretation $K : U \rightarrow V$ with parameters provides a parametrized set of inner models of U inside a model of V.

2.6. Notions of Theory Reduction based on Interpretations. Several notions of reduction between theories can be based on the notion of interpretations. These notions will not play an important role in this paper. We add them just for completeness.

- A theory U is *interpretable* in a theory V iff there is an interpretation K of U in V. We write $V \rhd U$ or $U \lhd V$ for U is interpretable in V.
- A theory U is *model-interpretable* in a theory V iff for each model $\mathcal{M}$ of V there is an internal model $\widetilde{\tau}(\mathcal{M})$ of U based on a translation τ. Equivalently, U is model-interpretable in V iff U is interpretable in each complete extension V^+ of V.
  We write $V \rhd_{\mathrm{mod}} U$ or $U \lhd_{\mathrm{mod}} V$ for U is model-interpretable in V.
- A theory U is *locally interpretable* in a theory V iff each finitely axiomatized subtheory U_0 of U is interpretable in V. We write $V \rhd_{\mathrm{loc}} U$ or $U \lhd_{\mathrm{loc}} V$ for U is locally interpretable in V.

One can show that $V \rhd U$ implies $V \rhd_{\mathrm{mod}} U$ and that $V \rhd_{\mathrm{mod}} U$ implies $V \rhd_{\mathrm{loc}} U$. There are sequential examples that illustrate that these arrows cannot be reversed.

We can refine these notions by restricting the allowable interpretations to a certain class of interpretations. E.g., the sequentiality of a theory U will be defined as the *1-direct interpretability* of a certain fixed theory in U.

3 Sequentiality

In Subsection 3.1 we provide the definition of sequentiality for one-sorted theories. We present the state of the art of the subject in Subsection 3.2. Finally, in Subsection 3.3, we give reasons why the present notion is not general enough.

3.1. Definitions of Sequentiality. We start with the primary definition in terms of a theory of sequences seq. This definition is in direct accordance with the intuitive picture we provided for sequential theories. The definition involves a 3-sorted theory seq which is given by 18 axioms. The primary definition is so involved in order to provide conceptual clarity about what is going on. From the technical point of view it is a ladder that we wish to throw away in order to replace it by much simpler definitions.

3.1.1 Sequences. We define the theory **seq** as follows. The language of **seq** has three sorts **o** (the sort of *objects*), **n** (the sort of *numbers*) and **s** (the sort of *sequences*). We have a unary predicate symbol Z (for: *zero*) of type **n**, a unary predicate symbol E (for: *empty sequence*) of type **s**, two binary predicate symbols S (for: *successor*) and $<$ of type **nn**, a binary predicate symbol L (for: *sequence length*) of type **sn**, a ternary predicate symbol Pu (for: *push*) of type **sos**, and a ternary predicate symbol Pr (for: *projection*) of type **sno**. The variables $x, x_0, x', y, z, u \ldots$ will be of sort **o**, the variables $n, n_0, n', m, k, \ldots$ will be of sort **n**, and the variables $s, s_0, s', t, \ldots$ will be of sort **s**. As usual we write $m \leq n$ for: $m = n \vee m < n$. The axioms are:

seq1 $\vdash (n < m \wedge m < k) \rightarrow n < k$
seq2 $\vdash \neg n < n$
seq3 $\vdash n \leq m \vee m < n$
seq4 $\vdash \exists n\, \mathtt{Z}(n)$
seq5 $\vdash \mathtt{Z}(n) \rightarrow n \leq m$
seq6 $\vdash \mathtt{S}(n, m) \leftrightarrow (n < m \wedge \forall k\, (n < k \rightarrow m \leq k))$
seq7 $\vdash \exists m\, \mathtt{S}(n, m)$
seq8 $\vdash \mathtt{Z}(n) \vee \exists m\, \mathtt{S}(m, n)$
seq9 $\vdash \exists n\, \mathtt{L}(s, n)$
seq10 $\vdash (\mathtt{L}(s, n) \wedge \mathtt{L}(s, m)) \rightarrow n = m$
seq11 $\vdash (\mathtt{L}(s, n) \wedge m < n) \rightarrow \exists x\, \mathtt{Pr}(s, m, x)$
seq12 $\vdash (\mathtt{L}(s, n) \wedge \mathtt{Pr}(s, m, x)) \rightarrow m < n$
seq13 $\vdash (\mathtt{Pr}(s, n, x) \wedge \mathtt{Pr}(s, n, y)) \rightarrow x = y$
seq14 $\vdash \exists s\, \mathtt{E}(s)$
seq15 $\vdash (\mathtt{E}(s) \wedge \mathtt{L}(s, n)) \rightarrow \mathtt{Z}(n)$
seq16 $\vdash \exists t\, \mathtt{Pu}(s, x, t)$
seq17 $\vdash (\mathtt{Pu}(s, x, t) \wedge \mathtt{L}(s, n) \wedge \mathtt{S}(n, m)) \rightarrow (\mathtt{L}(t, m) \wedge \mathtt{Pr}(t, n, x))$
seq18 $\vdash (\mathtt{Pu}(s, x, t) \wedge \mathtt{Pr}(s, m, y)) \rightarrow \mathtt{Pr}(t, m, y)$

We note that we may add extensionality for sequences by reinterpreting identity as extensional equivalence. There are many variants on **seq** that also would have served our purposes. Specifically, note that one can reduce the signature to just $<$ and Pr in the obvious way. Moreover, we could have avoided the sort **n** by working with a linear preorder on sequences for 'is a shorter sequence than'.

We remind the reader the reader that an (**o** : **o**)-direct interpretation is an interpretation that interprets the objects of the interpreted theory as the objects of the interpreting theory where the interpreted identity of objects is the identity of objects of the interpreting theory. For the other sorts we do allow relativization of domain and replacement of identity by a definable equivalence relation.

Definition 3.1. *We call a one-sorted theory U* sequential *iff there is a* $(\mathbf{o}:\mathbf{o})$*-direct interpretation of* `seq` *in* U*, that is one-dimensional for the three sorts of* `seq`.

Trivially, a one-sorted theory U is sequential iff there is a $(\mathbf{o}:\mathbf{o})$-direct interpretation of `seq` in U, that is one-dimensional for sorts $\mathbf{o}$ and $\mathbf{s}$, since we can compress the dimension of interpretation of the numbers using the sequences.

Remark 3.2. We note that the *random access* to the elements of sequences provided by the projection function is essential. Suppose we modify `seq` to a theory of stacks, say `stack`, where we omit the sort of numbers and replace the projection function by a function `Top` that reads the top element of the stack. Let's say that a one-sorted U is a *stack theory* iff there is a $(\mathbf{o},\mathbf{o})$-direct interpretation of `stack` in U, that is one-dimensional for the two sorts of `stack`. One can easily show that there are decidable stack theories,[11] where, in contrast, sequential theories are essentially undecidable.

3.1.2. Sets and Classes. We define the theories *adjunctive class theory* `ac` and *adjunctive set theory* `AS`.[12] The theory `ac` has two sorts $\mathbf{o}$, the sort of objects, and $\mathbf{c}$, the sort of classes. It has a binary predicate $\in$ of type $\mathbf{oc}$. We use $x, y, z, \ldots$ as variables for objects and $X, Y, Z, \ldots$ as variables for classes. We have the following axioms.

`ac1` $\vdash \exists X\, \forall x\; x \notin X$,
`ac2` $\vdash \forall X, x\, \exists Y\, \forall y\, (y \in Y \leftrightarrow (y \in X \vee y = x))$.

We note that we get extensionality for free in `ac` by replacing identity for classes by extensional identity. The theory `ac` allows finite models.

The theory `AS` is a one sorted theory with a binary relation $\in$.

`AS1` $\vdash \exists x\, \forall y\; y \notin x$,
`AS2` $\vdash \forall x, y\, \exists z\, \forall u\, (u \in z \leftrightarrow (u \in x \vee u = y))$.

The theory `AS` is much stronger than `ac`. E.g., it interprets Robinson's Arithmetic `Q`. Hence, it is essentially undecidable. Note that

[11]It is well known that there are decidable theories of non-surjective pairing. Any theory of non-surjective pairing is a stack theory.

[12]The use of capitals in the name of `AS` registers the fact that `AS` is a theory with 'iterated' elementhood.

both ac and AS have 'random access' to the objects contained in the classes/sets.

We have:

Theorem 3.3. *Let U be a one-sorted theory. The following are equivalent.*

i. U is sequential.

ii. There is an $(\mathbf{o} : \mathbf{o})$-direct interpretation of ac *in U that is one-dimensional for the sorts $\mathbf{o}$ and $\mathbf{c}$.*

iii. There is a direct interpretation of AS *in U.*

iv. There is an extension V of AS *in the same language, such that U is synonymous with V.*

Proof. The equivalence of (ii) and (iii) is easy and so is the implication from (i) to (iii). To prove the implication from (iii) to (i) one shows that seq has an $(\mathbf{o} : \mathbf{o})$-direct interpretation in AS that is one-dimensional for the sorts $\mathbf{s}$ and $\mathbf{n}$, using an elaborate bootstrap. We refer the reader to the literature. See, e.g., [21], [19], [10]. See also Subsection 3.2 for comments on ingredients of the proof. The equivalence of (iii) and (iv) is proved in [35]. ⊣

We note that AS is itself sequential. Moreover, sequentiality is preserved by synonymy. Of course, our definition does not work for many-sorted theories, but if we extend it in an appropriate way, it will follow that ac is not sequential, since it has finite models. ⊣

An alternative to AS was provided by Harvey Friedman. Here is Friedman's theory that we will call *reduced adjunctive subtractive set theory* $\mathtt{ASS}^-$. The axioms are as follows.

$\mathtt{ASS}^-\mathbf{1}$ $\vdash \forall x, y\, \exists z\, \forall u\, (u \in z \leftrightarrow (u \in x \vee u = y))$.

$\mathtt{ASS}^-\mathbf{2}$ $\vdash \forall x, y\, \exists z\, \forall u\, (u \in z \leftrightarrow (u \in x \wedge u \neq y))$.

A theory is sequential iff it directly interprets $\mathtt{ASS}^-$. To see this it is sufficient to show that AS and $\mathtt{ASS}^-$ are mutually directly interpretable.

Theorem 3.4. AS *and* $\mathtt{ASS}^-$ *are mutually directly interpretable.*

Proof. We can directly interpret $\mathtt{ASS}^-$ in AS using an argument in the style of the proof of Lemma 4.1. We can interpret AS in $\mathtt{ASS}^-$ by defining

$$x \in^*_z y :\leftrightarrow (x \in y \wedge x \notin z) \vee (x \notin y \wedge x \in z).$$

An $\in^*_z$-empty set is given by z. We define $\in^*_z$-adjunction of y as $\in$-adjunction of y in case $y \notin z$, and as $\in$-subtraction of y in case $y \in z$. ⊣

We note that the direct interpretation of AS in $\mathtt{ASS}^-$ uses a parameter. This leads to the following question.

Open Question 3.5. Is there a parameter-free direct interpretation of AS in $\mathtt{ASS}^-$?

A third alternative to AS can be obtained by adapting the definition of a much weaker theory given in [24]. Our theory, say RApp (Reset Applicative Theory), has a binary function symbol $\cdot$, for function application, that we write in infix notation.[13]

We have the following axioms.

RApp1 $\vdash \exists x, y\, x \neq y$

RApp2 $\vdash \forall x, y, z\, \exists w\, (w \cdot y = z \wedge \forall u\, (u \neq y \rightarrow w \cdot u = x \cdot u))$

We could also formulate a similar theory where application is a partial function. We have the following theorem.

Theorem 3.6. RApp *and* AS *are mutually directly interpretable.*

Proof. To interpret AS in RApp, we define $y \in_{x^*} w :\leftrightarrow w \cdot y \neq x^* \cdot y$, where x^* is a parameter. It is easily seen that this interpretation works.

In the other direction, we first interpret AS plus the subtraction axiom $\mathtt{ASS}^-2$ directly in AS. In this theory we can develop the non-extensional theory of pairing using Kuratowski's definition. We say that x is a function if x is a set of pairs that satisfies the following property: if p and q are in x and $(p)_0 = (q)_0$, then $p = q$. We define: $x \cdot y := z$ if x is a function and, for some $p \in x$, $(p)_0 = y$ and $(p)_1 = z$; $x \cdot y := y$, otherwise. It is easily seen that this definition does the job. $\dashv$

Open Question 3.7. Is there a parameter-free direct interpretation of AS in RApp?

3.2. State of the Art concerning Sequentiality. Richard Montague in his classical paper [18] introduces *strongly semantically closed theories* (surely not a very happy choice of naming). *Strong semantical closure* is in essence *sequentiality*, only where in the modern definition we have weak theories like Q or S^1_2, in Montague's definition we find Peano Arithmetic. Except for this, Montague's treatment is remarkably modern. He views incompleteness phenomena in a rigorous way from an interpretability-theoretic standpoint.[14] An illustrative example

[13] Officially, this is a ternary relation symbol and we have functionality axioms for it.

[14] See also Kreisel's review MR0150033 (27 #38). Kreisel seems to miss all conceptual progress in Montague's paper.

of application of Montague's framework is Stanislaw Krajewski's paper [14].

The modern notion of sequential theory was introduced by Pavel Pudlák in his paper [21]. Pudlák refers to the work of Vaught as a source of inspiration. Vaught uses a weaker set theory, Vaught set theory, to formulate a salient property of theories that is weaker than sequentiality. I call theories that satisfy Vaught's property: *Vaught theories*. See [28] and [40].

Pudlák uses his notion for the study of the degrees of local multi-dimensional parametric interpretability. He proves that sequential theories are prime in this degree structure. In [22], sequential theories provide the right level of generality for theorems about consistency statements. Let n-provability stand for provability from axioms with code below n and with a proof only involving formulas of complexity at most n, where the complexity measure is, e.g., *depth of quantifier alternations*. We write $\mathtt{con}_n(U)$ for an appropriate arithmetization of the consistency of U w.r.t. n-provability. For any sequential RE theory U, and any n, we can find a definable number system N such that $U \vdash \mathtt{con}_n^N(U)$. This is a formalization of the soundness theorem. On the other hand, if U is finitely axiomatized, for every definable number system N, we can find an n, such that $U \nvdash \mathtt{con}_n^N(U)$. This is a version of the second incompleteness theorem.

The notion of sequential theory was independently invented, modulo one detail, by Friedman who called it *adequate theory*. See Smoryński's survey [25]. Friedman's definition, as given by Smoryński, is still not quite *sequentiality*, because Elementary Arithmetic EA (aka EFA or $\mathrm{I}\Delta_0 + \mathtt{exp}$) is stipulated to be interpretable in adequate theories. This demand is too strong, even if it is already much closer than Montague's choice of Peano Arithmetic.

Friedman uses the notion to provide the Friedman characterization of interpretability among finitely axiomatized sequential theories. (See also [29] and [30].) Moreover, he shows that ordinary interpretability and faithful interpretability among finitely axiomatized sequential theories coincide. (See also [31] and [33].)

Examples of sequential theories are Adjunctive Set Theory AS, the theory $\mathtt{PA}^-$ (see: [12]), Buss' theory $\mathtt{S}_2^1$ (see: [2]) and synonymous variants of it like a theory of strings due to Ferreira (see: [6], [7], [8]) and a theory of sets and numbers due to Zambella (see: [42]), the theory $\mathrm{I}\Delta_0$, Wilkie and Paris' theory $\mathrm{I}\Delta_0 + \mho_1$ (see: [41]), PRA, Elementary Arithmetic EA, $\mathrm{I}\Sigma_1^0$, Peano Arithmetic PA, $\mathtt{ACA}_0$, ZF, GB.

We have provided several characterizations in Subsection 3.1. An important part of the story of the characterization involving AS is the interpretability of Robinson's Arithmetic Q in AS. Here are some historical notes about that characterization.

(1) In [26], Wanda Szmielew and Alfred Tarski announce the interpretability of Q in AS plus extensionality. See also [27], p. 34.
(2) A proof of the Szmielew-Tarski result is given by George Collins and Joseph Halpern in [3].
(3) Franco Montagna and Antonella Mancini, in [17], give an improvement of the Szmielew-Tarski result. They prove that Q can be interpreted in an extension of AS in which we stipulate the functionality of empty set and adjoining of singletons.
(4) In Appendix III of [19], Jan Mycielski, Pavel Pudlák and Alan Stern provide the ingredients of the interpretation of Q in AS.
(5) A new proof of the interpretability of Q in AS is given in [36].

A very nice presentation of the converse interpretability of (an extension of) AS plus extensionality in Q, is given in [20]. This is an interpretation with absolute identity.

For further work concerning sequential theories, see, e.g., [22], [25], [19], [10], [31], [32], [13], [33], [37], [38], [39].

Remark 3.8. An interesting paper that is closely related to, but as far as I can see entirely independent of, the work described above is Yuri Gurevich's and Saharon Shelah's paper [9]. In earlier work, Shelah had shown that true arithmetic reduces to the second order monadic theory of (the order structure of) the real line employing a reduction method that is not an interpretation. In [9], Gurevich and Shelah show that Shelah's reduction method cannot be replaced by an interpretation. They show that the theory AS cannot be interpreted in the monadic second order theory of a short colored chain.[15] Here the order structure of the real line is an example of a short chain. Their result implies that the monadic second order theory of a short chain is not polysequential.[16] In [16], Shmuel Lifsches and Saharon Shelah show that there is a forcing notion P such that in $\mathtt{V}^P$, the theory AS is not interpretable in the monadic second-order theory of chains.

[15] Gurevich and Shelah employ a version of AS in which union is total, but this is an inessential difference. They did not take the theory AS from the literature but invented it afresh.

[16] Monadic second order theories can be viewed as one-sorted theories via a standard reduction which makes the objects the atoms of the subset ordering. We apply the notion of polysequentiality to the thus flattened theories.

3.3. Why the Notion of Sequentiality is not Quite Right. The most direct argument that the notion of sequentiality is not quite right is to notice that there is an unmotivated and unnecessary component in the definition. Why do we demand that the interpretation of **seq** is one-dimensional for the sorts **s** and **o**? The fact that the interpretation is $(\mathbf{o} : \mathbf{o})$-direct, and, hence, one-dimensional for **o**, follows from our intuitive motivation: we want our sequences to be sequences of *all* objects of the original domain of the interpreting theory. However, there is no reason to put any constraint on our representation of sequences. We are just interested in the accessibility of the elements contained in the sequences, but not in the way the sequences are represented as long as it fulfills its desired function.

A second argument is that sequentiality fails to be preserved under the relation of bi-interpretability. The notion of *bi-interpretability* is a very good notion of sameness of theories. As we already mentioned in Subsubsection 2.3.2, it preserves such diverse properties of theories as finite axiomatizability and κ-categoricity. In the parameter-free case it preserves automorphism groups modulo isomorphism.

There are good reasons that sequentiality should be closed under bi-interpretability. If U is bi-interpretable with a sequential theory V, we have the following situation. For any model $\mathcal{M}$ of U, we have an internal model $\mathcal{N}$ of V, and *inside* this model we have again an internal model $\mathcal{M}'$ of U, definably isomorphic, say via F, with the original $\mathcal{M}$. Now we have sequences for $\mathcal{M}'$ provided by $\mathcal{N}$. These sequences are inherited by $\mathcal{M}$, in a definable way via F. So, $\mathcal{M}$ has sequences. Inspection of this argument shows that it fails as a proof precisely because of multi-dimensionality: the definition of the sequences in $\mathcal{M}$ may be more-dimensional, where we demanded it to be one-dimensional. Thus, the arbitrary demand of one-dimensionality blocks a very good closure condition of our theories.

In Section 6, we will show in detail that sequentiality is not closed under bi-interpretability. We will show in Subsection 4.3 that every polysequential theory is bi-interpretable with a sequential theory.

We claim that *polysequentiality* is the right notion. So, by combining the theorem and the claim, we only need to close off the sequential theories under bi-interpretability to arrive at the right notion.

Remark 3.9. It is easy to see that sequentiality is closed under synonymy.

Sequential theories are known not to be closed under mutual interpretability, however this does not constitute an argument against the definition. E.g., consider the disjoint union $U \uplus U$ of a sequential theory U with itself. This disjoint union is mutually interpretable with U but it is not sequential.[17] Yet, $U \uplus U$ proves, on different interpretations of number theory, its own restricted consistency statements. Shouldn't we also call $U \uplus U$ sequential? Well, in the light of our original motivation the answer should be *no*. The fact that we do have restricted consistency statements is inherited from U, which does have the means to verify these statements. The theory $U \uplus U$ itself does not have partial satisfaction for all objects in the domain of the theory.

A second example of a non-sequential theory that is mutually interpretable with a sequential one, is as follows. Robinson's arithmetic $\mathtt{Q}$ is mutually interpretable with $\mathtt{AS}$ but not itself sequential. See [35] and [12]. In Example 4.9, it is shown that $\mathtt{Q}$ is not even polysequential.

It seems to me that even for *iso-congruence* there is no strong reason that sequential theories should be closed under that equivalence relation. If we trace the argument above concerning the closure of sequentiality under bi-interpretability, we see that in a model $\mathcal{M}$ of U, we have sequences of elements of the isomorphic internal model $\mathcal{M}'$. However, we cannot make it visible in $\mathcal{M}$ that these sequences can also function as sequences of elements of $\mathcal{M}$. Moreover, where, by compactness, we were guaranteed uniformity of the isomorphisms, in the case of bi-interpretability, conceivably, for iso-congruence, the isomorphisms could be wildly different if we vary our models of U. (See also Question 4.13.)

4 What is Polysequentiality?

Our improved version of sequentiality is *polysequentiality*. In this section we give the official definition of polysequentiality and prove some equivalents. We define *polysequentiality* as follows.

Definition 4.1. *A one-sorted theory U is* polysequential *iff there is an $(\mathbf{o}, \mathbf{o})$-direct interpretation of* $\mathtt{seq}$ *in U.*

4.1. Some useful Results. As a preparation of our characterizations in the subsequent subsections, we prove some useful results about $\mathtt{seq}$. We distinguish between *internal sequences*, which are objects of type $\mathbf{s}$, and *external sequences*, which exist only because we can externally form sequences of terms (as in the definition of m-dimensional interpretations).

Theorem 4.2. *Consider any number m. Then, the theory* $\mathtt{seq}$ *interprets* $\mathtt{ac}$ *by an $(\mathbf{o} : \mathbf{o}^m)(\mathbf{c} : \mathbf{s^2})$-direct interpretation. In other words,*

[17] This follows e.g. from the results of [21] or [19].

there is a $\texttt{seq}$*-definable predicate* $(z_0, \ldots, z_{m-1}) \in (s_0, s_1)$*, where the* z_i *are of type* $\mathbf{o}$ *and the* s_j *are of type* $\mathbf{s}$*, such that:*

- $\texttt{seq} \vdash \exists \vec{s}\, \forall \vec{z}\ \vec{z} \notin \vec{s}$
- $\texttt{seq} \vdash \forall \vec{z}, \vec{s}\, \exists \vec{t}\, \forall \vec{u}\, (\vec{u} \in \vec{t} \leftrightarrow (\vec{u} \in \vec{s} \vee \bigwedge_{i<m} u_i = z_i))$

Here, of course, all (external) sequences are supposed to be of the right length.

Proof. We work in $\texttt{seq}$. It is easy to see that there are at least two objects. For arbitrary x and y with $x \neq y$, we define an *m, x, y-sequence* as follows. It is a sequence s such that

i. $\forall n, u\, (\texttt{Pr}(s, n, u) \to (u = x \vee u = y))$

ii. $\forall n\, (\texttt{Pr}(s, n, x) \to \bigvee_{i<m}((\texttt{Z}(n-i) \vee \texttt{Pr}(s, n-i-1, y)) \wedge$ $(\texttt{L}(s, n-i+m) \vee \texttt{Pr}(s, n-i+m, y)) \wedge \bigwedge_{j<m} \texttt{Pr}(s, n-i+j, x)))$

iii. $\forall n\, (\texttt{Pr}(s, n, y) \to \bigvee_{i<m}((\texttt{Z}(n-i) \vee \texttt{Pr}(s, n-i-1, x)) \wedge$ $(\texttt{L}(s, n-i+m) \vee \texttt{Pr}(s, n-i+m, x)) \wedge \bigwedge_{j<m} \texttt{Pr}(s, n-i+j, y)))$

Here, e.g., $\texttt{Z}(n-i)$ abbreviates $\exists k_0, \ldots k_{i-1}\, (\texttt{Z}(k_0) \wedge \texttt{S}(k_0, k_1) \wedge \ldots \wedge \texttt{S}(k_{i-1}, n))$.

We define $(z_0, \ldots, z_{m-1}) \in (s_0, s_1)$ by: for some x, y, we have $x \neq y$ and s_0 is an x, y, m-sequence and s_0 and s_1 have the same length and, for some n, $(\bigwedge_{i<m} \texttt{Pr}(s_0, n+i, x)$ or $\bigwedge_{i<m} \texttt{Pr}(s_0, n+i, y))$ and $\bigwedge_{i<m} \texttt{Pr}(s_1, n+i, z_i)$.

If we take s_0 and s_1 both the empty sequence, we see that (s_0, s_1) satisfies the empty class axiom. We fulfill the adjunction axiom by pushing m times x or m times y on top of s_0, where the choice of x and y depends on the last element of s_0, if any, and where we push subsequently $z_0, z_1, \ldots, z_{m-1}$ on s_1. ⊣

We provide a strengthening of Theorem 4.2 that will be only used in an alternative proof of Theorem 5.2. We start with a well-known lemma. Let $\texttt{ac}^+$ be $\texttt{ac}$ plus the following axioms.

$\texttt{ac3} \vdash \forall X, Y\, \exists Z\, \forall z\, (z \in Z \leftrightarrow (z \in X \wedge z \in Y))$

$\texttt{ac4} \vdash \forall X, x\, \exists Y\, \forall y\, (y \in Y \leftrightarrow (y \in X \wedge y \neq x))$

Lemma 4.1. *There is an* $\mathbf{o} : \mathbf{o}$*-direct interpretations, of* $\texttt{ac}^+$ *in* $\texttt{ac}$*, with profile* $(\mathbf{o} : \mathbf{o})(\mathbf{c} : \mathbf{c})$*, that is one-dimensional and identity preserving for* $\mathbf{c}$*.*

Proof. We work in $\texttt{ac}$. To simplify the presentation we replace identity on classes by extensional identity. Then we can justify the use of functional notations $\cap$ for intersection and $\setminus$ for subtraction. We define the virtual class $\mathcal{X}_0$ as consisting of all those classes X such that, for all Y,

$X \cap Y$ exists. We show that $\mathcal{X}_0$ is closed under empty class, adjunction, and intersection.

Clearly the empty class is in $\mathcal{X}_0$.

We have: $(X \cup \{x\}) \cap Z = (X \cap Z) \cup \{x\}$, if $x \in Z$, and $(X \cup \{x\}) \cap Z = X \cap Z$ if $x \notin Z$. In both cases $(X \cup \{x\}) \cap Z$ exists. Ergo, $X \cup \{x\}$ is in $\mathcal{X}_0$.

Suppose X and Y in $\mathcal{X}_0$. We note that $X \cap Y$ exists, since $X \in \mathcal{X}_0$. Consider any Z. We have $(X \cap Y) \cap Z = X \cap (Y \cap Z)$. Since $Y \in \mathcal{X}_0$, $Y \cap Z$ exists and since $X \in \mathcal{X}_0$, $X \cap (Y \cap Z)$ exists. Hence $X \cap Y$ is in $\mathcal{X}_0$.

By relativizing the classes to $\mathcal{X}_0$ we get an interpretation of ac+ac3. We now work in ac+ac3. As before, we add extensionality. Define $\mathcal{X}_1$ as the class of all X such that $X \setminus \{x\}$ exists, for all x. We show that $\mathcal{X}_1$ is closed under empty class, adjunction, intersection and subtraction.

It is clear that the empty class is in $\mathcal{X}_1$.

Suppose X is in $\mathcal{X}_1$. Consider x. We have $(X \cup \{x\}) \setminus \{y\} = (X \setminus \{y\}) \cup \{x\}$, if $x \neq y$, and $(X \cup \{x\}) \setminus \{y\} = X$, if $x = y$. In both cases $(X \cup \{x\}) \setminus \{y\}$ exists. So $X \cup \{x\} \in \mathcal{X}_1$.

Suppose X is in $\mathcal{X}_1$. Consider any Y in $\mathcal{X}_1$. We have $(X \cap Y) \setminus \{z\} = (X \setminus \{z\}) \cap Y$. So $X \cap Y$ is in $\mathcal{X}_1$.

Suppose X is in $\mathcal{X}_1$. Consider any x. We have

$$(\dagger) \quad (X \setminus \{x\}) \setminus \{y\} = (X \setminus \{x\}) \cap (X \setminus \{y\}).$$

The right hand side of $(\dagger)$ clearly exists. Hence, $X \setminus \{x\} \in \mathcal{X}_1$.

We relativize to $\mathcal{X}_1$ in ac+ac3. This gives us the desired interpretation. $\dashv$

Theorem 4.3. *Consider any number m. Then, the theory* seq *interprets* ac^+ *by an interpretation of profile* $(\mathbf{o} : \mathbf{o}^{\mathbf{m}})(\mathbf{c} : \mathbf{s}^{\mathbf{2}})$, *that is* $(\mathbf{o} : \mathbf{o}^{\mathbf{m}})$*-direct and* $(\mathbf{c} : \mathbf{s}^{\mathbf{2}})$*-identity preserving. In other words, there are* seq*-definable predicates $D(s_0, s_1)$ and $(z_0, \ldots, z_{m-1}) \in (s_0, s_1)$, where the z_i are of type* $\mathbf{o}$ *and the s_j are of type* $\mathbf{s}$*, such that:*

- $\mathsf{seq} \vdash \vec{x} \in \vec{s} \to D(\vec{s})$
- $\mathsf{seq} \vdash \exists \vec{s}\,(D(\vec{s}) \wedge \forall \vec{z}\ \vec{z} \notin \vec{s})$
- $\mathsf{seq} \vdash \forall \vec{z}, \vec{s}\,(D(\vec{s}) \to \exists \vec{t}\,(D(\vec{t}) \wedge \forall \vec{u}\,(\vec{u} \in \vec{t} \leftrightarrow (\vec{u} \in \vec{s} \vee \bigwedge_{i<m} u_i = z_i))))$
- $\mathsf{seq} \vdash$ $\forall \vec{s}, \vec{s}\,'\,((D(\vec{s}) \wedge D(\vec{s}\,')) \to \exists \vec{t}\,(D(\vec{t}) \wedge \forall \vec{z}\,(\vec{z} \in \vec{t} \leftrightarrow (\vec{z} \in \vec{s} \wedge \vec{z} \in \vec{s}\,'))))$

- $\mathsf{seq} \vdash \forall \vec{z}, \vec{s}\, (D(\vec{s}) \to \exists \vec{t}\, (D(\vec{t}) \wedge \forall \vec{u}\, (\vec{u} \in \vec{t} \leftrightarrow (\vec{u} \in \vec{s} \wedge \bigvee_{i<m} u_i \neq z_i))))$

Here all (external) sequences are supposed to be of the right length.

Proof. We compose the interpretations of Theorem 4.2 and Lemma 4.1.

4.2. m-Direct and Relaxed Direct Interpretations. Consider one-sorted theories U and V. An interpretation $M : U \to V$ is *relaxed direct* iff there is a V-definable relation R, such that:

- $V \vdash x\, R\, \vec{y} \to \delta_M(\vec{y})$,
- $V \vdash \forall x\, \exists \vec{y}\;\, x\, R\, \vec{y}$,
- $V \vdash (x\, R\, \vec{y} \wedge u\, R\, \vec{v} \wedge \vec{y} =_M \vec{v}) \to x = u$.

Here $=_M$ is the M-translation of identity in U.

If the interpretation contains parameters with domain α, the definition of R will also contain parameters. E.g., the first item becomes:

$$V \vdash (\alpha(\vec{z}) \wedge x\, R_{\vec{z}}\, \vec{y}) \to \delta_{M,\vec{z}}(\vec{y}),$$

etcetera. We have the following theorem.

Proposition 4.4. *Any m-direct interpretation is relaxed direct.*

Proof. Suppose $M : U \to V$ is m-direct. Let $x\, R\, \vec{y} :\leftrightarrow \bigwedge_{i<m} y_i = x$. Clearly, R witnesses that M is relaxed direct. ⊣

Here is our first characterization of polysequentiality.

Theorem 4.5. *The following are equivalent:* (i) *U is polysequential;* (ii) *for some m, there is an m-direct interpretation of* $\mathtt{AS}$ *in U,* (iii) *there is a relaxed direct interpretation of* $\mathtt{AS}$ *in U.*

Proof. We give the proof for the case without parameters. The case with parameters is an easy adaptation.

(i) $\Rightarrow$ (ii). Suppose $M : \mathsf{seq} \to U$ is $(\mathbf{o} : \mathbf{o})$-direct. Suppose the interpretation of type $\mathbf{s}$ is k-dimensional. Put $m := 2k$. We define:
$(z_1, \ldots, z_m) \in (x_0, \ldots, x_{k-1}, y_0, \ldots, y_{k-1}) :\leftrightarrow$
$\delta^{\mathbf{s}}_M(x_0, \ldots, x_{k-1}) \wedge \delta^{\mathbf{s}}_M(y_0, \ldots, y_{k-1}) \wedge$
$((z_1, \ldots, z_m) \in (x_0, \ldots, x_{k-1}, y_0, \ldots, y_{k-1}))^M$
where Theorem 4.2 provides the relation $\in$ inside M. (By coding the pair of internal sequences as one internal sequence, we can get the more efficient $m := k$.)

(ii) $\Rightarrow$ (iii). This is immediate from Proposition 4.4.

(iii) $\Rightarrow$ (i). Suppose $M : \mathtt{AS} \to U$ is relaxed direct. Let R be the witnessing relation. As we mentioned in the proof of Theorem 3.3, we can

$(\mathbf{o}:\mathbf{o})$-directly interpret seq in AS, say via K. See, e.g., [21], [19], [10]. Suppose $L := M \circ K : \mathtt{seq} \to U$. By the $(\mathbf{o}:\mathbf{o})$-directness of K, we find that $\delta^{\mathbf{o}}_L$ is, U-provably, equal to δ_M, and $=^{\mathbf{o}}_L$ is, U-provably, the same as $=_M$. We build the desired $(\mathbf{o}:\mathbf{o})$-direct interpretation, say, $Q : \mathtt{seq} \to U$, by defining:

- $\delta^{\mathbf{o}}_Q(x) :\leftrightarrow x = x$,
- $\delta^{\mathbf{s}}_Q(\vec{y}) :\leftrightarrow \delta^{\mathbf{s}}_L(\vec{y})$,
- $\delta^{\mathbf{n}}_Q(\vec{z}) :\leftrightarrow \delta^{\mathbf{n}}_L(\vec{z})$,
- $=^{\mathbf{s}}$, $=^{\mathbf{n}}$, Z, E, S, $<$ L, are interpreted according to Q as they are according to L,
- $x =^{\mathbf{o}}_Q y :\leftrightarrow x = y$,
- $\mathtt{Pu}_Q(\vec{s}, x, \vec{t}) :\leftrightarrow \exists \vec{y}\,(x\ R\ \vec{y} \wedge \mathtt{Pu}_L(\vec{s}, \vec{y}, \vec{t}))$,
- $\mathtt{Pr}_Q(\vec{s}, \vec{n}, x) :\leftrightarrow \exists \vec{y}\,(x\ R\ \vec{y} \wedge \mathtt{Pr}_L(\vec{s}, \vec{n}, \vec{y}))$.

By the U-provable, totality and $(=, =^{\mathbf{o}}_L)$-injectivity of R, we can easily verify that Q is indeed an interpretation of seq. ⊣

It is clear that we should have the following theorem.

Proposition 4.6. *Suppose K : AS $\to V$ is relaxed direct and one-dimensional. Then V is sequential.*

Proof. Suppose K : AS $\to V$ is relaxed direct and one-dimensional. Suppose R witnesses that K is direct relaxed. We define a new direct interpretation M of AS in V by setting $x \in_M y :\leftrightarrow \exists z\,(x\ R\ z \wedge z \in_K y)$. ⊣

Remark 4.7. Suppose that U is weak 1-pair-theory in the sense of Subsubsection 2.3.2 and Appendix B. We have seen that U is polysequential iff there is an m-dimensional interpretation of AS in U. Thus, it follows by Theorem 2.1, that U is *polysequential* iff U is sequential.

Remark 4.8. We call a theory U *conceptual* if there is an $(\mathbf{o}, \mathbf{o})$-direct interpretation of ac in U. It is easily seen that, if U is conceptual via a one-dimensional interpretation, then U is sequential. However, if the interpretation of the classes is allowed to be more-dimensional, then there are conceptual theories with finite models. So *conceptual* $\neq$ *polysequential.* As we hope to illustrate elsewhere the notion of conceptuality is useful in its own right.

We end this subsection with an example of a salient theory that fails to be polysequential.

Example 4.9 Robinson's arithmetic Q is not polysequential by an argument of Emil Jeřábek as an answer to a question of Andrés Cordón Franco (in private correspondence). The argument is as follows.

We show that Q is not even a weak m-pair theory. (See Subsubsection 2.3.2 for the definition of a weak m-pair theory.) To see this it is sufficient to show that there is no definable coding of $(m+1)$-tuples by m-tuples in Q. This can be proved using the following model: it consists of the standard model of arithmetic plus infinitely many other elements a_i, we define:

- $\mathsf{S}a_i = a_i$,
- $a_i + x = a_i$ for any x, and $n + a_i = a_i$, for standard n,
- $a_i \cdot 0 = 0$, and $a_i \cdot x = a_i$ for nonzero x, and $n \cdot a_i = a_i$ for standard n.

It is easily checked that this is a model of Q.

Clearly, every permutation of the a_i is an automorphism. Suppose we have a definable coding of $(k+1)$-tuples $(d_0, \ldots, d_k)$ by k-tuples with parameters $b_0, \ldots, b_{s-1}$. Without loss of generality we can assume that the parameters are non-standard. It follows that any code of $(d_0, \ldots, d_k)$ must involve all nonstandard elements that appear among $d_0, \ldots, d_k$ and are distinct from the parameters. Otherwise, we could move the tuple by an automorphism while leaving the code fixed. However, this is impossible if we take for $d_0, ..., d_k$ pairwise distinct nonstandard elements that are disjoint from the parameters.

4.3. bi-Interpretability. Any interpretation between one-sorted theories can be split up into first an identity-preserving interpretation and then a bi-interpretation, i.e., an isomorphism in the category $\mathtt{INT}_1$.

Theorem 4.10. *Suppose $K : U \to V$, where U and V are 1-sorted. Suppose V proves that there are at least two objects. Then, there is a theory V_K and there are interpretations $\imath_K^U : U \to V_K$ and $\jmath_K : V_K \to V$, such that $\imath_K^U$ is identity-preserving, $\jmath_K$ is a bi-interpretation, and $K = \jmath_K \circ \imath_K^U$.*

As a general methodology, one should strive —to quote Kreisel— *to say simple things simply.* Regrettably, the proof of the present theorem is a case where I failed miserably to fulfill this dictum. It seems to me that, with the right eyes, one can immediately see that it is true, but when you try to write the proof down with any precision ... My solution to the problem presented by this elaborate proof of a triviality is as follows. I provide a detailed proof of the theorem in Appendix A. Here I give a sketchy proof highlighting the main ideas of the construction.

Proof. We ignore the possibility of parameters. The theory V_K is formed as the disjoint sum of the theories U and V extended with a new predicate F. The obvious embeddings of U and V in V_K are respectively $\imath_K^U$

and $\imath_K^V$. The theory V_K also contains axioms stating that F is an isomorphism between U as interpreted via $\imath_K^V \circ K$ and U as interpreted via $\imath_K^U$ in V_K. In other words, $F : \imath_K^V \circ K \to \imath_K^U$ witnesses the $\mathtt{INT}_1$-identity between $\imath_K^V \circ K$ and $\imath_K^U$.

We want to construct an interpretation $\jmath_K : V \to V_K$, such that $\jmath_K$ and $\imath_K^V$ are inverses in the sense of $\mathtt{INT}_1$. Thus, V and V_K will be bi-interpretable. Moreover, since $\imath_K^V \circ K =_1 \imath_K^U$, we have $K = \jmath_K \circ \imath_K^V \circ K =_1 \jmath_K \circ \imath_K^U$. Here the standard embedding $\imath_K^U$ is clearly one-dimensional and identity-preserving.

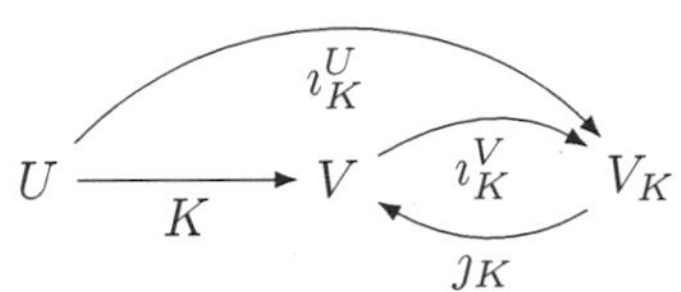

We turn to the construction of $\jmath_K$, i.e., of a uniform internal model of V_K in V. We want to reproduce the objects of V and the objects of δ_K as a disjoint sum. To do that more-dimensionality comes into play. The objects of δ_K are given as (external) sequences of length, say m. As a first move we also represent the objects of V as sequences of length m where we take the last component to represent the intended object. The rest is padding. We are given that V produces two distinct objects x and y, so one option is to make our domains disjoint by raising the dimension with 1 and represent the V-objects as $(x, \ldots, v)$ and the U-objects as $(y, u_0, \ldots, u_{m-1})$. The disadvantage of this construction is that we make use of parameters. This is an inessential use, however, and we can replace it by a construction where we do not use parameters. We raise the dimension with two and represent the V-objects by $(x, x, \ldots, v)$, where x is now arbitrary, and we represent the U-objects by $(x, y, u_0, \ldots, u_{m-1})$, where x and y are arbitrary and different.

Thus we produced disjoint copies of the objects of V and the objects of δ_U and the copies are given by sequences of the same length. It is now easy to extend this construction with define predicates to obtain an $m+2$-dimensional interpretation $\jmath_K$ of V_K in V.

The verification that $\jmath_K$ is inverse to $\imath_K^U$, finally, is just a matter of careful tracing of obvious details. ⊣

We use Theorem 4.10 to prove that any polysequential theory is bi-interpretable with a sequential theory.

Theorem 4.11. *Suppose V is polysequential. Then V is bi-interpretable with a sequential theory $V^\star$.*

Proof. Suppose $K : \mathtt{AS} \to V$ is a relaxed direct interpretation. Let R be the witness of relaxedness. We take $V^\star := V_K$ as provided by Theorem 4.10.

We start with an arbitrary object x from the domain of V_K, or, in other words, of $\mathtt{id}_{V_K}$. The bi-interpretation between V and V_K is given by $\imath_K^V : V \to V_K$ and $\jmath_K : V_K \to V$. We have an isomorphism $G : \imath_K^V \circ \jmath_K \to \mathtt{id}_{V_K}$. We relate x via G^{-1} to a sequence $\vec{y}$ of $\imath_K^V \circ \jmath_K$. Suppose $\imath_K^V$ is k-dimensional and $\jmath_K$ is ℓ-dimensional. We can view $\vec{y}$ as a sequence $\vec{y}_0, \ldots, \vec{y}_{\ell-1}$ of $\imath_K^V$-objects $\vec{y}_i$ of length k. Each $\imath_K^V$-object $\vec{y}_i$ is related via $R^{\imath_K^V}$ to an $\imath_K^V \circ K$-object $\vec{u}_i$. So we can relate $\vec{y}_0, \ldots, \vec{y}_{\ell-1}$ via $R^{\imath_K^V}$ pointwise to $\vec{u}_0, \ldots, \vec{u}_{\ell-1}$. The sequence $\vec{u}_0, \ldots, \vec{u}_{\ell-1}$ is an external sequence in $\imath_K^V \circ K : \mathtt{AS} \to V_K$. Inside $\imath_K^V \circ K$ we can work in $\mathtt{AS}$. Thus, we may relate the external sequence $\vec{u}_0, \ldots, \vec{u}_{\ell-1}$ to an internal one $\vec{v}$ in $\imath_K^V \circ K$. Let's call the internalizing relation $I^{\imath_K^V \circ K}$ (for the given length ℓ). We have $\vec{v} \in_{\imath_K^V \circ K} \vec{w}$ for some $\vec{w}$ in $\imath_K^V \circ K$. Finally let $H : \imath_K^V \circ K \to \imath_K^U$ be the isomorphism between $\imath_K^V \circ K$ and $\imath_K^U$. We can relate $\vec{w}$ to z via H.

Thus, we can define:

$$
\begin{aligned}
rlx \in^\star z :\leftrightarrow \exists \vec{y}_0, \ldots, \vec{y}_{\ell-1} \, (x \, G^{-1} \, (\vec{y}_0, \ldots, \vec{y}_{\ell-1}) \wedge \\
\exists \vec{u}_0, \ldots, \vec{u}_{\ell-1} \, (\bigwedge_{i<\ell} \vec{y}_i \, R^{\imath_K^V} \, \vec{u}_i \wedge \\
\exists \vec{v} \, ((\vec{u}_0, \ldots, \vec{u}_{\ell-1}) \, I^{\imath_K^V \circ K} \, \vec{v} \wedge \\
\exists \vec{w} \, (\vec{v} \in_{\imath_K^V \circ K} \vec{w} \wedge \vec{w} \, H \, z))))
\end{aligned}
$$

In this definition, we assume that domain and range are 'built in' in each relation involved. E.g., $\vdash x \, G^{-1} \, (\vec{y}_0, \ldots, \vec{y}_{\ell-1}) \to \delta_{\imath_K^V \circ \jmath_K} (\vec{y}_0, \ldots, \vec{y}_{\ell-1})$.

To see that the definition works, note that each of G^{-1}, $R^{\imath_K^V}$ and $I^{\imath_K^V \circ K}$ is total and injective (w.r.t. the appropriate equivalence relations) and that H is an isomorphism. ⊣

Finally we show that, retracts in the category $\mathtt{INT}_1$ of sequential theories are polysequential. A theory U is a retract in $\mathtt{INT}_1$ of a theory V if there are interpretations $K : U \to V$ and $M : V \to U$ and $F : M \circ K \to \mathtt{id}_U$, where F is a U-definable isomorphism.

We note that it follows that U is polysequential iff U is bi-interpretable with a sequential V iff U is a retract in $\mathtt{INT}_1$ of a sequential V.

Proposition 4.12. *If U is a retract in $\mathtt{INT}_1$ of a sequential theory V, then U is polysequential.*

Proof. Suppose $K : U \to V$ and $M : V \to U$ and $F : M \circ K \to \mathtt{id}_U$ witness that U is a retract of V in $\mathtt{INT}_1$. Let $N : \mathtt{AS} \to V$ be direct. Let

M be m-dimensional and let K be k-dimensional.

We have $M \circ N : \mathtt{AS} \to U$. Note that $M \circ N$ is m-dimensional. Let I be the relation between external and internal sequences in AS, for the given length k. We take:

$$x\ R\ \vec{z} :\leftrightarrow \exists \vec{y}_0, \ldots, \vec{y}_{k-1}\ x\ F^{-1}\ (\vec{y}_0, \ldots, \vec{y}_{k-1})\ I^{M \circ N}\ \vec{z}.$$

The relation R witnesses that $M \circ N$ is relaxed direct. ⊣

Open Question 4.13. It seems to me that there is no strong reason that polysequential theories should be closed under iso-congruence. However, it would be interesting to know whether they are. I conjecture that the answer is *no*.

A closely related question is as follows. Is Robinson's arithmetic Q iso-congruent to a sequential theory? I conjecture that Q is not even elementarily congruent to a sequential theory. (We know that Q is mutually interpretable with a sequential theory.)

4.4. Summarizing. We have the following characterization theorem.

Theorem 4.14. *The following are equivalent.*

char(i) *U is polysequential.*

char(ii) *For some m, there is an m-direct interpretation of* AS *in U.*

char(iii) *There is a relaxed direct interpretation of* AS *in U.*

char(iv) *U is bi-interpretable with a sequential theory V.*

char(v) *U is a retract in* $\mathtt{INT}_1$ *of a sequential theory V.*

char(vi) *U is bi-interpretable with an extension of* AS *in the same language.*

The last equivalence follows immediately using (iv) of Theorem 3.3.

5 Parameters

In this section, we will show that some sequential theories are not parameter-free sequential. In contrast, we show that all polysequential theories are parameter-free polysequential.

Example 5.1. Let us start with the Ackermann interpretation of the hereditarily finite sets in the natural numbers. We call this $\in_{\mathrm{ack}}$. Let α be a bijection between the integers and the natural numbers. We define, on the integers, $x \in^* y :\leftrightarrow \alpha(x) \in_{\mathrm{ack}} \alpha(y)$. Finally, we define, on the integers, $x \in_z y :\leftrightarrow (x - z) \in^* (y - z)$.

We consider the structure, say $\mathcal{M}$, on the set of the integers, given by $x \in_z y$. Clearly, $\mathcal{M}$ is sequential with one parameter.

Suppose a binary predicate $x \in y$ were definable in $\mathcal{M}$, without parameters, satisfying AS. Let m be a singleton. Since successor is an automorphism of $\mathcal{M}$ it follows that all elements are singletons, contradicting the assumption that we have $\mathtt{AS}$ for $\in$.

We prove that parameters can always be eliminated from the witnessing interpretations of polysequentiality.

Theorem 5.2. *Suppose V is polysequential. Then V is polysequential without parameters.*

Proof. Our proof is an adaptation of the proof that if V is polysequential via a k-dimensional interpretation of the sequences, then there is an $2k$-direct interpretation of $\mathtt{AS}$ in V.

Suppose U is polysequential. Let $M : \mathsf{seq} \to U$ be $(\mathbf{o} : \mathbf{o})$-direct. Suppose M has dimension k for the interpretation of sort $\mathbf{s}$. Suppose M has an ℓ-dimensional parameter domain α.

Put $m := 2k + l$. We define:
$(z_1, \ldots, z_m) \in (p_0, \ldots, p_{\ell-1}, x_0, \ldots, x_{k-1}, y_0, \ldots, y_{k-1}) :\leftrightarrow$
$\alpha_M(p_0, \ldots, p_{\ell-1}) \wedge \delta^{\mathbf{s}}_M(x_0, \ldots, x_{k-1}) \wedge \delta^{\mathbf{s}}_M(y_0, \ldots, y_{k-1}) \wedge$
$((z_1, \ldots, z_m) \in (x_0, \ldots, x_{k-1}, y_0, \ldots, y_{k-1}))^{M,\vec{p}}$
using Theorem 4.2 to provide the $\in$ inside $M, \vec{p}$. (By coding the triple of external sequences as one internal sequence, we can get the more efficient $m := k$.) $\dashv$

The above proof is a bit disappointing, since we have to detour over the characterization theorem. This means that, in a sense, we throw away the interpretations of the successor theory that we were given and have to regain it via a bootstrap. There is a more elegant and more interesting proof that we will give now. The construction shows how we can, in the presence of finite functions, glue together different systems of the ordering of the natural numbers.

Proof. Suppose $M : \mathsf{seq} \to U$ is $(\mathbf{o} : \mathbf{o})$-direct. Let M be k-dimensional for sort $\mathbf{n}$. Let $\vec{p}$ and $\vec{q}$ be in α_M. Using Theorem 4.3, we can define finite functions from $\delta^{\mathbf{n}}_{M,\vec{p}}$ to $\delta^{\mathbf{n}}_{M,\vec{q}}$ as finite sets of $2k$-sequences. Note that by Theorem 4.3, we can both adjoin pairs to a function and substract pairs from a function. We consider two such functions *equal*, when, whenever their inputs are $=_{M,\vec{p}}$-equal, then their outputs are $=_{M,\vec{q}}$-equal.

We say such a function f is a $\vec{p}, \vec{q}$-isomorphism on $\vec{n}$ (in $\delta^{\mathbf{n}}_{M,\vec{p}}$) iff the domain of f consists of all $\vec{n}'$ such that $\vec{n}' <_{M,\vec{p}} \vec{n}$, and the range is $<_{M,\vec{q}}$-downwards closed, and, if $\vec{n}'' <_{M,\vec{p}} \vec{n}' <_{M,\vec{p}} \vec{n}$, then $f(\vec{n}'') <_{M,\vec{q}} f(\vec{n}')$. Note that it follows that f commutes with successors. Clearly, we can find a $\delta^{\mathbf{n}}_{M,\vec{q}}$-number $\vec{n}^*$, such that we can consider f as function from $\vec{n}$ to $\vec{n}^*$.

Define $N_{\vec{p}}$ as the class of those $\vec{n}$ in $\delta^{\mathbf{n}}_{M,\vec{p}}$, such that, for all $\vec{n}' \leq_{M,\vec{p}} \vec{n}$, and for all $\vec{q}$ in α_M:

i. there is a $\vec{p}, \vec{q}$-isomorphism f on $\vec{n}'$, which is unique modulo equality. We will call the unique $\vec{p}, \vec{q}$ isomorphism on $\vec{n}$: $f_{\vec{n},\vec{p},\vec{q}}$.

ii. if $n'' <_{M,\vec{p}} n'$, then $f_{\vec{n}'',\vec{p},\vec{q}}$ is the restriction of $f_{\vec{n}',\vec{p},\vec{q}}$ to $\vec{n}'$.

iii. $f_{\vec{n}',\vec{p},\vec{p}}$ is the identity.

It is clear that $N_{\vec{p}}$ is downwards closed under $<_{M,\vec{p}}$ and that it contains the zero of $\delta^{\mathbf{n}}_{M,\vec{p}}$. We show that $N_{\vec{p}}$ is closed under successor. Consider any $\vec{n}$ in $N_{\vec{p}}$. Suppose $\mathtt{S}_{M,\vec{p}}(\vec{n}, \vec{n}^+)$. Either $\vec{n}$ is an $M, \vec{p}$-zero or an $M, \vec{p}$-successor. We leave the case that $\mathtt{Z}_{M,\vec{p}}(\vec{n})$ to the reader. Let the $<_{M,\vec{p}}$-predecessor of $\vec{n}$ be $\vec{n}^-$. Suppose $f_{\vec{n},\vec{p},\vec{q}}(\vec{n}^-) = \vec{n}^*$ and $\mathtt{S}_{M,\vec{q}}(\vec{n}^*, \vec{n}^\star)$. We obtain $f_{\vec{n}^+,\vec{p},\vec{q}}$ by adjoining $(\vec{n}, \vec{n}^\star)$ to $f_{\vec{n},\vec{p},\vec{q}}$. It is easy to see that $f_{\vec{n}^+,\vec{p},\vec{q}}$, thus defined, is a $\vec{p}, \vec{q}$-isomorphism for $\vec{n}^+$. Also property (ii) and (iii) are evident. Suppose g is a second $\vec{p}, \vec{q}$-isomorphism for $\vec{n}^+$. We subtract the pair corresponding to input $\vec{n}$ from g, thus obtaining a $\vec{p}, \vec{q}$-isomorphism h for $\vec{n}$. It follows that h is equal to $f_{\vec{n},\vec{p},\vec{q}}$. Hence g must be equal to $f_{\vec{n}^+,\vec{p},\vec{q}}$.

We define: $F_{\vec{p},\vec{q}}(\vec{n}) = \vec{n}^*$ iff $\mathtt{S}_{M,\vec{p}}(\vec{n}, \vec{n}^+)$ and $f_{\vec{n}^+,\vec{p},\vec{q}}(\vec{n}) = \vec{n}^*$. Clearly, $F_{\vec{p},\vec{q}}$ is an initial embedding of $N_{\vec{p}}$ in $\delta^{\mathbf{n}}_{M,\vec{q}}$. Also $F_{\vec{p},\vec{p}}$ is the identity on $N_{\vec{p}}$.

We want to improve the $N_{\vec{p}}$ and the $F_{\vec{p},\vec{q}}$ to a system of classes of 'numbers' with designated isomorphisms between them where the isomorphisms are closed under identity, inverse and composition. We proceed as follows. We define $N^\star_{\vec{p}}$ as the class of those $\vec{n}$ in $N_{\vec{p}}$, such that, for all $\vec{n}' \leq_{M,\vec{p}} \vec{n}$,

a. for all $\vec{q}$ in α_M, we have: $F_{\vec{p},\vec{q}}(\vec{n}') \in N_{\vec{q}}$,

b. for all $\vec{q}, \vec{r}$ in α_M, we have: $F_{\vec{q},\vec{r}}(F_{\vec{p},\vec{q}}(\vec{n}')) = F_{\vec{p},\vec{r}}(\vec{n}')$,

It is easy to see that $N^\star_{\vec{p}}$ is downwards closed w.r.t. $<_{M,\vec{p}}$, that it contains the zero of $N_{\vec{p}}$, and that it is closed under $\mathtt{S}_{M,\vec{p}}$.

Suppose $\vec{n}$ is in $N^\star_{\vec{p}}$. We show that $\vec{n}^\star := F_{\vec{p},\vec{q}}(\vec{n})$ satisfies the conditions (a) and (b). We note that (a) for $\vec{n}^\star$, follows from (b) and (a) for $\vec{n}$. We verify (b) for $\vec{n}^\star$. For $\vec{s}$ in α_M, we have:

$$\begin{aligned} llF_{\vec{r},\vec{s}}(F_{\vec{q},\vec{r}}(\vec{n}^\star)) &= F_{\vec{r},\vec{s}}(F_{\vec{q},\vec{r}}(F_{\vec{p},\vec{q}}(\vec{n}))) \\ &= F_{\vec{r},\vec{s}}(F_{\vec{p},\vec{r}}(\vec{n})) \\ &= F_{\vec{p},\vec{s}}(\vec{n}) \\ &= F_{\vec{q},\vec{s}}(F_{\vec{p},\vec{q}}(\vec{n})) \\ &= F_{\vec{q},\vec{s}}(\vec{n}^\star) \end{aligned}$$

Consider any $\vec{n}$ in $N^\star_{\vec{p}}$. Suppose $\vec{n}^* \leq_{M,\vec{q}} \vec{F}_{\vec{p},\vec{q}}(\vec{n})$. Since, $F_{\vec{p},\vec{q}}$ is an initial embedding, it follows that, for some $\vec{n}' \leq_{M,\vec{p}} \vec{n}$, we have $\vec{n}^* =$

$F_{\vec{p},\vec{q}}(\vec{n}')$. Clearly, $\vec{n}'$ is in $N^\star_{\vec{p}}$. Ergo by the above $\vec{n}^*$ satisfies conditions (a) and (b). We may conclude that $F_{\vec{p},\vec{q}}(\vec{n})$ is in $N^\star_{\vec{q}}$.

Let $F^\star_{\vec{p},\vec{q}}$ be $F_{\vec{p},\vec{q}}$ restricted to $N^\star_{\vec{p}}$. Clearly, we find $F^\star_{\vec{p},\vec{p}} = \mathtt{id}_{N^\star_{\vec{p}}}$, and $F^\star_{\vec{q},\vec{r}} \circ F^\star_{\vec{p},\vec{q}} = F^\star_{\vec{p},\vec{r}}$, and $F^\star_{\vec{p},\vec{q}} = (F^\star)^{-1}_{\vec{q},\vec{p}}$.

We assume that the $F^\star$ are defined in such a way that we have:

- $U \vdash \vec{n}\, F^\star_{\vec{p},\vec{q}}\, \vec{n}^* \to (N^\star_{\vec{p}}(\vec{n}) \wedge N^\star_{\vec{q}}(\vec{n}^*))$
- $U \vdash \vec{n} =_{M,\vec{p}} \vec{n}'\; F^\star_{\vec{p},\vec{q}}\, \vec{n}^* =_{M,\vec{q}} n^\star \to \vec{n}\, F^\star_{\vec{p},\vec{q}}\, \vec{n}^\star$

We define our parameter-free interpretation $M^\star : \mathtt{seq} \to U$.

$$\begin{aligned}
\delta^{\mathbf{o}}_{M^\star}(x) &\;:\leftrightarrow\; x = x,\\
x =^{\mathbf{oo}}_{M^\star} y &\;:\leftrightarrow\; x = y,\\
\delta^{\mathbf{n}}_{M^\star}(\vec{p},\vec{n}) &\;:\leftrightarrow\; \alpha_M(\vec{p}) \wedge N^\star_{\vec{p}}(\vec{n}),\\
\vec{p},\vec{n} =^{\mathbf{nn}}_{M^\star} \vec{q},\vec{n}^* &\;:\leftrightarrow\; \vec{n}\, F^\star_{\vec{p},\vec{q}}\, \vec{n}^*,\\
\mathtt{Z}_{M^\star}(\vec{p},\vec{n}) &\;:\leftrightarrow\; \mathtt{Z}_{M,\vec{p}}(\vec{n}),\\
\mathtt{S}_{M^\star}(\vec{p},\vec{n},\vec{q},\vec{n}^*) &\;:\leftrightarrow\; \exists \vec{n}'\,(S_{M,\vec{p}}(\vec{n},\vec{n}') \wedge \vec{n}'\, F_{\vec{p},\vec{q}}\, \vec{n}^*),\\
\vec{p},\vec{n} <_{M^\star} \vec{q},\vec{n}^* &\;:\leftrightarrow\; \exists \vec{n}'\,(\vec{n} <_{M,\vec{p}} \vec{n}' \wedge \vec{n}'\, F^\star_{\vec{p},\vec{q}}\, \vec{n}^*),\\
\delta^{\mathbf{s}}_{M^\star}(\vec{p},\vec{s}) &\;:\leftrightarrow\; \alpha_M(\vec{p}) \wedge \delta^{\mathbf{s}}_{M,\vec{p}}(\vec{s}),\\
\vec{p},\vec{s} =_{M^\star} \vec{q},\vec{t} &\;:\leftrightarrow\; \bigwedge_i p_i = q_i \wedge \vec{s} =_{M,\vec{p}} \vec{t},\\
\mathtt{E}_{M^\star}(\vec{p},\vec{s}) &\;:\leftrightarrow\; \mathtt{E}_{M,\vec{p}}(\vec{s}),
\end{aligned}$$

$$\begin{aligned}
\mathtt{L}_{M^\star}(\vec{p},\vec{s},\vec{q},\vec{n}^*) &\;:\leftrightarrow\; \exists \vec{n}\,(\vec{n}\, F^\star_{\vec{p},\vec{q}}\, \vec{n}^* \wedge \mathtt{L}_{M,\vec{p}}(\vec{s},\vec{n})),\\
\mathtt{Pr}_{M^\star}(\vec{p},\vec{s},\vec{q},\vec{n}^*,x) &\;:\leftrightarrow\; \exists \vec{n}\,(\vec{n}\, F^\star_{\vec{p},\vec{q}}\, \vec{n}^* \wedge \mathtt{Pr}_{M,\vec{p}}(\vec{s},\vec{n},x)),\\
\mathtt{Pu}_{M^\star}(\vec{p},\vec{s},x,\vec{q},\vec{t}) &\;:\leftrightarrow\; \bigwedge_i p_i = q_i \wedge \mathtt{Pu}_{M,\vec{p}}(\vec{s},x,\vec{t}).
\end{aligned}$$

It is not difficult to see that our interpretation delivers the goods. ⊣

6 Sequentiality is not preserved under Bi-interpretablity

We consider the following model $\mathcal{M}$ in a signature with unary predicate symbols A, B, S, C, a binary predicate symbol $\in$ and a ternary symbol app. The domain is partitioned in three infinite sets B, S and C. The set A is an infinite subset of S. Here A is the interpretation of A, etc. The union of the sets A and B forms the urelements. The set S consists of the hereditarily finite sets over A plus A. The relation $\in$ (that corresponds, par abus de langage, with the symbol $\in$) is the element relation on S. Here urelements are treated as empty sets. Note that we can define 'the true empty set' $\emptyset$: it is the unique empty set not in A. The set C consists of partial bijections from A to B. We treat the empty bijection in C, say $\triangle$, as distinct from the true empty set $\emptyset$ in S. Finally app stands for the application relation that we will again call app.

We use $a, a', a_0, \ldots$ to range over elements of A, and $b, b', b_0, \ldots$ for elements of B, and $s, s', s_0, t, \ldots$ for elements of S, and $\chi, \chi', \chi_0, \nu, \ldots$ for elements of C. Finally, $x, x', x_0, y, z, \ldots$ are general variables. There will be some ad hoc uses of other letters for variables.

We write $\chi(a) = b$ for $\mathtt{app}(\chi, a, b)$. We will use $\langle s, t \rangle$ for internally defined pairing on S and (x, y) for 'external pairing' as used in more-dimensional interpretations.

We first show that (the theory of) our model is polysequential. To do that we need some preparation. We start to work in S. Clearly, in S we have the familiar machinery available to implement certain inductive definitions. Let's say that the class of *substitution pairs* is the minimal class such that a pair $\langle c, s \rangle$ is a substitution pair, if c is a partial bijection from A to A, coded in the usual way as a set of pairs, and s is a set of substitution pairs such that for every element $\langle d, t \rangle$ of s we have $\mathtt{range}(d) \subseteq \mathtt{dom}(c)$.

We will prove that there is a relaxed direct interpretation of AS in (the theory of) $\mathcal{M}$ to show that (the theory of) $\mathcal{M}$ is polysequential. We first specify our interpretation of AS. The elements of our domain, say the class of set*s, are of the form (χ, s), where s is a set of substitution pairs for any $\langle d, t \rangle$ in s, we have $\mathtt{range}(d) \subseteq \mathtt{dom}(\chi)$.

We define: $(\nu, t) \in^* (\chi, s)$ iff, for some d, we have $\langle d, t \rangle \in s$ and $\nu = \chi \circ d$. In fact d is uniquely determined by this condition: we could also say $(\nu, t) \in^* (\chi, s)$ iff $\langle \chi^{-1} \circ \nu, t \rangle \in s$.

We note that for any $\langle d, t \rangle \in s$, we have that $(\chi \circ d, t)$ is a set*. Consider any $\langle e, u \rangle \in t$. Since $\langle e, u \rangle$ is a substitution pair, we have $\mathtt{range}(e) \subseteq \mathtt{dom}(d) = \mathtt{dom}(\chi \circ d)$.

We easily see that any pair $(\chi, \emptyset)$ will be an empty set*.

We define $\{(\chi, s)\}^* := (\chi, \{\langle j, s \rangle\})$, where j is the identity on $\mathtt{dom}(\chi)$, or, in other words, $j = \chi^{-1} \circ \chi$. We easily see that $\{(\chi, s)\}^*$ is indeed an $\in^*$-singleton containing (χ, s).

Next we want to show that —not necessarily unique— unions exist. We need the following lemmas.

Lemma 6.1. *Suppose $\chi \subseteq \rho$ and that (χ, s) is a set**. *Then, (ρ, s) is also a set**. *Moreover, (ρ, s) is $\in^*$-extensionally equal to (χ, s).*

Lemma 6.2. *Suppose that (χ, s) and (χ, t) are sets**, *then $(\chi, s \cup t)$ is a set**, *which is an $\in^*$-union of (χ, s) and (χ, t).*

Lemma 6.3. *Suppose χ and ρ have the same range. Consider (χ, s). We can find an u such that (ρ, u) is a set** *that is $\in^*$-extensionally equal to (χ, s).*

Lemmas 6.1 and 6.2 are easy. We prove Lemma 6.3.

Proof. Suppose χ and ρ have the same range and consider (χ, s). Define:

$$u = \{\langle \rho^{-1} \circ \chi \circ d, t\rangle \mid \langle d, t\rangle \in s\}.$$

We note that $\langle \rho^{-1} \circ \chi \circ d, t\rangle$ is a substitution pair, since $\mathtt{dom}(\rho^{-1} \circ \chi \circ d) = \mathtt{dom}(d)$. The pair (ρ, u) is a set*, since $\mathtt{range}(\rho^{-1} \circ \chi \circ d) \subseteq \mathtt{dom}(\rho)$. Finally, we have:

$$\begin{aligned} ll(\nu, t) \in^\star (\chi, s) &\leftrightarrow \langle \chi^{-1} \circ \nu, t\rangle \in s \\ &\leftrightarrow \langle \rho^{-1} \circ \chi \circ \chi^{-1} \circ \nu, t\rangle \in u \\ &\leftrightarrow \langle \rho^{-1} \circ \nu, t\rangle \in u \\ &\leftrightarrow (\nu, t) \in^* (\rho, u) \end{aligned}$$

This ends our proof. ⊣

Consider any set*s (χ_0, s_0) and (χ_1, s_1). Consider any ρ with as range the union of the ranges of the χ_i. Let ρ_i be the restriction of ρ to the range of χ_i. By Lemma 6.3, we can find a set u_i such that (ρ_i, u_i) is a set* and such that (ρ_i, u_i) is $\in^*$-extensionally equal to (χ_i, s_i). By Lemma 6.1, (ρ, u_i) is extensionally equal to (χ_i, s_i). Finally, by Lemma 6.2, $(\rho, u_0 \cup u_1)$ is a union of the (ρ, s_i) and, ipso facto, of the (χ_i, s_i).

We may conclude that we have indeed produced an interpretation of AS. We still need a witnessing relation R. We define by recursion the following mapping F from S to substitution pairs. $F(a) := \langle \emptyset, a\rangle$, $F(s) := \langle \emptyset, \{F(t) \mid t \in s\}\rangle$. For s in S, we set: $s\ R\ (\triangle, F(\{s\}))$. For b in B, we set $b\ R\ (\tau_{ab}, F(a))$, where τ_{ab} is the bijection with domain $\{a\}$ that maps a to b. For $\chi \in C$, we set $\chi\ R\ (\chi, F(\emptyset))$. It is easy to see that F is total, has set*s as values and is injective.

We conclude that our model is polysequential.

We show that our model is not sequential, i.e., that there is no one-dimensional direct interpretation, possibly with parameters, of AS in

$\mathcal{M}$. Suppose there was. Let the elementhood predicate be $\in^\star$. Since all elements of the model can be defined in terms of urelements, we may assume that all parameters are urelements. Suppose the number of parameters in A is n and the number of parameters in B is m.

Consider any object c representing a set* (modulo $\in^\star$-extensionality) of the form

$$\{a_0, \ldots, a_{n+m}, b_0, \ldots, b_n\}^\star$$

with a_i in A en b_j in B, where the a_i and b_j are not parameters.

Can c be in S? Consider any b' in B disjoint from the b_j and the parameters. Consider the automorphism σ generated by interchanging b_0 and b'. Since $b_0 \in^\star c$, we find $b' = \sigma(b_0) \in^\star \sigma(c) = c$. Quod non. By a similar argument c is not in B.

So c is a bijection χ. Suppose a_i is not in the domain of χ. Let σ be the automorphism generated by interchanging a_i and a', where a' is disjoint of the parameters and of the a_i. It follows that: $a' = \sigma(a_i) \in^\star \sigma(\chi) = \chi$. Quod non. So each of the a_i is in the domain of χ. By a similar argument, each of the b_j is in the range of χ.

By our choice of the number of a_i and b_j, we have, for some k, $\chi(a_k) = b'$, where b' is neither a parameter nor one of the b_j. Similarly, we have, for some ℓ, $\chi(a') = b_l$, where a' is not a parameter. Let τ be the automorphism generated by interchanging a_k and a', and b' and b_ℓ. We find: $b' = \tau(b_\ell) \in^\star \tau(\chi) = \chi$. A contradiction.

Summarizing we have the following table that tells us which class of theories is closed under which notion of sameness.

	synonymy	*bi-int.*	*iso-cong.*	*el. cong.*	*mut. int.*
sequentiality	+!	–	–	–	–
polysequentiality	+	+	?	?	–

References

[1] G. Ahlbrandt and M. Ziegler, Quasi finitely axiomatizable totally categorical theories, *Annals of Pure and Applied Logic*, vol. 30(1), 1986, pp. 63–82.

[2] S.R. Buss, *Bounded Arithmetic*, Bibliopolis, Napoli, 1986.

[3] G.E. Collins and J.D. Halpern, On the interpretability of Arithmetic in Set Theory, *Notre Dame Journal of Formal Logic*, vol. 11(4), 1970, pp. 477–483.

[4] K. L. de Bouvére, Logical synonymy, *Indagationes Mathematicae*, vol. 27, 1965, pp. 622–629.

[5] K. L. de Bouvére, Synonymous Theories. In J.W. Addison, L. Henkin, and A. Tarski, editors, *The Theory of Models, Proceedings of the 1963 International Symposium at Berkeley*, North Holland, Amsterdam, 1965, pp. 402–406.

[6] F. Ferreira, *Polynomial time computable arithmetic and conservative extensions*, Ph.D. Thesis, Pennsylvania State University, Pennsylvania, 1988.

[7] F. Ferreira, Polynomial time computable arithmetic. In W. Sieg, editor, *Logic and Computation*, volume 106 of *Contemporary Mathematics*, AMS, 1990, pp. 137–156.

[8] G. Ferreira and I. Oitavem, An interpretation of $\mathtt{S}_2^1$ in Σ_1^b-$\mathtt{NIA}$, *Portugaliae Mathematica*, vol. 63(4), 2006, pp. 427–450.

[9] Y. Gurevich and S. Shelah, On the strength of the interpretation method, *Journal of Symbolic Logic*, vol. 54(2), 1989 pp. 305–323.

[10] P. Hájek and P. Pudlák, *Metamathematics of First-Order Arithmetic*, Perspectives in Mathematical Logic, Springer, Berlin, 1993.

[11] W. Hodges, *Model theory*, Encyclopedia of Mathematics and its Applications, vol. 42, Cambridge University Press, Cambridge, 1993.

[12] E. Jeřábek, Sequence encoding without induction, *Mathematical Logic Quarterly*, vol. 58(3), 2012, pp. 244–248.

[13] J.J. Joosten and A. Visser, The interpretability logic of *all* reasonable arithmetical theories, *Erkenntnis*, vol. 53(1–2), 2000, pp. 3–26.

[14] S. Krajewski, Non-standard satisfaction classes, In *Set Theory and Hierarchy Theory; A Memorial Tribute to Andrzej Mostowski*, number 537 in Lecture Notes in Mathematics, Springer Verlag, Berlin, 1976, pp. 121–144.

[15] J. Krajíček, A note on proofs of falsehood, *Archiv für Mathematische Logik und Grundlagenforschung*, vol. 26(1), 1987, pp. 169–176.

[16] S. Lifsches and S. Shelah, Peano Arithmetic may not be interpretable in the monadic theory of linear orders, *Journal of Symbolic Logic*, vol. 62(3), 1997, pp. 848–872.

[17] F. Montagna and A. Mancini, A minimal predicative set theory, *Notre Dame Journal of Formal Logic*, vol. 35(2), 1994, pp. 186–203.

[18] R. Montague, Semantic closure and non-finite axiomatizability, In *Infinitistic Methods, Proceedings of the Symposium on Foundations*

of Mathematics, (Warsaw, 2-9 September 1959), Pergamon, Oxford, England, 1961, pp. 45–69.

[19] J. Mycielski, P. Pudlák and A.S. Stern, *A lattice of chapters of mathematics (interpretations between theorems)*, volume 426 of *Memoirs of the American Mathematical Society*, AMS, Providence, Rhode Island, 1990.

[20] E. Nelson, *Predicative arithmetic*, Princeton University Press, Princeton, 1986.

[21] P. Pudlák, Some prime elements in the lattice of interpretability types, *Transactions of the American Mathematical Society*, vol. 280, 1983, pp. 255–275.

[22] P. Pudlák, Cuts, consistency statements and interpretations, *The Journal of Symbolic Logic*, vol. 50(2), 1985, pp. 423–441.

[23] W.V. Quine, Concatenation as a Basis for Arithmetic, *The Journal of Symbolic Logic*, vol. 11(4), 1946, pp. 105–114.

[24] J. H. Schmerl, Recursively saturated models generated by indiscernibles, *Notre Dame Journal of Formal Logic*, vol. 26(2), 1985, pp. 99–105.

[25] C. Smoryński, Nonstandard models and related developments. In L.A. Harrington, M.D. Morley, A. Scedrov, and S.G. Simpson, editors, *Harvey Friedman's Research on the Foundations of Mathematics*, North Holland, Amsterdam, 1985, pp. 179–229.

[26] W. Szmielew and A. Tarski, Mutual Interpretability of some essentially undecidable theories, In *Proceedings of the International Congress of Mathematicians (Cambridge, Massachusetts, 1950)*, volume 1, page 734. American Mathematical Society, Providence, 1952.

[27] A. Tarski, A. Mostowski, and R.M. Robinson, *Undecidable theories*, North–Holland, Amsterdam, 1953.

[28] R.A. Vaught, Axiomatizability by a schema, *The Journal of Symbolic Logic*, vol. 32(4), 1967, pp. 473–479.

[29] A. Visser, Interpretability logic, In P.P. Petkov, editor, *Mathematical logic, Proceedings of the Heyting 1988 summer school in Varna, Bulgaria*, Plenum Press, Boston, 1990, pp. 175–209.

[30] A. Visser, An inside view of EXP, *The Journal of Symbolic Logic*, vol. 57(1), 1992, pp. 131–165.

[31] A. Visser, The unprovability of small inconsistency, *Archive for Mathematical Logic*, vol. 32(4), 1993, pp. 275–298.

[32] A. Visser, An Overview of Interpretability Logic. In M. Kracht, M. de Rijke, H. Wansing, and M. Zakharyaschev, editors, *Advances in Modal Logic*, volume 1, 87 of *CSLI Lecture Notes*, Center for the Study of Language and Information, Stanford, 1998, pp. 307–359.

[33] A. Visser, Faith & Falsity: a study of faithful interpretations and false Σ_1^0-sentences, *Annals of Pure and Applied Logic*, vol. 131(1–3), 2005, pp. 103–131.

[34] A. Visser, Prolegomena to the categorical study of interpretations. Logic Group Preprint Series 249, Faculty of Humanities, Philosophy, Utrecht University, Janskerkhof 13A, 3512 BL Utrecht, 2006.

[35] A. Visser, Pairs, sets and sequences in first order theories. *Archive for Mathematical Logic*, vol. 47(4), 2008, pp. 299–326.

[36] A. Visser, Cardinal arithmetic in the style of baron von Münchhausen, *Review of Symbolic Logic*, vol. 2(3), 2009, pp. 570–589.

[37] A. Visser, Growing commas–a study of sequentiality and concatenation, *Notre Dame Journal of Formal Logic*, vol. 50(1), 2009, pp. 61–85.

[38] A. Visser, The predicative Frege hierarchy, *Annals of Pure and Applied Logic*, vol. 160(2), 2009, pp. 129–153.
doi: 10.1016/j.apal.2009. 02.001.

[39] A. Visser, Can we make the Second Incompleteness Theorem coordinate free, *Journal of Logic and Computation*, vol. 21(4), 2011, pp. 543–560. First published online August 12, 2009.
doi: 10.1093/logcom/exp048.

[40] A. Visser, Vaught's theorem on axiomatizability by a scheme, *Bulletin of Symbolic Logic*, vol. 18(3), 2012, pp. 382–402.

[41] A. Wilkie and J.B. Paris, On the scheme of of induction for bounded arithmetic formulas, *Annals of Pure and Applied Logic*, vol. 35, 1987, pp. 261–302.

[42] D. Zambella, Notes on Polynomially Bounded Arithmetic, *The Journal of Symbolic Logic*, vol. 61(3), 1996, pp. 942–966.

Appendix A. Proof of Theorem 4.10

We repeat the theorem:

Theorem 4.10. *Suppose $K : U \to V$, where U and V are 1-sorted. Suppose V proves that there are at least two objects. Then, there is a theory V_K and there are interpretations $\imath_K^U : U \to V_K$ and $\jmath_K : V_K \to V$, such that $\imath_K^U$ is identity-preserving and $\jmath_K$ is a bi-interpretation and $K = \jmath_K \circ \imath_K^U$.*

Proof. We give the proof for the case without parameters and then sketch how it should be adapted if there are parameters.

Suppose $K : U \to V$, where U and V are 1-sorted and K is m-dimensional.

We define V_K as follows. The signature of V_K is the disjoint union of the signatures of U and V, plus a unary predicate D and a $m+1$-ary predicate F. As axioms we take the axioms of U relativized to D, the axioms of V relativized to the complement of D, plus axioms that express that F defines, in each model $\mathcal{M}$ of V_K, an isomorphism between the internal model $\widetilde{K}(\mathcal{P})$ of U in $\mathcal{M}$, where $\mathcal{P}$ is the internal model of V with domain $D^c_{\mathcal{M}}$, and the internal model $\mathcal{N}$ of U in $\mathcal{M}$ with domain $D_{\mathcal{M}}$.

We can view this as follows. There are the obvious identity-preserving interpretations $\imath_K^U : U \to V_K$ and $\imath_K^V : V \to V_K$. The axioms for F precisely make F a witness for the identity in $\mathtt{INT}_1$ of $\imath_K^V \circ K$ and $\imath_K^U$.

We define an $m+2$-dimensional interpretation $j_K : V_K \to V$ as follows.

- $\delta_{\jmath K}(x, y, \vec{z}) :\leftrightarrow x = y \vee \delta_K(\vec{z})$
- $(x_0, x_1, \vec{x}_2) =_{\jmath K} (y_0, y_1, \vec{y}_2) :\leftrightarrow$
  $(x_0 = x_1 \wedge y_0 = y_1 \wedge x_{2(m-1)} = y_{2(m-1)}) \vee$
  $(x_0 \neq x_1 \wedge y_0 \neq y_1 \wedge \vec{x}_2 =_K \vec{y}_2)$
- Suppose P is in (the disjoint version of) the signature of V:
  $P_{\jmath K}(x_{00}, x_{01}, \vec{x}_{02}, \ldots, x_{(n-1)0}, x_{(n-1)1}, \vec{x}_{(n-1)2}) :\leftrightarrow$
  $\bigwedge_{i<n} x_{i0} = x_{i1} \wedge P(x_{02(m-1)}, \ldots, x_{(n-1)2(m-1)})$
- Suppose P is in (the disjoint version of) the signature of U:
  $P_{\jmath K}(x_{00}, x_{01}, \vec{x}_{0,2}, \ldots, x_{(n-1)0}, x_{(n-1)1}, \vec{x}_{(n-1)2}) :\leftrightarrow$
  $\bigwedge_{i<n} x_{i0} \neq x_{i1} \wedge P_K(\vec{x}_{02}, \ldots, \vec{x}_{(n-1)2})$
- $D_{\jmath K}(x, y, \vec{z}) :\leftrightarrow x \neq y \wedge \delta_K(\vec{z})$
- $F_{\jmath K}(x_{00}, x_{01}, \vec{x}_{0,2}, \ldots, x_{(m-1)0}, x_{(m-1)1}, \vec{x}_{(m-1)2}, y_0, y_1, \vec{y}_2,) :\leftrightarrow$
  $\bigwedge_{i<m}(x_{i0} = x_{i1} \wedge x_{i2(m-1)} = y_{2i}) \wedge y_0 \neq y_1 \wedge \delta_K(\vec{y}_2)$

We want to show that $\jmath_K$ is an isomorphism in $\mathtt{INT}_1$ with inverse $\imath_K^V$. We first compute $M := \jmath_K \circ \imath_K^V$.

- $\delta_M(x, y, \vec{z}) \leftrightarrow x = y$.

- $(x_0, x_1, \vec{x}_2) =_M (y_0, y_1, \vec{y}_2) \leftrightarrow$
  $(x_0 = x_1 \wedge y_0 = y_1 \wedge x_{2(m-1)} = y_{2(m-1)})$
- $P_M(x_{00}, x_{01}, \vec{x}_{0,2}, \ldots, x_{(n-1)0}, x_{(n-1)1}, \vec{x}_{(n-1)2}) :\leftrightarrow$
  $\bigwedge_{i<n} x_{i0} = x_{i1} \wedge P(x_{02(m-1)}, \ldots, x_{(n-1)2(m-1)})$

It is clear that G with $G(x_0, x_1, \vec{x}_2, y) :\leftrightarrow x_0 = x_1 \wedge x_{2(m-1)} = y$ gives us the desired isomorphism $G : M \to \mathtt{id}_V$.

Next we compute $N := \imath_K^V \circ \jmath_K$. We write $\vec{x} : D^c$, where $\vec{x}$ has length k, for $\bigwedge_{i<k} \neg D(x_i)$. We write A^{D^c} for A with its quantifiers relativized to D^c.

- $\delta_N(x, y, \vec{z}) :\leftrightarrow x, y, \vec{z} : D^c \wedge (x = y \vee \delta_K^{D^c}(\vec{z}))$
- $(x_0, x_1, \vec{x}_2) =_N (y_0, y_1, \vec{y}_2) :\leftrightarrow x_0, x_1, \vec{x}_2, y_0, y_1, \vec{y}_2 : D^c \wedge$
  $(x_0 = x_1 \wedge y_0 = y_1 \wedge x_{2(m-1)} = y_{2(m-1)}) \vee$
  $(x_0 \neq x_1 \wedge y_0 \neq y_1 \wedge \vec{x}_2 =_K \vec{y}_2)$
- Suppose P is in (the disjoint version of) the signature of V:
  $P_N(x_{00}, x_{01}, \vec{x}_{0,2}, \ldots, x_{(n-1)0}, x_{(n-1)1}, \vec{x}_{(n-1)2}) :\leftrightarrow$
  $x_{00}, x_{01}, \vec{x}_{0,2}, \ldots, x_{(n-1)0}, x_{(n-1)1}, \vec{x}_{(n-1)2} : D^c \wedge$
  $\bigwedge_{i<n} x_{i0} = x_{i1} \wedge P(x_{02(m-1)}, \ldots, x_{(n-1)2(m-1)})$
- Suppose P is in (the disjoint version of) the signature of U:
  $P_N(x_{00}, x_{01}, \vec{x}_{0,2}, \ldots, x_{(n-1)0}, x_{(n-1)1}, \vec{x}_{(n-1)2}) :\leftrightarrow$
  $x_{00}, x_{01}, \vec{x}_{0,2}, \ldots, x_{(n-1)0}, x_{(n-1)1}, \vec{x}_{(n-1)2} : D^c \wedge$
  $\bigwedge_{i<n} x_{i0} \neq x_{i1} \wedge P_K^{D^c}(\vec{x}_{02}, \ldots, \vec{x}_{(n-1)2})$
- $D_N(x, y, \vec{z}) :\leftrightarrow x, y, \vec{z} : D^c \wedge x \neq y \wedge \delta_K^{D^c}(\vec{z})$
- $F_N(x_{00}, x_{01}, \vec{x}_{0,2}, \ldots, x_{(m-1)0}, x_{(m-1)1}, \vec{x}_{(m-1)2}, y_0, y_1, \vec{y}_2) :\leftrightarrow$
  $x_{00}, x_{01}, \vec{x}_{0,2}, \ldots, x_{(m-1)0}, x_{(m-1)1}, \vec{x}_{(m-1)2}, y_0, y_1, \vec{y}_2 : D^c \wedge$
  $y_0 \neq y_1 \wedge \delta_K^{D^c}(\vec{y}_2) \wedge \bigwedge_{i<m}(x_{i0} = x_{i1} \wedge x_{i2(m-1)} = y_{2i})$

We now may take to define our isomorphism $H : N \to \mathtt{id}_{V_K}$, the formula:

- $H(x_0, x_1, \vec{x}_2, y) :\leftrightarrow x_0, x_2, \vec{x}_2 : D^c \wedge$
  $((x_0 = x_1 \wedge x_{2(m-1)} = y)) \vee (x_0 \neq x_1 \wedge \delta_K^{D^c}(\vec{x}_2) \wedge F(\vec{x}_2, y))$

To add parameters, suppose the parameter domain of K is α and the number of parameters is k. We define the theory V_K as follows. For every predicate of V we have a copy in the predicates of V_K of the same arity. For every predicate of U of arity n we have a corresponding $k+n$-ary predicate. Again we keep the predicates corresponding to those of V and U disjoint. We have a unary predicate D_V and a $k+1$-ary predicate D_U. We have a $k+2$-ary predicate F. We have the following axioms.

- $\vdash D_U(\vec{p}, x) \to (\vec{p} : D_V \wedge \alpha^{D_V}(\vec{p}))$

- $\vdash \exists \vec{p}\, D_U(\vec{p}, x) \leftrightarrow \neg D_V(x)$
- $\vdash (D_U(\vec{p}, x) \wedge D_U(\vec{q}, x)) \to \vec{p} = \vec{q}$
- if A is an axiom of V: $\vdash A^{D_V}$
- $\vdash F(\vec{p}, x, y) \to (\alpha^{D_V}(\vec{p}) \wedge D_V(x) \wedge \delta^{D_V}_{K,\vec{p}}(x) \wedge D_U(\vec{p}, y))$
- $\vdash (\alpha^{D_V}(\vec{p}) \wedge D_V(x) \wedge \delta^{D_V}_{K,\vec{p}}(x)) \to \exists y\, F(\vec{p}, x, y)$
- $\vdash (\alpha^{D_V}(\vec{p}) \wedge D_U(\vec{p}, y)) \to \exists x\, F(\vec{p}, x, y)$
- $(F(\vec{p}, x, y) \wedge D_V(u) \wedge \delta^{D_V}_{K,\vec{p}}(u) \wedge x =^{D_V}_K u) \to F(\vec{p}, u, y)$
- $(F(\vec{p}, x, y) \wedge F(\vec{p}, u, v)) \to (x =^{D_V}_K u \leftrightarrow y = v)$
- If P is an n-ary predicate of U:
  $\vdash \bigwedge_{i<n} F(\vec{p}, x_i, y_i) \to (P^{D_V}_{K,\vec{p}}(x_0, \ldots, x_{n-1}) \leftrightarrow P(y_0, \ldots, y_{n-1}))$

Note that the axioms of U will take care of themselves because of the isomorphism. The rest of the development is precisely as expected. $\dashv$

Appendix B. i-Isomorphism meets Pairing

In this appendix, we provide some details of the proof of Theorem 2.1. For the reader's convenience, we repeat the definition of *weak pair theory*.

The theory of unordered pairing `UP` has as non-logical relation symbol the binary predicate $\in$.[18] It is axiomatized as follows.

`UP1` $\vdash \forall x, y\, \exists z\, \forall u\, (u \in z \leftrightarrow (u = x \vee u = y))$.

The theory U is a *weak m-pair theory* if there is an m-direct interpretation of `UP` in U. We prove Theorem 2.1 which states:

Consider any theory U. Then, U is a weak m-pair theory iff, for every n-dimensional interpretation $K : W \to U$, there is an m-dimensional interpretation $K^\star : W \to U$ that is i-isomorphic to K.

Proof. First suppose U is a weak m-pair theory. Let $K : W \to U$ be an n-dimensional interpretation. We want to construct an m-dimensional $K^\star : W \to U$ that is i-isomorphic to K.

In case $n \leq m$, we are easily done using padding. So, let's suppose that $n > m$. Using the unordered pairing we can develop ordered pairs using Cantor pairing. Let k be the smallest number such that $k \cdot m \geq n$. Then, in the usual way, we can define *internal* sequences of length k. Note that each *internal sequence* of length k will be represented by an *external sequence* of the form $(x_0, \ldots, x_{m-1})$. Moreover $(x_0, \ldots, x_{m-1})$ stands for a sequence of k external sequences of the form $(y_0, \ldots, y_{m-1})$. Say the projection function for the internal sequences of length k are π_i

[18]This theory is the theory $\texttt{PAIR}_{\text{uno}}$ of [35].

for $i < k$. We now define new projection functions for $j < n$ as follows. Suppose $j = p \cdot m + q$ and $q < m$. Then, $\pi^\star_j(x_0, \ldots, x_{m-1}) = z$ iff, for some $y_0, \ldots, y_{m-1}$, we have $\pi_p(x_0, \ldots, x_{m-1}) = (y_0, \ldots, y_{m-1})$ and $y_q = z$. Thus we have created sequences of length n with single-object projections.[19] (Note that we may have some don't care elements at the end of the sequence. These do not matter since we do not have extensionality for sequences anyway.) By the directness of the interpretation of UP we obtain all sequences of length n in this way. Now we take:

- $F(x_1, \ldots, x_{m-1}) = (\pi^\star_0(x_1, \ldots, x_{m-1}), \ldots, \pi^\star_{n-1}(x_1, \ldots, x_{m-1}))$
- $\delta_{K^\star}(x_1, \ldots, x_{m-1}) :\leftrightarrow \delta_K(F(x_1, \ldots, x_{m-1}))$
- $(x_1, \ldots, x_{m_1}) =_{K^\star} (y_1, \ldots, y_{m-1}) :\leftrightarrow$
  $F(x_1, \ldots, x_{m_1}) =_K F(y_1, \ldots, y_{m-1})$
- Similarly for the interpretation of the other predicate symbols.
- $F^+(x_1, \ldots, x_{m-1}, u_1, \ldots, u_{n-1}) :\leftrightarrow$
  $F(x_1, \ldots, x_{m-1}) =_K (u_0, \ldots, u_{n-1})$.
  (We assume that $=_K$ only holds between elements of δ_K.)

Clearly F^+ is a representation of the desired i-isomorphism between $K^\star$ and K.

Conversely, suppose that, for every W and every $K : W \to U$, there is an m-dimensional $K^\star : W \to U$. We show that U is a weak m-pair theory. Let EQ be the theory of equality and let $E_{2m} : \mathtt{EQ} \to U$ be the interpretation with as domain: all external sequences $(x_0, \ldots, x_{2m-1})$, and as interpretation of identity: component-wise identity. Let $E^\star_m : \mathtt{EQ} \to U$ be the promised m-dimensional interpretation and let $G : E^\star_{2m} \to E_m$ be the promised i-isomorphism. We define:

$$rl(x_0, \ldots, x_{m-1}) \in (z_0, \ldots, z_{m-1}) \text{ iff, there are } y_0, \ldots, y_{m-1}, \text{ such that } G(x_0, \ldots, x_{m-1}, y_0, \ldots, y_{m-1}, z_0, \ldots, z_{m-1}) \text{ or } G(y_0, \ldots, y_{m-1}, x_0, \ldots, x_{m-1}, z_0, \ldots, z_{m-1}).$$

It is easily seen that $\in$ does indeed satisfy the axiom UP1. ⊣

In the category $\mathtt{INT}_1$ we can assign to each interpretation, viewed as an object modulo i-isomorphism, *the i-dimension* of the interpretation which is the minimum of the dimensions of the translations embodying the interpretation. Using this terminology our theorem says that

[19] Our implementations is just one of many ways of building internal sequences of length n. Another strategy is to, first, use the pairing to form sequences of elements $(y_0, \ldots, y_{m-1})$ of length $n \cdot m$ and, then, use the function $\mathsf{d}_m(x) := (x, \ldots, x)$, where $(x, \ldots, x)$ has length m, to construct a sequence of single objects of length n from this.

a theory U is a weak m-pair theory iff all interpretations in U have i-dimension $\leq m$.

Part II

Some Problems in Logic and Number Theory, and their Connections

by Alan Robert Woods

A thesis submitted to the University of Manchester for the degree of Doctor of Philosophy in the Faculty of Science, Department of Mathematics, 1981.

Alan R. Woods (1953–2011)

Alan Robert Woods was born on June 21, 1953, in Brisbane, Australia, as the first child of Geoff and Jess Woods. After obtaining a B.Sc. (Honours, 1st class) in Mathematics at the University of Queensland (1971–1974), he continued his studies, completing first an M.Sc. by research at Monash University (1975–1978) and then a Ph.D. at the University of Manchester, UK (1978-1981). His Ph.D. thesis was entitled "Some problems in logic and number theory, and their connections" and was supervised by Jeff B. Paris.

Alan held academic appointments at the University of Malaya (Lecturer, 1982-1985), Yale University (Assistant Professor, 1985–1989) and at the University of Western Australia (Lecturer, 1990–1993, Honorary Research Fellow, 1994–2002, and Adjunct Associate Professor, 2003–2011). He also held visiting positions at Rutgers University (1989), Institute for Advanced Study, Princeton (2000), Université de Versailles Saint-Quentin (2003) and Mathematisches Forschungsinstitut Oberwolfach (2003). Alan gave about 30 invited lectures at international conferences, Universities and research centers all over the world.

Alan's research interests lay in Mathematical Logic, Computational Complexity and their connections with Algebra, Combinatorics and Number Theory. He (co-)authored 17 research papers which appeared in prestigious international journals, including the *Journal of Symbolic Logic*, *Theoretical Computer Science* and *Proceedings of the London Mathematical Society*, as well as in conference proceedings volumes. He supervised the Ph.D. thesis of J.M. Foy (Yale University), the M.Sc. thesis of K. Fergusson (University of Western Australia) and was alternate supervisor of the Ph.D. thesis of T.N. French (University of Western Australia).

Alan was very keen on discussing ideas with colleagues sharing his interests, ready to contribute highly innovative ideas in several research areas. He co-operated with many distinguished researchers all over the world, including C.G. Jockusch, J. Krajíček, A. Maciel, J.B. Paris, P. Pudlák, T. Pitassi, A.J. Wilkie and others, publishing papers that attracted hundreds of citations.

He was an amateur radio enthusiast. He served as president and committee member of the Western Australia VHF Group and was highly regarded for his technical knowledge. He was also an enthusiastic walker, especially enjoying long Sunday walks on Perth's beaches.

Alan was diagnosed with Parkinson's disease in 2006 and died at his home in Mt. Claremont (Western Australia) in mid-December 2011. He was survived by his son Robert and his daughter Sarah.

Costas Dimitracopoulos,
March 2013

Abstract

This thesis is primarily concerned with some problems related to both logic and number theory.

In chapter 1, Wilkie's problem,which asks whether the existence of arbitrarily large prime numbers can be proved using only induction on bounded quantifier (Δ_0) formulas together with the usual "algebraic" axioms (in the first order language with $\leq, +, \cdot$), is linked with a question of Macintyre about the provability of the pigeon hole principle in the same axiom system. It is shown that if an axiom schema asserting a version of the pigeon hole principle for Δ_0 formulas is added, then it is possible to prove Sylvester's theorem that for $y \geq x \geq 1$, some number among $y + l, y + 2, \ldots, y + x$ has a prime divisor $p > x$. Alternatively, if function symbols corresponding to Grzegorczyk's $\mathcal{E}^2$ functions are added (with their definitions) and induction allowed on bounded formulas involving them, then the pigeon hole principle can be proved for such formulas, and the existence of infinitely many primes follows.

The problem of defining addition and multiplication on the natural numbers $\mathbb{N}$ by first order formulas involving the predicate $x \perp y$ ("x and y are coprime") is considered in chapter 2. The predicates $z = x+y$, $z = x \cdot y$ are defined by bounded quantifier formulas involving only $\leq$ and $\perp$. As an application, a proof is given that the class of all rudimentary (i.e., Δ_0 definable) sets of positive numbers is contained in the class consisting of the spectra $S_\phi = \{|M| : M \textit{ finite}, M \models \phi\}$ of those sentences ϕ having only graphs as models, with equality between these classes if and only if S_ϕ, is rudimentary for *every* first order sentence ϕ. An analogous result holds with partial orderings in place of graphs. (It is noted in Appendix II that if every S_ϕ. is rudimentary, then $NP \neq co - NP$.)

Julia Robinson's question whether $+, \cdot$ are definable by formulas involving only $\perp$ and successor, is shown to be *equivalent* to an open problem in number theory, namely the conjecture that there is some k such that every $n \in \mathbb{N}$ is determined uniquely by the sequence of sets of distinct primes $S_0, S_1, \ldots, S_k$, where $S_i = \{p : p|n+i\}$. An unconditional proof is given that the theory of $\mathbb{N}$ in this language is undecidable.

In chapter 3 it is deduced from the linear case of Schinzel's hypothesis H that $z = x \cdot y$ is definable in the language with $=, +$ and the predicate, "x is a prime", but that the defining formula cannot be existential, since the conjecture provides an algorithm for deciding all existential sentences in this language.

Acknowledgements

I wish to express my gratitude to Jeff Paris who supervised the work and was a source of constant encouragement. Thanks are also due to all the members of the Manchester logic community who provided a stimulating environment with many interesting discussions on a wide range of topics, and especially to my fellow students, among them Jonathan Cox, Costas Dimitracopoulos, Adam Huby, Collin Seadon, Martin Merry, and Jose Ruiz. In particular several useful conversations related to the subject matter of part of this thesis were had with Costas Dimitracopoulos. I would also like to thank Jeff Paris and Alex Wilkie for sharing with me the knowledge gained from their recent work. In particular this contributed greatly to making my introduction to complexity theory relatively painless.

My sincere thanks go to Beryl Sweeney who typed the majority of the thesis and obviously put a great deal of effort and energy into it. Thanks too to Briget Iveson who typed the remainder.

Acknowledgement must also be made of the financial support received from the University of Manchester in the form of a research studentship in science and a Bishop Harvey Goodwin mathematical scholarship.

Finally I would like to thank my wife Ellisha for her understanding and help.

Alan Woods,
December, 1981.

Introduction

The guiding philosophy underlying this thesis is the conviction that significant connections between logic and number theory *do* exist, and that the only way to find them is by actively searching for them.

There are three chapters. Each is essentially independent of the others (except that some definitions and simple results from the early part of chapter 2 are assumed in chapter 3). However several themes run through the work as a whole. *One* is the important role that the prime divisors of consecutive integers play. Another (particularly in chapters 1 and 2) is the appearance of problems in computational complexity theory intimately interwoven with the at first sight seemingly unrelated questions under consideration.

A third theme—the goal of much of the work on definability and (un)decidability—is the desirability of achieving a fuller understanding of *logical redundancy* in the combined additive and multiplicative structure of the natural numbers, its finite initial segments, and nonstandard models of related axiom systems. Here by "logical redundancy" is meant the ability to recover the whole structure from seemingly (and in certain senses genuinely) weak parts of it, through the use of first order definitions. Certainly thinking about the structure of nonstandard models played a considerable role in the development of the proofs of the definability and decidability theorems in chapters 2 and 3.

The question of to what extent these "redundancy" properties are shared by other finite structures (including "machines"), and the related problem of "measuring" the relative strengths of different finite structure *theories*, are only touched on here, but in the author's view, they are potentially very important and worthy of intensive study in their own right.

Finally, some connections between logic and number theory have been found, and these will be described in the pages that follow, however more and deeper links no doubt still remain hidden. The author hopes that the present work will challenge its readers to actively seek them.

Logic notation

$\wedge$	(and),
$\vee$	(or),
$\neg$	(not),
$\rightarrow$	(implies),
$\Longleftrightarrow$	(if and only if),
$\forall$	(for every),
$\exists$	(there exists).
$\Sigma \vdash \phi$	(*ϕ is syntactically provable from the axioms Σ.*),

$M \models \phi(x_1, \ldots, x_n)[a_1, \ldots, a_n]$ or just $M \models \phi(a_1, \ldots, a_n)$	*$\phi(x_1, \ldots, x_n)$ is true in the model M at $x_1 = a_1, \ldots, x_n = a_n$. Here $a_1, \ldots, a_n \in M$.*

Note that there are some minor differences in notation between chapters. (For example the Δ_0 formulas of chapter 1 are called bounded $L_{\leq,+,\cdot}$ formulas in chapter 2.)

Other notation

[x] denotes the largest integer $n \leq x$.

1

Bounded induction, the pigeon hole principle, and the existence of arbitrarily large prime numbers

One of the benefits the study of logic can bring to other areas of mathematics is an understanding of which basic axioms are required to derive any given theorem. Although there has been considerable success in some areas (for example in general topology, infinite algebra and set theory with the axiom of choice, and more recently in finite combinatorics with variants of Ramsey's theorem—see Paris and Harrington [29]), very little of any significance is known about the axiomatic requirements of number theory, where by "number theory" we mean that body of theorems about the natural numbers which is *actually studied by number theorists.* Even the case of the very basic theorem that there exist arbitrarily large prime numbers, is not properly understood. Wilkie [47] highlighted the problem by asking whether this can be proved using an axiom system similar to Peano Arithmetic but having the restriction that the induction axiom is available only where the induction hypothesis is a bounded formula. (A formula $\phi(\vec{x})$ in the first order language with primitives $=, \leq, +, \cdot, 0, 1$ is *bounded* if every quantifier in $\phi(\vec{x})$ occurs either in the form $\forall u \leq v$ or in the form $\exists u \leq v$, where these are abbreviations for $\forall u(u \leq v \rightarrow \ldots)$ and $\exists u(u \leq v \wedge \ldots)$.) The class of all such formulas will be denoted by Δ_0, and for definiteness we will consider the axiom system $I\Delta_0$ consisting of:

$$0 \leq 0 \wedge \neg 1 \leq 0$$
$$\forall x(x + 0 = x \wedge x \cdot 0 = 0 \wedge x \cdot 1 = x)$$
$$\forall x \forall y(x + 1 = y + 1 \rightarrow x = y)$$
$$\forall x \forall y(x \leq y + 1 \iff x \leq y \vee x = y + 1)$$
$$\forall x \forall y(x + (y + 1) = (x + y) + 1)$$

$\forall x \forall y (x \cdot (y+1) = (x \cdot y) + x)$
$\forall \vec{x} (\theta(\vec{x}, 0) \wedge \forall y (\theta(\vec{x}, y) \to \theta(\vec{x}, y+1)) \to \forall y \theta(\vec{x}, y))$
for all Δ_0 formulas $\theta(\vec{x}, y)$,

together with the usual axioms for predicate calculus with identity.

Wilkie motivated his question by stating that

> "An affirmative answer here (which unfortunately seems unlikely) would, we believe, give an essentially new proof of the infinity of primes. This is because all current proofs, as far as we know, introduce functions of exponential growth and it is known that induction on bounded quantifier formulae, while natural and quite strong in many respects, cannot define functions of greater than polynomial growth".

The present author agrees that a proof using *only* Δ_0 would probably necessarily involve some new, and possibly powerful, idea (although he certainly does not think that a *proof* of a negative answer would be unfortunate). However as we shall see, the present lack of a proof has more to do with our inability to prove a certain combinatorial principle in $I\Delta_0$ than with the lack of functions of exponential growth.

The combinatorial principle referred to is a version of

The Pigeon Hole Principle. n pigeons cannot fit, one per hole, into fewer than n holes.

Specifically it will be shown that the existence of arbitrarily large primes can be proved if $I\Delta_0$ is augmented by the axiom schema:

Definition. $PHP(\Delta_0)$:

$$\forall \vec{x} \forall y (\forall u_1 \le y \forall u_2 \le y (\theta(\vec{x}, y, u_1, v) \wedge \theta(\vec{x}, y, u_2, v) \to u_1 = u_2) \wedge$$
$$\forall u \le y \exists v \le y \theta(\vec{x}, y, u, v) \to \forall v \le y \exists u \le y \theta(\vec{x}, y, u, v)),$$

Admittedly each instance of this schema can be proved if we add to $I\Delta_0$ an additional axiom asserting

$$\forall x (2^x \; exists).$$

However, $I\Delta_0 + PHP(\Delta_0) \nvdash \forall x (2^x \; exists)$, as can easily be seen by taking a nonstandard model $< M, \le, +, \cdot >$ of the theory of the natural numbers $< \mathbb{N}, \le, +, \cdot >$ and considering the model

$$< A, \le, +, \cdot > \models I\Delta_0 + PHP(\Delta_0) + \neg \forall x (2^x \; exists)$$

formed by choosing $a \in M \setminus N$ and letting $A = \{b \in M : \exists n \in \mathbb{N} (b < a^n)\}$.

With the axiom system $I\Delta_0 + PHP(\Delta_0)$ it is in fact possible to prove:

Theorem (Sylvester's Theorem). If $1 \le x \le y$ then some number among $y+1, y+2, ..., y+x$ has a prime divisor $p > x$.

The proof will be given in detail for the case $y \geq x^5$ since this case establishes the existence of arbitrarily large primes in $I\Delta_0 + PHP(\Delta_0)$ in a relatively simple way, while still requiring many of the essential ideas. §4 contains a sketch of how the method may be extended to the general case, essentially by recasting a variant of Sylvester's original argument as a proof in $I\Delta_0 + PHP(\Delta_0)$. Taking $y = x$ this case can be seen to include Chebychev's celebrated theorem:

Bertrand's Postulate. For every $x \geq 1$ there is a prime p satisfying $x < p \leq 2x$.

Presumably any of the standard "textbook" variants of the proof of Bertrand's Postulate can be similarly recast (with more or less difficulty).

As the reader may have surmised, it is not known whether $PHP(\Delta_0)$ can be proved from $I\Delta_0$. The question of whether the pigeon hole principle can be proved by bounded induction seems to have first been formulated by Angus Macintyre, who was led to it by the problem of proving that for each odd prime p there exist quadratic nonresidues modulo p, that is, residue classes $a\ (mod\ p)$ such that

$$\forall x(x^2 \not\equiv a\ (mod\ p)).$$

As this problem raises some interesting (and relevant) matters it is worth digressing for a moment to consider it.

As can easily be proved in $I\Delta_0$, for each prime p the $(p-1)/2$ numbers $1^2, 2^2, \ldots, ((p-1)/2)^2$ are incongruent $(mod\ p)$ and include a representative of each quadratic residue $a\ (mod\ p)$, that is, of each residue class $a \not\equiv 0\ (mod\ p)$ for which

$$\exists x(x^2 \equiv a\ (mod\ p)).$$

Thus (using the notation $[1, n] = \{1, 2, \ldots, n\}$) the function $f(x) = \min\{y :\ y \equiv x^2\ (mod\ p)\}$ is a one to one map from $[1, (p-1)/2]$ onto those elements of $[1, p-1]$ which are quadratic residues. Obviously this is contrary to $PHP(\Delta_0)$ if there are no quadratic nonresidues. Also the stronger result that there exist $(p-1)/2$ quadratic nonresidues follows immediately using the following combinatorial principle, in the statement of which $f : A \leftrightarrow B$ means that f is a one to one map from the set A onto the set B.

Bijection Extension Principle. If $A, B \subseteq [1, n]$ and $f : A \leftrightarrow B$ then there is an extension $f^\star$ of f such that $f^\star : [1, n] \leftrightarrow [1, n]$.

(More generally we could replace $[1, n]$ by a finite set C.) Clearly for the function f considered above,

$$f : [(p-1)/2 + 1, p-1] \leftrightarrow \{a \in [1, p-1] :\ \forall x(x^2 \not\equiv a\ (mod\ p))\}$$

(or rather the restriction of f^* to $[(p-1)/2+1, p-1]$ has this property) and hence there is a one to one map from $[1, (p-1)/2]$ onto the quadratic nonresidues.

At first sight the bijection extension principle would appear to be quite powerful, since as far as the author is aware, it is not known whether there is a Δ_0 formula $\theta(x, y, p)$ which (on the natural numbers $\mathbb{N}$) defines for each odd prime p a function $y = g_p(x)$ with

$$g_p : [1, (p-1)/2] \leftrightarrow \{a \in [1, p-1] : \forall x(x^2 \not\equiv a \ (mod\ p))\}.$$

Certainly there is such a predicate $\theta(x, y, p)$ in the class $\mathcal{E}^2_\star$ defined by Grzegorczyk [16], but although every predicate on $\mathbb{N}$ which can be defined by a Δ_0 formula is in $\mathcal{E}^2_\star$, it is not known whether these classes are equal, that is, whether $\Delta_0 = \mathcal{E}^2_\star$. (Ritchie [33] showed that for $\mathbb{N}$, $\mathcal{E}^2_\star$ is identical with the class of predicates whose characteristic functions can be computed by deterministic Turing machines working in linear space, and in fact there is a function $y = g_p(x)$ satisfying the above which can be computed *nondeterministically* in linear space and polynomial time (where these are measured, as usual, with respect to the number of digits in x and p), but even in this case it is not known whether all such functions have Δ_0 definable graphs.)

$\mathcal{E}^2_\star$ is defined to be the class of all predicates having characteristic functions, where a function is in $\mathcal{E}^2$ if it can be *defined* by the satisfaction of a finite sequence of the following schemes:

(1) $y = 0$, $z = 1$, $z = x + y$, $z = x \cdot y$ are in $\mathcal{E}^2$

(2) *Substitution:* If $z = f(x_1, \ldots, x_k, \ldots, x_m)$ and $z = g(y_1, \ldots, y_n)$ are in $\mathcal{E}^2$ then $z = f(x_1, \ldots, , g(y_1, \ldots, y_n), x_{k+1}, \ldots, x_m)$ is in $\mathcal{E}^2$.

(3) *Limited recursion:* If $z = g(\vec{y})$, $z = h(x, \vec{y}, w)$ and $z = b(x, \vec{y})$ are in $\mathcal{E}^2$ then so also is the function $z = f(x, \vec{y})$ satisfying:
   (i) $f(0, \vec{y}) = g(\vec{y})$
   (ii) $f(x + 1, \vec{y}) = h(x, \vec{y}, f(x, \vec{y}))$
   (iii) $f(x, \vec{y}) \leq b(x, \vec{y})$.

(Note that it is easy to prove by induction on the complexity of definition, that *every $\mathcal{E}^2$ function has at most polynomial growth*).

Now suppose we associate a function symbol f (say) with each $\mathcal{E}^2$ definition of a function, and turn each such definition into an axiom DEF(f). If we add these axioms to $I\Delta_0$ and allow induction on bounded formulas $\theta(\vec{x}, y)$ involving the new function symbols as well as $\leq, +, \cdot$ we obtain a new axiom system which will be denoted by $I\mathcal{E}^2_\star$. Similarly we can formulate a schema $PHP(\mathcal{E}^2_\star)$ analogous to $PHP(\Delta_0)$ by allowing $\theta(\vec{x}, y, u, v)$ to have these new function symbols, and ask the obvious question whether $\mathcal{E}^2_\star \vdash PHP(\mathcal{E}^2_\star)$. The answer is affirmative and of course it follows that $I\mathcal{E}^2_\star \vdash$ *there exist arbitrarily large prime numbers.* Thus *if* one could show that every $\mathcal{E}^2$ definition (defining a function $y = g(\vec{x})$, say) corresponds to a Δ_0 formula $\phi_g(\vec{x}, y)$ in such a way that replacing every function symbol in the axioms DEF(f) using the ϕ_g's yields sentences provable in $I\Delta_0$, then it would follow that $I\Delta_0 \vdash PHP(\Delta_0)$ and hence $I\Delta_0 \vdash$ *there exist arbitrarily large primes.*

Although it seems unlikely that this approach can succeed, it does suggest analysing $I\mathcal{E}^2_\star$ proofs of the existence of arbitrarily large primes to see exactly which $\mathcal{E}^2$ functions are involved.

A standard example of an $\mathcal{E}^2$ function not known to be definable by a bounded formula is $\pi(x)$, the number of primes not exceeding x. (A proof that $y = \pi(x)$ cannot be defined by a bounded formula would help explain why no computationally efficient formulas for $\pi(x)$ or its "inverse" $y = p_n$, the nth prime number, have been found.)

This leads the author to conjecture that there should be a proof of the existence of arbitrarily large primes in which the only $\mathcal{E}^2$ function required is $\pi(x)$, that is that

$$I\Delta_0(\pi) + \mathrm{def}(\pi) \models \textit{there exist arbitrarily large primes,}$$

where $I\Delta_0(\pi)$ denotes $I\Delta_0$ with induction allowed on bounded formulas involving $\leq, +, \cdot, \pi$, and def(π) is an axiom asserting that $\pi(0) = 0$ and for every x,

$$\pi(x+1) = \begin{cases} \pi(x) + 1, & \text{if } x+1 \text{ is prime,} \\ \pi(x), & \text{otherwise.} \end{cases}$$

Before proceeding further, a few words on the interdependence of the sections of this chapter would seem to be in order. §1 contains a resume of Sylvester's method and its history. This may or may not help the reader understand the proof of the existence of arbitrarily large primes in $I\Delta_0 + PHP(\Delta_0)$ given in §3, which technically depends only on some basic properties of $I\Delta_0$ dealt with in §2. It is therefore possible to omit §1 and go straight on to §2 and §3, coming back only to provide the background for the $I\Delta_0 + PHP(\Delta_0)$ proof of Sylvester's theorem sketched in §4, or "to see where it all comes from". §5 is devoted to a discussion of census functions with the theorem that $I\mathcal{E}^2_\star \vdash PHP(\mathcal{E}^2_\star)$ as a corollary. (§5 is independent of all other sections.) §6 contains the proof of a lemma about the sums which can be defined by Δ_0 formulas in models of $I\Delta_0$. This lemma is used in §4 but its proof is deferred in order to be able to make use of the results in §5.

1 Historical perspective

The number theoretic idea underlying the $I\Delta_0 + PHP(\Delta_0)$ proof of the existence of arbitrary large primes to be given in §3 comes from Sylvester [44]. This paper (which admittedly is somewhat eccentric) seems to have been sadly neglected by most 20th century number theorists even to the extent that this fundamental idea is often credited to a living mathematician. However the principle is clearly stated there in a footnote as follows:

> "The author was wandering in an endless maze in his attempts at a general proof of his theorem, until in an auspicious hour when taking a walk on the Banbury road (which leads out of Oxford) the Law of Ademption flashed upon his brain: meaning thereby the law (the nerve, so to say, of the preceding investigation) that *if all the terms of a natural arithmetical series be increased by the same quantity so as to form a second such series, no prime number can enter in a higher power as a factor of the product of the terms in this latter series, when a suitable term has been taken away from it, than the highest power which enters as a factor into the product of the terms of the original series*"

Sylvester goes on to explain:

> "The whole matter is thus made to rest on (Tschebyscheff's) .. superior limit to the sum of the logarithms of the primes not exceeding a given number, from which ... a superior limit may be deduced to the number of such primes."

To understand this in the case of the arithmetic progressions $1, 2, \ldots, x$ and $y+1, y+2, \ldots, y+x$, let $[z]_p$ denote the largest power of the prime p which divides z, and $\{z\}_p$ denote exponent of p in z, so

$$z = \prod_{p|z} [z]_p = \prod_{p|z} p^{\{z\}_p}.$$

Now consider the products:

$$\prod_{1 \leq s \leq x} s = \prod_{p \leq x} \prod_{1 \leq s \leq x} [s]_p, \quad \prod_{1 \leq i \leq x} (y+i) = \prod_{p \leq y+x} \prod_{1 \leq i \leq x} [y+i]_p.$$

Sylvester's *crucial* observation was that for each prime p,

$$w_p \prod_{1 \leq i \leq x} [s]_p \geq \prod_{1 \leq i \leq x} [y+i]_p \tag{1}$$

where w_p is the largest single factor on the right hand side of this inequality (and thus will not occur if the corresponding term $y+m$ is deleted from the arithmetic progression).

This was proved using the fact that

$$\left\{\prod_{1\leq s\leq x} s\right\}_p = \sum_{k\geq 1} |\{s \leq x : p^k|s\}| = \sum_{k\geq 1} [x|p^k]. \qquad (2)$$

By the maximality of w_p, if $w_p|y+m$ then for all k,

$$p^k|y+m+j \iff p^k|j \ (for\ 1\leq j\leq x-m)$$

and

$$p^k|y+m-j \iff p^k|j \ (for\ 1\leq j\leq m-1).$$

Thus

$$\left\{\prod_{m\leq i\leq x}(y+i)\right\}_p = \left\{\prod_{1\leq j\leq x-m} j\right\}_p = \sum_{k\geq 1}[(x-m)/p^k]$$

$$\left\{\prod_{1\leq i\leq m}(y+i)\right\}_p = \left\{\prod_{1\leq j\leq m-1} j\right\}_p = \sum_{k\geq 1}[(m-1)/p^k]$$

and therefore

$$\begin{aligned}\left\{\prod_{1\leq i\leq x}(y+i)\right\}_p &= \{w_p\}_p + \sum_{k\geq 1}\left([(m-1)/p^k] + [(x-m)/p^k]\right) \\ &\leq \{w_p\}_p + \sum_{k\geq 1}\left[\frac{x}{k}\right] = \{w_p\}_p + \left\{\prod_{1\leq x\leq x} s\right\}_p\end{aligned}$$

which implies (1).

From (1) it follows that if no prime divisor of the numbers $y+1, y+2, \ldots, y+x$ exceeds x then

$$\left(\prod_{p\leq x} w_p\right)\cdot\left(\prod_{1\leq s\leq x} s\right) \geq \prod_{1\leq i\leq x}(y+i) \qquad (3)$$

and hence

$$\left(\prod_{x-\xi(x)<i\leq x}(y+i)\right)\cdot\left(\prod_{1\leq s\leq x} s\right) \geq \prod_{1\leq i\leq x}(y+i) \qquad (4)$$

for any function $\xi(x)\geq\pi(x)$ (where $\pi(x) = |\{p\leq x : p\ prime\}|$). (4) can be rewritten as

$$x!y! \geq (x+y-\xi(x))!.$$

But Chebychev had already shown how to prove $\pi(x)\leq Ax(\log x)^{-1}$ via the bound $\sum_{p\leq x}\log p\leq Bx$ (where A, B are constants and log denotes the logarithm to base e). Taking x sufficiently large and $\xi(x) = Ax(\log x)^{-1}$ with a "good" value of $A(1 < A < 2\log 2)$ one can show that (4) fails for $y \geq x$ by applying Stirling's asymptotic formula $x!$, or alternatively just the weak version $\log(x!) = x\log x - x + o(x)$, (The latter can easily be proved by comparing $\sum_{1\leq n\leq x} \log n$ with $\int_1^x \log t dt$. The notation $g(x) = o(f(x))$ means $g(x)/f(x) \to 0$ as $x \to \infty$.) This proves Sylvester's theorem for all but finitely many values of x. (Actually Sylvester used a slightly different argument aimed at making the checking of these cases easier).

If $y \geq x^m$ for m a sufficiently large constant then the same result can be obtained by means of much simpler bounds. For example suppose we take the trivial upper bound $\xi(x) = [x/2] + 1 \geq \pi(x)$ and simplify (4) to:

$$y^{[x/2]+1}x^x \geq y^x \tag{5}$$

or in logarithmic form:

$$([x/2] + 1)\log y + x\log x \geq x\log y. \tag{5a}$$

Clearly (5a) fails for all $y > x^m$, provided $m > 2$ and x is I sufficiently large, and thus establishes Sylvester's theorem in this case.

Superficially (that is, as written) the above proofs appear to involve numbers greater than 2^x (as the products in (1) to (4) can have values of this order), and as was noted in the introduction the existence of numbers of this size for all x cannot be proved in $I\Delta_0 + PHP(\Delta_0)$.

However there is still some hope since the logarithmic version of (1) namely:

$$\log w_p + \sum_{1\leq s\leq x} \{s\}_p \log p \geq \sum_{1\leq s\leq x} \{y+i\}_p \log p \tag{1a}$$

and also the logarithmic version of (4) involve "quantities" whose values can be bounded by a polynomial in x. This suggests using "*approximate logarithms*", that is functions taking values which are *rational numbers* in the sense of the model under consideration, and which behave in the way one would expect approximations to logarithms to behave. For the logician it should perhaps be mentioned that the making of approximations which are most naturally viewed at the logarithmic level (because the error can then be written as a term rather than as a factor) is crucial for many arguments in prime number theory. The built in tolerance to errors of the magnitude is the reason we will be able to make do with "approximations" which are *not* very good. This is important for our work with $I\Delta_0$ because it does not seem to be known even whether the predicate $y = [x\log z]$ can be defined on $\mathbb{N}$ by a Δ_0 formula although it is in $\mathcal{E}^2_\star$. In other words, the approximations to the function $y = \log x$ which are known to be definable by Δ_0 formulas are rather inaccurate. For this and other reasons, the use of approximate logarithms (which is a fairly obvious technique for Peano Arithmetic—see for example Woods [50]) appears to deserve some care when applied to $I\Delta_0$.

Fortunately however, we will be able to postpone these considerations to §4, since for the special case of Sylvester's theorem with $y > x^m$ (m sufficiently large) the crudeness of the bounds which suffice allows the use of *integer* valued approximate logarithms which can be obtained relatively easily (provided we use base 2 rather than e).

There is still one further difficulty. The logarithmic version of the proof sketched above involves establishing inequalities by manipulating sums having a large number of terms (interchanging double summations, etc.), and as yet there is no known way of defining the predicate $z = \sum_{i \leq x} f(i)$ on $\mathbb{N}$ by a Δ_0 formula, given an arbitrary function $y = f(i)$ defined by a Δ_0 formula, let alone a proof that this can be done using only the axioms of $I\Delta_0 + PHP(\Delta_0)$. (For $\mathbb{N}$ this is another special case of the $\Delta_0 = \mathcal{E}^2_\star$ problem.) We will side step this problem by "unravelling" the inequalities to find the underlying "*comparison map*". This is possible because the inequalities were proved in the first place *by comparing terms in some order*, and is useful because in the cases considered below the comparison map is Δ_0 definable.

2 Some functions available in $I\Delta_0$

Although it is *not* possible to prove in $I\Delta_0$ that

$$\forall x \forall y \exists z (z = x^y),$$

it *is* possible to define the predicate $z = x^y$ by a Δ_0 formula which provably satisfies the usual inductive definition whenever x^y does exist.

Proposition 2.1. *There is a Δ_0 formula $z = x^y$ such that*

$$I\Delta_0 \vdash z = x^y \text{ is a partial function,}$$

and

$$I\Delta_0 \vdash \forall x (1 = x^0 \wedge \forall y (z = x^y \rightarrow z \cdot x = x^{y+1})).$$

For a proof see Dimitracopoulos [9]. Note that any other Δ_0 formula with the same property is $I\Delta_0$ provably equivalent to $z = x^y$. The other properties of exponentiation which could reasonably be expected to hold in $I\Delta_0$ can fairly obviously also be proved so we will use these freely.

In particular we can define by a Δ_0 formula an "inverse" function $y = [\log_x z]$ (which *is* provably total) by taking y to be the largest number such that $x^y \leq z$. It is very important here to realise that [log] should be considered as a single symbol as it does *not* seem to be at all clear that even $\log_2 z$ can given a reasonable meaning for an arbitrary model of $I\Delta_0$ (unlike the situation for, say, Peano Arithmetic). Thus notation such as $[\log_2 z]$, although intended to be suggestive, should not be taken too literally.

To make it easier to indicate what techniques are available in $I\Delta_0$ we will now consider a fixed model $M \models I\Delta_0$, although the procedures described are quite uniform and could be handled syntactically. *The further down we go in M (that is, towards 0) the more the behaviour of the initial segments is compelled to approach that of the initial segments in models of Peano Arithmetic.* (In fact, M will always have an initial segment which is a model of Peano Arithmetic—see for example Lessan [26], proposition 4.1.7.) Thus we will be concerned with the "top" section of M.

The ability to code "finite" sequences of numbers by a single number is rather limited in $I\Delta_0$ (since in general it is to be expected that giving a distinct code number to each sequence of 0's and 1's of length x will require code numbers up to at least $2^x - 1$). However the technique can be applied if the length of the sequence and the size of its elements are sufficiently small. By a Δ_0 *sequence* we will mean a function taking a value $u_i \in M$ for each $i \in [1, d]$ (for some $d \in M$) which is defined on M by a Δ_0 formula (possibly involving parameters). A Δ_0 *function* will be a function defined on M (or some obvious subset) by a Δ_0 formula (again possibly with parameters).

Lemma 2.2. *For each $n \in N$ there is a Δ_0 function $g_n(c, i, a)$ with the property that if $a \in M$ and $u_1, u_2, \ldots u_d$, is any Δ_0 sequence with $d \leq [\log_2 a]/([\log_2[\log_2 a]]+1)$ and each $u_i < [\log_2 a]^n$, then there exists $c \in M$ such that*

$$\forall i \in [1, d](u_i = g_n(c, i, a)).$$

Proof. Let $b = [log_2 a]^n$. c will be the number having $u_d u_{d-i} \ldots u_1$ as the *digits* of its representation to base b, so we take

$$g_n(c, i, a) = \left[\frac{c}{b^{i-1}}\right] - b\left[\frac{c}{b^i}\right].$$

That c exists can be established by using induction on j to prove that for each $j \in [1, d]$ there is a number $c_j < b^j$ with digits $u_j u_{j-1} \ldots u_1$, to base b. Since

$$\begin{aligned} b^j \leq b^d &\leq [\log_2 a]^{n[\log_2 a]/([\log_2[\log_2 a]]+1)} \\ &< 2^{n[\log_2 a]} \leq a^n \end{aligned}$$

the quantifiers in the induction hypothesis can be bounded by a^n (which exists in M since $n \in \mathbb{N}$ and M is closed under multiplication). Thus the proof only requires $I\Delta_0$. ⊣

Of course coding is important not only as a means of quantifying over known Δ_0 sequences but also as a technique for constructing *new* Δ_0 sequences. In particular *the coding technique used in lemma 2.2 enables us to define sums of terms from an arbitrary Δ_0 sequence having considerably more elements than the sequences considered in its statement.* (Later in section 6 it will be shown that sums of sequences having much *larger* elements can also be handled.)

Lemma 2.3. *If $u_1, u_2 \ldots u_d$ is a Δ_0 sequence in M with $d \leq [\log_2 a]^k$ and each $u_i < [\log_2 a]^n$ for some $n, k \in \mathbb{N}$, $a \in M$, then there is a Δ_0 sequence (with elements denoted by) $\sum_{1 \leq i \leq j} u_i, j \in [1, d]$, such that $\sum_{1 \leq i \leq l} u_i = u_1$, and for all $j \in [1, d-1]$,*

$$\sum_{1 \leq i \leq j+1} u_i = \left(\sum_{1 \leq i \leq j} u_i \right) + u_{j+1}.$$

Proof. Let $d_0 = [\log_2 a]/([\log_2[\log_2 a]] + 1)$. If $d \leq d_0$ we simply prove the existence of a number coding a sequence $v_1, v_2, \ldots, v_d$ with $v_1 = u_1$ and $v_{j+1} = v_j + u_{j+1}$ for all $j \in [1, d-1]$. This can be done since if each $u_i \leq [log_2 a]^n$ then each $v_j < [\log_2 a]^{n+1}$. ⊣

The result is extended to the case $d \leq d_0^k$ for $k = 2^m \in \mathbb{N}$ by induction on m using a "divide and conquer" argument. Suppose the lemma has been proved for *all* Δ_0 sequences $u_1, u_2, \ldots, u_d$, with $d \leq d_1$ (where $d_1 \leq [\log_2 a]^k$ for some $k \in \mathbb{N}$). If $w_1, w_2, \ldots w_d$, is a Δ_0 sequence of length $d \leq d_1^2$ and there is some $n \in \mathbb{N}$ such that each $w_i < [log_2 a]^n$, then $\sum_{1 \leq i \leq j} w_i$ can be defined by:

$$\sum_{1 \leq i \leq j} w_i = \sum_{0 \leq s \leq [j/d_1]} \sum_{1 \leq i \leq d_1} w_{sd_1+i} + \sum_{1 \leq i \leq j-d_2} w_{d_2+i}$$

where $d_2 = [j/d_1]d_1$. Clearly the inner and final sums can be handled by the induction hypothesis. So also can the sum $\sum_{0 \leq s \leq [j/d_1]}$, since each of its terms is of the form:

$$\sum_{1 \leq i \leq d_1} w_{sd_1+i} < d_1[\log_2 a]^n \leq [\log_2 a]^{k+n},$$

and there are at most d_1 of them.

The properties $\sum_{1 \leq i \leq l} w_i = w_1$ and $\sum_{1 \leq i \leq j+1} w_i \sum_{1 \leq i \leq j} w_i + w_{j+1}$ are inherited from the similar properties possessed by the sums with at most d_1 terms.

A different sort of coding is used to give Δ_0 definitions of the functions:

$$p_n(x) = \textit{the nth prime divisor of } x \textit{ (in order of magnitude)}$$
$$\nu(x) = \textit{the number of (distinct) prime divisors of } x.$$

Lemma 2.4. *For every $x \in M$ there is some $c \in M$ such that*

(i) *If p is the least prime divisor of x then $c \equiv 1 \pmod p$.*
(ii) *If $p < q$ are prime divisors of x, $c \equiv n \pmod p$, and there is no prime $r|x$ with $p < r < q$, then $c \equiv n+1 \pmod q$.*

Proof. Develop enough of the Chinese remainder theorem in $I\Delta_0$ to prove the existence of c. ⊣

Clearly, if $n < p$ and $p|x$ then $p = p_n(x) \iff c \equiv n \pmod p$. Also,

$$n = \nu(x) \iff p_n(x) \textit{ is the largest prime divisor of } x.$$

It should be remarked that, subject to the limitations imposed by the fact that models need not be closed under functions of faster than polynomial growth, the development of the properties of congruences and primes (up until the problem of the existence of "infinity" many) proceeds quite smoothly in $I\Delta_0$. For example (essentially) the standard proofs that the usual definitions of the primes are equivalent, and that every $x > 1$ has at least one prime divisor, work. Some models of weaker axiom systems (notably the system with induction restricted to quantifier free formulas studied by Wilkie [47] and Shepherdson [41])do not have these properties. (See for example the appendix to Woods [50].)

3 The pigeon hole principle and arbitrarily large primes

We are now in a position to give an $I\Delta_0 + PHP(\Delta_0)$ proof of the existence of arbitrarily large primes. In fact the argument below shows in $I\Delta_0 + PHP(\Delta_0)$, that for all x, y at least one of the numbers $y+1, y+2, \ldots, y+x$ is divisible by some prime $p > x$, *provided* that $y > x^5$ and $x > C$ where C is some fixed (standard) natural number. The condition $x > C$ can be removed by:

Lemma 3.1. *Suppose $M \models I\Delta_0$ and $x \in \mathbb{N}$. Then for every $y \in M$ with $y \geq x \geq 1$, at least, one of the numbers $y+1, y+2, \ldots, y+x$ has a prime divisor $p > x$.*

Proof. If $y \in \mathbb{N}$ this is just Sylvester's theorem for $\mathbb{N}$, so suppose $y \in M \setminus \mathbb{N}$ (that is, y is *nonstandard*). If $\nu(y+i) \leq x$ for some $i \in [1, x]$ then we are done, while if $\nu(y+i) \leq x$ for all $i \in [1, x]$, then assign to each $y+i$ the least prime p such that $p^e | y+i$ for some nonstandard power p^e. (Some such p must exist since the product of a standard number of standard numbers is standard.) The x primes p obtained in this way are distinct, since otherwise $p^e | (y+i_1) - (y+i_2) = i_1 - i_2 \in \mathbb{N}$ for some $p^e \in M \setminus \mathbb{N}$. Thus at least one of them satisfies $p > x$. $\dashv$

(This argument is adapted from Grimm [15].)

Theorem 3.2. $I\Delta_0 + PHP(\Delta_0) \vdash \forall x \exists p (p > x \wedge p \textit{ is prime})$.

Proof. We will construct a Δ_0-formula $\theta(u, v, x, y)$ such that it can be proved in $I\Delta_0$ that if $p \leq x$ for all prime divisors p of $y+1, y+2, \ldots, y+x$, then $\{< u, v >: \theta(u, v, x, y)\}$ is the graph of a one-to-one map from $[1, a]$ into $[1, b]$, where

$$a = x[\log_2 y]$$
$$b = 2x[\log_2 x] + ([x/2] + 1)[\log_2 y].$$

If $a > b$, this is clearly contrary to $PHP(\Delta_0)$. But for $y \geq x^5$,

$$\begin{aligned}(x - [x/2] - 1)[\log_2 y] &\geq 5(x - [x/2] - 1)[\log_2 x] \\ &> 2x[\log_2 x]\end{aligned}$$

provided $x > C$, where $C \in \mathbb{N}$ is a suitably chosen constant, and therefore $a > b$.

To construct the map from $[1, a]$ to $[1, b]$ we first subdivide $[1, a]$ into x (disjoint) intervals A_i, $i = 1, 2, \ldots, x$, each of *length* $|A_i| = [\log_2 y]$, and $[1, b]$ into x intervals B_r, $r = 1, 2, \ldots, x$, each of length $|B_r| = 2[\log_2 x]$ followed by $[x/2] + 1$ intervals C_i, $i = 1, 2, \ldots, [x/2] + 1$, each of length $|C_i| = [\log_2 y]$.

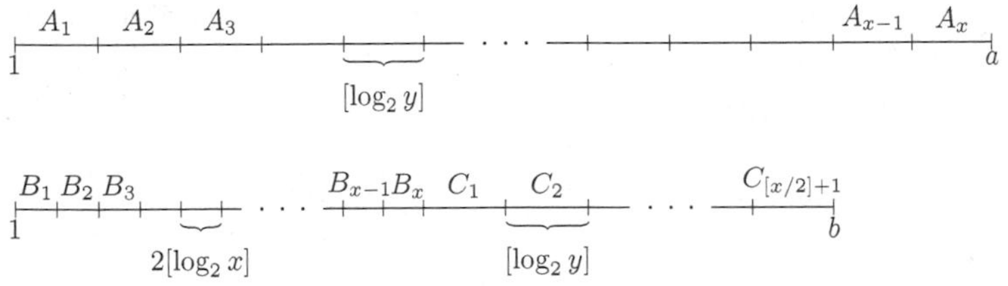

A_1 A_2 A_3 … A_{x-1} A_x
1 … a
$[\log_2 y]$

$B_1 B_2 B_3$ … $B_{x-1} B_x$ C_1 C_2 … $C_{[x/2]+1}$
1 … b
$2[\log_2 x]$ $[\log_2 y]$

Each A_i is now subdivided in a manner determined by the prime power decomposition $y+i = \prod_{1 \leq j \leq \nu(y+i)} p_j^{e_j}$ the idea being to represent factors $p_j^{e_j}$ of this product by lengths corresponding roughly to their logarithms. (The indexing P_j, $j = 1, 2, \ldots, \nu(y+i)$ of the prime divisors of $y+i$ is available in $I\Delta_0$ by lemma 2.4.) A_i is divided into nonempty subintervals A_{ij}, $j = 1, 2, \ldots, h$, with

$$|A_{ij}| = \begin{cases} e_j([\log_2 p_j] + 1) & \text{for } j < h, \\ \leq e_h([\log_2 p_h] + 1) & \text{for } j = h. \end{cases}$$

By lemma 2.3 the sum $\sum_{1 \leq t \leq j} e_t([\log_2 p_t]+1)$ makes sense in $I\Delta_0$, and the endpoints of the A_{ij}'s form a Δ_0 sequence. Also it can easily be proved in $I\Delta_0$ that

$$\sum_{1 \leq j \leq \nu(y+1)} e_j([\log_2 p_j] + 1) > [\log_2 y]$$

so it is possible to "carry out" this construction for some $h \leq \nu(y+i)$.

Each A_{ij} is then further subdivided into subintervals A_{ijk} with $|A_{ijk}| = [\log_2 p_j]+1$, corresponding to the factors p_j of $p_j^{e_j}$. (In the case of A_{ih} we construct as many disjoint subintervals of length $[\log_2 p_h]+1$ as there is space for, allowing the final A_{ihk} to have $|A_{ijk}| \leq [\log_2 p_h] + 1$).

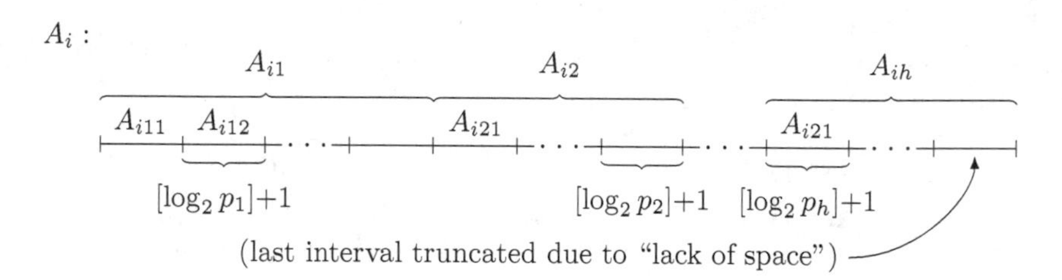

Similarly each B_r is subdivided using the prime power decomposition of $r = \prod_{1 \leq s \leq \nu(r)} p_s^{e_s}$, first into intervals B_{rs}, $s = 1, 2, \ldots, \nu(r)$, with $|B_{rs}| = e_s([\log_2 p_s] + 1)$, and a "left over" interval $B_{r\nu(r)+1}$ (possibly empty), and then (for $s \leq \nu(r)$) into subintervals B_{rst}, $t = 1, 2, \ldots, e_s$ with $|B_{rst}| = [\log_2 p_s] + 1$. The first of these subdivisions can always be carried out since

$$\begin{aligned} \sum_{1 \leq s \leq \nu(r)} e_s([\log_2 p_s] + 1) &\leq \sum_{1 \leq s \leq \nu(r)} 2e_s[\log_2 p_s] \\ &\leq 2[\log_2 r] \leq 2[\log_2 x]. \end{aligned}$$

The elements (if any) of the "left over" interval $B_{r\nu(r)+1}$ will not occur in the range of the function to be constructed.

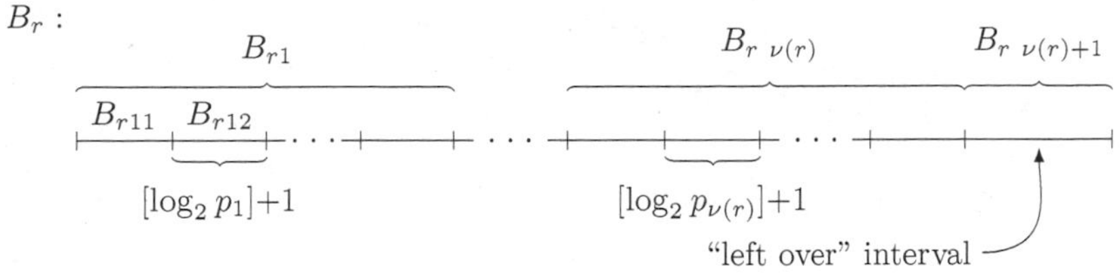

B_r :

B_{r1} $B_{r\ \nu(r)}$ $B_{r\ \nu(r)+1}$

B_{r11} B_{r12}

$[\log_2 p_1]+1$ $[\log_2 p_{\nu(r)}]+1$

"left over" interval

Notice that each interval $A_{ijk}, B_{rst}(s \leq \nu(r))$ is associated with a prime by this construction, so these primes can be used as *labels* on the elements in these intervals. Give labels $2, 3, 5, ..., 2i-1, ..., 2[x/2]+1$ to the elements in the intervals $C_1, C_2, C_3, \ldots, C_i, \ldots, C_{[x/2]+1}$ respectively. Only elements having prime labels $\leq x$ will occur in the range and domain of the one-to-one function f which we will now construct. f will preserve labels and can therefore be defined separately for each label $p \leq x$.

Fix $p \leq x$ and consider the least number $i^\star$ with the property:

$$\forall i \in [1, x] \forall m (p^m | y + i \rightarrow p^m | y + i^\star).$$

Define f on the interval $A_{i^\star j}$ with label p to be a one-to-one map into the interval C_i with label p. This can be done uniformly using a Δ_0 formula since we can define the endpoints of the intervals and then map consecutive elements to consecutive elements. There are enough of these in C_i because

$$|A_{i^\star j}| \leq |A_{i^\star}| = [\log_2 y] = |C_i|.$$

To define f on the remaining elements of $[1, a]$ with label p, it suffices to define a one-to-one map taking each $A_{ijk}(i \neq i^\star)$ with label p to some B_{rst} with the same label. B_{rst} has enough elements because

$$|A_{ijk}| \leq [\log_2 p] + 1 = |B_{rst}|.$$

In fact we can map A_{ijk} to some B_{rst} with $t = k$. To see this observe that for any given i, k, if there exists j such that A_{ijk} has label p, then $p^k | y + i$. But then $p^k | y + i^\star$ so $p^k || i - i^\star |$, and therefore either $i = i^\star + zp^k$ for some $z \leq [(x - i^\star)/p^k]$, or $i = i^\star - zp^k$ for some $z \leq [i^\star/p^k]$. Also

$$\begin{aligned} \exists s\ (B_{rsk}\ \textit{has label}\ p) &\iff p^k | r \\ &\iff \exists z \leq [x/p^k](r = zp^k), \end{aligned}$$

and $[(x - i^\star)/p^k] + [i^\star/p^k] \leq [x/p^k]$, so there is an obvious one-to-one map from $\{i \mid i \neq i \wedge \exists j\ (A_{ijk}\ \textit{has label}\ p)\}$ into $\{r \mid \exists s\ (B_{rsk}\ \textit{has label}\ p)\}$.

This completes the Δ_0 definition of a function f mapping those elements of $[1, a]$ which have labels $p \leq x$, one-to-one into $[1, b]$. But every element of $[1, a]$ has a label which is a prime, divisor of some number among $y+1, y+2, \ldots, y+x$, so if these have no prime divisor greater than x, then f is defined on all of $[1, a]$, which is contrary to $PHP(\Delta_0)$ if $a > b$. $\dashv$

4 Approximate logarithms and Sylvester's theorem

To obtain a proof of the general case of Sylvester's theorem using only $I\Delta_0 + PHP(\Delta_0)$ we will need:

(I) more "accurate" approximate logarithms than those used in §3,

(II) some way of getting around the requirement (indicated in §1) for a good upper bound on $\pi(x)$.

Fix $M \models I\Delta_0$, $a \in M \setminus \{0, 1, 2\}$, $n \in \mathbb{N} \setminus \{0\}$, and let $Q^+(M)$ denote the set of non negative rational numbers in the sense of M. We will now construct a Δ_0 function $\log^\star : \{x \in Q^+(M) : \ x \geq 1\} \to Q^+(M)$. More precisely, the construction gives a Δ_0 formula with parameter a, which defines a map taking each ordered pairs $< b, c >$ with $b, c \in M$, $b \geq c$, to some $< u, v > \in M \times M$ with $\log^\star \frac{b}{c} = \frac{u}{v}$.

The intention is that $\log^\star x$ should "behave like" an approximation to $\log x$ with error less than a fraction $K[\log_2 a]^{-n}$ of its value for x large, where $K \in \mathbb{N}$ is a constant. The definition could be extended to all of $Q^+(M) \setminus \{0\}$ by taking $\log^\star \frac{b}{c} = -\log^\star \frac{c}{b}$ for some $b < c$.

We first define a "Δ_0" function $\log^\star x$ for $x \in Q^+(M)$ with $1 \leq x < 4$, by considering a grid of squares "under the graph" of the function $y = \frac{1}{x}$, the sides of the squares being parallel to the axes and of length $h = 1/2^k$, where $k \in M$ satisfies $2^{k-1} < [\log_2 a]^n \leq 2^k$. (Recall that $\log x = \int_1^x \frac{1}{t} dt$.) For $1 \leq x < 4$ we put:

$$\log^+ x = \left(\sum_{1 \leq j \leq [(x-1)/h]} \left[\frac{1}{(1+jh)h} \right] \right) \cdot h^2,$$

that is, we count the *number* of complete squares of area h^2 which lie *entirely* in the region under the graph. Since $[(x-1)/h] < 6[\log_2 a]^n$ and $\left[\frac{1}{(1+jh)h}\right] \leq 2[\log a]^n$, the sum in brackets can be handled by lemma 2.3.

Heuristically we expect that $\log^+ x$ will behave like an approximation to "$\log x$" accurate to within $C[\log_2 a]^{-n}$ for some $C \in \mathbb{N}$, since the number of squares which lie *partially* in the region under the graph is $\leq C[\log_2 a]^n$ for some $C \in \mathbb{N}$, and these therefore represent an area $\leq C[\log_2 a]^{-n}$. What we will *actually* show in $I\Delta_0$ is that for $1 \leq x \leq 2$, $1 \leq y < 2$,

$$\log^+(x \cdot y) \doteqdot log^+ x + log^+ y, \tag{1}$$

where $\doteq$ *means that the two sides differ by at most* $C[\log_2 a]^{-n}$ *for some constant* $C \in \mathbb{N}$ *independent of* x, y, a, M.

Note that it is trivial to verify directly from the definition of $\log^+$ that if $x_1 = u/2^k \leq x < (u+l)/2^k$ with $u \in M$, then $\log^+ x = \log^+ x_1$ and $\log^+(x \cdot y) \doteq \log^+(x_1 \cdot y)$ (for $1 \leq y < 2$). Therefore we may suppose that $x = u/2^k$ for some $u \in M$, where obviously $u \leq 2^{k+1}$ for $1 \leq x \leq 2$.

Since $\int_1^{xy} \frac{1}{t} dt = \int_1^x \frac{1}{t} dt + \int_x^{xy} \frac{1}{t} dt$, proving (1) corresponds to showing

$$\int_1^{xy} \frac{1}{t} dt = \int_1^y \frac{1}{t_1} dt_1$$

in the standard case, which is of course done by putting $t_1 = t/x$. The effect of this change of variable is to transform the square grid (with sides $= h$) under the graph of $\frac{1}{t}$ into a rectangular grid (with vertical sides $= xh$ and horizontal sides $= h/x$) under the graph of $\frac{1}{t_1}$.

The area ($\doteq \log^+(x \cdot y) - \log^+ x$) represented by the complete rectangles of this sort which lie entirely in the region corresponding to $\int_1^y \frac{1}{t_1} dt_1$ can be compared with the approximation $\log^+ y$ to $\int_1^y \frac{1}{t_1} dt_1$ obtained by counting complete squares in a square grid with sides $= h$, *without going beyond* $I\Delta_0$. To do this consider a *common refinement* of these square and rectangular grids consisting of a grid of squares with sides $h_1 = h/(u \cdot 2^k)$, where $x = u/2^k$. Since $h_1 \geq 1/2^{3k+1} > 1/(2^4[\log_2 a]^{3n})$, the *number* of squares in this common refinement is small enough to enable lemma 2.3 to be used to count the number of these "captured" when the square grid with side h is used and compare this with the number captured when using the rectangular grid. The details of the formal proof are more appropriate to a tedious first course in integral calculus so are omitted.

Now to define $\log^\star x$ for all $x \in Q^+(M)$, $x \geq 1$, write $x = 2^m \cdot w$ where $m \in M$ and $w = x/2^m \in Q^+(M)$ with $1 \leq w < 2$, and put

$$\log^\star x = m \cdot \log^+ 2 + \log^+ w = [\log_2 x] \cdot \log^+ 2 + \log^+ w.$$

Lemma 4.1. $\log^\star 1 = 0$, *and for all* $x, y \in Q^+(M)$, $x, y \geq 1$,

$$\log^\star x + \log^\star y \doteq \log^\star(x \cdot y).$$

(That is, $|\log^\star x + \log^\star y = \log^\star(x \cdot y)| < K[\log_2 a]^{-n}$ *for some constant* $K \in \mathbb{N}$*).*

Proof. Suppose $x = 2^b \cdot u$, $y = 2^c \cdot v$ where $b, c \in M$ and $1 \leq u < 2$, $1 \leq v < 2$, so $x \cdot y = 2^{b+c} \cdot u \cdot v$.

If $2 \leq u \cdot v < 4$, then

$$\log^\star(x \cdot y) = (b+c+1) \cdot \log^+ 2 + log^+ \frac{u \cdot v}{2}.$$

But by (1), $\log^+ 2 + \log^+ \frac{u \cdot v}{2} \doteq \log^+(u \cdot v)$, so

$$\log^\star(x \cdot y) \doteq (b+c) \cdot \log^+ 2 + \log^+(u \cdot v).$$

The same is trivially true if $1 \leq u \cdot v < 2$, and therefore by (1),

$$\begin{aligned}\log^\star(x \cdot y) &\doteq (b \cdot \log^+ 2 + log^+ u) + (c \cdot \log^+ 2 + \log^+ v) \\ &= \log^\star x + \log^\star y.\end{aligned}$$ ⊣

Notice also that for every $i \in \mathbb{N}$ and every standard rational $\varepsilon > 0$ there is some $j \in \mathbb{N}$ (independent of M) such that $\log^\star i$ can be made to approximate $\log i$ to within ε simply by taking $a > j$. (More precisely the cut in the standard rationals determined by $\log^\star i$ can be brought to within ε of the real number $\log i$.) *We will assume in future that a is large enough to validate any standard arithmetic we do with* $\log^\star$.

At this stage it is convenient to state the following improvement of lemma 2.3, the proof of which will be deferred until §6.

Lemma 4.2. *If $u_1, u_2, \ldots, u_d$, is a Δ_0 sequence of elements of M where $d \leq [log_2 a]^n$ and each $u_i \leq u$ for some $a, u \in M$, $n \in \mathbb{N}$, then there is a Δ_0 sequence $\sum_{1 \leq i \leq j} u_i$, $j = 1, 2, \ldots, d$ such that $\sum_{1 \leq i \leq j} u_i = u$ and*

$$\sum_{1 \leq i \leq j+1} u_i = \sum_{1 \leq i \leq j} u_i + u_{j+1}$$

for all $j \in [1, d-1]$.

Lemma 4.2 will enable us to define $\sum_{1 \leq i \leq x} \log^\star i$. Consider the points $x \in Q^+(M)$ at which there is a "jump" in the value of the "step function" $\log^\star x$. From the definitions of $\log^\star$ and $\log^+$ it can be seen that these jumps occur at:

$$\begin{aligned}1 + \frac{1}{2^k}, 1 + \frac{2}{2^k}, 1 + \frac{3}{2^k}, \ldots, 2, 2 + \frac{2}{2^k}, 2 + \frac{2 \cdot 2}{2^k}, 2 + \frac{3 \cdot 2}{2^k}, \ldots, 4, \\ \ldots, 2^m, 2^m + \frac{2^m}{2^k}, 2^m + \frac{2 \cdot 2^m}{2^k}, 2^m + \frac{3 \cdot 2^m}{2^k}, \ldots, 2^{m+1}, \ldots\end{aligned} \tag{2}$$

where $m \in M$, and $k \in M$ satisfies $2^{k-1} < [\log_2 a]^n \leq 2$. Let $v_1, v_2, \ldots, v_d$ be a list of the points $v \in M$, $v < x$, at which $\log^\star(v+1) > \log^\star v$. From (2) it is easily seen that:

(i) $v_j = j$ for $j \leq 2^k$.

(ii) The numbers $v_j + 1, 2^k \leq j \leq d$ form a Δ_0 sequence comprised of that part of list (2) which lies between $2^k + 1$ and x. Therefore $d \leq 2^k([\log_2 x] + 1) \leq 2[\log_2 a]^n \cdot ([\log_2 x] + 1)$.

Recalling that each $\log^\star v_j = u_j/2^{2k}$ for some $u_j \in M$ we see that we can make the definition:

$$\sum_{1\le i\le x} \log^\star i = 2^{-2k} \sum_{1\le j\le d} (v_j - v_{j-1})u_j + (x - v_d)\log^\star x,$$

where $v_0 = 0$. The sum on the right can be handled by lemma 4.2, so it is obvious that $\sum_{1\le i\le x} \log^\star i$ will have the desired inductive properties.

Next, define a partial function $\exp^\star w$ for $Q^+(M)$ by taking $\exp^\star w$ to be the least $x \in M$ such that $\log^\star x \ge w$. Observe that $\exp^\star \log^\star x = v_d + 1$.

Now suppose $x \ge 2^k$, $x \in M$, and $\log^\star(x+1) > \log^\star x$ (or equivalently, $\exp^\star \log^\star(x+1) = x+1$). Then the distance between $x+1 = v_d+1$ and $\exp^\star \log^\star x = v_{d-1} + 1$ is

$$\exp^\star \log^\star(x+1) - \exp^\star \log^\star x = 2^{[\log_2 x]-k}, \tag{3}$$

and from the definitions of $\log^\star$ and $\log^+$ it can be seen that

$$\log^\star(x+1) - \log^\star x = \left\lceil \frac{2^{[\log_2 x]+k}}{x+1} \right\rceil \cdot 2^{-2k},$$

so

$$2^{[\log_2 x]-k} - (x+1)\cdot 2^{-2k} < (x+1)\cdot(\log^\star(x+1) - \log^\star x) \le 2^{[\log_2 x]-k}. \tag{4}$$

We are now able to prove a $\log^+$ analogue of the relation $\sum_{1\le i\le x} \log i = x\log x - x + o(x)$. By $r = s + O(t)$ we will mean that the terms r, s, t satisfy $|r - s| \le Kt$ for some constant $K \in \mathbb{N}$ (independent of M and the variables occurring in r, s, t).

Lemma 4.3. *For all $x > [\log_2 a]^{2n}$,*

$$\sum_{1\le i\le x} \log^\star i = x\log^\star x - x + O\left(\frac{x[\log_2 x]}{[\log_2 a]^n}\right).$$

Proof. Let $\sigma(x) = (x+1)\log^\star x - \exp^\star \log^\star x$ and let

$$\gamma(x) = \sum_{1\le i\le x} \log^\star i - \sigma(x).$$

If $x \ge 2^k$ then by (3) and (4),

$$\sigma(x+1) - \sigma(x) = \begin{cases} \log^\star(x+1), & \text{if} \log^\star(x+1) = \log^\star x \\ \log^\star(x+l) + O(x\cdot 2^{-2k}), & \text{if} \log^\star(x+l) > log^\star x. \end{cases}$$

Using this it can be shown induction on $x \ge 2^k$ that

$$\begin{aligned} \gamma(x) - \gamma(2^k) &= O(|\{j:\ 2^k \le v_j < x\}| \cdot x \cdot 2^{-2k}) \\ &= O\left(\frac{x[\log_2 x]}{[\log_2 a]^n}\right). \end{aligned}$$

Therefore,

$$\begin{aligned}\gamma(x) &= O\left(2^k \log^\star(2^l) + \frac{x[\log_2 x]}{[\log_2 a]^n}\right)\\ &= O\left([\log_2 a]^n [\log_2 [log_2 a]] + \frac{x[\log_2 x]}{[\log_2 a]^n}\right)\\ &= O\left(\frac{x[\log_2 x]}{[\log_2 a]^n}\right)\end{aligned}$$

for $x \geq [\log_2 a]^{2n}$. But by (3)

$$\sigma(x) = x \log^\star x - x + O\left([\log_2 x] + 2^{[\log_2 x]-k}\right),$$

so for all $x \geq [\log_2 a]^{2n}$,

$$\sum_{1\leq i\leq x} \log^\star i = \sigma(x) + \gamma(x) = x \log^\star x - x + O\left(\frac{x[\log_2 x]}{[\log_2 a]^n}\right). \qquad \dashv$$

Our aim now will be to sketch an argument showing the existence of an $I\Delta_0 + PHP(\Delta_0)$ proof of:

Theorem (Sylvester's Theorem). If $1 \leq x \leq y$ then some number among $y+1, y+2, \ldots, y+x$ has a prime divisor $p > x$.

In view of lemma 3.1 we may suppose that $x \geq K$ for any fixed $K \in \mathbb{N}$. Also, the method of §3 extends readily to the case $y \geq x^{2-\varepsilon}$ where $\varepsilon > 0$ is any sufficiently small standard rational number. (This will be left to the reader. Hint: use a better, but still trivial, bound on $\pi(x)$, for example $\pi(x) < x/3$, together with more "accurate" approximate logarithms than those used in §3. Actually at the expense of doing, some extra number theory for $\mathbb{N}$, any $\varepsilon < 1$ can be shown to work). *We can therefore assume that* $y + x \geq x^{2-\varepsilon}$.

We now turn to the question of the need for a "good" upper bound on $\pi(x)$ in the case $y + x < x^{2-\varepsilon}$. Before considering $I\Delta_0 + PHP(\Delta_0)$ we will investigate how good an upper bound on $\pi(x)$ is required for the standard proof of Sylvester's theorem for $\mathbb{N}$ described in §1, and how this can be obtained. Actually we will work with the closely related function $\theta(x) = \sum_{p\leq x} \log p$ rather than $\pi(x)$ since this will be more convenient later. (It is relatively easy to prove $\pi(x) = \frac{\theta(x)}{\log x} + o(\pi(x))$).

Under the assumption that $y+1, y+2, \ldots, y+x$ all have no prime divisor $p > x$, inequality (3) of §1 asserts (in logarithmic form):

$$\sum_{p\leq x} \log w_p + \sum_{1\leq i\leq x} \log i \geq \sum_{y<i\leq y+x} \log i \tag{5}$$

where w_p is the largest power of p which divides some $y+i$. Since $w_p \le y+x < x^{2-\varepsilon}$ we see that if $p > x^{1-\varepsilon/2}$ (say) then either $w_p = p$ or $w_p = 1$. Therefore

$$\sum_{x_0<p\le x} \log p + x^{1-\varepsilon/2}\log(x^{2-\varepsilon}) + \sum_{1\le i\le x} \log i \ge \sum_{y<i\le y+x} \log i$$

so

$$\sum_{p\le x} \log p + \sum_{1\le i\le x} \log i + o(x) \ge \sum_{y<i\le y+x} \log i \ge \sum_{x<i\le 2x} \log i \tag{6}$$

Using $\sum_{1\le i\le x} \log i = x\log x - x + o(x)$ it follows that:

$$\sum_{p\le x} \log p \ge (2\log 2)x + o(x),$$

so a contradiction will be obtained if it can be shown that $\sum_{p\le x} \log p \le Ax + o(x)$ for some constant $A < 2\log 2$.

Remark 4.4. *If we do not make the assumption that no prime divisor of $y+1, y+2, \ldots, y+x$ exceeds x then instead of (5) we obtain:*

$$\sum_{p\le x+y} \log w_p + \sum_{1\le i\le x} \log i \ge \sum_{y<i\le y+x} \log i.$$

Taking $y = x$ it follows from this (as above) that

$$\sum_{p\le x} \log p \ge (\log 2)x + o(x).$$

This proves proposition 4.2 of chapter 2 of this thesis.

Sylvester's method can also be used to give an upper bound on $\sum_{p\le x} \log p$. Our starting point is the method used to prove inequality (1) of §1. Recall that this asserted that

$$w_p \prod_{1\le s\le x} [s]_p \ge \prod_{1\le i\le x} [y+i]_p.$$

A close examination of the proof shows that we also have

$$\prod_{1\le s\le x} [s]_p \le \prod_{1\le i\le x} [y+i]_p.$$

Furthermore, if $p > x$ and $p|y+i$ for some $i \in [1,x]$, then

$$p \cdot \left(\prod_{1\le s\le x} [s]_p\right) \le \prod_{1\le i\le x} [y+i]_p.$$

since each $[s]_p = 1$. Taking $y = x$ we see that

$$\left(\prod_{x \le p \le 2x} p\right) \cdot \left(\prod_{1 \le s \le x} s\right) \le \prod_{1 \le i \le x} (x+i).$$

or in logarithmic form:

$$\sum_{x \le p \le 2x} \log p + \sum_{1 \le i \le x} \log i \le \sum_{x < i \le 2x} \log i,$$

from which it can easily be deduced that

$$\sum \log p \le (2\log 2)x + o(x).$$

This, of course, is *not* good enough for our purposes since we require a constant *strictly less than* $2\log 2$.

There are several ways to produce the slight improvement required. One approach is to take heed of the implication in the quotation from Sylvester given in §1 that the method applies to arithmetic progressions other than $y+1, y+2, \ldots, y+x$. Indeed if y is odd then all the primes $p \in [y+1, y+2x]$ will lie in the arithmetic progression:

$$y+2, y+4, \ldots, y+2i, \ldots, y+2x,$$

while as before

$$\prod_{1 \le s \le x} [s]_p \le \prod_{1 \le i \le x} [y+2i]_p$$

for all primes $p \ne 2$. Thus for $y = x$ (and odd) it follows that

$$\left(\prod_{x < p \le 3x} p\right) \cdot \left(\prod_{1 \le s \le x} s\right) \le \left(\prod_{1 \le i \le x} (x+2i)\right) \cdot \left[\prod_{1 \le s \le x} s\right]_2 .$$

But from equation (2) of §1 we know that

$$\log \left[\prod_{1 \le s \le x} s\right]_2 = \sum_{1 \le k \le [\log_2 x]} [x/2^k] \log 2 \le x \log 2,$$

so

$$\sum_{x < p \le 3x} \log p + \sum_{1 \le i \le x} \log i \le \sum_{1 \le i \le x} \log(x+2i) + x\log 2 \qquad (7)$$

But

$$\begin{aligned} \sum_{1 \le i \le x} \log(2i-1) &= \sum_{1 \le i \le 2x} \log i - \sum_{1 \le i \le x} \log 2i, \\ &= \sum_{1 \le i \le 2x} \log i - \sum_{1 \le i \le x} \log i - x\log 2, \end{aligned} \qquad (8)$$

so

$$\sum_{1\le i\le x} \log(x+2i) = \sum_{x<i\le 3x} \log i - \sum_{x/2<i\le 3x/2} \log i - x\log 2,$$

Putting

$$\sigma_1(x) = \sum_{1\le i\le x/3} \log i$$

and

$$\sigma_2(x) = \sum_{x/3<i\le x} \log i - \sum_{x/6<i\le x/2} \log i,$$

it follows that

$$\sum_{x<p\le 3x} \log p + \sigma_1(3x) \le \sigma_2(3x).$$

Hence for *arbitrary* x

$$\sum_{x/3^{j+1}<p\le x/3^j} \log p + \sigma_1(x/3^j) \le \sigma_2(x/3^j) + O(\log x),$$

and thus

$$\sum_{p\le x} \log p + \sum_{j\le[\log_3 x]} \sigma_1(x/3^j) \le \sum_{j\le[\log_3 x]} \sigma_2(x/3^j) + O(\log x) \tag{9}$$

But since

$$\sum_{1\le i\le x} \log i = x\log x - x + o\left(\frac{x}{\log x}\right), \tag{10}$$

$$\sigma_2(x) - \sigma_1(x) = \left(\frac{1}{2}\log 3 + \frac{1}{3}\log 2\right)x + o\left(\frac{x}{\log x}\right),$$

so

$$\begin{aligned}\sum_{j\le[\log_3 x]} (\sigma_2(x/3^j) - \sigma_1(x/3^j)) &= \left(\sum_j \frac{1}{3^j}\right)\cdot\left(\frac{1}{2}\log 3 + \frac{1}{3}\log 2\right)x \\ &\quad + o(x) \\ &= \frac{3}{2}\left(\frac{1}{2}\log 3 + \frac{1}{3}\log 2\right)x + o(x)\end{aligned}$$

Therefore, $\sum_{p\le x}\log p \le \left(\frac{3}{4}\log 3 + \frac{1}{2}\log 2\right)x + o(x)$, and as the reader may check, $\frac{3}{4}\log 3 + \frac{1}{2}\log 2 < 2\log 2$ (note the use of an error bound better than $o(x)$ in (10). This is of course possible since the *actual* error is $O(\log x)$ by Stirling's formula).

Theorem 4.5. *$I\Delta_0 + PHP(\Delta_0) \vdash$ Sylvester's theorem.*

Proof. (Sketch) Let $M \models I\Delta_0 + PHP(\Delta_0)$, $a \in M$, and $n \in \mathbb{N}$. (We will choose a, n explicitly later). Define $\log^\star$ as above and let $\sigma_1^\star, \sigma_2^\star$ be the $\log^\star$ analogues of σ_1, σ_2. Then the "starred" version of (9) is

$$\sum_{p \leq x} \log^\star p + \sum_{j \leq [\log_3 x]} \sigma_1^\star(x/3^j) \leq \sum_{j \leq [\log_3 x]} \sigma_2^\star(x/3^j) + o(x). \qquad (9)^\star$$

There is a way of *interpreting* this inequality which is viable in $I\Delta_0$. Observe first that multiplying by 2^k, where as before $2^{k-1} < a \leq 2^k$, converts the inequality into one involving sums of terms of the form $2^{2k} \log^\star i$ and $2^{2k} \log^\star p$, and by the definition of $\log^\star$ all such terms are elements of M (rather than just $Q^+(M)$). For the same reason the sums

$$b = \sum_{j \leq [\log_3 x]} 2^{2k} \sigma_1^\star(x/3^j), \; c = \sum_{j \leq [\log_3 x]} 2^{2k} \sigma_2(x/3^j).$$

can be defined using lemma 4.2. For suitably chosen b_1, c_1 approximately equal to b, c, the proof of $(9)^\star$ (as indicated for (9) above) can be turned into the construction of a *one-to-one comparison map*:

$$f_1 : [1, b_1] \to [1, c_1] \setminus \bigcup_{p \leq x} A_p$$

where $\{A_p \mid p \; prime \wedge p \leq x\}$ *is a set of disjoint subintervals of* $[1, c_1]$ *with lengths* $|A_p| = 2^k \log^\star p$ *and* f_1 *is* Δ_0 *definable.*

The construction of f_1 is very similar to the construction of the map f used in §3. $[1, b_1]$ is first subdivided into intervals of length roughly $2^{2k}\sigma_1^\star(x/3^j) = \sum_{1 \leq i \leq x/3^{j+1}} 2^{2k} \log^\star i$, then into intervals of length approximately $2^{2k} \log^\star i$, and these are then subdivided into intervals of length *exactly* $2^{2k} \log^\star p_m$ (except where we "run out of space") corresponding to the prime factors of $i = \prod_{1 \leq m \leq \nu(i)} p_m^{e_m}$. Since the starred version of the equation $\log(w \cdot z) = \log w + log z$ is only approximate, namely $|\log^\star(w \cdot z) - \log^\star w - \log^\star z| < \frac{K}{[\log_2 a]^n}$ for some constant $K \in \mathbb{N}$, we will actually start from intervals slightly *smaller* than $2^{2k} \log^\star i$ so as to ensure that we always use up all the space available in the interval. Since $\sum_{1 \leq m \leq \nu(i)} e_m \leq [\log_2 x]$ we have

$$\sum_{1 \leq m \leq \nu(i)} e_m \log^\star p_m \geq \log^\star i - K[\log_2 x] \cdot [\log_2 a]^{-n},$$

so it suffices to reduce the length from $2^{2k} \log^\star i$ to

$$2^{2k} \log^\star i - 2^{k+1} K[\log_2 x],$$

(that is, to use approximately $\log^\star i - K[\log_2 x] \cdot [\log_2 a]^{-n}$ in place of $\log^\star i$). This reduction corresponds to taking

$$b_1 = b - 2^{k+1} K[\log_2 x] \sum_{j \leq [\log_3 x]} [x/3^{j+1}].$$

Clearly the reduced length subintervals can be "marked out" using a Δ_0 function, and

$$b_1 = 2^{2k}\left(\sum_{j\le[\log_3 x]} \sigma_1^\star(x/3^j) + O\left(\frac{x[\log_2 x]}{[\log_2 a]^n}\right)\right).$$

Similarly $[1, c_1]$ is subdivided into intervals of length either $2^k \log_2$ or approximately $2^{2k}\log^\star(x_j+2i)$, where x_j is an odd number divisible by 3 approximating $x/3^{j+1}$. (These can be marked off by a Δ_0 function since the function $\sum_{1\le i\le z}\log^\star(2i-1)$ can be defined, either directly in the same way as $\sum_{1\le i\le x} log^\star i$, or alternatively via (8) at the expense of introducing an $O(x\cdot[\log_2 a]^{-n})$ error). This time the intervals of length $2^{2k}\log^\star(x_j+2i)$ will have to be expanded to ensure that there is "enough space" for subintervals of length $2^{2k}\log^\star p$ corresponding to *all* of the prime factors of x_j+2i. However for $x \ge [\log_2 a]^n$ we can still choose:

$$c_1 = 2^{2k}\left(\sum_{j\le[\log_3 x]} \sigma_2^\star(x/3^j) + O\left(\frac{x[\log_2 x]}{[\log_2 a]^n}\right)\right).$$

The function f_1 maps subintervals of $[1, b_1]$ with length $2^{2k}\log^\star p$ to subintervals of $[1, c_1]$ with the same length in an almost completely analogous fashion to the function constructed in §3 (with intervals of length $[\log_2 p]+1$).

In a similar way, under the assumption that no prime $p > x$ is a divisor of $x+1, x+2, \ldots, x+y$, the proof of a starred version of (6), namely:

$$\sum_{1\le i\le x}\log^\star p \le \sum_{1\le i\le x}\log^\star(y+i) \le \sum_{p\le x}\log^\star p + \sum_{1\le i\le x}\log^\star i + O\left(\frac{x}{[\log_2 x]}\right) \tag{6*}$$

can be interpreted as defining a one-to-one Δ_0 comparison map

$$f_2 : [1, b_2] \to [1, c_2] \setminus \bigcup_{p\le x} B_p,$$

where

$$b_2 = 2^{2k}\left(\sum_{1<i<x}\log^\star(x+i)] + O\left(\frac{x[\log_2 x]}{[\log_2 a]^n}\right)\right),$$

$$c_2 = 2^{2k}\left(\sum_{1\le i\le x}\log^\star i + O\left(\frac{x}{[\log_2 x]} + \frac{x[\log_2 x]}{[\log_2 a]^n}\right)\right)$$

and $\{B_p \mid p\ prime \wedge p \leq x\}$ is a set of disjoint intervals with $B_p \cap [1, c_1] = \phi$ and lengths $|B_p| = 2^{2k} \log^\star p$. As with the A_p's there are Δ_0 functions mapping $\{p \leq x \mid p\ prime\}$ to the end points of the B_p's, but note that we are *not* assuming that the A_p's or B_p's can be Δ_0 indexed by any *interval* $[1, m]$. Obviously however there *is* a one-to-one Δ_0 function

$$g : \bigcup_{p \leq x} B_p \to \bigcup_{p \leq x} A_p \text{ with } g : B_p \to A_p \text{ for each } p \leq x.$$

Now it is possible, in effect, to add inequalities $(6)^\star$ and $(9)^\star$. More precisely, the inequality:

$$\sum_{j \leq [\log_3 x]} \sigma_1^\star(x/3^j) + \sum_{1 \leq i \leq x} \log^\star(x+i) \tag{6}$$

$$\leq \sum_{j \leq [\log_3 x]} \sigma_2^\star(x/3^j) \tag{7}$$

$$+ \sum_{1 \leq i \leq x} \log^\star(x+i) + O\left(\frac{x}{[\log_2 x]} + \frac{x[\log_2 x]}{[\log_2 a]^n}\right) \tag{8}$$

follows by $PHP(\Delta_0)$ once we construct a one-to-one Δ_0 function $f : [1, b_1 + b_2] \to [1, c_1 + c_2]$ by taking

$$f(x) = \begin{cases} f_1(x) & \text{if } x \in [1, b_1], \\ c_1 + f_2(x - b_1) & \text{if } x \in [b_1 + 1, b_1 + b_2] \\ & \wedge f_2(x - b_1) \in [1, c_2] \setminus \bigcup B_p, \\ g(f_2(x - b_1)) & \text{if } x \in [b_1 + 1, b_1 + b_2] \\ & \wedge f_2(x - b_1) \in \bigcup_{p \leq x} B_p. \end{cases}$$

But now choose a = x, n = 2, so that the error term in (11) is

$$O\left(\frac{x[\log_2 x]}{[\log_2 a]^n}\right) = O\left(\frac{x}{[\log_2 x]}\right)$$

which is negligible compared with x. Estimating both sides of (11) using the relation

$$\sum_{1 \leq i \leq x} \log^\star i = x \log^\star x - x + O\left(\frac{x}{[\log_2 x]}\right)$$

obtained from lemma 4.3 (for the same values of a, n) now shows (by arithmetic already done above) that

$$\left(2 \log^\star 2 - \frac{3}{4} \log^\star 3 - \frac{1}{2}\right) x \leq C \frac{x}{[\log_2 x]}$$

for some $C \in \mathbb{N}$.

But this is false for all $x > C_0$, where $C_0 \in \mathbb{N}$ is a constant, contrary to our earlier observation that x can be assumed larger than any such C_0. ⊣

5 Counting the pigeons—census functions and the pigeon hole principle

The problem of proving the pigeon hole principle in $I\Delta_0$ suggests introducing a new (undefined) function symbol f and then considering the axiom system $I\Delta_0(f)$ obtained from $I\Delta_0$ by allowing induction on $\Delta_0(f)$ formulas, that is, on bounded formulas in which f as well as $+$ and $\cdot$ may occur. Specifically, let $PHP(f)$ be a sentence saying:

$$\forall n(f \textit{ does not map } [1, n] \textit{ one-to-one into } [1, n-1]).$$

Does $I\Delta_0(f) \vdash PHP(f)$?

Obviously an affirmative answer here would imply $I\Delta_0 \vdash PHP(\Delta_0)$, however a negative answer seems more likely in view of a conditional result due to Alex Wilkie based on a conjecture which appears (implicitly) in Cook and Reckhow [6]. Consider one of the standard axiomatisation of the propositional calculus and a doubly indexed set of propositional variables $A_{ij}, i, j \in \mathbb{N}$.

Conjecture 5.1. *For each $m \in \mathbb{N}$ there exist arbitrarily large numbers $n \in \mathbb{N}$ such that every proof in proposition calculus of the tautology*

$$\bigwedge_{i \leq n} \bigvee_{j \leq n-1} A_{ij} \rightarrow \bigvee_{k \leq n-1} \bigvee_{i < j \leq n} (A_{ik} \wedge A_{jk})$$

has more than n^m symbols.

In other words, the conjecture states that the time required to write out the shortest proof of this formula for a given value of n cannot be bounded by a polynomial in n (or alternatively by a polynomial in the length of the formula). Note that the formula embodies the pigeon hole principle for $[0, n]$ and is therefore a tautology by virtue of the truth of the principle for $\mathbb{N}$.

Proposition 5.2. *(Wilkie). The above conjecture implies*

$$I\Delta_0(f) \nvdash PHP(f).$$

Now consider any predicate $\theta(x, \vec{y})$ in some language for $\mathbb{N}$. The *census function c_θ* for $\theta(x, \vec{y})$ (with respect to x) is defined by

$$c_\theta(x, \vec{y}) = |\{j \in [1, x] : \theta(j, \vec{y})\}|,$$

or alternatively (with models other than $\mathbb{N}$ in mind) by:

$$c_\theta(0,\vec{y}) = 0 \tag{9}$$

$$c_\theta(x+1,\vec{y}) = \begin{cases} c_\theta(x,\vec{y})+1 & \text{if } \theta(x+l,\vec{y}) \text{ holds,} \\ c_\theta(x,\vec{y}) & \text{otherwise.} \end{cases} \tag{10}$$

Let $def(c_\theta)$ be an axiom asserting that this definition is satisfied. In what follows, bounded formulas involving $f, g, +, \cdot, \leq$ will be called $\Delta_0(f,g)$ formulas, and $I\Delta_0(f,g)$ will denote the axiom system similar to $I\Delta_0$ but with induction allowed on $\Delta_0(f,g)$ formulas.

Theorem 5.3. *Let $\theta(x,y)$ be the $\Delta_0(f)$ formula $f(x) \leq y$ and let $c_\theta(x,y)$ be a function symbol for the corresponding census function with respect to x. Then*

$$I\Delta_0(f,c_\theta) + \mathrm{def}(c_\theta) \vdash PHP(f).$$

Proof. Working in $I\Delta_0(f,c_\theta)+\mathrm{def}(c_\theta)$ suppose to the contrary that $f : [1,n] \to [1,n-1]$ and is one-to-one. The idea is to remove $f(1), f(2), \ldots$ from $[1,n-1]$ one step at a time, closing up the gaps so caused as we go, until after $n-2$ steps we are left with a one-to-one map from $[n-1,n]$ into $\{1\}$ which can trivially be proved to be impossible. At the ith step we will have the function $f_i^\star$ defined on $[i+1,n]$ by:

$$f_i^\star(x) = f(x) - c_\theta(i, f(x)).$$

(Intuitively $f_i^\star(x) = f(x) - |\{j \in [1,i] : f(j) \leq f(x)\}|$). Clearly the predicate $y = f_i^\star(x)$ can be defined by a $\Delta_0(f,c_\theta)$ formula with variables x, y, i, provided that we can show $c_\theta(i, f(x)) \leq f(x)$ for all $i \leq n-2$, $x \in [i+1,n]$. *Intuitively* we should have $c_\theta(i,y) \leq y$ for all y, since as f is one-to-one we would expect that

$$c_\theta(x,v) = |\{j \in [1,x] : f(j) \leq y\}| \tag{11}$$

$$= |\{j \in [1,y] : \exists z \in [1,x](f(z) = j)\}|, \tag{12}$$

in other words, that $c_\theta(x,y)$ is the census function *with respect to* y of the formula $\exists z \in [1,x](f(z) = y)$. We now show that $c_\theta(x,y)$ has the inductive properties defining that census function.

Firstly, $c_\theta(x,0) = 0$ by a trivial induction on x. We will also use induction on x to prove:

$$c_\theta(x,y+1) = c_\theta(x,y) + \begin{cases} 1 & \text{if } \exists z \in [1,x](f(z) = y+1), \\ 0 & \text{if } \exists z \in [1,x](f(z) \neq y+1). \end{cases}$$

This is trivially true for $x = 0$ since both sides are zero. Suppose the equation holds for x. We want to show

$$c_\theta(x+1, y+1) - c_\theta(x+1, y) = \begin{cases} 1 & \text{if } \exists z \in [1, x+1](f(z) = y+1), \\ 0 & \text{if } \exists z \in [1, x+1](f(z) \neq y+1). \end{cases}$$

From the definition of c_θ we know that

$$c_\theta(x+1, y) = c_\theta(x, y) + \begin{cases} 1 & \text{if } f(x+1) \leq y, \\ 0 & \text{if } f(x+1) > y \end{cases}$$

and

$$c_\theta(x+1, y+1) = c_\theta(x, y+1) + \begin{cases} 1 & \text{if } f(x+1) \leq y+1, \\ 0 & \text{if } f(x+1) > y+1. \end{cases}$$

Therefore

$$\begin{aligned} c_\theta(x+1, y+1) - c_\theta(x+1, y) &= c_\theta(x, y+1) \\ &\quad - c_\theta(x, y) + \begin{cases} 1 & \text{if } f(x+1) = y+1, \\ 0 & \text{if } f(x+1) \neq y+1. \end{cases} \end{aligned}$$

and hence by the induction hypothesis,

$$\begin{aligned} & c_\theta(x+1, y+1) - \\ & c_\theta(x+1, y) = \begin{cases} 1 & \text{if } \exists z \in [1, x](f(z) = y+1), \\ 0 & \text{if } \forall z \in [1, x](f(z) \neq y+1). \end{cases} \\ & \qquad + \begin{cases} 1 & \text{if } f(x+1) = y+1, \\ 0 & \text{if } f(x+1) \neq y+1. \end{cases} \end{aligned}$$

Since f is one-to-one the terms on the right cannot both be nonzero, so

$$c_\theta(x+1, y+1) - c_\theta(x+1, y) = \begin{cases} 1 & \text{if } \exists z \in [1, x+1](f(z) = y+1), \\ 0 & \text{if } \forall z \in [1, x+1](f(z) \neq y+1). \end{cases}$$

as required. Thus $c_\theta(x, y)$ is the census function with respect to y for the formula $\exists z \leq x(f(z) = y)$.

That $c_\theta(x, y) \leq y$ for all x, y now follows by a trivial induction on y. More generally we can show that for all x, y_1, y_2,

$$y_1 \leq y_2 \to c_\theta(x, y_2) - c_\theta(x, y_1) \leq y_2 - y_1. \tag{1}$$

(Let $y_2 = y_1 + k$ and use induction on k. Equality holds for $k = 0$, so suppose $c_\theta(x, y_1 + k) - c_\theta(x, y_1) \leq k$. Then there $c_\theta(x, y_1 + k + 1) \leq c_\theta(x, y_1 + k) + 1$ it follows that $c_\theta(x, y_1 + k + 1) - c_\theta(x, y_1) \leq k + 1$).

Similarly,

$$y_1 < y_2 \wedge \forall z \in [1, x](f(z) \neq y_2) \to c_\theta(x, y_2) - c_\theta(x, y_1) \leq y_2 - y_1.$$

Since f is one-to-one, $\forall z \in [1,i](f(z) \neq f(i+1))$, so it follows that for all i, y,

$$f(i+1) > y \rightarrow f(i+1) - c_\theta(i, f(i+1)) > y - c_\theta(i,y).$$

But by (1),

$$f(i+1) \leq y \rightarrow f(i+1) - c_\theta(i, f(i+1)) \leq y - c_\theta(i,y),$$

so

$$f(i+1) \leq y \iff f(i+1) - c_\theta(i, f(i+1)) \leq y - c_\theta(i,y), \qquad (2)$$

Using this we can now show that $f_1^\star$ can be "produced" in the way stated at the beginning, namely that $f_0^\star = f$, and on $[i+2, n]$,

$$f_{i+1}^\star(x) = \begin{cases} f_i^\star(x) - 1 & \text{if } f_i^\star(x) \geq f_i^\star(i+1), \\ f_i^\star(x) & \text{if } f_i^\star(x) < f_i^\star(i+1), \end{cases} \qquad (3)$$

From the way $f_i^\star$ was *actually* defined, this is equivalent to

$$c_\theta(i{+}1, f(x)) = \begin{cases} c_\theta(i, f(x)) + 1 & \text{if } f(x) - c_\theta(i, f(x)) \geq f(i+1) \\ & \qquad\qquad -c_\theta(i, f(i+1)), \\ c_\theta(i, f(x)) & \textit{otherwise.} \end{cases} \qquad (4)$$

But by definition

$$c_\theta(i+1, f(x)) = \begin{cases} c_\theta(i, f(x)) + l & \text{if } f_i(i+1) \leq f(x), \\ c_\theta(i, f(x)) & \text{otherwise} \end{cases}$$

so (4) follows immediately from (2).

To complete the proof we will prove the following hypothesis by induction on $i \leq n-2$:

$IH(i)$: $f_i^\star$ *is one-to-one and its range is contained in* $[1, n-i-1]$.

(Recall that the domain of definition of $f_i^\star$ is $[i+1, n]$).

As $f_0^\star = f$, $IH(0)$ holds, so suppose $IH(i)$ is satisfied. Since $f_i^\star$ is one-to-one, we know from (3) that for all $x \in [i+2, n]$,

$$f_{i+1}^\star(x) = \begin{cases} f_i^\star(x) & \text{if } f_i^\star(x) < f_i^\star(i+1), \\ f_i^\star(x) - 1 & \text{if } f_i^\star(x) > f_i^\star(i+1), \end{cases}$$

so we see immediately that for $x \in [i+2, n]$, either

$$f_{i+1}^\star(x) = f_i^\star(x) \geq 1 \text{ or } f_{i+1}^\star(x) = f_i^\star(x) - 1 \geq f_i^\star(i+1) \geq 1.$$

Also, either

$$f_{i+1}^\star(x) = f_i^\star(x) \leq f_i^\star(i+1) - 1 \leq n - (i+1) - 1,$$

or

$$f_{i+1}^{\star}(x) = f_i^{\star}(x) - 1 \leq n - (i+1) - 1.$$

Thus the range of $f_{i+1}^{\star}$ is contained in $[1, n-(i+1)-1]$.

Now suppose $f_{i+1}^{\star}(x) = f_{i+1}(y)$ for some $x, y \in [i+2, n]$ with $x \neq y$. Since $f_i^{\star}$ is one-to-one we may assume without loss of generality that $f_i^{\star}(x) < f_i^{\star}(y)$. Clearly we must then have

$$f_{i+1}^{\star}(x) = f_i^{\star}(x) < f_i^{\star}(i+1) \leq f_i^{\star}(y) - 1 = f_{i+1}^{\star}(y),$$

so $f_{i+1}^{\star}(x) < f_{i+1}^{\star}(y)$ contrary to the assumption that $f_{i+1}^{\star}(x) = f_{i+1}(y)$. Thus $f_{i+1}^{\star}$ is one-to-one and $IH(i+1)$ holds.

By $IH(n-2)$, $f_{n-2}^{\star}$ has range $\{1\}$ and $f_{n-2}^{\star}$ is one-to-one. But the domain of $f_{n-2}^{\star}$ is $[n-1, n]$ so this is impossible, completing the proof of $PHP(f)$. ⊣

Remark. A similar argument shows that the theorem remains true if $\theta(x, y)$ is taken to be $\exists z \in [1, y](f(z) = x)$, instead of $f(x) \leq y$.

Corollary 5.4. $I\mathcal{E}_{\star}^2 \vdash PHP(\mathcal{E}_{\star}^2)$

Proof. Using essentially the standard argument (see Grzegorczyk [16]) it can be shown that for each bounded formula $\phi(\vec{x})$ involving (function symbols for) $\mathcal{E}^2$ functions there is an $\mathcal{E}^2$ function χ_ϕ, which can be proved in $I\mathcal{E}_{\star}^2$ to be the characteristic function for ϕ. Therefore we will refer to such formulas ϕ as $\mathcal{E}_{\star}^2$ formulas.

Recall that $PHP(\mathcal{E}_{\star}^2)$ states:

$$\forall x \forall y (\forall u_1 \leq y \forall u_2 \leq y \forall v \leq y(\theta(\vec{x}, y, u_1, v) \wedge \theta(\vec{x}, y, u_2, v) \rightarrow u_1 = u_2)$$
$$\wedge \forall u \leq y \exists v \leq y \theta(\vec{x}, y, u, v) \rightarrow \forall v \leq y \exists u \leq y \theta(\vec{x}, y, u, v)),$$

for each formula $\theta(\vec{x}, y, u, v)$.

Working in $I\mathcal{E}_{\star}^2$, suppose $\theta(\vec{x}, y, u, v)$ is an $\mathcal{E}_{\star}^2$ formula and that, contrary to $PHP(\mathcal{E}_{\star}^2)$, for some values of the parameters x, y,

$$g(u) = \min\{v : \ \theta(\vec{x}, y, u, v)\}$$

defines a function $g : [0, y] \rightarrow [0, y]$ which is one-to-one but *not* onto. Let $f(u) = g(u)+1$ and $n = y+1$ so $f : [1, n] \rightarrow [1, n]$ is also one-to-one but not onto. We may suppose (by redefining f at $f^{-1}(n)$ if this exists) that $f : [1, n] \rightarrow [1, n-1]$.

Let $\phi(\vec{x}, y, u, v)$ be an $\mathcal{E}_{\star}^2$ formula defining the predicate $f(u) \leq v$. The census function $c_\theta(\vec{x}, y, u, v)$ for ϕ with respect to u has the $\mathcal{E}^2$ definition:

$$c_\phi(\vec{x}, y, 0, v) = 0$$
$$c_\phi(x, y, u+1, v) = c_\phi(x, y, u, v) + \chi_\phi(x, y, u+1, v)$$

where χ_ϕ is the $\mathcal{E}^2$ characteristic function of ϕ. Thus in $I\mathcal{E}^2_\star$ we can apply induction to $\Delta_0(f, c_\phi)$ formulas and deduce by theorem 5.3 that $PHP(\mathcal{E}^2_\star)$ holds, contradicting our assumption that $PHP(\mathcal{E}^2_\star)$ fails. $\dashv$

In $I\mathcal{E}^2_\star$ we can also prove the following version of the pigeon hole principle provided A, B and f are defined by $\mathcal{E}^2_\star$ formulas (with parameters allowed):

If $B \subset A \subseteq [1, n]$, $B \neq A$, and $f : A \to B$, then f is not one-to-one.

This is because we have $\mathcal{E}^2$ census functions $c_A(x), c_B(x)$ for A and B, so taking c_A^{-1} to be the one-to-one function with range A defined by:

$$c_A^{-1}(y) = \min\{x : \ c_A(x) = y\},$$

we can prove in $I\mathcal{E}^2_\star$ that if f is one-to-one then for some k, m with $k > m$, the function $c_B f c_A^{-1} : [1, k] \to [1, m]$ (formed by composition of functions) is one-to-one (and onto) which is contrary to corollary 5.4.

Similarly the bijection extension principle, namely:

If $A, B \subseteq [1, n]$ and $f : A \leftrightarrow B$ then there is an extension $f^\star$ of f such that $f^\star : [1, n] \leftrightarrow [1, n]$, is available in $I\mathcal{E}^2_\star$ for A, B, f with $\mathcal{E}^2_\star$ definitions. For (working in $I\mathcal{E}^2_\star$) suppose there is some bijection $f : A \leftrightarrow B$ where $A, B \subset [1, n]$. Then (as above) there is some k such that

$$c_B f c_A^{-1} : [1, k] \leftrightarrow [1, k]$$

(Intuitively $k = |A| = |B|$). Denoting the complements of A, B by A', B', it can be proved that the $\mathcal{E}^2$ functions

$$c_{A'}(x) = x - c_A(x), \ c_{B'}(x) = x - c_B(x)$$

have the property that $c_{A'}^{-1} : [1, n-k] \leftrightarrow [1, n] \setminus A$ and $c_{B'}^{-1} : [1, n-k] \leftrightarrow [1, n] \setminus B$, and therefore

$$c_{B'}^{-1} c_{A'} : [1, n] \setminus A \leftrightarrow [1, n] \setminus B.$$

Thus we can take

$$f^\star(x) = \begin{cases} f(x) & \text{if } x \in A, \\ c_{B'}^{-1} c_{A'}(x) & \text{if } x \in [1, u] \setminus A. \end{cases}$$

As indicated in the introduction, a special case of the $\Delta_0 = \mathcal{E}^2_\star$ problem asks whether for each predicate $\theta(\vec{x}, y)$ defined on $\mathbb{N}$ by a Δ_0 formula, the corresponding census function $c_\theta(x, \vec{y})$ has the property that the predicate $z = c_\theta(x, y)$ can also be defined by a Δ_0 formula. One can "strengthen" this question to asking whether there is some *uniform* way of defining the census function. For example is there some *bounded* formula $\phi(x, y, A)$ in a new unary predicate variable A (as well as $+, \cdot, \leq$)

such that for all $A \subseteq \mathbb{N}, x, z \in \mathbb{N}$,

$$z = c_A(x) \iff \phi(x, z, A)?$$

The following corollary gives a conditional answer to an axiomatic version of this problem (in obvious notation).

Corollary 5.5. *If conjecture 5.1 is true then*

$I\Delta_0(A) \not\vdash \phi(0,0,A) \wedge \forall x \forall z(\phi(x,z,A) \rightarrow$
$(\phi(x+1, z+1, A) \iff A(x+1)) \wedge (\phi(x+1, z, A) \iff \neg A(x+1)))$
for any $\Delta_0(A)$ *formula* $\phi(x, z, A)$.

Proof. Suppose the contrary held. Then replacing $A(v)$ in $\phi(x, z, A)$ by the formula $f(v) \leq y$ would yield a $\Delta_0(f)$ definition of $z = c_\theta(x, y)$ (where $\theta(x, y)$ is $f(x) \leq y$), and we would have $I\Delta_0(f) \vdash def(c_\theta)$.

Therefore by theorem 5.3, $I\Delta_0(f) \vdash PHP(f)$ in contradiction to proposition 5.2. ⊣

6 On summing sequences

In this section we will consider the problem of summing a Δ_0 sequence $u_1, u_2, \ldots, u_d$ in a model $M \models I\Delta_0$ with $d \leq [\log_2 a]^n$ for some $a \in M, n \in \mathbb{N}$. But first we will prove the following number theoretic lemma:

Lemma 6.1. *There is some* $C \in \mathbb{N}$ *such that*

$$I\Delta_o \vdash \forall x \exists y (y = \prod_{p \leq C[\log_2 x]} \wedge y > x).$$

Proof. Let $x_0 \in M \models I\Delta_0$. (The constant C produced by the proof will clearly be independent of the choice of M.) In order to prove the lemma for $x = x_0$ we may suppose that $x_0 \in M \setminus \mathbb{N}$, since if $x_0 \in \mathbb{N}$ then the result follows immediately by standard number theory (for example, remark 4.4).

Define an approximate logarithm function $\log^\star$ as in §4 using $a = [\log_2 x_0]^2$ and $n = 2$. Then for $x \leq a$ there is a Δ_0 definition of the sum $\sum_{p \leq x} \log^\star p$ by lemma 2.3, and using the method described in the proof of theorem 4.5 we can prove a $\star$ analogue of the result of remark 4.4, namely that $\sum_{p \leq x} \log^\star p \geq C_0 x$ for all $x \in [2, a]$, where C_0 is some fixed standard rational number. Similarly a $\star$ analogue of inequality (9) of §4 can be proved leading to the result:

$$\textstyle\sum_{p \leq x} \log^\star p < C_1 x \text{ for some fixed } C_1 \in \mathbb{N}, \text{ and all } x \leq a.$$

In both cases the pigeon hole principle is used to establish the relevant inequality (for example, inequality $(9)^\star$ of §4 in the second case) but each of these applications of the pigeon hole principle applies to a Δ_0 function

$$f : [1, m_1] \to [1, m_2] \text{ with } m_1, m_2 \le a^n \le [\log_2 x_0]^{2n}$$

for some $n \in \mathbb{N}$, and can therefore be handled by theorem 5.3, since the census function for the formula $f(x) \le y$ can obviously be defined using lemma 2.3. Similarly all uses of lemma 4.4 in the original arguments can be replaced by uses of lemma 2.3, since taking $a \le [\log_2 x_0]^2$ ensures that the size of the terms to be summed will be sufficiently small.

Now choose $C \in \mathbb{N}$ large enough so that $\sum_{p \le Cx} \log^\star p > (1+\varepsilon)x$ for all $x \le a/C$, where $\varepsilon > 0$ is some standard rational number. In particular this inequality will be satisfied for $x = [\log_2 x_0]$. Thus

$$(1+\varepsilon)[\log_2 x_0] < \sum_{p \le C[\log_2 x_0]} \log^\star p < D \log^\star x_0$$

where $D > C\, C_1/\log 2$, $D \in \mathbb{N}$.

Using the fact that $|\log^\star(u \cdot v) - \log^\star u - log^\star v| \le \frac{A}{[\log_2 a]^2}$ for some constant $A \in \mathbb{N}$, it follows by induction on $i \le C[\log x_0]$ that:

$$\exists y \le x_0^{D+\varepsilon}(y \text{ is squarefree} \wedge \forall p(p \text{ is prime} \to (p|y \iff p \le i))$$

$$\wedge(1+\varepsilon)[\log_2 x_0] - \frac{Ai}{[\log_2 a]^2} \le \log^\star y + \sum_{i<p \le C[\log_2 x_0]} \log^\star p$$

$$< D \log^\star x_0 + \frac{Ai}{[\log_2 a]^2}$$

Taking $i = [\log_2 x_0]$, it follows that a number $y = \prod_{p \le C[\log_2 x]} p$ *exists*, and that $[\log_2 x_0] < \log^\star y$ so $x_0 < y$. ⊣

Remark. Notice that the product version of Sylvester's method (described in §1) does not seem to be applicable, since it would involve numbers of size $[\log_2 x_0]!$ (that is roughly $x_0^{[\log_2[\log_2 x]]}$) which need not exist in M.

We can now prove:

Lemma 6.2. *If $M \models I\Delta_0$ and $u_1, u_2, \ldots, u_d$ is a Δ_0 sequence of elements of M with $d \le [\log_2 a]^n$ and each $u_i \le u$ for some $a, u \in M, n \in \mathbb{N}$, then there is a Δ_0 sequence $\sum_{1 \le i \le j} u_i$, $j = 1, 2, \ldots, d$, such that $\sum_{1 \le i \le 1} u_i = u_1$, and for all $j \in [1, d-1]$,*

$$\sum_{1 \le i \le j+1} u_i = \sum_{1 \le i \le j} u_i + u_{j+1}.$$

Proof. Clearly if the sum exists its value will be $\le [\log_2 a]^n u = v$ (say). By lemma 6.1 there is some $C \in \mathbb{N}$ such that there exists

$y \in M$ with $y = \prod_{p \le C[\log_2 v]} p > v$. The idea is to do the summation *mod p* for each of these primes p. Let $(x)_p$ denote the least number w such that $x \equiv w$ (*mod p*). Then for each $p \le C[\log_2 v]$, the numbers $(u_1)_p, (u_2)_p, \ldots, (u_d)_p$ form a Δ_0 sequence having p as a parameter in its definition. Since each $(u_i)_p \le C[\log_2 v]$ the sum $\sum_{1 \le i \le j} (u_i)_p$ can be defined for $j \in [1, d]$ using lemma 2.3. Now write:

$$z = \sum_{1 \le i \le j} u_i \iff$$

$$z \le u_j \wedge \forall p (p \textit{ is prime} \wedge p | y \to \sum_{1 \le i \le j} (u_i)_p \equiv z (\mod p)).$$

More precisely a unique z satisfying the right hand side of this equivalence can be proved to exist by induction on j, and a similar induction shows

$$\sum_{1 \le i \le j+1} u_i = \sum_{1 \le i \le j} u_i + u_{j+1}. \qquad \dashv$$

The lemma fails if the condition that there be some upper bound u on the elements of the Δ_0 sequence is omitted.

Obviously if the Δ_0 sequence $u_1, u_2, \ldots, u_d$ has no maximum element (that is, if $M = \{x \in M : \exists i \in [1, d](x \le u_i)\}$) then $\sum_{1 \le i \le j} u_i$ cannot exist for all j, since if it did we could prove by a Δ_0 induction on $d - j$ that $u_j \le \sum_{1 \le i \le j} u_i \le \sum_{1 \le i \le d} u_i$ for all $j \in [1, d]$, which is impossible. Therefore the problem amounts to showing that such Δ_0 sequences do exist in some model M (and that it is possible to have $d \le [\log_2 a]^n$ for some $a \in M, n \in \mathbb{N}$). This has been done by:

Paris, J.B.; Kirby, L.A.S., Σ_n – *Collection schemas in arithmetic.* Logic Colloquium '77, North Holland, 1978.

7 The future?

Wilkie's question of whether

$$I\Delta_0 \vdash \textit{there exist arbitrarily large prime numbers}$$

remains an intriguing problem. On the one hand there is the theorem of Alex Wilkie that the existence of arbitrarily large primes cannot be proved by induction on open formulas (together with the usual "algebraic" axioms) which has recently been extended by Zofia Adamowicz (unpublished) to induction on certain formulas with simple quantifier prefixes. On the other hand there is the possibility that the number theoretic properties of $I\Delta_0$ (which as we have seen are quite strong in some ways) may be powerful enough to prove the theorem. However even if this does transpire, the story cannot end there. For consider

Linnik's theorem (see for example Prachar [31]) that there is some constant C such that for ail $a, b \in \mathbb{N}$ with a coprime to b, there exists a prime $p \equiv a(\mod b)$, $p < b^C$. In view of this theorem it is just as sensible to ask whether

$$I\Delta_0 \vdash \text{Dirichlet's theorem}$$

that is, the theorem which states that if a, b are coprime then there exist arbitrarily large primes $p \equiv a(\mod b)$.

Of course there are numerous other examples of number theoretic properties which hold in every structure of the form $\{b \in M : \exists n \in \mathbb{N}(b \leq a^n)\}$, where M is a model of (say) Peano arithmetic and $a \in M$, but for which no proof in $I\Delta_0 \vdash$ is known. *The really important problem would seem to be to find a proof that some "natural" property of this sort cannot be proved in* $I\Delta_0$. Other possible examples include additive basis theorems such as that of Schnirelmann [38] which states that there is some constant C such that every $x \in \mathbb{N}$, $x \geq 2$, can be written as a sum of primes $x = p_1 + p_2 + \ldots + p_m$ with $n \leq C$, or even the analogous result with squarefree numbers instead of primes. (There do exist arbitrarily large squarefree numbers in any model of $I\Delta_0$).

The proof that $I\Delta_0 \vdash$ Sylvester's theorem also suggests some lines of further enquiry. As the reader may have noticed, much of the proof hinged on avoiding explicit use of the bijection extension principle.

Problem. Is the bijection extension principle a conservative extension of $I\Delta_0 + PHP(\Delta_0)$ in the sense that adding new function symbols $f_1, f_2, \ldots$ for the functions whose existence is demanded by the principle, and allowing induction on $\Delta_0(f_1, f_2, \ldots)$ formulas, cannot prove any extra theorems in the original language?

Of course this would follow from a proof of $\Delta_0 = \mathcal{E}^2_\star$ which did not go beyond the capabilities of $I\Delta_0 + PHP(\Delta_0)$, but perhaps this is more than is needed.

The $I\Delta_0 + PHP(\Delta_0)$ proofs of the existence of arbitrarily large primes also seem to make essential use of argument by contradiction, since if we do not assume that $y + 1, y + 2, \ldots, y + x$ have no prime divisor greater than x then we have no way of proving that the function to which we wish to apply the pigeon hole principle is properly defined. This motivates the following question:

Problem. Can classical $I\Delta_0 + PHP(\Delta_0)$ be interpreted in intuitionistic $I\Delta_0 + PHP(\Delta_0)$?

(Where intuitionistic $I\Delta_0 + PHP(\Delta_0)$ is defined in an obvious manner). Of course we have the usual (Gödel) interpretation of classical $I\Delta_0$ in intuitionistic $I\Delta_0$, so this question may have to await the solution of the $I\Delta_0 \vdash PHP(\Delta_0)$ problem. On the other hand it might conceivably provide valuable insight into that problem.

2

Definability properties of the coprimenes predicate

This chapter describes some of the logical properties of the two place predicate $\perp$ defined for natural numbers x, y by:

$$x \perp y \leftrightarrow x \textit{ and } y \textit{ have no common prime divisor.}$$

Throughout $L_{\alpha,\beta,\ldots,\gamma}$ will denote the language consisting of all formulas which can be constructed (according to the usual rules) from the symbols $\forall, \exists, \neg, \wedge, \vee, \rightarrow$; variables $x, y, z, \ldots$ intended to range over the natural numbers $\mathbb{N}$; and the predicates or operations $\alpha, \beta, \ldots, \gamma$. For example, $L_{=,',|}$ is the language with equality ($x = y$) and divisibility ($x|y$) predicates, and the successor operation ($x' = x + 1$).

Given a language L for $\mathbb{N}$, a basic "first question" is whether addition and multiplication are L definable, that is, are there L formulas $\Phi_A(x, y, z)$, $\Phi_M(x, y, z)$ such that for all $x, y, z \in \mathbb{N}$, $\Phi_A(x, y, z) \leftrightarrow x + y = z$ and $\Phi_M(x, y, z) \leftrightarrow x \cdot y = z$? If so, then a second basic question is: what sort of defining formulas are possible? Julia Robinson [35] made a fundamental contribution to this subject when she investigated these questions for various languages, and in particular proved that addition and multiplication are $L_{',|}$ definable. If we regard $'$ and $|$ as weak primitives of an additive and multiplicative kind respectively, then we see that in a certain sense this theorem is stronger than the corresponding results for $L_{\leq,|}, L_{=,+,|}, L_{=,',\cdot}$ and $L_{\leq,\cdot}$. For successor can be defined from $\leq$ which in turn is $L_{=,+}$ definable, but $\leq$ is not $L_{=,'}$ definable, nor is addition $L_{\leq}$ definable. Similarly divisibility is $L_{=,\cdot}$ definable, but multiplication is not $L_{|}$ definable. Thus a natural way to attempt to strengthen the theorem is to replace $|$ by a new "weaker" primitive which is $L_{|}$ definable, but which, by itself, is not strong enough to allow $|$ to be defined. The comprimeness predicate being an obvious candidate, Julia Robinson asked in her paper whether multiplication (and therefore addition) can be defined in $L_{=,+,\cdot}$ or even whether multiplication is $L_{=,+,\perp}$ definable. The present chapter is primarily devoted to these questions and the intermediate problem of defining multiplication in $L_{\leq,\perp}$.

In §1 two simple proofs of the $L_{=,+,\perp}$ definability of multiplication are given. Correspondence with Professor Robinson has revealed that the second of these was discovered independently (and much earlier) by her but has not previously been published. A stronger theorem, namely that addition and multiplication can be defined by *bounded* $L_{\leq,\perp}$ formulas, is proved in §4 by a more complicated argument. In fact, a relation on $\mathbb{N}$ can be defined by a bounded $L_{\leq,+,\cdot}$ formula if and only if it can be defined by a bounded $L_{\textcircled{\le}}$ formula, where $\textcircled{\le}$ is the preordering defined by:

$$x \textcircled{\le} y \leftrightarrow x = y \vee (x \leq y \wedge x \perp y).$$

(A *preordering* is a binary relation which can be extended to a linear ordering. Note that $\textcircled{\le}$ is not transitive.) As the relations on $\mathbb{N}$ which can be defined by bounded $L_{\leq,+,\cdot}$ formulas were shown by Bennett [3] to be precisely the *rudimentary predicates* of computational complexity theory, this theorem yields a new characterisation of these. In particular it follows that every rudimentary set of positive natural numbers is the spectrum of a sentence ϕ of the predicate calculus with a single binary predicate symbol γ (say). (The *spectrum* of a sentence ϕ is taken here to be the set $\{|M| : M \models \phi\}$ comprised of the cardinalities of the domains of all finite normal models M of ϕ). Furthermore it is shown that ϕ may be chosen in such a way that all finite normal models of ϕ are graphs (that is, have γ symmetric and antireflexive) and that similar results hold for partial orderings and quasiorderings.

Concerning the original problem about the properties of $L_{',\perp}$ it is proved in §2 that there cannot be any algorithm for deciding whether arbitrary sentences of this language are true. Also, Julia Robinson's question about the $L_{=,',\perp}$ definability of multiplication turns out to be equivalent to an open problem in number theory. Indeed all of the following statements are equivalent:

(i) $z = x \cdot y$ is $L_{=,',\perp}$ (or $L_{',\perp}$) definable.

(ii) $z = x + y$ is $L_{=,',\perp}$ (or $L_{',\perp}$) definable.

(iii) $x \leq y$ is $L_{=,',\perp}$ (or $L_{',\perp}$) definable.

(iv) $x = y$ is $L_{',\perp}$ definable.

(v) *There is some $k \in \mathbb{N}$ such that every natural determined uniquely by the sequence $S_0, S_1, \ldots, S_k$ of sets of (distinct) prime numbers defined by $S_i = \{p : p|x + i\}$.*

In view of the classical theorem of Størmer [34] which states that for any x there are at most finitely many numbers which produce the same sets S_0, S_1 as x, and the (admittedly limited) numerical evidence which can be extracted from the tables compiled by Lehmer [25], it seems plausible to conjecture that (v) may be true for quite small values of k. This conviction is strengthened by the fact that statement (v) (with $k = 20$ for x sufficiently large) is a consequence of a general conjecture of Hall and Schinzel about the magnitude of the solutions of certain diophantine equations. The truth of (i) - (v) would also follow from an affirmative answer to a question posed recently by Erdös [11], who asked whether there is some k such that every natural number x is determined uniquely by $\bigcup_{i \leq k} S_i$.

0 Two facts about languages for number theory

As a preliminary we will list two "pieces of folklaw" which have consequences for many languages which describe the natural numbers. The first is a corollary of Matiiasevic's negative solution to Hilbert's "tenth problem" the existence of an algorithm for deciding whether a polynomial equation with integer coefficients has a solution in $\mathbb{N}$. (An account of this work may be found in Davis [7]).

An *existential* formula is one of the form $\exists \vec{w} Q(\vec{w})$ where $Q(\vec{w})$ is quantifier free. For any language L, the existential L theory of $\mathbb{N}$ consists of all existential sentences of L which are true of $\mathbb{N}$. A theory T is *decidable* if there is an algorithm for deciding whether an arbitrary sentence is in T.

Proposition 0.1. *If the existential $L_{=,\alpha,\beta,\dots,\gamma}$ theory of $\mathbb{N}$ is decidable then at least one of the predicates $z = x + y$, $z = x \cdot y$ cannot be defined by an existential $L_{=,\alpha,\beta,\dots,\gamma}$ formula.*

Proof. Set $A(x, y, z) \leftrightarrow z = x + y$ and $M(x, y, z) \leftrightarrow z = x \cdot y$. For each $L_{=,+,\cdot,'}$ sentence ψ of the form

$$\exists \vec{w}(P(\vec{w}) = Q(\vec{w})) \qquad (\star)$$

there is an effectively found equivalent $L_{=,A,M}$ sentence of the form $\exists \vec{y} \bigwedge_{h=1}^{m} \psi_h(\vec{y})$ where each $\psi_h(\vec{y})$ is of one of the forms $A(y_i, y_j, y_k)$, $M(y_i, y_j, y_k)$, $y_i = y_j$. (The reason for including this last form is that $x = 1 \leftrightarrow \exists u \exists v(\neg u = v \wedge x \cdot u = u \wedge x \cdot v = v)$).

Now if $\phi_A(x, y, z)$ and $\phi_M(x, y, z)$ are existential $L_{=,\alpha,\beta,\dots,\gamma}$ definitions for $A(x, y, z)$ and $M(x, y, z)$ respectively, then each $\psi_h(\vec{y})$ of the form $A(y_i, y_j, y_k)$ can be replaced by $\phi_A(y_i, y_j, y_k)$, and similarly each $\psi_h(\vec{y})$ of the form $M(y_i, y_j, y_k)$ can be replaced by $\phi_M(y_i, y_j, y_k)$. The resulting sentence is clearly equivalent both to ψ and to an effectively found existential $L_{=,\alpha,\beta,\ldots,\gamma}$ sentence. Thus if the existential $L_{=,\alpha,\beta,\ldots,\gamma}$ theory of $\mathbb{N}$ is decidable then there is a decision procedure for $L_{=,+,\cdot,'}$ sentences of form $(\star)$. But since the terms $P(\vec{w})$, $Q(\vec{w})$ are (essentially) arbitrary polynomials, this contradicts Matijasevič's theorem. $\dashv$

Consider any language with a symbol ($\leq$, say) intended to denote an ordering or preordering. A *quantified variable* $\forall x$ or $\exists x$ is *bounded* by y (with respect to $\leq$) in a formula ψ if it begins a subformula of ψ of the form $\forall x(x \leq y \to \phi)$ or $\exists x(x \leq y \wedge \psi)$. (These will be abbreviated to $\forall x \leq y\psi$ and $\exists x \leq y\psi$.) A *formula* ψ is *bounded* if every quantified variable in ψ is bounded by some *variable*. If only one of the primitive symbols of the language denotes an ordering or preordering, "bounded" will, of course, mean bounded with respect to that symbol.

The basic idea behind the next proposition already appears in the literature, for example in work of Paris and Dimitracopoulos [28]. However the proof will be sketched here as we will later make use of the technique and it is desirable to emphasise one of its subtleties.

Proposition 0.2. *If the predicates $z = x + y$, $z = x \cdot y$ are definable by bounded $L_{\leq,\alpha,\beta,\ldots,\gamma}$ formulas then every bounded $L_{\leq,+,\cdot}$ formula is equivalent to an effectively found bounded $L_{\leq,\alpha,\beta,\ldots,\gamma}$ formula.*

Proof. Taking $A(x, y, z)$, $M(x, y, z)$ as in the previous proof, it suffices to show that each bounded $L_{\leq,+,\cdot}$ formula $\psi(\vec{w})$ is equivalent to an effectively found bounded $L_{\leq,A,M}$ formula $\psi^\star(\vec{w})$.

Note that $\psi(w_1, w_2, \ldots, w_m) \leftrightarrow \bigvee_{i=1}^{m} \theta_i(\vec{w})$ where

$$\theta_i(\vec{w}) \leftrightarrow \bigwedge_{\substack{j=1 \\ j\neq i}}^{m} (w_j \leq w_i) \wedge \psi(w_1, w_2, \ldots, w_m),$$

so we may consider each $\theta_i(\vec{w})$ separately. Clearly all terms occuring in $\theta_i(\vec{w})$ can be bounded by a polynomial in w_i, and indeed (at the expense of treating the first few values of w_i separately) it can be assumed that the bound is w_i^n for some constant $n \in \mathbb{N}$.

Replacing the operations $+$ and $\cdot$ in $\theta_i(\vec{w})$ by the predicates A, M using quantified variables bounded by a new variable c, yields a bounded $L_{\leq,A,M}$ formula $\rho_i(\vec{w}, c)$ such that

$$\forall\vec{w}\forall c \geq w_i^n(\rho_i(\vec{w}, c) \leftrightarrow \theta_i(\vec{w})).$$

But each number less than $([\sqrt{w_i}] + 1)^{2n}$ has a representation $a_1, a_2, \ldots a_{2n}$ to base $[\sqrt{w_i}] + 1$, the digits of which are numbers $a_k \leq [\sqrt{w_i}]$, and it is easy to see that there are bounded $L_{\leq,A,M}$ formulas $\phi_A(\vec{x}, \vec{y}, \vec{z}, w_i)$, $\phi_M(\vec{x}, \vec{y}, \vec{z}, w_i)$, $\phi_{\leq}(\vec{x}, \vec{y}, w_i)$ such that for any three numbers less than $([\sqrt{w_i}]+1)^{2n}$ with representations $x_1, x_2, \ldots, x_{2n}$, $y_1, y_2, \ldots, y_{2n}$, $z_1, z_2, \ldots, z_{2n}$,

$$\phi_A(x_1, x_2, \ldots, x_{2n}, y_1, y_2, \ldots, y_{2n}, z_1, z_2, \ldots, z_{2n})$$
$$\leftrightarrow x_1, x_2, \ldots, x_{2n} + y_1, y_2, \ldots, y_{2n} = z_1, z_2, \ldots, z_{2n}, \text{ etc..}$$

Also (and this point is *very* important) there is a bounded formula $\eta(x, \vec{a}, w_i)$ such that

$$\eta(x, \vec{a}, w_i) \leftrightarrow x \leq w_i \wedge x = a_1 a_2 \ldots a_{2n}.$$

By quantifying over $2n$-tuples $a_1 a_2 \ldots a_{2n}$ in place of numbers less than $([\sqrt{w_i}]+1)^n$ and using ϕ_A, ϕ_M, $\phi_\leq$ instead of A, M, $\leq$ where appropriate, we can now obviously construct a bounded $L_{\leq,A,M}$ formula equivalent to $\rho_i(\vec{w}, ([\sqrt{w_i}]+1)^{2n}-1)$, and therefore equivalent to $\theta_i(\vec{w})$. ⊣

1 Addition and coprimeness

We now give two different ways of defining multiplication by an $L_{=,+,\perp}$ formula. The observation underlying the first of these is:

Lemma 1.1. *Given a sequence of numbers $v_0 < v_1 < \ldots < v_n$ there exist numbers x, c such that for all y, if $x + v_0 \leq y \leq x + v_n$ then*

$$y \perp c \leftrightarrow \exists i(y = x + v_i).$$

Proof. Let $w_0, w_1, \ldots, w_k$, be the numbers between v_0 and v which are not v_i's, and choose a sequence of $k+1$ distinct primes p_j such that $p_j \not| \prod_{i \leq n}(v_i - w_j)$. By the Chinese remainder theorem there exists a number $x \equiv -w_j (\text{mod} p_j)$ for all j. But it is easy to check that if $c = \prod_{j \leq k} p_j$ then x, c have the required property. ⊣

Theorem 1.2. *$z = x \cdot y$ is $L_{=,+,\perp}$ definable.*

Proof. Since $z = x \cdot y$ is $L_{=,+,|}$ definable (see Tarski [45] or Robinson [35]) it suffices to show that the divisibility predicate $a|b$ is $L_{=,+,\perp}$ definable.

It is claimed that a/b if and only if there exist x, c such that

(i) $x \perp c \wedge (x+b) \perp c$

(ii) If $x \leq u < u_\star \leq x + b$ and $u, u_\star$ are consecutive numbers coprime to c then $u_\star = u + a$.

To see this note that (i) and (ii) describes the pattern:

$$\underbrace{\overbrace{x = u_0 \;\star\!\!-\!\!-\!\!-\!\!\star\; u_1}^{a}\;\overbrace{\star\!\!-\!\!-\!\!-\!\!\star\; u_2}^{a}\;\cdots\;\overbrace{u_{n-1}\;\star\!\!-\!\!-\!\!-\!\!\star\; u_n = x+b}^{a}}_{b} \qquad u_i - u_{i-1} = a$$

where the numbers between x and $x+b$ which are coprime to c are $x = u_0 < u_1 < u_2 < \ldots < u_n = x+b$. Obviously if such a pattern exists then $a|b$. Conversely, if $a|b$ then lemma (1.1) ensures that such a pattern does exist.

Since $\leq$ is $L_{=,+}$ definable (i) and (ii) can clearly be expressed by $L_{=,+,\perp}$ formulas so the theorem follows. $\dashv$

Analysing the above argument shows that

$$x \cdot y = z \leftrightarrow \exists w \psi(x, y, z, w)$$

for some bounded $L_{\leq,+,\perp}$ formula $\psi(x, y, z, w)$. Our second proof of theorem 1.2 will do better than this.

Theorem 1.3. *$z = x \cdot y$ is definable by a bounded $L_{\leq,+,\perp}$ formula.*

Proof. (Also found by J. Robinson):

Note first that

$$x = 0 \leftrightarrow \forall y \leq x (y \geq x)$$
$$x = 1 \leftrightarrow \neg x = 0 \wedge \forall y \leq x (y = 0 \vee y = x).$$

x is a prime $\leftrightarrow \neg x = 0 \wedge \neg x = 1 \wedge \forall y \leq x (y = 0 \vee y \perp x \vee y = x)$.

By Schnirelmann's theorem [38] there is a natural number n such that every number $x \geq 2$ is the sum of fewer than n primes. Thus the product of any two numbers $x, y \geq 2$ can be expressed in the form

$$x \cdot y = (p_1 + \ldots + p_m) \cdot (q_1 + \ldots + q_k) = \sum_{i,j} p_i \cdot q_j$$

where $x = p_1 + \ldots + p_m$, $y = q_1 + \ldots + q_k$, the p_i's and q_j's are primes, and $m, k < n$. Since the sum $\sum_{i,j} p_i \cdot q_j$ has fewer than n^2 terms the theorem will follow if it can be shown that there is a bounded $L_{\leq,\perp}$, formula which defines $z = p \cdot q$ for all primes p, q. But $z = p \cdot q \leftrightarrow$ *z is the least number greater than both p and q such that*

$$\neg z \perp p \wedge \neg z \perp q \wedge \forall y \leq z (y \perp p \wedge y \perp q \rightarrow y \perp z). \qquad \dashv$$

Recalling proposition (0.2) and noting that $x \perp y$ can be defined by a bounded $L_{\leq,+,\cdot}$ formula we have:

Corollary 1.3. *A relation on $\mathbb{N}$ can be defined by a bounded $L_{\leq,+,\cdot}$ formula if and only if it can be defined by a bounded $L_{\leq,+,\perp}$ formula.*

In the course of settling Hilbert's tenth problem Matijasevič proved the following theorem. (See, for example, Davis [7]).

Proposition 1.4. (Matijasevič). *Every bounded $L_{\leq,+,\cdot}$ formula is equivalent to an existential $L_{\leq,+,\cdot}$ formula.*

However the analogue of this fails for $L_{\leq,+,\perp}$ since $z = x \cdot y$ cannot be defined by an existential $L_{\leq,+,\perp}$ formula. This follows from a result of Bel'tyukov [2] and Lipshitz [27]. (See also Wilkie [49]).

Proposition 1.5. (Bel'tyukov, Lipshitz). *The existential $L_{\leq,=,',|}$ theory of $\mathbb{N}$ is decidable.*

Corollary 1.6. *The existential $L_{\leq,+,\perp}$ theory of $\mathbb{N}$ is decidable.*

Proof. We transform the decision problem for existential $L_{\leq,+,\perp}$ sentences into a decision problem for certain existential $L_{\leq,+,',|}$ sentences. Given any existential $L_{\leq,+,\perp}$ sentence ϕ, there is an effectively found equivalent sentence of the form $\exists \vec{y} \bigvee_{i=1}^{k} \psi_i(\vec{y})$, where each $\psi_i(\vec{y})$ is a finite conjunction of atomic and negated atomic formulas. If all unnegated atomic formulas of the form $s \perp t$ (s and t terms) are replaced by $\exists a(s|a \wedge t|a')$ and all negated atomic formulas of the form $\neg s \perp t$ are replaced by $\exists a \exists b(\neg a|b \wedge a|s \wedge a|t)$, then the resulting $L_{\leq,+,',|}$, sentence (equivalent to ϕ) is clearly equivalent to an effectively found existential $L_{\leq,+,',|}$, sentence. $\dashv$

Applying proposition (0.1) gives immediately:

Corollary 1.7. *$z = x \cdot y$ cannot be defined by an existential $L_{\leq,+,\perp}$ formula.*

2 Successor and coprimeness

We begin by defining some notation which will be used throughout this section.

P is the set of all prime numbers. (p and q will always denote elements of P).

$\nu(x)$ is the number of distinct prime divisors of x.

$x \approx y \leftrightarrow \nu(x) = \nu(y)$

$x \sim y \leftrightarrow x$ and y have the same (distinct) prime divisors.

$x \underset{n}{\sim} y \leftrightarrow x \sim y \wedge x+1 \sim y+1 \wedge \ldots \wedge x+n \sim y+n.$

(Thus $x = y \to \ldots \to x \underset{n+1}{\sim} y \to x \underset{n}{\sim} y \to \ldots \to x \underset{1}{\sim} y \to x \sim y \to x \approx y$).

For $x \neq 0$, $\overline{x}$, $\overline{x}^n$ and $\overline{\overline{x}}$ will denote, respectively, the $\sim$, $\underset{n}{\sim}$ and $\approx$ equivalence classes of which x is a representative, and we will write $\overline{x}|y$ or $\overline{x}|\overline{y}$ if $\forall p \in P(p|x \to p|y)$. Let $\overline{P} = \{\overline{p} : p \in P\}$. Elements of $\overline{P}$ will be denoted by the capital letters U, V, W, X, Y, Z. Thus, for example,

$$Z = \{p^m : m \in \mathbb{N} \setminus \{0\}\}$$

for some prime p, and $Z|x \leftrightarrow \overline{p}|x \leftrightarrow p|x$.

The reason for introducing these concepts is that with the exception of $\approx$, $\overline{\overline{x}}$, $\nu(x)$ and P, they are all easy to define in $L_{\prime,\perp}$. In particular we have:

(i) $x \sim y \leftrightarrow \forall z(z \perp x \leftrightarrow z \perp y)$

(ii) $\overline{x}|y \leftrightarrow \forall z(z \perp y \rightarrow z \perp x)$

(iii) $\overline{x} \in \overline{p} \leftrightarrow \forall y \forall z(y \perp z \rightarrow y \perp x \vee z \perp x) \wedge \exists y(\neg y \perp x)$

(iv) $x = 0 \leftrightarrow \forall y(\neg y \perp x \vee \forall z(y \perp z))$.

Consequently when constructing "$L_{\prime,\perp}$ formulas" we may use the expressions on the left of (i), (ii), (iii), terms $0, 1, 2, \ldots$ (that is, $0, 0', 0'', \ldots$), the capital letter notation for elements of $\overline{P}$, and, for each fixed $n \in \mathbb{N}$, the predicates $x \underset{n}{\sim} y$ and $\overline{x}^n = \overline{y}^n$.

Let $\overline{\overline{\mathbb{N}}} = \{\overline{\overline{x}} : x \in \mathbb{N} \setminus \{0\}\}$, $\overline{\mathbb{N}} = \{\overline{x} : x \in \mathbb{N} \setminus \{0\}\}$, $\overline{\mathbb{N}}^n = \{\overline{x}^n : x \in \mathbb{N} \setminus \{0\}\}$. Notice that the map $\nu^{-1} : \mathbb{N} \rightarrow \overline{\overline{\mathbb{N}}}$ defined by $\nu^{-1}(\nu(x)) = \overline{\overline{x}}$ is a bijection and may thus be used to induce operations $+, \cdot$ on $\overline{\overline{\mathbb{N}}}$ so that

$$< \mathbb{N}, =, +, \cdot > \cong < \overline{\overline{\mathbb{N}}}, =, +, \cdot >$$

under ν^{-1}. This is the basic observation underlying the proofs in this section. It will be shown below that when considered as predicates on $\mathbb{N}$, the expressions $\overline{\overline{x}} = \overline{\overline{y}}$, $\overline{\overline{x}} + \overline{\overline{y}} = \overline{\overline{z}}$ and $\overline{\overline{x}} \cdot \overline{\overline{y}} = \overline{\overline{z}}$ can be defined by $L_{\prime,\perp}$ formulas, and hence there is an effective way of transforming any sentence ψ of $L_{=,+,\cdot}$ into an equivalent $L_{\prime,\perp}$ sentence $\psi^\star$ asserting $< \overline{\overline{\mathbb{N}}}, =, +, \cdot > \models \psi$. Thus we will be able to deduce the undecidability of the $L_{\prime,\perp}$ theory of $\mathbb{N}$ from the known undecidability of the $L_{=,+,\cdot}$ theory. Also, the existence of these formulas reduces the problem of the $L_{\prime,\perp}$ definability of $z = x \cdot y$, etc., to the question of whether the isomorphism ν^{-1} (or equivalently the predicate $y = \nu(x)$) is $L_{\prime,\perp}$ definable.

To ease notational problems we will adopt the following conventions:

(i) If $\theta(y_1, \ldots, y_n)$ is an $L_{=,+,\cdot}$ definable predicate and $\overline{\overline{x}}_1, \ldots, \overline{\overline{x}}_n \in \overline{\overline{\mathbb{N}}}$ then $\overline{\overline{\mathbb{N}}} \models \theta(\overline{\overline{x}}_1, \ldots, \overline{\overline{x}}_n)$ will be used as an abbreviation for

$$< \overline{\overline{\mathbb{N}}}, =, +, \cdot > \models \theta(y_1, \ldots y_n)[\overline{\overline{x}}_1, \ldots, \overline{\overline{x}}_n].$$

(ii) If $y = \gamma(x_1, \ldots, x_n)$ is an $L_{=,+,\cdot}$ definable function, then $\gamma(\overline{\overline{x}}_1, \ldots, \overline{\overline{x}}_n)$ will denote the element $\overline{\overline{y}} \in \overline{\overline{\mathbb{N}}}$ which satisfies $\overline{\overline{\mathbb{N}}} \models \overline{\overline{y}} = \gamma(\overline{\overline{x}}_1, \ldots, \overline{\overline{x}}_n)$. (For example, $\overline{\overline{x}} + 1$ is the element $\overline{\overline{y}}$ for which $\overline{\overline{\mathbb{N}}} \models \overline{\overline{y}} = \overline{\overline{x}} + 1$).

(iii) We will write $\overline{\overline{x}} \leq \overline{\overline{y}}$ instead of $\overline{\overline{\mathbb{N}}} \models \overline{\overline{x}} \leq \overline{\overline{y}}$.

Lemma 2.1. *$\overline{\overline{x}} \leq \overline{\overline{y}}$ is $L_{\prime,\perp}$ definable.*

Proof. Write $x \mathbin{\textcircled{\le}} y$ if there exists u such that for

$$f_u = \{< X, Y >: \ X|x \wedge Y|y \wedge X \neq \overline{2} \wedge Y \neq \overline{2} \\ \wedge \exists z(\overline{z} \in \overline{p} \wedge \overline{z}|u \wedge X|z' \wedge Y|z')\}$$

one of the following is satisfied:

(i) $\neg\overline{2}|x \vee (\overline{2}|x \wedge \overline{2}|y)$ *and* f_u *is the graph of a one-to-one function mapping* $\{X: \ X|x \wedge X \neq \overline{2}\}$ *into* $\{Y: \ Y|y \wedge Y \neq \overline{2}\}$.

(ii) $\overline{2}|x \wedge \neg\overline{2}|y$ *and* f_u *is the graph of a one-to-one function mapping* $\{X: \ X|x \wedge X \neq \overline{2}\}$ *onto a proper subset of* $\{Y: \ Y|y\}$.

(iii) $\neg\overline{2}|x \wedge \overline{2}|y$ *and* f_u *is the graph of a one-to-one function with range contained in* $\{Y: \ Y|y \wedge Y \neq \overline{2}\}$ *and domain equal to* $\{X: \ X|x\} \setminus \{W\}$ *for some* $W|x$.

Obviously $x \mathbin{\textcircled{\le}} y$ can be defined by an $L_{',\perp}$ formula. Also, for any numbers x, y if $x \mathbin{\textcircled{\le}} y$ then $x \neq 0$, $y \neq 0$ and there is a one-to-one map from $\{\overline{q}: \ q|x\}$ into $\{\overline{q}: \ q|y\}$, so $\nu(x) \leq \nu(y)$ and therefore $\overline{\overline{x}} \leq \overline{\overline{y}}$.

On the other hand, if $\overline{\overline{x}} \leq \overline{\overline{y}}$ then $x \neq 0$, $y \neq 0$ and $\nu(x) \leq \nu(y)$, so it is possible to choose a one-to-one map taking each prime $q|x$ to a prime $r_q|y$, and having the additional property that if $q|y$ then $r_q = q$. Now choose for each $q|x$ a prime p_q satisfying

$$\begin{aligned} p_q &\equiv -1(\text{mod} q) \\ &\equiv -1(\text{mod} r_q) \\ &\equiv 1(mods) \end{aligned}$$

for all other primes $s|x \cdot y$. Such primes p_q exist by virtue of the Chinese remainder theorem (which is used to combine these congruences into a single congruence of the form $p_q \equiv a(\mod b)$ with $a \perp b$) and Dirichlet's theorem that if $a \perp b$ then there are infinitely many primes $p \equiv a(\text{mod} b)$. (A proof of Dirichlet's theorem may be found in Shapiro [40]). Noting that $p \equiv 1(\text{mod} s) \Rightarrow p^m \not\equiv -1(\text{mod} s)$ for all m and all $s > 2$, it is easy to check that $u = \prod_{q|x} p_q$ has the property required to make $x \mathbin{\textcircled{\le}} y$.

Thus, $\overline{\overline{x}} \leq \overline{\overline{y}} \leftrightarrow x \mathbin{\textcircled{\le}} y$. ⊣

Lemma 2.2. $\overline{\overline{x}} + \overline{\overline{y}} = \overline{\overline{z}}$ *and* $\overline{\overline{x}} \cdot \overline{\overline{y}} = \overline{\overline{z}}$ *are* $L_{',\perp}$ *definable.*

Proof. We simply extend the trick of coding one-to-one mappings of prime divisors developed in lemma 2.1 to define $+, \cdot$ on $\overline{\overline{N}}$ in the same way that one defines $+, \cdot$ for cardinal arithmetic in set theory. For example, $\overline{\overline{x}} \cdot \overline{\overline{y}} = \overline{\overline{z}}$ if and only if there exist *some* representatives x, y, z of these equivalence classes such that

$$\{< X, Y, Z >: \ X|x \wedge Y|y \wedge Z|z \wedge \exists w(\overline{w} = Z \wedge X|w' \wedge Y|w')\}$$

is the graph of a one-to-one function from $\{< X, Y >: X|x \wedge Y|y\}$ onto $\{Z : Z|z\}$.

In fact, since $\overline{\overline{x}} \leq \overline{\overline{y}}$ is $L_{',\perp}$ definable, and addition is $L_{\leq,\cdot}$ definable, it is sufficient just to define $\overline{\overline{x}} \cdot \overline{\overline{y}} = \overline{\overline{z}}$. ⊣

It follows immediately from these lemmas that for each $L_{=,+,\cdot}$ formula $\psi(x_1, \ldots, x_n)$ there is an effectively found $L_{',\perp}$ formula $\psi^\star(x_1, \ldots, x_n)$ such that

$$\psi^\star(x_1, \ldots, x_n) \leftrightarrow \overline{\overline{\mathbb{N}}} \models \psi(\overline{\overline{x}}_1, \ldots, \overline{\overline{x}}_n)$$

In particular, since $< \overline{\overline{\mathbb{N}}}, =, +, \cdot > \cong < \mathbb{N}, =, +, \cdot >$, we have thus proved:

Theorem 2.3. *For every sentence ψ of $L_{',\perp}$ there is an effectively found sentence $\psi^\star$ of $L_{',\perp}$ such that $\psi \leftrightarrow \psi^\star$.*

Corollary 2.4. *The $L_{',\perp}$ theory of $\mathbb{N}$ is undecidable.*

It seems prudent at this stage to eliminate any further "coding" difficulties by proving a general coding lemma (lemma 2.6) which will be more than adequate for the purposes of this paper. As a preliminary we define $\delta : \overline{P} \to \overline{\mathbb{N}}$ by taking $\delta(\overline{p}) = \overline{p+1}$ for all primes p, and prove:

Lemma 2.5. *$\overline{y} = \delta(Z)$ is $L_{',\perp}$ definable. (More precisely, the predicate $\overline{x} \in \overline{P} \wedge \overline{y} = \delta(\overline{x})$ is $L_{',\perp}$ definable).*

Proof. Let $\zeta(x, y, Z)$ be the formula:

$$Z = \overline{x'} \wedge \overline{y} = \overline{x''} \wedge \forall w (Z = \overline{w'} \to \overline{x}|w \wedge (\overline{w}|x \to \overline{x''}|w'')).$$

It is claimed that $\overline{y} = \delta(Z) \leftrightarrow \exists x \zeta(x, y, Z)$.

To check this, suppose first that for some Z, $\zeta(x, y, Z)$ is satisfied by $< x, y > = < x_1, y_1 >, < x_2, y_2 >$. Then (taking $w = x_2$ in $\zeta(x_1, y_1, Z)$) we see that $\overline{x}_1 | x_2$ and (taking $w = x_1$ in $\zeta(x_2, y_2, Z)$) it follows that $\overline{x_2''}|x_1''$. Similarly, $\overline{x_1''}|x_2''$ so $\overline{x_1''} = \overline{x_2''}$, that is, $y_1 \sim y_2$.

Thus if $\exists x \zeta(x, y_1, Z)$ holds then $\{y : \exists x \zeta(x, y, Z)\} = \overline{y}_1$, so it suffices to show $\exists x \zeta(x, p+1, Z)$ for the prime p with

$$Z = \overline{p} = \{p^m : m > 0\}.$$

But $x = p - 1$ has this property. For using the fact that $\overline{w'} = Z$ if and only if $w = p^m - 1$ for some $m > 0$, it can be seen that $\zeta(p-1, p+1, Z)$ is equivalent to $\forall m > 0 \left(\overline{p-1}|p^m - 1 \wedge (\overline{p^m - 1}|p - 1 \to \overline{p+1}|p^m + 1)\right)$, and as $p - 1 | p^m - 1$, this in turn is equivalent to the formula:

$$\forall m > 0 (\forall q \in P(p^m \equiv 1 \ (mod\ q) \to p \equiv 1 \ (mod\ q)) \to$$
$$\forall q \in P(p \equiv -1 \ (mod\ q) \to p^m \equiv -1 \ (mod\ q))),$$

which is true. ⊣

Lemma 2.6. *For each* $n \in \mathbb{N}$ *there is an* $L_{\prime,\perp}$ *formula* $\theta_n(x_1, \ldots, x_n, u)$ *with the property that for every finite set* $R \subseteq \overline{\mathbb{N}} \times \overline{\mathbb{N}} \times \ldots \times \overline{\mathbb{N}}$ *(n copies of* $\overline{\mathbb{N}}$*) there is some* $\overline{u} \in \overline{\mathbb{N}}$ *such that for every* $v \in \overline{u}$*,*

$$\forall x_1 \ldots \forall x_n (< \overline{x}_1, ..., \overline{x}_n > \in R \leftrightarrow \theta_n(x_1, \ldots, x_n, v)).$$

Proof. It suffices to prove the lemma for $n = 2$ since the general case then follows by induction, defining θ_{n+1} by:

$$\theta_{n+1}(x_1, \ldots, x_{n+1}, u) \leftrightarrow \exists v(\theta_n(x_1, \ldots, x_n, v) \wedge \theta_2(x_{n+1}, v, u)).$$

The proof has two parts. The first describes how to "decode" an arbitrary $\overline{u} \in \overline{\mathbb{N}}$ to obtain a finite set $R_{\overline{u}} \subset \overline{\mathbb{N}} \times \overline{\mathbb{N}}$ in such a way that when considered as a predicate in x_1, x_2, u, the expression $< \overline{x}_1, \overline{x}_2 > \in R_{\overline{u}}$ is $L_{\prime,\perp}$ definable. In the second part it is shown that if $R \subset \overline{\mathbb{N}} \times \overline{\mathbb{N}}$ is finite then a code $\overline{u}$ with $R_{\overline{u}} = R$ can always be found.

As an aid to decoding $\overline{u}$, define $\overline{a} \cap \overline{b}$ for $\overline{a}, \overline{b} \in \overline{\mathbb{N}}$ by $\overline{c} = \overline{a} \cap \overline{b} \leftrightarrow \forall Z \in \overline{P}(Z|c \leftrightarrow Z|\overline{a} \wedge Z|\overline{b})$. (If a, b are regarded as denoting subsets of $\overline{p}$ then $\overline{a} \cap \overline{b}$ denotes their intersection). For $\overline{z} \in \overline{\mathbb{N}}$, $W \in \overline{P}$ define $(\overline{z})_W \in \overline{\mathbb{N}}$ by:

$$\forall Z \in \overline{P}(Z|(\overline{z})_W \leftrightarrow (Z \neq \overline{2} \wedge Z \neq W \wedge Z|\overline{z}) \vee (Z = \overline{2} \wedge W|\overline{z})).$$

($(\overline{z})_W$ corresponds to the subset of $\overline{P}$ obtained by replacing W by $\overline{2}$ in the set denoted by $\overline{z}$, if W occurs there, and by suppressing any occurrence of $\overline{2}$ otherwise).

We also introduce an $L_{\prime,\perp}$ definable partial ordering $\overset{\star}{<}$ on $\overline{P}$ by taking

$$Z_1 \overset{\star}{<} Z_2 \leftrightarrow \overline{\overline{\delta(Z_1)}} < \overline{\overline{\delta(Z_2)}},$$

where $\overline{\overline{a}} = \overline{\overline{\delta(Z)}} \leftrightarrow \exists b(a \approx b \wedge \overline{b} = \delta(Z))$. In other words, if $Z_1 = \overline{p}_1$, $Z_2 = \overline{p}_2$ with p_1, p_2 prime, then $Z_1 \overset{\star}{<} Z_2$ if and only if $p_1 + 1$ has fewer (distinct) prime divisors than $p_2 + 1$.

The first step in decoding an arbitrary $\overline{u} \in \overline{\mathbb{N}}$ is to check whether there exist $W, X \in \overline{p}$ and $\overline{a}, \overline{b} \in \overline{\mathbb{N}}$ such that $\overline{a} = \overline{b}$,

$$\{Z \in \overline{P} : Z|\overline{u}\} = \{W, X\} \cup \{(Y \in \overline{P} : Y|\overline{a}\} \cup \{Z \in \overline{P} : Z|\overline{b}\},$$

and this set is totally ordered by $\overset{\star}{<}$ with $W \overset{\star}{<} X \overset{\star}{<} Y \overset{\star}{<} Z$ for all $Y|\overline{a}$, $Z|\overline{b}$. If not, put $R_{\overline{u}} = \phi$. If so, then regard W as a surrogate for $\overline{2}$ (this is needed because, as in the proof of lemma 2.1, it is necessary to treat $\overline{2}$ as a special case) and define $R_{\overline{u}}$ by: $< \overline{x}_1, \overline{x}_2 > \in R_{\overline{u}} \leftrightarrow$ *there exist* $Y|\overline{a}$, $Z|\overline{b}$ *such that*

$$Y|\delta(Z) \wedge \overline{x}_1 = (\delta(X) \cap \delta(Y)))_W \wedge \overline{x}_2 = (\delta(X) \cap \delta(Z))_W.$$

In view of lemma 2.5 it is obvious that there is an $L_{\prime,\perp}$ formula

$\theta_2(x_1, x_2, u)$ such that

$$\forall x_1 \forall x_2 (< \overline{x}_1, \overline{x}_2 > \in R_{\overline{u}} \leftrightarrow \theta_2(x_1, x_2, u)).$$

To complete the proof it suffices to show that for every finite $R \subset \overline{\mathbb{N}} \times \overline{\mathbb{N}}$ there is some $\overline{u} \in \overline{\mathbb{N}}$ such that $R = R_{\overline{u}}$. If elements of $\overline{\mathbb{N}}$ are thought of as denoting sets of primes then any such R may be regarded as a binary relation between subsets of some finite set $S \subset P$. Choose a prime $w \notin S$ as the surrogate for 2 and form a new relation $R^\star$ by replacing 2 by w at each occurrence of 2 in the subsets related by R. Thus $R^\star$ is a relation between "subsets" $\overline{y}$ of $S^\star = S \cup \{w\} \setminus \{2\}$. Enumerate all of these as $\overline{y}_1, \overline{y}_2, \ldots, \overline{y}_m$. Now using the Chinese remainder theorem and Dirichlet's theorem choose, *in turn*, odd primes

$$r < s_1 < s_2 < \ldots < s_m < t_1 < t_2 < \ldots < t_m$$

satisfying:

(i) $r \equiv -1 \ (mod\ q)$ for all primes $q \in S^\star$.

(ii) For $i = 1, 2, \ldots, m$,

$$\begin{aligned} s_i &\equiv -1 \ (mod\ q) \text{ for all primes } q | y_i, \\ &\not\equiv -1 \ (mod\ q) \text{ for all other primes } q | r+1. \end{aligned}$$

(iii) For $j = 1, 2, \ldots, m$,

$$\begin{aligned} t_j &\equiv -1 \ (mod\ q) \text{ for all primes } q | y_j, \\ &\not\equiv -1 \ (mod\ q) \text{ for all other primes } q | r+1, \\ &\equiv -1 (\text{mod} s_i) \text{ for all } i \text{ such that } < \overline{y}_i, \overline{y}_j > \in R, \\ &\not\equiv -1 \ (mod\ s_i) \text{ for all } i \text{ such that } < \overline{y}_i, \overline{y}_j > \notin R. \end{aligned}$$

(iv) $\overline{w} \overset{\star}{<} \overline{r} \overset{\star}{<} \overline{s}_1 \overset{\star}{<} \overline{s}_2 \overset{\star}{<} \ldots \overset{\star}{<} \overline{s}_m \overset{\star}{<} \overline{t}_1 \overset{\star}{<} \overline{t}_2 \overset{\star}{<} \ldots \overset{\star}{<} \overline{t}_m.$ ⊣

This last condition can be satisfied by adding, *at each stage*, sufficiently many extra congruences of the form:

$$r, s_i, \text{ or } t_j \equiv -1 \ (mod\ q), q \text{ prime}.$$

Taking $W = \overline{w}$, $X = \overline{r}$, $a = \prod_{1 \le i \le m} s_i$, $b = \prod_{1 \le j \le m} t_j$ it is now easy to check that $R = R_{\overline{u}}$ for $u = w \cdot r \cdot a \cdot b$.

Lemma 2.7. *For each $n \in \mathbb{N}$ there is an $L_{',\perp}$ formula $\chi_n(x, y, u)$ with the property that for any finite set $R \subset \overline{\overline{\mathbb{N}}} \times \overline{\mathbb{N}}^n$ there is some $u \in \mathbb{N}$ such that $\forall x \forall v (< \overline{\overline{x}}, \overline{y}^n > \in R \leftrightarrow \chi_n(x, y, u))$.*

Proof. With θ_{n+2} as in lemma 2.6, take $\chi_n(x, y, u)$ to be the formula:

$$\exists z \exists w (x \approx z \wedge \bigwedge_{i=0}^{n} (w_i \sim y + i) \wedge \theta_{n+2}(z, w_0, w_1, \ldots, w_n, u)).$$

This works because of the obvious correspondence between $\overline{y}^n$ and $< \overline{y}$, $\overline{y+1}, \ldots, \overline{y+n} >$ and the fact that $\approx$ equivalence classes are unions of $\sim$ equivalence classes so $\overline{x}$ can be used as a "representative" for $\overline{\overline{x}}$. $\dashv$

To complete the preparations for our final theorem about definability in $L_{',\perp}$ we require a simple number theoretic lemma.

Lemma 2.8. *If* $a \underset{k}{\sim} b$ *and* $a \neq b$ *then*

$$\prod_{p \leq k+1} p \leq \prod_{p|a(a+1)\ldots(a+k)} p \leq |a-b|.$$

Proof. In fact $\leq$ can be replaced by $|$. If $p \leq k+1$ then p divides at least one of the $k+1$ numbers $a, a+1, \ldots, a+k$. Also if $p|a+i$ for some $i \leq k$, then since $a \underset{k}{\sim} b$, it follows that $p|b+i$ and hence $p\,\big||a-b|$. $\dashv$

Theorem 2.9. *The following statements are equivalent:*

(i) $x = y$ *is* $L_{',\perp}$ *definable.*

(ii) $z = x \cdot y$ *is* $L_{',\perp}$ *(or* $L_{=,',\perp}$*) definable.*

(iii) $z = x + y$ *is* $L_{',\perp}$ *(or* $L_{=,',\perp}$*) definable.*

(iv) $z = x \leq y$ *is* $L_{',\perp}$ *(or* $L_{=,',\perp}$*) definable.*

(v) $\exists k \forall x \forall y (x \underset{k}{\sim} k \rightarrow x = y)$.

Proof. Obviously (v) $\rightarrow$ (i), while (ii) $\rightarrow$ (iii) follows from the $L_{=,',\cdot}$ definability of addition (Robinson [35]), and (iii) $\rightarrow$ (iv) is a consequence of the $L_{=,+}$ definability of $\leq$. Thus it will suffice to show (i) $\rightarrow$ (v) and (iv) $\rightarrow$ (v) $\rightarrow$ (ii).

(i) $\rightarrow$ (v): Suppose $\phi(x, y)$ is an $L_{',\perp}$ definition of equality, and let k be the largest number for which the variable y followed by k successor symbols occurs in $\phi(x, y)$. Since all atomic subformulas of $\phi(x, y)$ (and in particular all those containing y) are of the form $u^{''\cdots'} \perp v^{''\cdots'}$ (u and v variables) it is clear that

$$\forall y \forall z (y \underset{k}{\sim} z \rightarrow \forall x (\phi(x, y) \leftrightarrow \phi(x, z))).$$

Replacing ϕ by $=$ in this formula we see that

$$\forall y \forall z (y \underset{k}{\sim} z \rightarrow y = z).$$

(iv) $\rightarrow$ (v): This can be proved by an extension of the argument used to show (i) $\rightarrow$ (v), however it seems easier to consider a nonstandard model $< M, =, +, \cdot >$ of the $L_{=,',\cdot}$ theory of $\mathbb{N}$. Suppose (v) fails. Then

$$M \models \forall k \exists x \exists y (x < y \wedge x \underset{k}{\sim} y).$$

Take a nonstandard number $c \in M$ and let $a, b \in M$ have the property that

$$M \models a < b \wedge a \underset{2c}{\sim} b.$$

By lemma 2.8, $b - a$ is nonstandard. This allows us to define a bijection $f : M \to M$ by:

$$\left.\begin{array}{l} f(a+c+j) = b+c+j \\ f(b+c+j) = a+c+j \end{array}\right\} \text{for all standard (positive, negative, or zero) integers } j,$$

$$f(d) = d, \text{otherwise.}$$

$$\left.\begin{array}{l} f(a+c+j) = b+c+j \\ f(b+c+j) = a+c+j \end{array}\right\} \text{for all standard (positive, negative, or zero) integers } j,$$

$$f(d) = d, \text{otherwise.}$$

Clearly f is an automorphism of $< M, =, ', \perp >$ but f does not preserve $\leq$, so this relation cannot be $L_{=,',\perp}$ definable.

(v) $\to$ **(ii)**: Suppose that $\forall y \forall z (y \underset{k}{\sim} z \to y = z)$. Then for every $y \neq 0$ the $\underset{k}{\sim}$ equivalence class $\overline{y}^k$ contains only the single element y, so by lemma 2.7 the $L_{',\perp}$ formula $\chi_k(x, y, u)$ has the property that for any finite set $R \subset \overline{\overline{\mathbb{N}}} \times (N \setminus \{0\})$ there is some $u \in \mathbb{N}$ such that

$$\forall x \forall y (< \overline{\overline{x}}, y > \in R \leftrightarrow \chi_k(x, y, u)).$$

It follows that the isomorphism v^{-1} or more precisely, the predicate $\overline{\overline{x}} = v^{-1}(y)$ (which is the same as $v(x) = y$) is $L_{',\perp}$ definable. For $\overline{\overline{x}} = v^{-1}(y)$ if and only if $x = 1 \wedge y = 0$ or there is some finite set $R \subseteq \overline{\overline{\mathbb{N}}} \times \mathbb{N}$ such that

(1) *R is the graph of a one-to-one function,*
(2) $< \overline{\overline{2}}, 1 > \in R$ (that is, $< v^{-1}(1), 1 > \in R$)
(3) $\forall \overline{\overline{z}} < \overline{\overline{x}} \forall w (< \overline{\overline{z}}, w > \in R \to < \overline{\overline{z}}', w' > \in R)$
(4) $< \overline{\overline{x}}, y > \in R$, Appealing to lemma 2.2 where necessary, it is clear that the existence of an R satisfying (1)–(4) can be asserted by an $L_{',\perp}$ formula.

But

$$z = x \cdot y \leftrightarrow \exists u \exists c \exists w (\overline{\overline{u}} = v^{-1}(x) \wedge \overline{\overline{v}} = v^{-1}(y) \wedge \overline{\overline{w}} = v^{-1}(z) \wedge \overline{\overline{w}} = \overline{\overline{u}} \cdot \overline{\overline{v}})$$

so multiplication is L definable. ⊣

This last theorem shows that settling the question of the $L_{',\perp}$ definability of multiplication is tantamount to establishing the truth or otherwise of the number theoretic problem (v). It seems unreasonable to expect that the latter problem (and therefore the former) can be handled without the use of nontrivial number theoretic methods.

Postscript. *Recently the author has learned that Denis Richard has independently obtained a different proof of the undecidability of the* $L_{=,',\perp}$ *theory of* $\mathbb{N}$ *(cf. corollary 2.4). His proof is based on the following number theoretic fact:*

Proposition (Birkhoff and Vandiver [4]). *Suppose* $a > b$, $a \perp b$, *and* $n > 2$. *Then except for the single case* $a = 2$, $b = 1$, $n = 6$, *there is always some prime* $p | a^n - b^n$ *such that* $p \not| a^m - b^m$ *for all* $m \in [1, n-1]$.

This also has the interesting consequence that all numbers x of the form $x = p^n - 1$, $p \in P$ have the property:

$$\forall y(x \underset{2}{\sim} y \to x = y).$$

3 A number theoretic interlude[1]

It is timely now to survey what is currently known (for k fixed) about how large $|a - b|$ must be if $a \underset{k}{\sim} b$ with $a \neq b$. If a, b have this property then from lemma 2.8,

$$|a - b| \geq \prod_{p|a(a+1)\ldots(a+k)} p. \tag{1}$$

which suggests seeking good lower bounds on the product

$$\prod_{p|a(a+1)\ldots(a+k)} p. \tag{2}$$

As will be seen below, it follows from a conjecture of Hall and Schinzel that this approach should actually succeed in proving the conjecture

$$\exists k \forall a \forall b (a \underset{k}{\sim} b \to a = b).$$

For $k > 1$, the bounds on (2) discussed here are all due to Langevin [22], [23] and [24], however as it seems desirable to collect the relevant details in one place, their derivation will be described.

The following lemmas will enable us to concentrate on bounding $\prod_{p|a(a+1)} p$ and $\prod_{p|a(a+2)} p$. For any $A \subset \mathbb{N}$ let $S(A) = \{p : \exists a \in A(p|a)\}$.

Lemma 3.1. *Suppose* $A = \bigcup_i A_i \subseteq \{a, a+1, \ldots, a+k\}$ *where the* A_i*'s are all pairwise disjoint. Then*

$$\prod_{p \in S(A)} p \geq k^{-k+1} \prod_i \prod_{p \in S(A_i)} p.$$

[1] The only subsequent use made of the material in this section is in the proof of the $L_{\leq,\perp}$ definability of addition and multiplication given in §4. An alternative proof of that theorem which does not rely on the present section nay be found in appendix I.

Proof. Any prime p divides at most $[k/p]+1$ of the numbers $a, a+1, \ldots, a+k$ and at least $[k/p]$ of the numbers $2, 3, 4, \ldots, k$, so

$$\prod_i \prod_{p \in S(A_i)} p \leq \left(\prod_{p \leq k} p^{[k/p]}\right) \cdot \left(\prod_{p \in S(A)} p\right)$$
$$\leq k! \prod_{p \in S(A)} p \leq k^{k-1} \prod_{p \in S(A)} p. \qquad \dashv$$

Lemma 3.2. *Suppose $g(a)$ is a nondecreasing function such that for all a,*

$$\prod_{p|a(a+1)} p > (g(a))^2 \text{ and } \prod_{p|a(a+2)} p > (g(a))^2$$

Then for all $k \geq 1$,

$$\prod_{p|a(a+1)\ldots(a+k)} p > k^{-k}(g(a))^{k+1}, \tag{3}$$

and for $k = 2$ or $k \geq 4$, $i \leq k$,

$$\prod_{p|\frac{a(a+1)\ldots(a+k)}{(a+i)}} p > k^{-k}(g(a))^k. \tag{4}$$

Proof. Since 2 is the only prime which can divide more than one of the numbers a, $a+1$, $a+2$,

$$2\left(\prod_{p|a(a+1)(a+2)} p\right)^2 \geq \left(\prod_{p|a(a+1)} p\right) \cdot \left(\prod_{p|a(a+1)(a+2)} p\right) \cdot \left(\prod_{p|a(a+2)} p\right),$$

so

$$\prod_{p|a(a+1)(a+2)} p > \frac{1}{\sqrt{2}}(g(a))^3 \geq \frac{1}{\sqrt{k}}(g(a))^3 \text{ for } k > 1.$$

Writing

$$\{a, a+1, \ldots, a+k\} = \begin{cases} \bigcup_{i \leq m}\{a+2i, a+2i+1\}, & \textit{if } \mathrm{k = 2m+1}, \\ \bigcup_{i<m} \{a+2i, a+2i+1\} \\ \cup \quad \{a+2m, a+2m+1, \\ \quad a+2m+2\} & \textit{if } \mathrm{k = 2m+2}, \end{cases}$$

(3) follows by lemma 3.1. The proof of (4) is similar. $\dashv$

For $n = 1, 2$, bounds on $\prod_{p|a(a+n)} p$ are given by theorems 7 and 8 of Lehmer [25]. These yield:

$$\sum_{p|a(a+n)} \log p > \frac{2}{3}(1 - o(1)) \log\log a, \text{ for } n = 1, 2. \tag{5}$$

(Lehmer's work also gives the corresponding result for $n = 4$). $o(1)$ here denotes some real valued function of a which tends to zero as $a \to \infty$. Similarly, $o_k(1)$ will denote some function of a, k which, for each fixed k, tends to zero as $a \to \infty$.

By lemma 3.2 we obtain for all $k \geq 1$, Langevin's result (6):

$$\sum_{p|a(a+1)\dots(a+k)} \log p > \frac{k+1}{3}(1 - o_k(1)) \log\log a - k \log k, \tag{6}$$

$$> \frac{k+1}{3}(1 - o_k(1)) \log\log a,$$

and for $i \leq k$ with $k = 2$ or $k \geq 4$, the analogous inequality (7):

$$\sum_{p|\frac{a(a+1)\dots(a+k)}{(a+i)}} \log p > \frac{k}{3}(1 - o_k(1)) \log\log a - k \log k, \tag{7}$$

$$> \frac{k}{3}(1 - o_k(1)) \log\log a,$$

Combining (1) and (6) gives:

Proposition 3.3. *For every real number $\varepsilon > 0$, and every $k \geq 1$, there exists $a_k(\varepsilon)$ such that for all $a > a_k(\varepsilon)$ and all $b \neq a$,*

$$a \underset{k}{\sim} b \to \log|a - b| > \frac{k+1}{3}(1 - \varepsilon) \log\log a.$$

Lehmer deduced his inequalities (5) from properties of the Pell equation:

$$x^2 - dy^2 = 1 \tag{8}$$

using Størmer's observation that if $x = 2a + 1$ then for some y, and some $d| \prod_{p|a(a+1)} p$,

$$(x-1)(x+1) = 4a(a+1) = dy^2$$

so x, y, d satisfy (8). (For $\prod_{p|a(a+2)} p$ take $x = a + 1$). An alternative approach used by Langevin is to consider the Mordell equation:

$$y^2 = x^3 + m. \tag{9}$$

Let v be the least number such that for some w,

$$4a(a+n) = (2a+n)^2 - n^2 = vw^3 \tag{10}$$

Then $v \mid \left(2\prod_{p|a(a+n)} p\right)^2$ and multiplying (10) by v^2 yields:

$$(v(2a+n))^2 = (vw)^3 + (vn)^2. \tag{11}$$

That is, $x = vw$, $y = v(2a+n)$ is a solution of (9) with $m = (vn)^2 \neq 0$.

Theorem 3.4. *If there exist positive real valued constants C_0, D such that*

$$\forall y \forall z \neq 0(\exists x(y^2 - z^2 = x^3) \to z^D \geq C_0 y) \tag{12}$$

then there is a constant a_0 such that for $k > 4D - 5$,

$$\forall a > a_0 \forall b(a \underset{k}{\sim} b \to a = b).$$

Proof. Applying (12) to (11) shows $v^{D-1} \geq C_0(2a+n)n^{-D}$, so

$$\prod_{p|a(a+n)} p \geq \frac{1}{2}(C_n(2a+n))^{1/(2D-2)}$$

where $C_n > 0$ depends only on n. Taking $n = 1, 2$ it follows by lemma 3.2 that for each $k \geq 1$ there is a constant $C_k^\star > 0$ such that

$$\prod_{p|a(a+1)\dots(a+k)} p \geq C_k^\star a^{(k+1)/(4D-4)}$$

(This inequality is implicit in Langevin [24]). Thus there is some a_o such that for all $a > a_0$,

$$\prod_{p|a(a+1)\dots(a+k)} p > a \text{ for } k = [4D-5]+1.$$

But if $a \underset{k}{\sim} b$ with $b \neq a$ and $a > a_0$ then we may sunpose without loss of generality that $a > b$ and therefore $a - b > \prod_{p|a(a+1)\dots(a+k)} p > a$ which is impossible. $\dashv$

The hypothesis of theorem 3.4 (with $D = 6$) is a consequence of the following conjecture attributed by Langevin [23] to Hall and Schinzel:

Conjecture (Hall and Schinzel). *For any pair of natural numbers $n > 1$, $m > 1$, there is a constant C, such that for all integers x, y with $x^n \neq y^m$,*

$$|x^n - y^m|^6 \geq C \max\{|x|^n, |y|^m\}.$$

Thus it follows from this conjecture (with $n = 3$, $m = 2$) that if a is sufficiently large then a is determined uniquely by the sequence $S_0, S_1, \dots, S_{20}$ where $S_i = \{p : p|a+i\}$. Although at present the hypothesis of theorem 3.4 remains unproved, Stark [42] has shown that for every real number $\varepsilon > 0$ there is a constant $C(\varepsilon)$ such that all integer solutions of (9) with $m \neq O$ satisfy

$$\log|m| + C(\varepsilon) > (1-\varepsilon)\log\log\max\{|x|, |y|\}.$$

Applying this to (11) yields:

$$2\log(vn) + C(\varepsilon) > (1-\varepsilon)\log\log(v(2a+n)),$$

from which it follows that

$$\sum_{p|a(a+n)} \log p > \frac{1}{4}(1-o(1))\log\log(2a+n) - \frac{1}{2}\log n. \qquad (13)$$

As noted by Langevin [23], if $b > a$ and we take $n = b - a$ then (13) becomes

$$\sum_{p|ab} \log p + \frac{1}{2}\log(b-a) > \frac{1}{4}(1-o(1))\log\log(a+b)$$

and hence by (1),

$$a \sim b \to \log(b-a) > \frac{1}{6}(1-o(1))\log\log(a+b),$$

giving a bound for the case $k = 0$ which was omitted from proposition 3.3. (The existence of a constant $C > 0$ such that for $a \neq b$,

$$a \sim b \to \log|b-a| > C\log\log(a+b)$$

was also proved by Erdös and Shorey [12] using a different method).

(13) also provides an alternative starting point for obtaining weaker versions (with $\frac{1}{3}$ replaced by $\frac{1}{8}$) of the inequalities (6) and (7). This version of (7) is, nevertheless, quite suitable for use in the proofs of the definability results given in the next section. However the reader is warned that the $L_{\leq,\perp}$ defining formulas for addition and multiplication produced in this way may be *much* longer, not only because larger values of k must be used, but more importantly because it would seem a must be taken much larger in order to reduce the $o(1)$ term to a reasonable size if one starts with Stark's work rather than Lehmer's.

4 Order and coprimeness

We now turn to the problem of giving $L_{\leq,\perp}$ defintions of addition and multiplication. The basic idea is to exploit the fact that if $|a-b|$ is *small* with respect to a suitably chosen function of a and k then we certainly do have:

$$a \underset{k}{\sim} b \to a = b.$$

For technical reasons it will be desirable to exclude certain "trouble-some" primes from the discussion by using a slightly weaker equivalence relation than $\sim$, namely that defined between integers a and b (for a given q) by:

$$a \overset{q}{\sim} b \leftrightarrow \{p:\ p|a \wedge p \nmid q\} = \{p:\ p|b \wedge p \nmid q\}.$$

We will also want an analogue of $\underset{k}{\sim}$ which applies to arithmetic progressions other than $a, a+1, \ldots, a+k$. (To make it easier to obtain bounded formulas this will be defined using $a, a-d, a-2d, \ldots, a-kd$ rather than $a, a+d, a+2d, ..., a+kd$). Put:

$$a \underset{k,d,q}{\sim} b \leftrightarrow a \overset{q}{\sim} b \wedge a-d \overset{q}{\sim} b-d \wedge \ldots \wedge a-kd \overset{q}{\sim} b-kd.$$

Lemma 4.1. *For $k=2$ or $k \geq 4$ and any real number $\varepsilon > 0$, there is some $a_k(\varepsilon)$ such that for all primes q, and all a, b with $a > a_k(\varepsilon)$,*

$$a \underset{k,l,q}{\sim} b \rightarrow \log|a-b| > \frac{k}{3}(1-\varepsilon)\log\log a \vee a = b.$$

Proof. Suppose $a \underset{k,l,q}{\sim} b$ with q prime. The argument used to prove lemma 2.8 shows that if $a \neq b$ then

$$\begin{aligned}\log|a-b| &\geq \sum_{\substack{p \mid a(a-1)\ldots(a-k)\\ p \neq q}} \log p,\\ &\geq \sum_{p \mid \frac{a(a-1)\ldots(a-k)}{(a-i)}} \log p - \log k\end{aligned}$$

for some $i \leq k$, since either $q \leq k$ or q divides at most one of the numbers $a+i$ $i \leq k$. The result now follows by inequality (7) of §3. ⊣

On several occassions we will have cause to use Chebychev's weak version of the Prime Number theorem (see for example, Hardy and Wright [18] or §4 of chapter 1 of this thesis).

Proposition 4.2 (Chebychev). *There is a constant $A > 0$ such that for all $x \geq 2$, $\sum_{p \leq x} \log p > Ax$.*

Lemma 4.3. *There is a constant B such that for all primes q and all a, b, k with $k > B\log a$,*

$$a \geq b \wedge a \underset{k,l,q}{\sim} b \rightarrow a = b.$$

Proof. Suppose that $a > b$ and $a \underset{k,l,q}{\sim} b$ for $k = [B\log a] + 1$. Then

$$\begin{aligned}\log(a-b) &\geq \sum_{\substack{p \mid a(a-1)\ldots(a-k)\\ p \neq q}} \log p,\\ &\geq \sum_{p \leq k} \log p - \log k \qquad \text{(as in lemma 2.8)}\\ &> Ak - \log k > (A/2)k\end{aligned}$$

for B (and therefore k) sufficiently large, where A is the constant in proposition 4.2. Thus

$$(A/2)k < \log(a - b) < \log a < B^{-1}k$$

which is false for $B > 2/A$. ⊣

Remark. The statement of lemma 4.3 remains true if $\log a$ is replaced by $(\log a) \cdot (\log\log a)^{-1}$ (The extra details may be found in Langevin [24], theorem 11).

Let $\overline{w}^{k,d,q}$ denote the $\underset{k,d,q}{\sim}$ equivalence class to which w belongs.

Lemma 4.4. *There exists $k \in \mathbb{N}$ such that for any prime q and any number y, if u is the smallest number for which the elemnts of $W = \{w : u \leq w \leq y\}$ belong to distinct $\underset{k,l,q}{\sim}$ equivalence classes, and $\leq$ denotes the ordering induced on $Y = \{\overline{w}^{k,l,q} : w \in M\}$ by the ordering $\leq$ on W, then the triple $< Y, \leq, q >$ determines y uniquely, that is, no other number y can give rise to the same triple in this way.*

Proof. We will assume throughout that y is sufficiently large, since if the lemma is true for all $y > y_0$ with $k = k_0$, then it is true for all y with $k = \max\{k_0, y_0\}$. Under this assumption, it will be shown that the lemma holds for $k \geq 4$.

Suppose $y_1 < y_2$ both give rise to $< Y, \leq, q >$ for some prime q, with u_1, W_2 and u_2, W_2 corresponding to y_1, y_2 respectively. Since $|W_1| = |Y| = |W_2|$, it follows that $y_1 - u_1 = y_2 - u_2 = m$ (say) and thus

$$W_1 = \{y_1, y_1 - 1, \ldots, y_1 - m\}, W_2 = \{y_2, y_2 - 1, \ldots, y_2 - m\}$$

As the orderings on W_1, W_2 induce the same ordering on Y, we see that for each $i \leq m$,

$$\overline{y_1 - i}^{k,l,q} = \overline{y_2 - i}^{k,l,q}$$

so $y_1 - i \overset{q}{\sim} y_2 - i$. Hence $y_1 \underset{m,l,q}{\sim} y_2$.

To obtain a contradiction by lemma 4.3 we need only show $m > B \log y_2$. This is the case for $k \geq 4$, since if $m < y_2$ then there is some $w \in W_2$ such that $w \underset{k,l,q}{\sim} u_2 - 1$, and therefore by lemma 4.1 (with $\varepsilon < \frac{1}{4}$),

$$u_2 - 1 < w - (\log w)^{4(l-\varepsilon)/3} \leq y_2 - (\log y_2)^{4(l-\varepsilon)/3}$$

so $m = y_2 - u_2 > (\log y_2)^{4(l-\varepsilon)/3} - 1 > B \log y_2$. ⊣

We are now ready to investigate the definability properties of $L_{\leq,\perp}$. Clearly (cf. §1 and §2) each of the following predicates can be defined by bounded $L_{\leq,\perp}$ formula:

$$x \text{ is squarefree}$$
$$x \in P \text{ (that is, } x \text{ is prime)}$$
$$x \text{ is squarefree} \wedge x|y.$$

Observe also that if $d|a$ and $kd \leq a$ then $\{a, a-d, a-2d, \ldots, a-kd\}$ consists of the largest $k+1$ elements of $\{w : w \leq a \wedge d|w\}$. Using this fact it is easily seen that for each *fixed* $k \in \mathbb{N}$ there is a bounded $L_{\leq,\perp}$ formula which defines the predicate:

$$d \text{ is squarefree} \wedge d|a \wedge a \underset{k,d,q}{\sim} b.$$

Similarly, for each fixed $k \in \mathbb{N}$ we can construct: a bounded $L_{\leq,\perp}$ formula which defines the predicate:

$$d \text{ is squarefree} \wedge d|a \wedge f_{k,d,q}(a) =< v_0, v_1, \ldots, v_k >,$$

where $f_{k,d,q}$, is the mapping which "identifies" the equivalence $\overline{a}^{k,d,q}$ class with the $k+1$ tuple

$$< v_0, v_1, \ldots, v_k >=< \prod_{\substack{p|a \\ p\nmid q}} p, \prod_{\substack{p|a-d \\ p\nmid q}} p, \ldots \prod_{\substack{p|a-kd \\ p\nmid q}} p >$$

the elements of which are either squarefree or zero (adopting the convention that $\prod_{p\nmid 0} p = 0$).

Lemma 4.5. *The predicate* $q \in P \wedge q \cdot y = z$ *can by a bounded* $L_{\leq,\perp}$ *formula.*

Proof. Fix k and suppose the prime $q|z$. Consider the triple $< Z, \leq, q >$ constructed by taking $u_\star$ to be the least multiple of q such that all elements of

$$W_\star = \{w : u_\star \leq w \leq z \wedge q|w\} = \{z, z-q, z-2q, \ldots, u_\star\}$$

belong to distinct $\underset{k,q,q}{\sim}$ equivalence classes, setting

$$Z = \{\overline{w}^{k,q,q} : w \in W_\star\},$$

and letting $\leq$ denote the ordering induced on Z by the natural ordering $\leq$ on $W_\star$.

Now if $z = q \cdot y$ and $< Y, \leq, q >$ is the triple constructed from y as described in lemma 4.4, then identifying Y and Z with the corresponding sets of $k+1$ tuples of squarefree (or zero) numbers, it is clear that $< Y, \leq, q >=< Z, \leq, q >$. Also for each fixed k, there is obviously a bounded $L_{\leq,\perp}$ formula $\psi_k(q, y, z)$ which holds if and only if $q \in P$, $q|z$, and the triples $< Y, \leq, q >$, $< Z, \leq, q >$ constructed from y and z in the ways described above are equal.

But by lemma 4.4, y is uniquely determined by $< Y, \leq, q >$ provided k was chosen sufficiently large. Thus for some $k \in \mathbb{N}$,

$$\forall q \forall y \forall z(\psi_k(q, y, z) \leftrightarrow q \in P \wedge q \cdot y = z). \qquad \dashv$$

Recall that our goal is to prove:

Theorem 4.6. *$z = x \cdot y$ and $z = x + y$ can be defined by bounded $L_{\leq,\perp}$ formulas.*

In view of lemma 4.5 this is easily seen to be equivalent to the following theorem (the proof of which does not depend on the previous lemmas).

Let $\alpha(q, y, z)$ be the predicate defined by

$$\alpha(q, y, z) \leftrightarrow q \in P \wedge q \cdot y = z.$$

Theorem 4.7. *$z = x \cdot y$ and $z = x + y$ can be defined by bounded $L_{\leq,\alpha}$ formulas.*

Theorem 4.7 will be proved via several lemmas. The first two of these show that there are $L_{\leq,\alpha}$ formulas having all variables bounded by a parameter w, which define $x \cdot y = z$ and $x + y = z$ provided w is somewhat larger than z, and which cannot be satisfied by any triple x, y, z not possessing the properties $x \cdot y = z$, $x + y = z$ respectively.

Lemma 4.8. *There is a bounded $L_{\leq,\alpha}$ formula $\phi_M(x, y, z, w)$ and constants $m, C > 0$ such that*

$$\forall w \forall x \forall y \forall z(\phi_M(w, x, y, z) \rightarrow x \cdot y = z) \qquad (1)$$

$$\forall w \forall x \forall y \forall z(x \cdot y = z \wedge z \leq w \cdot (C \log w)^{-2m} \rightarrow \phi_M(x, y, z, w)). \qquad (2)$$

Proof. Clearly there is a bounded $L_{\leq,\alpha}$ formula $\phi_M(x, y, z, w)$ which is satisfied if and only if either

$$((x = 0 \vee y = 0) \wedge z = 0) \vee (x = 1 \wedge y = 1 \wedge z = 1)$$

or $x \leq z$, $y \leq z$ and there exist squarefree numbers $d_1, d_2 \leq w$ such that

(i) $z < d_1 \wedge z < d_2 \wedge d_1 \perp d_2$

(ii) *For each prime $p | d_1 d_2$ there exist primes q, r with*

$$q \cdot x \leq w \wedge q \cdot x \equiv \ (mod\ p),$$
$$r \cdot y \leq w \wedge r \cdot y \equiv 1\ (mod\ p),$$
$$\textit{and } q \cdot r \cdot z \leq w \wedge q \cdot r \cdot z \equiv 1\ (mod\ p).$$

For example, for p, q, r primes,

$$q \cdot r \cdot z \leq w \wedge q \cdot r \cdot z = 1\ (mod\ p) \leftrightarrow$$
$$\exists u \leq w \exists v \leq w(p | u - 1 \wedge \alpha(q, v, u) \wedge \alpha(r, z, v)).$$

Now suppose $\phi_M(x, y, z, w)$ holds. Then for each prime $p|d_1 d_2$,

$$q \cdot r \cdot x \cdot y \equiv 1 \equiv q \cdot r \cdot z \ (mod\ p)$$

and therefore $x \cdot y \equiv z\ (mod\ p)$ (since $q \cdot r \perp p$). Hence $x \cdot y \equiv z\ (mod\ d_1 d_2)$ and as both sides of this congruence are less than $d_1 d_2$ it follows that $x \cdot y = z$. This establishes (1).

On the other hand, suppose $x \cdot y = z > 1$. By proposition 4.2, if C is sufficiently large then

$$Cz^3 \log z < \prod_{p < C \log z} p \text{ and hence } Cz^2 \log z < \prod_{\substack{p < C \log z \\ p \perp z}} p.$$

Therefore it is possible to choose coprime squarefree numbers d_1, d_2 such that $d_1 d_2 \perp z$, $z < d_i < Cz \log z$ (for $i = 1, 2$) and all primes $p|d_1 d_2$ satisfy $p < C \log z$.

But by a theorem of Linnik (see, for example, Prachar [30]) there is a constant $m > 1$ such that for any pair of numbers a, b with $a \perp b$ there is a prime $q < b^m$ satisfying $q \equiv a\ (mod\ b)$. Thus for each $p|d_1 d_2$ there exist primes $q, r < p^m < (C \log z)^m$ such that $q \equiv x^{-1}\ (mod\ p)$ and $r \equiv y^{-1}\ (mod\ p)$ where x^{-1} and y^{-1} denote natural numbers satisfying $x^{-1} \cdot x \equiv 1\ (mod\ p)$, $y^{-1} \cdot y \equiv 1\ (mod\ p)$, and hence $q \cdot x \equiv 1\ (mod\ p)$, $r \cdot y = 1\ (mod\ p)$, and $q \cdot r \cdot z \equiv 1\ (mod\ p)$.

Clearly if $z \leq w \cdot (C \log w)^{-2m}$ then (provided C is large enough) $q \cdot x$, $r \cdot y$ and $q \cdot r \cdot z$ are all less than w, so $\phi_M(x, y, z, w)$ is satisfied. $\dashv$

Readers familiar with Robinson [35] will notice that the proof of the above lemma is an adaption of the proof of the $L_{','|}$ definability of $x \cdot y = z$ found in that paper. The essential difference is that here q and r are chosen to be primes so that only the predicate $\alpha(q, y, z)$ is required. A second difference is that to get a good upper bound on the least w which will work, the definition has been used modulo p for a sufficient number of small primes p. (Although it does simplify matters later, this second step is not absolutely necessary - without it theorems 4.6 and 4.7 could still be proved by iterating the construction used in lemma 4.10 below).

Notice also that $z = x \cdot y \leftrightarrow \exists w \phi_M(x, y, z, w)$. In view of Julia Robinson's observation that $z = x + y$ is $L_{=,',\cdot}$ definable since for $z \neq 0$,

$$x + y = z \leftrightarrow (x \cdot z + 1) \cdot (y \cdot z + 1) = (x \cdot y + 1) \cdot z^2 + 1,$$

we have thus already proved (by a somewhat longer route than is necessary) that both $z = x \cdot y$ and $z = x + y$ are $L_{\leq,\alpha}$ and hence $L_{\leq,\perp}$ definable. In order to proceed towards obtaining *bounded* definitions, we now adapt the $L_{=,',\cdot}$, definition of addition just mentioned using much the same techniques as for the previous lemma.

Lemma 4.9. *There is a bounded $L_{\leq,\alpha}$ formula $\phi_A(x,y,z,w)$ and constants C and m (the same as in lemma 4.8) such that*

$$\forall w \forall x \forall y \forall z(\phi_A(x,y,z,w) \to x+y=z) \tag{3}$$

$$\forall w \forall x \forall y \forall z(x+y=z \wedge z \leq w \cdot (C \log w)^{-2m-1} \to \phi_A(x,y,z,w)). \tag{4}$$

Proof. Note first that there is a bounded $L_{\leq,\alpha}$ formula $\psi(s,t,p,w)$ such that

$$\forall p \in P \forall s \forall t \forall w(\psi(s,t,p,w) \to s \equiv t \pmod p) \tag{5}$$

and if $s,t \leq w \cdot (C \log w)^{-2m-1}$ then for all primes $p \leq C \log w$,

$$s \equiv t \pmod p \to \psi(s,t,p,w). \tag{6}$$

For if p is prime then

$$s = t \pmod p \leftrightarrow (p|s \wedge p|t) \vee \exists v < p(p|v \cdot s - 1 \wedge p|v \cdot t - 1),$$

and replacing the two occurrences of $\cdot$ in this second formula by the use of $\phi_<(\ldots,w)$ yields a predicate $\psi(s,t,p,w)$ which can obviously be defined by a bounded $L_{\leq,\alpha}$ formula. Applying lemma 4.8 shows that $\psi(s,t,p,w)$ has properties (5) and (6).

Now consider the expression:

$$\exists a < p \exists b < p \exists c < p(a \equiv x \pmod p) \wedge b \equiv y \pmod p) \wedge c \equiv z \pmod p)$$
$$\wedge (a \cdot c + 1) \cdot (b \cdot c + 1) \equiv (a \cdot b + 1) \cdot c \cdot c + 1 \pmod p)). \tag{7}$$

Let $\theta(x,y,z,p,w)$ be the bounded $L_{\leq,\alpha}$ formula obtained from this by replacing $\cdot$ by $\phi_M(\ldots,w)$, and $\equiv$ by $\psi(\ldots,w)$. Take $\phi_A(x,y,z,w)$ to be a bounded $L_{\leq,\alpha}$ formula which asserts that there exists a squarefree number $d < w$ such that $d \perp z$, $2 \cdot x < d$, $2 \cdot y < d$, $z < d$, and $\forall p \in P(p|d \to \theta(x,y,z,p,w))$.

If $\phi_A(x,y,z,w)$ holds then for every prime $p|q$,

$$(x \cdot z + 1) \cdot (y \cdot z + 1) \equiv (x \cdot y + 1) \cdot z^2 + 1 \pmod p$$

and therefore since $z \perp p$, $x + y \equiv z \pmod p$. Hence $x + y \equiv z \pmod d$ and thus $x + y = z$.

On the other hand if $x + y = z$ then (7) is clearly satisfied for all primes p, and thus $\theta(x,y,z,p,w)$ holds for all w,p with $p < C \log z$, $z < w(C \log w)^{-2m-1}$. Also (as in the proof of the previous lemma) there is a squarefree number $d| \prod_{\substack{p < C \log z \\ p \perp z}} p$ such that $2 \cdot z < d < w$, so $\phi_A(x,y,z,w)$ is satisfied. ⊣

Let $\underset{w}{+}$, $\underset{w}{\cdot}$ denote the partial functions (extendible to $+,.$) defined by

$$x \underset{w}{+} y = z \leftrightarrow \phi_A(x, y, z, w)$$
$$x \underset{w}{\cdot} y = z \leftrightarrow \phi_M(x, y, z, w)$$

where ϕ_A, ϕ_M formulas constructed in lemmas 4.8 and 4.9. Denote by $h(w)$ the largest number $h \leq w$ such that for all $x, y, z \leq h$,

$$(x + y = z \rightarrow x \underset{w}{+} y = z) \wedge (x \cdot t = z \rightarrow x \underset{w}{\cdot} y = z).$$

For each w, the interval $[0, h(w)]$ is the largest initial segment of $[0, w]$ on which ϕ_A, ϕ_M correctly define the addition and multiplication predicates A, M. Our problem is to "extend" these definitions to the whole of $[0, w]$.

The predicate $v = h(w)$ can be defined by a bounded $L_{\leq,\alpha}$ formula. Indeed if $\underset{w}{+}, \underset{w}{\cdot}$ are commutative and $0 \underset{w}{\cdot} y = 0$ for all y (as is the case with the definitions ϕ_A, ϕ_M actually constructed in the proofs above), then it is obvious that $h(w)$ is the largest number $h \leq w$ such that

(i) $\forall z \leq h \forall x \leq z \exists y \leq z (z = x \underset{w}{+} y)$

(ii) $\forall z \leq h \forall x \leq z (x \neq 0 \rightarrow \exists q \leq z \exists r < x (z = x \underset{w}{\cdot} q \underset{w}{+} r))$.

((i) and (ii) say that subtraction and division work).

Also by lemmas 4.8 and 4.9, $h(w) \geq w(C \log w)^{-2m-1}$. Applying Bertrand's postulate (see for example Hardy and Wright [18] or chapter 1 of this thesis) that for every $x \geq 2$ there is a prime p with $x < p < 2x$, it follows that if w is sufficiently large then there is a prime p such that

$$h(w) \geq w(C \log w)^{-2m-1} \geq 2\sqrt{w} > p > \sqrt{w}.$$

But for w less than any *fixed* bound, it is clearly possible to "write out" the addition and multiplication tables up to w as bounded $L_{\leq}$ formulas and incorporate these in ϕ_A and ϕ_M, so it can be assumed that for every $w > 1$ there is a prime p with $\sqrt{w} < p \leq h(w)$.

The technique used in the proof of proposition 0.2 allows us to define the addition and multiplication relations on an isomorphic copy $[0, a^2]^\star$ of the interval $[0, a^2]$ by means of $L_{\leq,A,M}$ formulas in which all quantified variables are bounded by a. (The elements of $[0, a^2]^\star$ are 4 tuples of numbers $\leq [\sqrt{a}]$ treated as the digits of the base $[\sqrt{a}] + 1$ representation of a single number.) Replacing A, M by ϕ_A, ϕ_M we see that there are bounded $L_{\leq,\alpha}$ formulas $\phi_A^\star(X, Y, Z, w, a)$, $\phi_M^\star(X, Y, Z, w, a)$ (where X, Y, Z denote 4 tuples) such that

$$\phi_A^\star(X, Y, Z, w, h(w)) \leftrightarrow [0, h(w)^2]^\star \models X + Y = Z$$
$$\phi_M^\star(X, Y, Z, w, h(w)) \leftrightarrow [0, h(w)^2]^\star \models X \cdot Y = Z.$$

Similarly, if for $u \in [0, w]$ we use $u^\star$ to denote the corresponding element of $[0, h(w)^2]^\star$ then there is a bounded $L_{\leq,\alpha}$ formula $\eta(u, U, w)$ such that for all $u \in [0, h(w)^2]^\star$,

$$\eta(u, U, w) \leftrightarrow U = u^\star.$$

Thus η defines the isomorphism $u \to u^\star$ for $u \leq h(w)$. (See Figure 1)

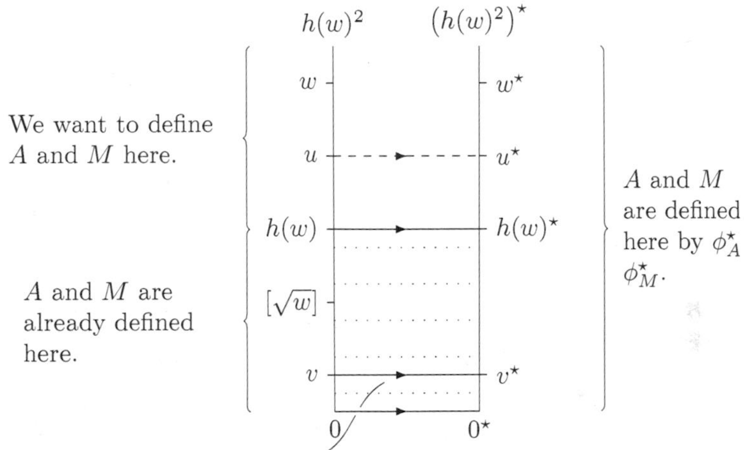

Isomorphism (for A, M) defined by $\eta(u, U, w)$.

FIGURE 1

Our strategy is to "extend" the definition given by $\eta(u, U, w)$ by constructing a new bounded $L_{\leq,\alpha}$ formula $\eta^\star(u, U, w)$ which defines the isomorphism $u \to u^\star$ for all $u \leq w$.

Lemma 4.10. *There is a bounded $L_{\leq,\alpha}$ formula $\eta^\star(u, U, w)$ such that*

$$\forall w \forall u \in [0, w] \forall U \in [0, h(w)^2]^\star (U = u^\star \leftrightarrow \eta^\star(u, U, w)).$$

Proof. Essentially what we will do is to use bounded $L_{\leq,\alpha}$ formulas to associate certain elements and subsets of $[0, h(w)]$ with each $u \in [0, w]$ in such a way that these elements and subsets of $[0, h(w)]$ determine u uniquely in the interval $[0, h(w)^2]$, and thus their images under $\star$ (which is already defined on $[0, h(w)]$ by $\eta(u, U, w)$) will completely determine $u^\star$ in $[0, h(w)2]^\star$. (The novelty of the argument lies in the use of sets, rather than just a fixed finite number of elements, to determine u).

It is claimed that for all $u \in [0, w]$, $U \in [0, h(w)^2]^\star$, $U = u^\star$ if and only if U is the least element of $[0, h(w)^2]^\star$ possessing the following three properties:

(I) *If p is the largest prime $\leq h(w)$ and t is the largest number such that $p \cdot t < u$, then $t^\star$ is the largest element of $[0, h(w)^2]^\star$ for which $[0, h(w)^2]^\star \models p^\star \cdot t^\star < U$.*

(II) $\exists s \leq u(2 \cdot s = u) \leftrightarrow [0, h(w)^2]^\star \models \exists s \leq U(2 \cdot S = U)$.

(III) *For* $i = 0, 1, 2$ *and all primes* $q \leq h(w)$,

$$\exists s \leq u(q \cdot (3 \cdot s + 1) < u \wedge \neg q \cdot (3 \cdot s + i + 1) < u) \leftrightarrow$$
$$[0, h(w)^2]^\star \models \exists s \leq U(q^\star \cdot (3^\star \cdot S + i^\star) < U$$
$$\wedge \neg q^\star \cdot (3^\star \cdot S + i^\star + 1^\star) < U)$$

Taking $U^\star = u^\star$ obviously satisfies (I), (II), (II), so it suffices to show that no $U < u^\star$ has these three properties. Suppose $U < u^\star$ satisfies (I), (II) and (III). Since $U < u^\star$, we must have $U = u^\star$ for some $v < u$. By (II), $2 \leq u - v$, while by (I), $pt < v < u \leq p(t+1)$ so $u - v < p \leq h(w)$. Hence by Bertrand's postulate there is some prime $q \leq h(w)$ such that $q \leq u - v < 2q$.

But now suppose $qs_1 < u \leq q(s_1 + 1)$ and $qs_2 < v \leq q(s_2 + 1)$. Then

$$qs_2 + q < v + (u - v) < q(s_2 + 1) + 2q$$

so $q(s_2 + 1) < u < q(s_2 + 3)$ and thus $s_1 = s_2 + 1$ or $s_1 = s_2 + 2$. In either case $s_1 \not\equiv s_2(\text{mod}3)$ so (III) fails.

Note that in (I), (II), (III) all terms of the form $x^\star$ are restricted to $x \leq h(w)$. (In this case of (I) to get $t \leq h(w)$ we make use of the assumption, justified above, that $\sqrt{w} < p \leq h(w)$ and therefore $t < \sqrt{w} < h(w)$). Consequently since $U = u^\star$ is defined on $[0, h(w)]$ by the formula $\eta(u, U, w)$, and addition and multiplication predicates are defined on $[0, h(w)^2]^\star$ by expressions in (I), (II), (Ill) beginning with $[0, h(w)^2]^\star] \models$ can be replaced by bounded $L_{\leq,\alpha}$ formulas. Since all occurrences of $\cdot$ in (I), (II), (III) involve multiplication by a prime, and there is no nontrivial use of $+$, it is now clear that a suitable bounded $L_{\leq,\alpha}$ formula $\eta^\star(u, U, w)$ can be constructed. ⊣

Completion of the proof of theorems 4.6 and 4.7:

To complete the proof that $x + y = z$ and $x \cdot y = z$ can be defined by bounded $L_{\leq,\alpha}$ formulas (and therefore also by bounded $L_{\leq,\perp}$ formulas) it is only necessary to observe that for all $x, y \leq z$,

$$x + y = z \leftrightarrow$$
$$\exists X \leq x \exists Y \leq y \exists Z \leq z(X = x^\star \wedge Y = y^\star \wedge Z = z^\star \wedge \phi_A^\star(X, Y, Z, z))$$
$$x \cdot y = z \leftrightarrow$$
$$\exists X \leq x \exists Y \leq y \exists Z \leq z(X = x^\star \wedge Y = y^\star \wedge Z = z^\star \wedge \phi_M(X, Y, Z, z))$$

where $U = u^\star$ is being used as an abbreviation for $n^\star(u, U, z)$, and X, Y, Z are 4 tuples of variables each of which is bounded by z.

Remark. The use of Linnik's (or some weaker) bound on the least prime $p \equiv a(\text{mod}b)$ in proving the above theorems can be avoided by appealing to the analogous result for squarefree numbers, which, at least at present, is easier to prove. (See Pracher [31] or Erdos [10]). The argument is modified as follows: having first proved lemma 4.5, that lemma is used to give a similar proof that the predicate

$$q \textit{ is squarefree} \wedge q \cdot y = z$$

is definable by a bounded $L_{\leq,\perp}$ formula. The proofs of theorems 4.7 and 4.8 can then be reworked using this predicate instead of a, and squarefree numbers instead of primes at the appropriate places. (The original theorem 4.8 can of course be recovered as a corollary of theorem 4.7).

Applying proposition 0.2 to theorem 4.7 yields:

Corollary 4.11. *A relation on $\mathbb{N}$ can be defined by a bounded $L_{\leq,+,\cdot}$ formula is and only if it can be defined by a bounded $L_{\leq,\perp}$ formula.*

In order to deduce the same result with a single binary relation in place of $\leq, \perp$ we will use the following approximation to Goldbach's conjecture obtained by the use of "sieve" methods. (a readable introduction to these can be found in Cel'fond and Linnik [14]).

Proposition 4.12 (Brun [5]). *There is a number $k \in \mathbb{N}$ with the property for every $n \in \mathbb{N}$ there exist numbers q, r with $2n = q + r$, such that every prime $p|qr$ satisfies $p \geq (2n)^{1/k}$.*

Define the preordering $\overset{\star}{<}$ by:

$$x \overset{\star}{<} y \leftrightarrow x < y \wedge x \perp y.$$

The reader is reminded that $\overset{\star}{<}$ is *not* transitive. ($x \overset{\star}{<} y \overset{\star}{<} z$ should be read as $x \overset{\star}{<} y \wedge y \overset{\star}{<} z$).

Lemma 4.13. *There is a number $g \in \mathbb{N}$ such that for all x, y with $x \overset{\star}{<} y$ there exist numbers $z_0, z_1, \ldots, z_m$ with $m \leq g$ satisfying:*

$$x = z_0 \overset{\star}{<} z_1 \overset{\star}{<} z_2 \overset{\star}{<} \ldots \overset{\star}{<} z_m = y.$$

Furthermore, the numbers z_i may be chosen to be alternately even and odd.

Proof. Take $g = (2k-2)^k + 2k$ where k satisfies proposition 4.12. If $y - x = h \leq g$ we have $x \overset{\star}{<} x+1 \overset{\star}{<} x+2 \overset{\star}{<} \ldots \overset{\star}{<} x+h = y$ so we may suppose $y - x \geq g = (2k-2)^k + 2k$. Let

$$2n = \begin{cases} (y-x) - (2k-2) & \text{if this is even,} \\ (y-x) - (2k-2) - 1, & \text{otherwise,} \end{cases}$$

so $2n > (2k-2)^k$. By proposition 4.12 there exist q, r such that $2n = q + r$ and all primes $p|qr$ satisfy $p \geq (2n)^{1/k} > 2k - 2$. (In particular, since $k \geq 2$ we have $p > 2$, so q, r are both odd). Observe that each prime $p|q$ can divide at most one of the numbers $x, x+1, \ldots, x+2k-2$, and that q has fewer than k prime divisors. It follows that $x+i \perp q$ for at least k of the numbers $i = 0, 1, \ldots, 2k-2$. Similarly, $x+2n+i \perp r$ for at least k of these i's. Hence there is some $i \leq 2k-2$ such that $x+i \perp q$ *and* $x+2n+i \perp r$. Taking $z = x+i+q = x+2n+i-r$ exhibits a number satisfying $x+1 \overset{\star}{<} z \overset{\star}{<} x+2n+i$ and having the opposite parity to $x+i$. Thus the sequence

$$x \overset{\star}{<} x+1 \overset{\star}{<} x+2 \overset{\star}{<} \ldots \overset{\star}{<} x+i \overset{\star}{<} z \overset{\star}{<} x+2n+i \overset{\star}{<} x+2n+i+1 \overset{\star}{<} \ldots$$
$$\ldots \overset{\star}{<} y-2 \overset{\star}{<} y-1 \overset{\star}{<} y$$

of at most $2k+2$ elements has the required properties. ⊣

Problem. What is the smallest value of g for which this lemma is true? (The author conjectures: If $y - x > 1$ then some z in the interval $x < z < y$ is coprime to both x and y).

Theorem 4.14. *A relation* $\mathbb{N}$ *can be defined by a bounded* $L_{\leq,+,\cdot}$ *formula if and only if it can be definded by a bounded* $L_{\overset{\star}{\leq}}$ *formula, where* $\overset{\star}{\leq}$ *is the preordering defined by:*

$$x \overset{\star}{\leq} y \leftrightarrow x = y \vee (x \leq y \wedge x \perp y).$$

Proof. Obviously,
$$x = y \leftrightarrow x \overset{\star}{\leq} y \wedge y \overset{\star}{\leq} x$$
$$x \perp y \leftrightarrow \neg x = y \wedge (x \overset{\star}{\leq} y \vee y \overset{\star}{\leq} x)$$
and for g satisfying lemma 4.13,

$$x \leq y \leftrightarrow \exists z_{g-1} \overset{\star}{\leq} y \exists z_{g-2} \overset{\star}{\leq} z_{g-1} \ldots \exists z_1 \overset{\star}{\leq} z_2 (x \overset{\star}{\leq} z_1)$$
$$\forall x \leq y \phi \leftrightarrow \forall z_{g-1} \overset{\star}{\leq} y \forall z_{g-2} \overset{\star}{\leq} z_{g-1} \ldots \forall z_1 \overset{\star}{\leq} z_2 \forall x \overset{\star}{\leq} z_1 \phi$$
$$\exists x \leq y \phi \leftrightarrow \exists z_{g-1} \overset{\star}{\leq} y \exists z_{g-2} \overset{\star}{\leq} z_{g-1} \ldots \exists z_1 \overset{\star}{\leq} z_2 \exists x \overset{\star}{\leq} z_1 \phi$$

where the variables $z_1, z_2, \ldots, z_{q-1}$ are chosen from those which do not occur in ϕ.

Hence the theorem follows from corollary 4.11 by induction on the complexity of bounded $L_{\leq,\perp}$ formulas. ⊣

As mentioned in the introduction, a relation on $\mathbb{N}$ is *rudimentary* if and only if it can be defined by a bounded $L_{\leq,+,\cdot}$ formula. By a *rudimentary graph* G we will mean a structure $< N, \gamma >$ where γ is a two place, symmetric, antireflexive, rudimentary relation.

(Vertices $x, y \in \mathbb{N}$ are joined by an edge if and only if $x\gamma y$). Let G_n denote the restriction of G to $[0, n]$.

Now consider the rudimentary graph $G =< N, \gamma >$ where γ is the symmetric relation defined by requiring $x\gamma y$ and $y\gamma x$ to hold between numbers x, y *with* $x \leq y$ if and only if one of the following is satisfied:

(i) x *and* y *are both odd and* $x \perp y$.

(ii) x *is odd,* y *is even, and* $x \perp y$.

(iii) $x = 0, y \neq 0$, *and* y *is even.*

Theorem 4.15. *The rudimentary graph G defined above has the property that for every rudimentary predicate* $\rho(x_1, \ldots, x_k)$ *there is an effectively found* L_γ *formula* $\phi(x_1, \ldots, x_k)$ *such that for all* $n \geq 5$, *and all* $x_1, \ldots, x_k \in [0, n]$,

$$\rho(x_1, \ldots, x_k) \leftrightarrow G_n \models \phi(x_1, \ldots, x_k).$$

Proof. By corollary 4.11, a predicate $\rho(x_1, \ldots, x_k)$ is rudimentary if and only if it can be defined by a bounded $L_{\leq,\perp}$ formula, so the theorem will follow by induction on the complexity of such formulas if it can be shown that there are L_γ, formulas $\phi_\leq(x, y)$, $\phi_\perp(x, y)$ such that for all $n \geq 5$, and all $x, y \in [0, n]$,

$$x < y \leftrightarrow G_n \models \phi_\leq(x, y)$$
$$x \perp y \leftrightarrow G_n \models \phi_\perp(x, y).$$

To avoid continually writing "$G_n \models$" for the rest of this proof variables x, y, z etc. will range over the elements of some *fixed* initial segment of $\mathbb{N}$ containing $\{0, 1, 2, 3, 4, 5\}$. (These initial segments are the intervals $[0, n], n \geq 5$, and $\mathbb{N}$ itself). However the formulas produced will be independent of which initial segment is used. We begin with some less ambitious definitions:

(i) $x = y \leftrightarrow \forall z(x\gamma z \leftrightarrow y\gamma z)$

(ii) $x = 1 \leftrightarrow \exists y(\neg y = x \wedge \forall z(\neg z\gamma x \leftrightarrow z = y \vee z = x))$

(iii) $x = 0 \leftrightarrow \neg x\gamma 1 \wedge \neg x = 1$.

To verify (i) check that for all x and y with $x < y$ (say), $(x\gamma z \leftrightarrow y\gamma z)$ fails for $z = 0$ (if x, y have opposite parity), for $z = y - 1$ (if x, y are both even), or for $z = x + 1$ (if x, y are both odd). (i) and (ii) follow from the fact that for $x = 1$ there is a unique number $z \neq x$ (namely $z = 0$) such that $\neg z\gamma x$, but this cannot happen for $x \neq 1$, since if x is odd then $\neg z\gamma x$ holds for $z = 0, 2$, while if x is even then $\neg z\gamma x$ holds (with $z \neq x$) for $z = 2, 4$ except for the cases $x = 0, 2, 4$ which can be dealt with by examining the picture in Figure 2.

Note now that

$$x \textit{ is even} \leftrightarrow x = 0 \vee x\gamma 0$$

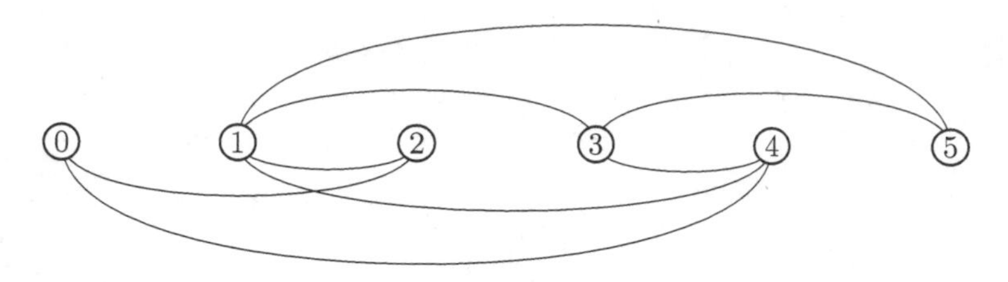

FIGURE 2

and if x, y are both odd then

$$x \perp y \leftrightarrow x\gamma y.$$

Also,

$$x \textit{ is a power of an odd prime } \leftrightarrow$$
$$x \textit{ is odd} \wedge \forall y \forall z(y, z \textit{ are odd} \wedge y \perp z \rightarrow y \perp z \vee z \perp x),$$

and if y is odd then

$$x \sim y \leftrightarrow x \textit{ is odd} \wedge \forall z(z \textit{ is odd} \rightarrow (z \perp x \leftrightarrow z \perp y)).$$

Using these equivalences we can define the odd primes, since

$$x \textit{ is an odd prime} \leftrightarrow$$
$$x \textit{ is a power of an odd prime } \wedge x \neq 1 \wedge$$
$$\forall z(z \neq x \wedge z \sim x \rightarrow \exists w(x\gamma w \wedge \neg z\gamma w)).$$

(To see this take w to be the even number $z-1$).

Note that the predicates $y = 2$, $x = 3$ can be defined since $x = 3$ if and only if x is odd and there exists a *unique* even number $y \neq 0$ (namely $y = 2$) such that $x\gamma y$. Also *for x even,*

$$3 \perp x \leftrightarrow x = 2 \vee 3\gamma x.$$

Now suppose x is even, $3 \perp x$, and p is an odd prime $\neq 3$. Then

$$p \leq x \leftrightarrow p\gamma x \vee \psi(p, x),$$

where $\psi(p, x)$ is the formula:

$$\exists z \exists w(z \textit{ is odd} \wedge w \textit{ is even} \wedge p\gamma w \wedge \neg z\gamma w \wedge z\gamma x$$
$$\wedge \forall q(q \textit{ is an odd prime} \wedge \neg q \perp z \rightarrow q\gamma w))$$

Clearly $p\gamma x \Rightarrow p \leq x$. Also if z, w are as required by $\psi(p, x)$ then $z \perp w$ but $\neg z\gamma w$ so $w < z$, and since $p\gamma w$ and $z\gamma x$ it follows that $p < w < z < x$.

On the other hand, suppose $p \leq x$. If $p \not| x$ then $p\gamma x$. *If* $p|x$ we have two possibilities:

Case 1: $p < 3^n < x$ for some $n > 1$. Then either $p\gamma(3^n - 1)$ or $p\gamma(3^n - 5)$ so choosing $w = 3^n - 1$ or $3^n - 5$ as appropriate, and $z = 3^n$ satisfies $\psi(p, x)$.

Case 2: $3^n < p < x < 3^{n+1}$ for some $n \geq 1$. Then $\frac{x}{p} < 3$, but since $p|x$ it follows that $x = 2p$. Either $2p - 1$ or $2p - 5$ is divisible by 3 (and if $p = 5$ then it is the former) so choosing z to be that one of these, and

$$w = z - 1 \geq \begin{cases} 2p - 6 > p & \text{if } p \geq 7, \\ 2p - 2 > p & \text{if } p = 5, \end{cases}$$

satisfies $\psi(p, x)$ since $p < w < 2p$, and thus $p \perp w$.

Observe that

$$x \textit{ is a power of } 2 \leftrightarrow x \textit{ is even} \wedge 3 \perp x \wedge$$
$$\forall q(q \textit{ is an odd prime } \neq 3 \wedge q \leq x \rightarrow q\gamma x).$$

Now if x *is even and* p *is an odd prime*, then

$$p \leq x \leftrightarrow p\gamma x \vee$$
$$\exists z \exists w(w \textit{ is a power of } 2 \wedge z \textit{ is odd} \wedge p\gamma w \wedge z\gamma x \wedge \neg z\gamma w).$$

For clearly $p\gamma x \rightarrow p \leq x$, while if the second disjunct of this definition is satisfied then $p < 2^n < z < x$ for some n, z. Conversely if $p \leq x$ and $p \not| n$ then $p\gamma x$, while if $p|x$ then $2p \leq x$ so there exists n with $p < 2^n < 2p \leq x$, and taking $z = x - 1$, $w = 2^n$ satisfies the second disjunct above.

Also for x even and p an odd prime,

$$p \perp x \leftrightarrow p\gamma x \vee \neg p \leq x.$$

Finally if x (say) is even and y is odd, then

$$x \perp y \leftrightarrow \forall p(p \textit{ is an odd prime} \rightarrow p \perp x \vee p \perp y),$$

and since the case where x, y are both odd has already been dealt with, the existence of an L_γ formula $\phi_\perp(x, y)$ satisfying

$$\forall x \forall y(x \perp y \leftrightarrow \phi_\perp(x, y))$$

clearly follows.

We next define for each fixed natural number $k \geq 1$ a preordering $\underset{k}{\leq}$ on the *even* natural numbers by

$$x \underset{1}{\leq} y \leftrightarrow x = y \vee \exists z(x \perp z \wedge z\gamma y \wedge \neg z\gamma x),$$
$$x \underset{k+1}{\leq} y \leftrightarrow x \underset{k}{\leq} y \vee \exists z(z \textit{ is even} \wedge x \underset{1}{\leq} z \wedge z \underset{k}{\leq} y).$$

In the notation of lemma 4.11, for x, y both even we have $x \underset{k}{\leq} y$ if and only if for some $m \leq k$ there exists a sequence of alternatively even and odd numbers $z_0, z_1, \ldots, z_{2m}$ such that $x = z_0 \overset{\star}{<} z_1 \overset{\star}{<} \ldots \overset{\star}{<} z_{2m} = y$. Thus if k is sufficiently large then for all even numbers x, y,

$$x \leq y \leftrightarrow x \underset{k}{\leq} y.$$

Now for x odd and y even,

$$x \leq y \leftrightarrow \exists z(z \textit{ is even} \wedge x\gamma z \wedge z \leq y)$$

(as can be seen by taking $z = x + 1$ if this exists) and $y \leq x \leftrightarrow \neg x \leq y$.
Finally, if x and y are both odd,

$$x \leq y \leftrightarrow x = y \vee \exists z(z \textit{ is even} \wedge x \leq z \wedge z \leq y).$$

Since all cases have now been covered it is clear that there is an L_γ formula $\phi_\leq(x, y)$ such that

$$\forall x \forall y(x \leq y \leftrightarrow \phi_\leq(x, y)). \qquad \dashv$$

Remark. With the exception of the trivial case $n = 0$, the restriction $n \geq 5$ in the above theorm is unavoidable since every graph with at least 2 and at most 5 vertices admits a nontrivial automorphism which obviously cannot preserve $\leq$.

To complete this section we will sketch the proof of the analogue of theorem 4.15 for a rudimentary partial ordering.

Theorem 4.16. *There is a rudimentary partial ordering $H =< \mathbb{N}, \circledleq >$ such that $\circledleq$ is extendible to $\leq$, and if H_n denotes the restriction of H to $[0, n]$ then for every rudimentary predicate $\rho(x_1, \ldots, x_m)$ there is an $L_{\circledleq}$ formula $\phi(x_1, \ldots, x_m)$ with the property that for all $n \in \mathbb{N}$ and all $x_1, \ldots, x_m \leq n$*

$$\rho(x_1, \ldots, x_m) \leftrightarrow H_m \models \phi(x_1, \ldots, x_m).$$

Proof. As the reader should by now be well versed in the sort of techniques used, only a sketch will be provided.

To define $\circledleq$ we will construct a rudimentary directed graph having $\mathbb{N}$ as its set of points,with the property that there is a directed edge from y to x if and only if x is an immediate $\circledleq$-predecessor of y. Then we will have $z \circledleq y$ if and only if $x = y$ or there is a directed path from y to x. (Some care is needed here as "path" is a second order concept. However the paths in the directed graph constructed will be "trivial enough" to allow the existence of a path from y to x to be described easily by an $L_{\leq,+,\cdot}$ formula, thus making $\circledleq$ rudimentary).

The basic constituents of the directed graph will be of the form in Figure 3 and will be called *sextets*. The points in the $4, 5, \ldots, 9$ positions

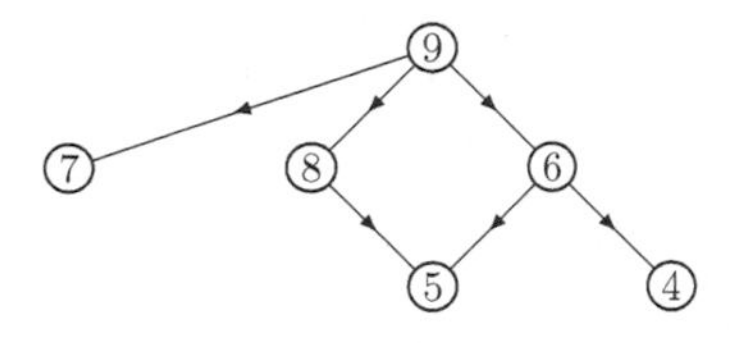

FIGURE 3

(called i-points, $i = 4, 5, \ldots, 9$) will be consecutive numbers (greater than 3) congruent to $4, 5, \ldots, 9(\text{mod}6)$ respectively.

The idea is to ensure that $\leq, \perp$ can be recovered from $\textcircled{\leq}$ on $[0, n]$. For example, for $n \equiv 9(\text{mod}6)$ it will be possible to define $\leq$ since it will be arranged that:

(i) The predicate "x is a 4-point" is $L_{\textcircled{\leq}}$ definable in H_m.

(ii) $\textcircled{\leq}$ linearly orders the 4-points.

(iii) This fact can be used to linearly order the 9-points (which will be the maximal points of the $\textcircled{\leq}$ ordering).

(iv) The sextet to which a point x belongs is then determined uniquely by the smallest 9-point above x.

As the asymmetry of the sextet allows $\leq$ to be defined on the sextet, conditions (i)-(iv) will make it possible (for $n \equiv 9(\text{mod}6)$) to give an $L_{\textcircled{\leq}}$ definition of $\leq$ on $[0, n]$ which is independent of n.

Now consider the partial ordering $H^\circ = < \mathbb{N}, \textcircled{\leq} >$ corresponding to the following directed graph in Figure 4.

Clearly this partial ordering satisfies (i)-(iv) for $n \equiv 9(\text{mod}6)$, The addition of the special bottom section makes it easy to comply with (i). (Hint: first define the predicate $x = 0$).

If $n \equiv 9(\text{mod}6)$ then only part of the uppermost sextet is included in H°_m. These "partial sextets" are of the forms in Figure 5. $\leq$ can still be defined on H°_n in these cases (this is left for the reader to check) and hence there is a single $L_{\textcircled{\leq}}$ formula which defines $\leq$ on H°_n in all cases.

However $\perp$ *cannot* be defined on all H°_n by a single $L_{\textcircled{\leq}}$ formula (as can easily be seen by considering nonstandard models), so to produce a partial ordering $H = < \mathbb{N}, \textcircled{\leq} >$ such that $\perp$ can be defined on all H_n by a single $L_{\textcircled{\leq}}$ formula we will add new edges to the directed graph for H° while at the same time preserving the definition of $\leq$ (for example, in the case of $n \equiv 9(\text{mod}6)$ preserving (i)-(iv)).

If p ($\neq 2, 3$) is a prime less than x and $p|x$ then we will add an edge from x's sextet (or two edges from the preceeding sextet) to the point p in such a way that information about exactly which element(s) of x's sextet is (are) multiple(s) of p is coded into the corresponding partial

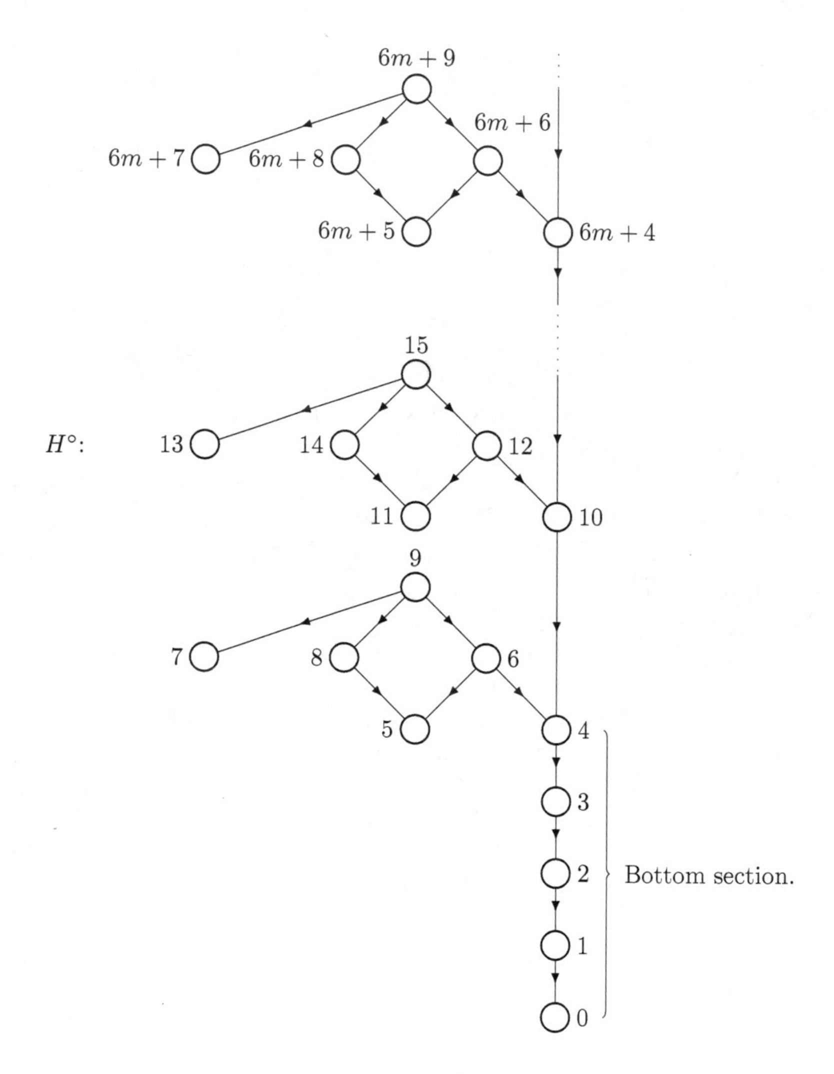

$6m+9$
$6m+7$
$6m+8$
$6m+6$
$6m+5$
$6m+4$
15
H°:
13
14
12
11
10
9
7
8
6
5
4
3
2
1
0
Bottom section.

FIGURE 4

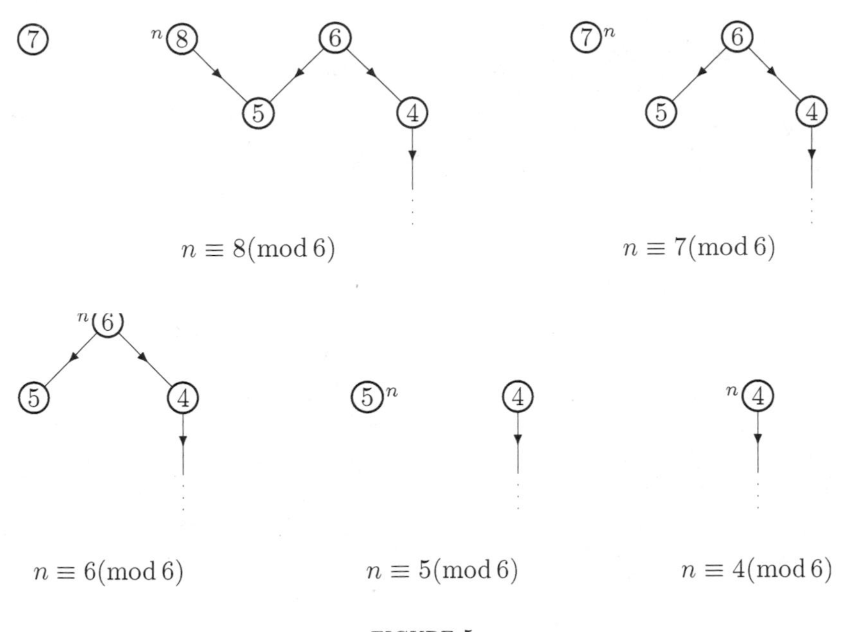

FIGURE 5

ordering ⩽ (circled).

The only points in a sextet which can be on the "receiving" end of these additional edges will be those which are primes, and these can only be 5-points or 7-points (since if $p \geq 5$ is prime then $p \equiv 5(\mathrm{mod} 6)$ or $p \equiv 7(\mathrm{mod} 6)$). These prime points will not be on the "transmitting" end of any additional edges and will constitute *exactly* the minimal points of the ⩽ (circled) partial ordering (aside from 0). (Figure 6)

Thus given x, y it will be possible to decide whether $x \perp y$ from the minimal (prime) points lying under the sextets of x and y (and/or the sextets immediately preceeding these). Note that it is trivial to decide whether 2 or 3 divides x from x's position in the sextet.

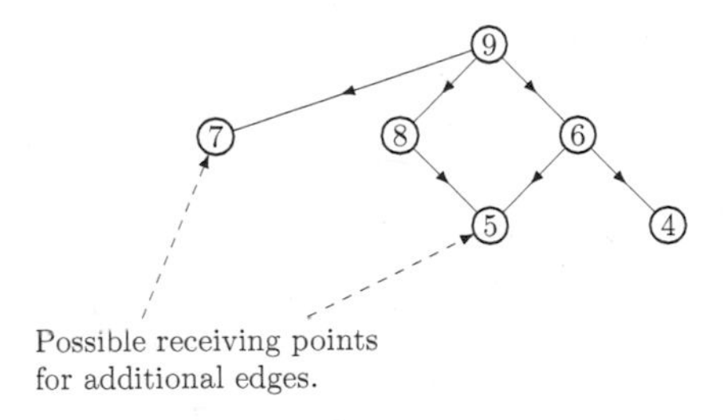

FIGURE 6

Suppose $p|x$ where $5 \leq p < x$. Then unless x is a 4-point, add an edge from x to p. If x is a 4-point then add edges from the 6- and 8-points of the *preceeding* sextet (that is from the numbers $x-2$, $x-4$) to p. No confusion between these two cases can arise, because if $p > 5$ then p can divide at most one of the numbers $x, x-1, x-2, \ldots, x-6$, while if $p = 5$ and $p|x$ where x is a 4-point, then p will divide $x - 5$, that is the 5-point of the preceeding sextet, and (see the example below) the resulting partial ordering is the same as if the two additional edges from $x - 2$, $x - 4$ to p were left out.

Apart from this case, none of the additional edges is redundant (that is, an omission would change $\circledleq$) so the requirement that information be coded into the partial ordering about which element(s) of x's sextet is (are) divisible by p is satisfied. (To check this use the minimality of primes and the fact that if $p > 5$ then p can divide at most one element of a sextet).

Examples.

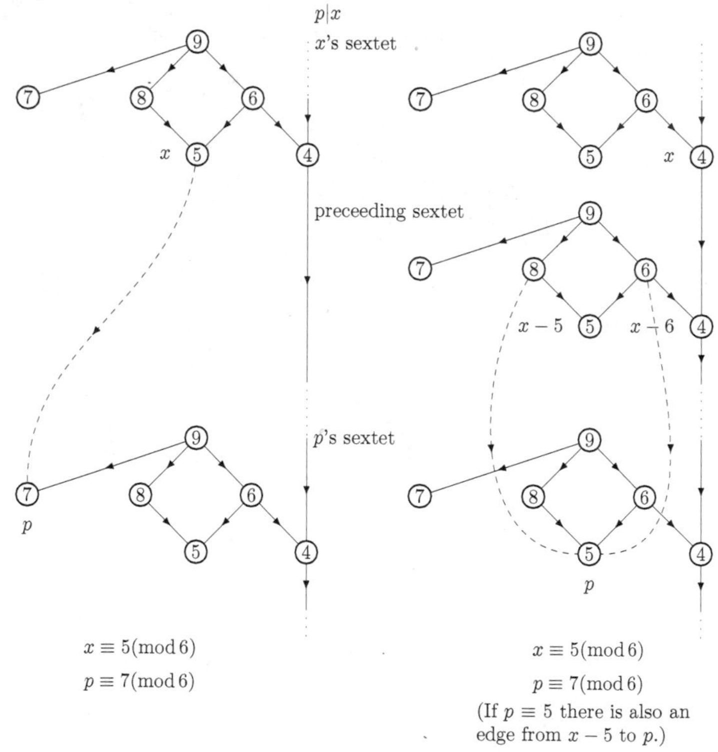

Let $H = < \mathbb{N}, \circledleq >$ be the partial ordering obtained from H° by adding edges as described above. Then there is an $L_{\circledleq}$ formula which defines $\perp$ in all H_n's. It is left to the reader to verify that the definition of $\leq$ on $[0, n]$ has been preserved. $\dashv$

5 Applications to spectrum problems.

Consider any language L of first order predicate calculus (with or without identity). A model M will be said to be *normal* if for each pair $a, b \in M$ with $a \neq b$ there is some formula $\psi(u, v_1, \ldots, v_k)$ of L and elements $c_1, \ldots, c_k \in M$ such that

$$M \models \psi(a, c_1, \ldots, c_k) \text{ and } M \models \neg\psi(b, c_1, \ldots, c_k).$$

Obviously if the identity relation $=$ can be defined on M by an L formula then M is normal. (Conversely if M is a *finite* normal model then $=$ can be defined on M by some L formula, but in general this formula will of course depend on M).

The *spectrum* of a sentence ϕ of L is the set of positive natural numbers:

$$S_\phi = \{|M| : \ M \text{ is a finite normal model of } \phi\}.$$

Let

$$S = \{S_\phi : \ \phi \text{ is a sentence of some first order language}\}.$$

The *spectrum problem* (Scholz [39]) is to characterise S. Some characterisations are known. For example, $S = \mathrm{NEXP}$ where NEXP is the class consisting of all sets $S \subset \mathbb{N} \setminus \{0\}$ with the property that there are a nondeterministic Turing machine T_S and a constant c (dependent on S) such that T_S will accept the binary representation of any element of S with n digits (and no others) in time 2^{cn}. (See Jones and Selman [21]). However although it has long been known that the class

$$\mathrm{RUD} = \{R \subseteq \mathbb{N} \setminus \{0\} : \ x \in R \text{ is a rudimentary predicate}\}$$

is contained in S, it is not known whether this containment is strict. A proof that $S = \mathrm{RUD}$ would settle, in the negative, the $\mathrm{P} = \mathrm{NP}$ problem in computational complexity theory. In fact it would follow not only that $\mathrm{NP} \neq \mathrm{co-NP}$, but also that the polynomial time hierarchy of Stockmeyer [43] would collapse at some level higher than this. (For details see chapter 4. A general discussion of the $\mathrm{P} = \mathrm{NP}$ problem may be found in Hopcroft and Ullman [20]).

Analogues of the spectrum problem can be obtained by starting with some theory T in a first order language L, and defining the *T-spectrum* of a sentence ϕ to be:

$$S_\phi^T = \{|M| : \ M \text{ is a finite normal model of } T \cup \{\phi\}\}.$$

The problem is to characterise

$$S_T = \{S_\phi^T : \ \phi \text{ is an } L \text{ sentence}\}.$$

For example, suppose L is the language $L_{=.\gamma}$ with two binary predicate symbols and that T is axiomatized by the axioms of equality (for $=$) plus the sentence

$$\forall x \forall y (x\gamma y \leftrightarrow y\gamma x) \wedge \forall x(\neg x\gamma x) \qquad (\dagger)$$

Then $< V, =, \gamma >$ is a normal model of T if and only if $< V, \gamma >$ is a graph. (Normal here simply implies $=$ is interpreted by identity). Thus $n \in S_\phi^T$ if and only if some graph with n vertices has the first order "property" and the spectrum problem is to characterise the sets S_ϕ^T which can be obtained m this way.

If instead we use the language L_γ and take T to be the theory with ($\dagger$) as its only axiom, then $< V, \gamma >$ is a normal model of T if and only if $< V, \gamma >$ is a graph in which no two vertices are joined by edges to exactly the same set of vertices. (We will call $< V, \gamma >$ a *normal graph* in these circumstances).

Similarly other examples yielding nontrivial spectrum problems are obtained by taking L to be L and T to be the theory of partial orderings, or the theory of quasiorderings. (a *quasiordering* is a preordering which can be extended to just one linear ordering, that is, one whose transitive closure is a linear ordering).

As a final example consider the language $L_{\leq,A,M}$ (where $A(x, y, z) \leftrightarrow x + y = z$ and $M(x, y, z) \leftrightarrow x \cdot y = z$) and let T be the theory FA of *finite arithmetic*:

$$\text{FA} = \{\psi : \ \psi \text{ is an } L_{\leq,A,M} \text{ sentence} \wedge \forall n \in \mathbb{N}(< [0, n], \leq, A, M > \models \psi)\}.$$

In this case S_T = RUD by virtue of the second part of the following lemma which states that FA is categorical in every finite cardinality.

Lemma 5.1. *There is a finite set Σ of axioms in the language $L_{\leq,A,M}$ such that*

(i) *For each $n \in \mathbb{N}$, $< [0, n], \leq, A, M > \models \Sigma$.*

(ii) *Every finite normal model $< X, \leq^\star, A^\star, M^\star >$ of Σ with cardinality n is isomorphic to $< [0, n-1], \leq, A, M >$.*

Proof. The axioms in Σ describe the basic "algebraic" properties of $\leq$, A, M, and assert that A and M satisfy the inductive definitions of addition and multiplication as far as is possible in a finite initial segment of $\mathbb{N}$. (ii) is proved by induction on n. $\dashv$

Now if $R \in$ RUD then (by proposition 0.2) there is a bounded $L_{\leq,A,M}$ formula $\psi_R(x)$ such that

$$\forall n(n \in R \leftrightarrow \psi_R(n-1)),$$

and it follows by the lemma that $R = S_\phi^{\text{FA}}$, where ϕ is the sentence

$$\exists x(\forall y(y \leq x) \wedge \psi_R(x)).$$

On the other hand, if ϕ is an $L_{\leq,A,M}$ sentence and $\phi^\star(x)$ is the bounded formula obtained from ϕ by bounding all quantifiers in ϕ by x (where x is a variable not occurring in ϕ) then

$$n \in S_\phi^{\text{FA}} \leftrightarrow \exists x \leq n(n = x+1 \wedge \phi^\star(x)).$$

Thus $S_{\text{FA}} =$ RUD.

(Note also that, as mentioned earlier, RUD $\subseteq S$, since for all ϕ, $S_\phi^{\text{FA}} = S_{\phi\wedge\sigma} \in S$ where σ is the conjunction of the axioms in Σ).

An obvious question now is to ask how RUD is related to the other theories T cited above as examples.

Theorem 5.2. *If T is the theory of graphs, the theory of normal graphs, the theory of quasiorderings, or the theory of partial orderings, then* RUD $\subseteq S_T$.

Proof. Since RUD $= S_{\text{FA}}$ it will suffice to show that in each case $S_{\text{FA}} \subseteq S_T$. Therefore suppose ψ is an $L_{\leq,A,M}$ sentence and take ψ° to be $\psi \wedge \sigma$ where σ is the conjunction of the axioms Σ for FA given by lemma 5.1. Fix three formulas $\phi_\leq(x,y)$, $\phi_A(x,y,z)$, $\phi_M(x,y,z)$ in the language L corresponding to T, and let $\psi^\star$ be the L sentence obtained by replacing all occurrences of $\leq, A, M$ in ψ° by $\phi_\leq$, ϕ_A, ϕ_M in the obvious way. By lemma 5.1, $S_{\psi^\star}^T \subseteq S_\psi^{\text{FA}}$, since if $< X, \ldots > \models \psi^\star$ then taking $\leq^\star, A^\star, M^\star$ to be the predicates defined on X by $\phi_\leq$, ϕ_A, ϕ_M yields a model $< X, \leq^\star, A^\star, M^\star >$ of FA (with the same cardinality) in which ψ is satisfied.

Thus we will have $S_{\psi^\star}^T = S_\psi^{\text{FA}}$, if we can choose $\phi_\leq$, ϕ_A, ϕ_M having the property that for each $n \in \mathbb{N} \vee \{0\}$ there is some model $< X, \ldots > \models T$ with $|X| = n$ such that $< X, \leq^\star, A^\star, M^\star > \cong < [0, n-1], \leq, A, M >$. But for T the theory of graphs (or normal graphs), partial orderings or quasiorderings, the existence of formulas $\phi_\leq$, ϕ_A, ϕ_M possessing this property is guaranteed by theorems 4.15, 4.16 and 4.14 respectively. (Note that the preordering $\overset{\star}{\leq}$ in theorem 4.14 is a quasiordering by lemma 4.13. Also strictly speaking in the case of (normal) graph theory the construction fails because it is impossible to define $\phi_\leq, \phi_A, \phi_M$ which will work for $n = 2, 3, 4, 5$. However theorem 4.15 does allow the construction of a sentence $\psi^\star$ such that $S_{\psi^\star}^T$ and S_ψ^{FA}, differ only on $[1, 5]$ and it is easy to do a "finite modification" on this sentence so that for $n \leq 5$ the normal models of $\psi^\star$ with exactly n distinct elements will be all the (normal) graphs with n vertices if $n \in S_\psi^{\mathrm{FA}}$, and nonexistent otherwise. Since there *are* normal graphs with $1, 2, 3, 4$ and 5 vertices, we will then have $S_{\psi^\star}^T = S_\psi^{\mathrm{FA}}$). ⊣

Corollary 5.3. *Every $R \in \mathrm{RUD}$ can be represented in the form $R = S_\phi$ where ϕ is a sentence of first order predicate calculus with only a single binary predicate symbol.*

This corollary shows that the RUD $= \mathcal{S}$ problem would be settled in the negative if it could be shown that there is some $\mathrm{S} \in \mathcal{S}$ which cannot be represented in the form $\mathrm{S} = \mathrm{S}_\phi$, for any sentence involving only a single binary predicate symbol. Similarly it follows from theorem 5.2 that RUD $\neq \mathcal{S}$ if $\mathcal{S}_T \neq \mathcal{S}$ where T is the theory of graphs, normal graphs, partial orderings or quasiorderings, since in each case $\mathcal{S}_T \subseteq \mathcal{S}$. In fact to prove the existence of a nonrudimentary spectrum it would suffice to show that one of these theories T has the property that for each $n \in \mathbb{N}$ there is a sentence ϕ such that $\mathrm{S}_\phi^{\mathrm{T}} \neq \mathrm{S}_\psi^{\mathrm{T}}$ for all sentences ψ of the form $Q_1 Q_2 \ldots Q_n X$ where X is quantifier free and each Q_i denotes a *block* of similar quantifiers of the form $\exists\exists \ldots \exists$ or $\forall\forall \ldots \forall$. This is because (as shown in Harrow [19], theorem 3) the assumption that $\mathcal{S} = \mathrm{RUD}$ implies the existence of an absolute bound on the number of *changes* of quantifier (but not the total number of quantifiers,[2] see Wilkie [48]) required in giving a bounded $L_{\leq,+,\cdot}$ (or $L_{\leq,A,M}$) definition $\psi_R(x)$ for any $R \in \mathrm{RUD}$. This bound implies the existence of a similar (but larger) bound on the number of blocks of quantifiers which need appear in the prenex normal forms of the sentences $\psi^\star$ constructed in the proof of theorem 5.2.

It should also be mentioned that Fagin [13] has conjectured that for each k there is some element of $\mathcal{S}$ which is not the spectrum of any sentence having only k-ary predicate symbols. He highlighted the nature of the problem by proving:

Proposition 5.4. *(Fagin [13]). For every $\mathrm{S} \in \mathcal{S}$ there is some $k \in \mathbb{N} \setminus \{0\}$ such that $\{n^k : n \in \mathrm{S}\} \in \mathcal{S}_T$ where T is the theory of graphs.*

An easy modification of Fagin's argument shows that the same is true if T is the theory of normal graphs. Furthermore, the theories of quasiorderings and partial orderings have similar properties since:

Lemma 5.5. *If T_1, T_2, T_3 are the theories of graphs, quasiorderings, and partial orderings respectively, then for all* S,

(i) $\mathrm{S} \in \mathcal{S}_{T_1} \to \{2n : n \in S\} \in \mathcal{S}_{T_2}$

---

[2] at least if the *matrix* is an $L_{=,+,\cdot}$ formula.

(ii) $S \in S_{T_1} \rightarrow \{n^2 : n \in S\} \in S_{T_3}$.

Proof. (i) It will suffice to show that there are L formulas $\phi_G(x)$, $\phi_\gamma(x, y)$ such that:

(I) For any graph $< G, \gamma >$ it is possible to construct a quasiordering $< Q, \leq >$ with $|Q| = 2|G|$ on which ϕ_G, ϕ_γ define a subset $G^\star \subseteq Q$ and a binary relation $\gamma^\star$ on $G^\star$ with $< G^\star, \gamma^\star > \cong < G, \gamma >$.

(II) There is an L sentence which if satisfied by a quasiordering $< Q^\star, \leq^\star >$ will ensure that the structure $< G^\star, \gamma^\star >$ defined by ϕ_G, ϕ_γ is a graph with $2|G^\star| = |Q^\star|$.

Given $< G, \gamma >$ to construct $< Q, \preceq >$ let $Q = G \cup H$ where $G \cap H = \emptyset$ and $|H| = |G| = n$ (say). Linearly order Q as

$$h_1 \leq g_1 \leq h_2 \leq g_2 \leq \ldots \leq h_n \leq g_n$$

where $h_1, \ldots, h_n$ and $g_1, \ldots, g_n$ are the elements of H and G respectively (in some order). Now define a quasiordering $\preceq$ on Q by:

$$h_i \preceq h_j \leftrightarrow i \leq j$$
$$g_i \preceq g_j \leftrightarrow (i \leq j \wedge g_i \gamma g_j) \vee i = j$$
$$h_i \preceq g_j \leftrightarrow i = j$$
$$g_i \preceq h_j \leftrightarrow j \leq i + 1.$$

Clearly for all $x, y \in Q$,

$$x \leq y \leftrightarrow \exists z \exists w (x \preceq z \wedge z \preceq w \wedge w \preceq y) \qquad (\star)$$

so the $\leq$-least and $\leq$-least-but-one elements h_1, g_1 are L definable. But

$$x \in G \leftrightarrow x = g_1 \vee \neg h_1 \preceq x \qquad (\dagger)$$

and for $x, y \in G$,

$$x \gamma y \leftrightarrow (x \preceq y \vee y \preceq x) \wedge x = y, \qquad (\dagger\dagger)$$

which proves (I).

Also we can define the predicate $y = x'$, that is, y is the immediate $\leq$-successor of x, by an L formula, so using the definitions for G and g_1 given above, there is an L sentence which says:

$$g_1 \in G \wedge \forall x (\neg x \in G \leftrightarrow \exists y \in G (y = x')).$$

Clearly if a quasiordering $< Q^\star, \preceq^\star >$ satisfies the conjunction of this sentence with the sentence saying that the formula on the right of $(\star)$ defines a linear ordering, then the formula $\phi_G(x)$ given by $(\dagger)$ will define a subset $G^\star \subseteq Q^\star$ with $2|G^\star| = |Q|$, and on $G^\star$ the formula on the right of $(\dagger\dagger)$ will define a relation $\gamma^\star$ such that $< G^\star, \gamma^\star >$ is a graph.

(ii) The second part of the lemma can be proved in a similar way by considering the following method for coding an arbitrary graph

$< G, \gamma >$ with $|G| = n$ into a partial ordering $< Q, \preceq >$ with $Q = G \cup H$, where $G \cap H = \emptyset$ and $|H| = n^2 - n$. The partial ordering $\preceq$ is defined so as to associate the elements of H in a two-to-one fashion with the (unordered) pairs of elements of G. Each pair $\{a, b\} \subseteq G$ is associated with two distinct elements $h_1, h_2 \in H$ in one of the following two ways in Figure 7. $\preceq$ is taken to be the (reflexive) partial ordering generated

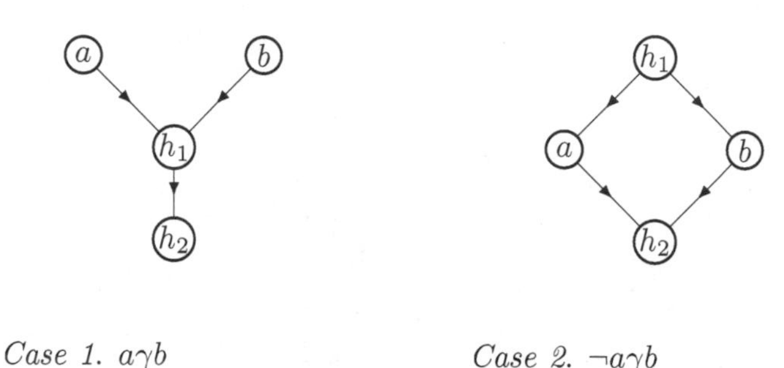

Case 1. $a\gamma b$ *Case 2.* $\neg a\gamma b$

FIGURE 7

by the directed graph containing just the edges required by these two cases. Since there are $n(n-1)/2$ unordered pairs from G, this is possible if and only if $|H| = n(n-1)$.

The rest of the details are left to the reader. ⊣

Now suppose T is one of the theories considered in theorem 5.2. Then

$$S_T \subseteq \text{RUD} \leftrightarrow S_T = \text{RUD}.$$

For if $\text{S} \in S$, then by proposition 5.4 and lemma 5.5 there are non-zero natural numbers j, k such that $\{jn^k : n \in \text{S}\} \in S_T$. Thus if $S_T \subseteq \text{RUD}$ then $\{jn^k : n \in \text{S}\} \in \text{RUD}$. But it is easy to see that if $\{jn^k : n \in \text{S}\}$ can be defined by a bounded $L_{\leq,+,\cdot}$, formula, then S can also be defined by a bounded $L_{\leq,+,\cdot}$ formula, so $\text{S} \in \text{RUD}$.

Combining this observation with theorem 5.2 yields:

Theorem 5.6. $S = \text{RUD} \leftrightarrow S_T = \text{RUD}$, *for* T *the theory of graphs, normal graphs, quasiorderings, or partial orderings.*

Thus (using the equivalence of RUD and S_{FA}) problem of whether $S \neq \text{RUD}$ can be viewed as a question asking whether graph theory, the theory of partial orderings, etc., are each more complex than the theory of finite arithmetic, in the sense that they generate a broader class of spectra.

Obviously there is much more to be learnt about the classes of spectra generated by the various theories of finite structures, and the relationships between these classes. It might be interesting to know, for example, what can be said about the spectra generated by field theory, group theory, the theory of trees, etc.

3

Some logical consequences of the linear case of Schinzel's hypothesis H

Let $L_{=,+,P}$ be the first order language for $\mathbb{N}$ (or more generally for some model of Peano arithmetic, PA) with primitive symbols $=, +$, and the predicate $P(x)$ defined by:

$$P(x) \leftrightarrow x \text{ is a prime number}$$

Obviously several famous open problems in number theory can be expressed by very simple formulas in this language. For example noting that the predicates

$$x \le y, x = 0, x = 1, x = 2, \ldots, y = 2x, y = 3x, \ldots$$

all have existential (or quantifier free) $L_{=,+,P}$ definitions we have:

$$\forall x \exists p (p \ge x \wedge P(p) \wedge P(p+2))$$
(there exist infinitely many twin primes)

$$\exists x \forall y \exists p (p \ge y \wedge P(p) \wedge P(p+2x))$$
($\liminf(p_{i+1} - p_i) < \infty$, *where* p_i *denotes the ith prime number*)

$$\forall x (x \ge 2 \rightarrow \exists p \exists q (P(p) \wedge P(q) \wedge 2x = p + q))$$
(Goldbach's conjecture).

This suggests the following question:

Open problem. Is there some $L_{=,+,P}$ sentence ϕ such that PA $\models \phi$?

(Recall that by a well known theorem of Presburger [32] the $L_{=,+}$ theory of $\mathbb{N}$ is identical with the $L_{=,+}$ consequences of PA and therefore *decidable*).

In this chapter a partial answer to this question is deduced from the following conjecture (in which $\mathbb{Z}[x]$ denotes the ring of all polynomials over the integers $\mathbb{Z}$):

Hypothesis H **(Schinzel).** *Let $f_1(x), f_2(x), \ldots, f_n(x)$ be irreducible polynomials in $\mathbb{Z}[x]$ with positive leading coefficients and suppose that*

$$\forall y \neq 1 \exists x \left(y \not| \prod_{1 \leq i \leq n} f_i(x) \right),$$

then there exist infinitely many numbers x for which $f_1(x)$, $f_2(x), \ldots,$ $f_n(x)$ are all primes.

Actually we will require only the linear case where $f_1(x)$, $f_2(x), \ldots,$ $f_n(x)$ are all polynomials of degree 1:

Conjecture H_L **(Dickson [8]).** *If $a_1, a_2, \ldots, a_n, b_1, b_2, \ldots, b_n \in \mathbb{Z}$ with all $a_i > 0$ and*

$$\forall y \neq 1 \exists x \left(y \not| \prod_{1 \leq i \leq n} (a_i x + b_i) \right)$$

then there exist infinitely many numbers x such that $a_i x + b_i$ is prime for all i.

Remark. Note that in each case the conjecture states that an obviously necessary condition for the existence of primes is also sufficient. Also the words "infinitely many" are actually redundant - see Schinzel and Sierpinski [37]. Special cases of H_L include Dirichlet's theorem on the existence of infinitely many primes $p \equiv a (\text{mod} b)$ provided a and b are coprime, the twin primes conjecture, and the conjecture that there exist arbitrarily long arithmetic progressions consisting entirely of prime numbers.

Assuming H_L has the following consequence for $\mathbb{N}$:

Theorem 1. *If H_L is true then the predicate $z = x \cdot y$ can by defined by an $L_{=,+,P}$ formula.*

However the definition cannot be an existential formula since:

Theorem 2. *If H_L is true then the existential $L_{=,+,P}$ theory of $\mathbb{N}$ is decidable.*

In fact it will be easy (and left to the reader) to check that the proof of theorem 1 shows that there is an $L_{=,+,P}$ formula which defines $z = x \cdot y$ on any model of the $L_{=,+,P}$ consequences of PA$+H_L$ (or indeed of $I\Delta_0 + \forall x(2^x$ *exists*$+H_L)$ where H_L is a *single sentence* expressing the conjecture H_L. (Note that here $\mathbb{Z}$ will be interpreted as "integers in the sense of the model"). Therefore turning any undecidable $L_{=,+,\cdot}$ sentence into an $L_{=,+,P}$ sentence in the obvious way gives:

Corollary 3. *If* PA + H_L *is consistent then there is some* $L_{=,+,P}$ *sentence* ϕ *such that* PA $\nvdash \phi$.

Similarly the proof of theorem 2 will show that the existential $L_{=,+,P}$ theory of $\mathbb{N}$ is the set of existential sentences ϕ such that

$$\mathrm{PA} + H_L^1 + H_L^2 + H_L^3 + \ldots \vdash \phi$$

(assuming this axiom system is sound, where H_L^i, $i \in \mathbb{N}$, is an enumeration of the instances of H_L with $n, a_1, a_2, \ldots, a_n, b_1, b_2, \ldots, b_n$ *standard* integers. (Actually PA can be replaced here by an axiom system considerably weaker than $I\Delta_0$).

The proofs of theorems 1 and 2 depend on the following lemma (in which $\equiv$ denotes the identity relation between polynomials).

Proposition 4 (Schinzel [36]). *Suppose* $a_1x+b_1, a_2x+b_2, \ldots, a_nx+b_n$ *satisfy the conditions of* H_L *and let* $c_1x+d_1, c_2x+d_2, \ldots, c_mx+d_m$ *be in* $\mathbb{Z}[x]$ *with* $c_j > 0$ *and* $c_jx+d_j \not\equiv a_ix+b_i$ *for all* i, j. *Then* H_L *implies the existence of infinitely many numbers* x *such that*

(i) $a_ix + b_i$ *is a prime for each* $i \in [1, n]$.

(ii) $c_jx + d_j$ *composite for each* $j \in [1, m]$.

Using this we now prove:

Lemma 5. H_L *implies that for all* $a > 1$, $n \in \mathbb{N}$ *there exist* $n+1$ consecutive *prime numbers* $p_0, p_1, \ldots, p_n$ *with* $p_i - p_{i-1} = 2ia$ *for all* $i \in [1, n]$.

Proof. Take the polynomials $x, x+2a, \ldots, x+i(i+1)a, \ldots, x+n(n+1)a$ as the a_ix+b_i's and the polynomials $x+m$ with $0 < m < n(n+1)a$, $m \neq i(i+1)a$ for all i, as the c_jx+d_j's. We need only check that for every prime p there is some x such that $\forall i \le n(p \nmid (x+i(i+1)a))$, in other words that there is some residue class $x(\bmod p)$ such that $x \not\equiv -i(i+1)a(\mathrm{mod} p)$ for all i. But clearly the residue class of $-i(i+1)a(\mathrm{mod} p)$ is determined by that of $i(\mathrm{mod} p)$, so since $i \equiv 0(\mathrm{mod} p)$ and $i \equiv -1(\mathrm{mod} p)$ both produce $-i(i+1)a \equiv 0(\mathrm{mod} p)$ we see that $-i(i+1)a$ takes fewer than p values (mod$p$). Therefore such an x exists. ⊣

Theorem 1. H_L *implies that* $z = x \cdot y$ *can be defined by an* $L_{=,+,P}$ *formula.*

Proof. Since $z = x \cdot y$ can be defined by an $L_{+,|}$ formula (see Robinson [35] or chapter 2 of this thesis) it suffices to show that the predicate $a|b$ is $L_{=,+,P}$ definable. Suppose $a|b$ with $n = \frac{a}{b}$ (and $b \neq 0$). Then assuming H_L, it follows by lemma 5 that a pattern of the following sort exists:

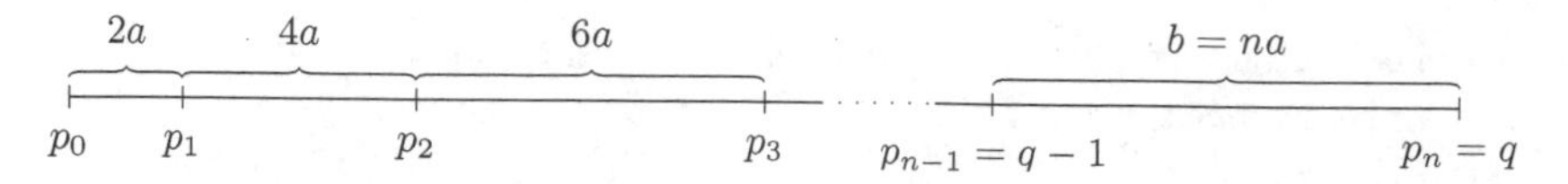

where $p_0, p_1, \ldots, p_n$ are consecutive prime numbers with $p_i - p_{i-1} = 2ia$. Conversely if such a pattern exists then obviously $a|b$. But we can construct an $L_{=,+,P}$ formula $\phi(a,b)$ asserting that there exist primes $p_0 < p_1 < q_{-1} < q$ such that

(i) p_0, p_1 *are consecutive primes*$\wedge p_1 = p_0 + 2a$,

(ii) $q_{-i} < q$ *are consecutive primes*$\wedge q = q_{-1} + 2b$,

(iii) $\forall p \forall r \forall s(p_0 < p < r < s < q \wedge p, r, s$ *are consecutive primes*$\rightarrow s - r = r - p + 2a)$. Clearly, $a|b \leftrightarrow \phi(a,b)$. ⊣

Theorem 2. *H_L implies that the existential $L_{=,+,P}$ theory of $\mathbb{N}$ is decidable.*

Proof. Suppose ϕ is an existential $L_{=,+,P}$ sentence. Using the equivalences:

$$\begin{array}{ll} (\psi \rightarrow \chi) \leftrightarrow \chi \vee \neg\psi, & \neg(\psi \vee \chi) \leftrightarrow \neg\psi \wedge \neg\chi, \\ \neg(\psi \wedge \chi) \leftrightarrow \neg\psi \vee \neg\chi, & \chi \wedge (\psi \vee \rho) \leftrightarrow (\chi \wedge \psi) \vee (\chi \wedge \rho), \end{array}$$

we can effectively find an equivalent sentence in *disjunctive normal form:*

$$\exists \vec{x}(\psi_1(\vec{x}) \vee \psi_2(\vec{x}) \vee \ldots \vee \psi_m(\vec{x})),$$

where each $\psi_j(\vec{x})$ is a conjunction of atomic and negated atomic formulas of the forms:

$$\begin{array}{r} P(a_0 + a_1x_1 + a_2x_2 + \ldots + a_nx_n) \\ \neg P(b_0 + b_1x_1 + b_2x_2 + \ldots + b_nx_n) \\ c_0 + c_1x_1 + c_2x_2 + \ldots + c_nx_n = 0 \\ \neg d_0 + d_1x_1 + d_2x_2 + \ldots + d_nx_n = 0 \end{array}$$

where we have taken the liberty of replacing terms by linear polynomials with coefficients in $\mathbb{Z}$, so $a_i, b_i \in \mathbb{N}$ and $c_i, d_i \in \mathbb{Z}$.

But since

$$\exists \vec{x}(\psi_1(\vec{x}) \vee \ldots \vee \psi_m(\vec{x})) \leftrightarrow \exists \vec{x}\psi_1(\vec{x}) \vee \ldots \vee \exists \vec{x}\psi_m(\vec{x}),$$

the individual disjuncts $\exists \vec{x}\psi_j(\vec{x})$ can be considered separately, so it suffices to show that there is an algorithm for deciding whether or not a system of relations of the form:

$$P\left(a_0 + \sum_{1\leq i\leq n} a_i x_i\right), \quad \vec{a} \in A,$$

$$\neg P\left(b_0 + \sum_{1\leq i\leq n} b_i x_i\right), \quad \vec{b} \in B,$$

$$c_0 + \sum_{1\leq i\leq n} c_i x_i = 0, \quad \vec{c} \in C,$$

$$\neg d_0 + \sum_{1\leq i\leq n} d_i x_i = 0, \quad \vec{d} \in D,$$

with $A, B \subset \mathbb{N}^{n+1}, C, D \subset \mathbb{Z}^{n+1}$ *finite sets*, has a simultaneous solution $\vec{x} \in \mathbb{N}^n$.

At the expense of some more "separating out" of disjuncts we may replace the inequalities

$$d_0 + \sum_{1\leq i\leq n} d_i x_i = 0$$

using the equivalence:

$$d_0 + \sum_{1\leq i\leq n} d_i x_i = 0 \leftrightarrow$$
$$\left(d_0 + \sum_{1\leq i\leq n} d_i x_i > 0\right) \vee \left((-d_0) + \sum_{1\leq i\leq n} (-d_i) x_i > 0,\right)$$

and thus obtain systems of the form:

$$\begin{aligned}
&P\left(a_0 + \sum_{1\leq i\leq n} a_i x_i\right), && \vec{a} \in A,\\
&\neg P\left(b_0 + \sum_{1\leq i\leq n} b_i x_i\right), && \vec{b} \in B,\\
&c_0 + \sum_{1\leq i\leq n} c_i x_i = 0, && \vec{c} \in C,\\
&d_0 + \sum_{1\leq i\leq n} d_i x_i > 0, && \vec{d} \in D,
\end{aligned} \tag{1}$$

We will actually consider systems of this sort where A, B (as well as C, D) are allowed to be finite subsets of $\mathbb{Z}^{n+1}$ but with the following added requirement:

Restriction R. For every $\vec{a} \in A$, $\vec{b} \in B$, the system (1) must contain:

(i) $(a_0 - k_{\vec{a}}) + \sum_{1 \leq i \leq n} a_i x_i > 0$ for some $k_{\vec{a}} \in \mathbb{N}$,
(ii) $(b_0 - m_{\vec{b}}) + \sum_{1 \leq i \leq n} b_i x_i > 0$ for some $m_{\vec{b}} \in \mathbb{N}$.

It will be left to the reader to verify that this requirement can be preserved through all the reductions made below.

We will also assume that if any $\vec{a}$, $\vec{b}$, $\vec{c}$, or $\vec{d}$ is zero apart (possibly) from the first term a_0, b_0, c_0 or d_0 (respectively) then the truth or falsity of the relevant formula is determined immediately. If it is false then the system has no solution. If it is true then the formula is deleted.

Now consider two cases:

Case 1 $C \neq \emptyset$.
Multiplying equations by -1 and renumbering variables (if necessary) we may suppose that the system contains an equation

$$c_0^\star + \sum_{1 \leq i \leq n} c_i^\star x_i = 0$$

with $c_n^\star > 0$, for some $\vec{c}^\star \in C$. Thus if $\vec{x}$ is a solution of (1), then

$$c_0^\star + \sum_{1 \leq i \leq n-1} c_i^\star x_i \equiv 0 (\text{mod } c_n^\star) \tag{2}$$

But we can test effectively whether this congruence has a solution, and if so, find a finite set E of solutions $\vec{X} =< X_1, X_2, \ldots, X_{n-1} >$ such that E includes each solution $\vec{X}(\mod c^\star)$ with $0 \leq X_i < c_n^\star$ for all i. Now if $\vec{x}$ is a solution of (1) then there is some $\vec{t} \in \mathbb{Z}^{n-1}$, $\vec{X} \in E$, such that:

$$x_i = X_i + c_n^\star t_i \text{ for } i \in [1, n-1],$$

$$x_n = \frac{-1}{c_n^\star} \left(c_0^\star + \sum_{1 \leq i \leq n-1} c_i^\star X_i \right) - \sum_{1 \leq i \leq n-1} c_i^\star t_i. \tag*{$(3)_{\vec{X}}$}$$

Thus system (1) has a solution if and only if for some $\vec{X} \in E$ there exists a solution of the system (called $(1)_{\vec{X}}$) obtained from (1) by substituting the expressions on the right of $(3)_{\vec{X}}$ for the variables $x_1, \ldots, x_n$ in (1), and adjoining the inequality:

$$1 - \frac{-1}{c_n^\star} \left(c_0^\star + \sum_{1 \leq i \leq n-1} c_i^\star X_i \right) - \sum_{1 \leq i \leq n-1} c_i^\star t_i > 0$$

Clearly the systems $(1)_{\vec{X}}$ in the variables $t_1, t_2, \ldots, t_{n-1}$ may now be considered individually, and each is of the form (1) but *with one less variable.*

Continuing in this way we either decide that the system has or does not have a solution, or eventually arrive at:

Case II $C = \emptyset$.
Here the system is of the form:

$$\begin{aligned} &P\left(a_0 + \sum_{1\le i\le n} a_i x_i\right), && \vec{a} \in A, \\ \neg &P\left(b_0 + \sum_{1\le i\le n} b_i x_i\right), && \vec{b} \in B, \\ &d_0 + \sum_{1\le i\le n} d_i x_i > 0, && \vec{d} \in D, \end{aligned} \tag{4}$$

We may assume that there is no $a_0 = 0$. For since $\vec{0} \notin A$, the system of linear inequalities

$$a_0 + \sum_{1\le i\le n} a_i x_i \neq 0, \quad \vec{a} \in A$$

certainly has an (effectively found) solution $\vec{X} \in \mathbb{N}^n$ (since some $\vec{X} \in \mathbb{N}^n$ lies outside the union of the $n-1$ dimensional "subspaces" comprised of the rational solutions of the equations $a_0 + \sum_{1\le i\le n} a_i x_i = 0$), and clearly all $\vec{x} \in \mathbb{N}$ satisfy

$$(\bigwedge_i x_i = X_i + t_i) \bigvee \bigvee_i (x_i = 0 \vee x_i = 1 \vee \ldots \vee x_i = X_i - 1).$$

Thus the problem reduces to considering:

(i) the systems obtained from (4) by replacing some x_i with a number $< X_i$,

(ii) the system with variables $\vec{t}$ obtained from (4) by putting $x_i = X_i + t_i$ for $i \in [1, x]$.

But the systems in (i) have fewer than n variables, while the system in (ii) satisfies the requirement that all constant terms be nonzero, since

$$a_0 + \sum_{1\le i\le n} a_i x_i = \left(a_0 + \sum_{1\le i\le n} a_i X_i\right) + \sum_{1\le i\le n} a_i t_i$$

where $a_0 + \sum_{1\le i\le n} a_i X_i \neq 0$. We can also assume that every $d_0 \le 0$, since if $d_0 > 0$ then

$$d_0 + \sum_{1 \le i \le n} d_i x_i \leftrightarrow \left(\sum_i d_i x_i > 0\right) \vee \left(\sum_i d_i x_i = 0\right) \vee \left(1 + \sum_i d_i x_i = 0\right) \vee \ldots \vee \left((d_0 - 1) + \sum_i d_i x_i = 0\right).$$

so we may reduce the problem to considering the system with

$$d_0 + \sum_{1 \le i \le n} d_i x_i > 0$$

replaced by $\sum_{1 \le i \le n} d_i x_i > 0$, and d_0 systems of the form to which case I applies. (Note that each time we return to Case I the number of variables is *reduced* by the procedure used there, so the "complexity" of the systems which need be considered is decreased).

We now have only systems of the form:

$$\begin{aligned} &P\left(a_0 + \sum_{1 \le i \le n} a_i x_i\right), && \vec{a} \in A, \quad (a_0 \ne 0), \\ &\neg P\left(b_0 + \sum_{1 \le i \le n} b_i x_i\right), && \vec{b} \in B, \\ &d_0 + \sum_{1 \le i \le n} d_i x_i > 0, && \vec{d} \in D, \quad (d_0 \le 0), \end{aligned} \tag{5}$$

left to consider, where by the restriction R introduced earlier there are inequalities present which imply $a_0 + \sum_{1 \le i \le n} a_i x_i > 0$ and $b_0 + \sum_{1 \le i \le n} b_i x_i \ge 0$ for all $\vec{a} \in A$, $\vec{b} \in B$. Nov let

$$S = \{p : \ P(p) \wedge \exists \vec{a} \in A(p | a_0)\}.$$

Consider the system of relations:

$$\begin{aligned} &a_0 + \sum_{1 \le i \le n} a_i x_i \not\equiv 0 (\mathrm{mod} p), && p \in S, \quad \vec{a} \in A, \\ &a_0 + \sum_{1 \le i \le n} a_i x_i \ne b_0 + \sum_{1 \le i \le n} b_i x_i, && \vec{a} \in A, \quad \vec{b} \in B, \\ &d_0 + \sum_{1 \le i \le n} d_i x_i > 0, && \vec{d} \in D. \end{aligned} \tag{6}$$

Since

$$x \equiv 0 \ (\mathrm{mod}\ p) \leftrightarrow \exists z (x = \underbrace{z + z + \ldots + z}_{p \text{ times}}),$$

there is clearly an $L_{=,+}$ sentence which expresses the statement that system (6) has a solution. But using Presburger's algorithm we can decide effectively whether a given $L_{=,+}$ sentence is true and thus whether (6) has a solution. If it does not have a solution, then the only way (5) can have a solution is if some $a_0 + \sum_{1 \le i \le n} a_i x_i = p$ for some prime $p \in S$. Therefore the problem can be reduced to considering the $|S| \cdot |A|$ systems obtained by adding one of the possible equations of the form

$$(a_0 - p) + \sum_{1 \le i \le n} a_i x_i = 0$$

to (5), and of course case I applies to all of these.

If (6) does have a solution, then some such solution $\vec{X}$ can obviously be found effectively. It is claimed that in this case (5) will also have a solution (provided H_L is true). To prove this we will apply proposition 4 to the polynomials (in x):

$$\begin{aligned} &a_0 + \left(\sum_{1 \le i \le n} a_i X_i \right) x, \qquad \vec{a} \in A, \\ &b_0 + \left(\sum_{1 \le i \le n} b_i X_i \right) x, \qquad \vec{b} \in B. \end{aligned} \tag{7}$$

The restriction R introduced earlier and the requirement that $d_0 \le 0$ ensure that $\sum_{1 \le i \le n} a_i X_i > 0$ and $\sum_{1 \le i \le n} b_i X_i > 0$. To check that the product of the polynomials in the first line of (7) has no fixed divisor (other than 1), suppose that some prime

$$p \left| \prod_{\vec{a} \in A} \left(a_0 + \left(\sum_{1 \le i \le n} a_i X_i \right) x \right) \right.$$

for every $x \in \mathbb{N}$. Taking $x = 0$ we see that $p \in S$, while taking $x = 1$ shows that $p | \prod_{\vec{a} \in A} \left(a_0 + \sum_{1 \le i \le n} a_i X_i \right)$, that is, that $a_0 + \sum_{1 \le i \le n} a_i X_i \equiv 0 (\mathrm{mod} p)$ for *some* $\vec{a} \in A$, which is impossible since X is a solution of (6).

Thus assuming H_L it follows from proposition 4 that $x > 0$ can be chosen satisfying:

$$P\left(a_0 + \sum_{1\le i\le n} a_i(X_i x)\right), \qquad \vec{a} \in A,$$

$$\neg P\left(b_0 + \sum_{1\le i\le n} b_i(X_i x)\right), \qquad \vec{b} \in B.$$

Taking $x_i = X_i x$ gives a solution of (5) since

$$\begin{aligned} d_0 + \sum_{1\le i\le n} d_i x_i &= d_0 + \left(\sum_{1\le i\le n} d_i X_i\right) x \\ &\ge d_0 + \sum_{1\le i\le n} d_i X_i > 0, \end{aligned}$$

(because $\vec{X}$ is a solution of (6) and thus $\sum_{1\le i\le n} d_i X_i > -d_0 \ge 0$). $\dashv$

Having proved theorem 2 it now follows immediately (by proposition 0.1 of chapter 2) that:

Corollary 6. *If H_L is true then $z = x \cdot y$ cannot be defined by an existential $L_{=,+,P}$ formula.*

Remark. *The argument above is easily expended to give a decision procedure for all $L_{=,+,P}$ sentences of the form $\tilde{\exists}\vec{x}\exists\vec{y}\phi(\vec{x},\vec{y})$ where $\phi(\vec{x},\vec{y})$ is quantifier free and $\tilde{\exists}x_i$ denotes $\forall z_i \exists x_i \ge z_i$ (where z_i does not occur in $\phi(\vec{x},\vec{y})$, that is, $\tilde{\exists}$ is the quantifier:*

there exist infinitely many x_i.

Corollary 6 can also be proved for $L_{=,+,P}$ formulas with only $\exists$ and $\tilde{\exists}$ quantifiers, by analysing the properties of predicates defined by such formulas.

The theorems proved above suggest asking what the situation regarding decidability is for the languages $L_{=,',P}$, $L_{\le,P}$ (in the notation of chapter 2). In the first case we have

Theorem 7. *H_L implies that the $L_{=,',P}$ theory of $\mathbb{N}$ is decidable.*

This can be proved directly by quantifier elimination, but in view of theorem 2 it also follows from the general result:

Proposition 8 (Hanf [17], Thomas [46]). *Let $P(x)$ be any unary predicate on $\mathbb{N}$. Then the $L_{',P}$ theory of $\mathbb{N}$ is decidable if and only if the existential $L_{',P}$ theory of $\mathbb{N}$ is decidable.*

Open Question. Is the $L_{\leq,P}$ theory of $\mathbb{N}$ undecidable?

As observed by Thomas [46], this question has a negative answer if H_L, fails so badly that $\liminf(p_{i+1} - p_i) = \infty$ (where p_i denotes the i th prime).

It also seems worthwhile to remark that although it is obviously hopeless to seek unconditional proofs of the decidability results above using only currently known number theory, it does not seem quite so implausible that a proof of the *undecidability* of the $L_{=,+,P}$ theory of $\mathbb{N}$ might be obtained without going very far beyond what is currently known about primes. Certainly there are other ways of deducing the undecidability result from H_L. For example, it is a consequence of the conjecture:

If r is divisible by all prime numbers $\leq n$ then there exists a sequence of n consecutive prime numbers in arithmetic progression with common difference r. (This states that arithmetic progressions of consecutive primes of all conceivable sorts exist, and is a consequence of H_L - see Schinzel and Sierpinski [37]).

Finally there is the question of what natural axiom systems it is reasonable to expect H_L (or H for that matter) to be provable in (assuming it is true). Let $I\Sigma_1$ be the axiom system similar to the system $I\Delta_0$ defined in chapter 1, but with induction allowed for induction hypotheses expressed by Σ_1 formulas, that is, formula of the form $\exists\vec{x}\theta(\vec{x}, y)$, where $\theta(\vec{x}, y)$ is a Δ_o formula. (Intuitively $I\Sigma_1$ corresponds to doing inductive arguments with recursively enumerable predicates as the induction hypotheses - a lar proportion of known number theory can be developed this way.)

Conjecture. $I\Sigma_1 \nvdash H_L$

(where H_L denotes the single sentence expressing the conjecture, as described above).

Appendix I

An alternative proof that addition and multiplication can be defined by bounded $L_{\leq,\perp}$ formulas.

Since the proof of theorem 1.3 shows that $z = x \cdot y$ can be defined by a bounded $L_{\leq,A,\perp}$ formula where $A(x, y, z) \leftrightarrow x + y = z$, it suffices to prove that $z = x + y$ can be defined by a bounded $L_{\leq,\perp}$ formula.

Recall that for all real numbers a, b the integer part sign [ ] has the property:

$$[a] + [b] \leq [a + b] \leq [a] + [b] + 1.$$

Lemma. *$z = x + y$ if and only if z is the* least *number such that:*

(I) $x \leq z \wedge y \leq z$

(II) $z \equiv x + y (mod 6)$

(III) *for every prime* $p \leq z$, $p \neq 7$, *either*

$$\left[\frac{z}{p}\right] \equiv \left[\frac{z}{p}\right] + \left[\frac{z}{p}\right] (mod\ 7)$$

$$or \left[\frac{z}{p}\right] \equiv \left[\frac{z}{p}\right] + \left[\frac{z}{p}\right] + 1 (mod\ 7)$$

Proof. Obviously $z = x+y$ satisfies (I), (II), (III), so it is only necessary to show that if $z < x + y$ at least one of these must fail. Suppose $z < x + y$ satisfies (I) and (II). Then by (II), $x + y - z = 6m$ for some $m \geq 1$. But for any $n \geq 1$ there is a prime p with $n < p \leq 2n$ by Bertrand's postulate (a theorem of Chebychev - see, for example, Hardy and Wright [18] or chapter 1 of this thesis). Hence for any $m \geq 1$ there is a prime $p \neq 7$ with $\frac{3}{2}m \leq p \leq 3m$, and thus some prime $p \neq 7$ satisfies $2p \leq x + y - z \leq 4p$.

But then $2 \leq \left[\frac{x}{p} + \frac{y}{p} - \frac{z}{p}\right] \leq 4$ so

$$2 + \left[\frac{z}{p}\right] \leq \left[\frac{x}{p} + \frac{y}{p}\right] \leq 5 + \left[\frac{z}{p}\right]$$

and

$$1 + \left[\frac{z}{p}\right] \leq \left[\frac{x}{p}\right] + \left[\frac{y}{p}\right] \leq 5 + \left[\frac{z}{p}\right]$$

In other words

$$\left[\frac{z}{p}\right] - \left[\frac{x}{p}\right] - \left[\frac{y}{p}\right] = -1,\ -2,\ -3,\ -4 \text{ or } -5.$$

Hence

$$\left[\frac{z}{p}\right] - \left[\frac{x}{p}\right] - \left[\frac{y}{p}\right] \not\equiv 0,\ 1 \text{ (mod 7)},$$

and since $p \leq x + y - z \leq z + z - z = z$ (by (I)), condition (III) fails. ⊣

Theorem. $z = x + y$ can be defined by a bounded $L_{\leq,\perp}$ formula.

Proof. We show that conditions (I), (II), (III) in the lemma can be defined by bounded $L_{\leq,\perp}$ formulas.

As the predicates $y = x'$, $x = 0$ are definable by bounded $L_{\leq}$ formulas, $x \equiv i$ (mod 6) can be defined for $i = 0, 1, \ldots, 5$ using:

$$x \equiv i \text{ (mod 6)} \leftrightarrow x = i \vee (x \geq 6 \wedge \neg(2 \perp x - i) \wedge \neg(3 \perp x - i)).$$

Thus $x + y \equiv z(\text{mod} 6)$ can be defined by a bounded $L_{\leq,\perp}$ formula, since

$$x + y = z \text{ (mod 6)} \leftrightarrow \bigvee_{\substack{i,j,k \leq 5 \\ i+j \equiv k \ (mod\ 6)}} (x \equiv i \text{ (mod 6)} \wedge y \equiv j \text{ (mod 6)} \wedge z \equiv k \text{ (mod 6)})$$

$\dashv$

Similarly condition (III) will be expressible by a bounded $L_{\leq,\perp}$ formula if it can be shown that for $i = 0, 1, ..., 6$,

$$p \textit{ is prime} \wedge p \neq 7 \wedge \left[\frac{x}{p}\right] \equiv i \text{ (mod 7)}$$

can be defined by a bounded $L_{\leq,\perp}$ formula. But this is the case since the predicate p *is prime* can obviously be handled (as in §1), and $\left[\frac{x}{p}\right] \equiv i$ (mod 7) for a prime $p \neq 7$ if and only if there exists $u \leq x$ with $\neg p \perp u \wedge \neg 7 \perp u$ such that there are *exactly* i distinct numbers v satisfying $u < v \leq x \wedge \neg p \perp v$. (To see this consider the numbers: $u = 7pw < p(7w+1) < p(7w+2) < \ldots < p(7w+i) \leq x < p(7w+i+1)$).

$\dashv$

Remark. *This method can be refined to show that the full stength of* $\perp$ *(with* $\leq$*) is not required to define the addition and multiplication predicates on* $[0, n]$. *Write* $p|_S x$ *for* $p \in S \wedge p|x$, *where* S *is a set of primes. Then there are* $L_{\leq,|_S}$ *formulas* $\phi_A(x, y, z)$, $\phi_M(x, y, z)$ *such that for all* n *and all* $x, y, z \in [0, n]$,

$$+y = z \leftrightarrow < [0, n], \leq, |_S > \models \phi_A(x, y, z)$$
$$x \cdot y = z \leftrightarrow < [0, n], \leq, |_S > \models \phi_M(x, y, z)$$

provided $\{p : p \leq C \log n\} \subseteq S$ *(where* C *is a suitably large constant, for example* $C > 1$.

This is because for $y \leq 2C \log n$ the definition of $x + y = z$ given above requires only a knowledge of which numbers are divisible by primes $p \leq C \log n$. Thus there is an $L_{\leq,|_S}$, formula $\theta(z, y, p)$ such that for all primes $p \leq C \log n$ and all z, y,

$$\theta(z, y, p) \leftrightarrow y < p \wedge z \equiv y \text{ (mod } p).$$

Also since the predicates $x + y = z$ and thus $x \cdot y = z$ can be defined on $[0, 2C \log n]$, it is possible to define $x + y \equiv z$ (mod p) and $x \cdot y \equiv z$ (mod p) (the latter in two steps) for all $x, y \in [0, n]$, $p \leq C \log n$ and hence obtain definitions for $x + y = z$ and $x \cdot y = z$ on $[0, n]$ from the equivalences:

$$x + y = z \leftrightarrow \forall p \leq C \log n(x + y \equiv z \pmod{p})$$
$$x \cdot y = z \leftrightarrow \forall p \leq C \log n(x \cdot y \equiv z \pmod{p}).$$

Using these ideas, the method can also be extended to show that there are bounded $L_{\leq,\perp}$ formulas $\psi_A(x,y,z)$, $\psi_M(x,y,z)$ with the property that if a suitable axiom ϕ defining $\perp$, for example:

$$\forall x \forall y (x \perp y \leftrightarrow \forall z(\exists u(x = z \cdot u) \wedge \exists v(y = z \cdot v) \rightarrow z = 1)),$$

is added to the axiom system $I\Delta_0$ considered in chapter 1, then

$$I\Delta_0 + \phi \models \forall x \forall y \forall z(x + y = z \leftrightarrow \psi_A(x,y,z)),$$
$$I\Delta_0 + \phi \models \forall x \forall y \forall z(x \cdot y = z \leftrightarrow \psi_M(x,y,z)).$$

Appendix II

Some problems in computational complexity

As mentioned in chapter 2 a positive solution to the RUD $= S$ problem would solve several open problems in computational complexity theory. For a start the class of all sets S of positive integers such that S can be accepted or rejected by some deterministic Turing machine using linear working space (or equivalently such that S has an $\mathcal{E}^2_\star$ definition) is easily seen to include RUD and to be contained in S. Thus we have:

$$\text{RUD}(= \Delta_0 \textit{ sets}) \subseteq \textit{linear space}(= \mathcal{E}^2_\star \textit{ sets}) \subseteq S(= \text{NEXP}),$$

where it is not known whether *linear space*= NEXP or whether RUD = *linear space*.

A proof that RUD $=$ S would also settle the P = NP problem which asks whether every S $\subseteq \mathbb{N}$ which can be accepted in polynomial time by some nondeterministic Turing machine (that is, every S $\in$ NP) is a member of the class P of all sets which can be accepted (or rejected) in polynomial time by a deterministic Turing machine. Let co $-$ NP $= \{\mathbb{N} \setminus \text{S} : \text{S} \in \text{NP}\}$. At present it is not known whether NP $=$ co $-$ NP, although P $=$ NP $\Rightarrow$ NP $=$ co $-$ NP since if S $\in$ P then obviously $\mathbb{N} \setminus$ S $\in$ P. These problems were generalised by Stockmeyer [43]. The *Stockmeyer polynomial time hierarchy* is defined by taking

$\Sigma^P_0 = \Pi^P_0 = P,$

$\text{S} \in \Sigma^P_{n+1} \leftrightarrow$

S *is accepted in polynomial time by some nondeterministic Turing machine using an oracle* $A \in \Pi^P_n$

$\Pi^P_{n+1} = \{\mathbb{N} \setminus \text{S} : \text{S} \in \Sigma^P_{n+1}\}.$

The problem of whether the classes $\Sigma_0 \subseteq \Sigma_1^P \subseteq \Sigma_2 \subseteq \ldots$ form a *strict* hierarchy is open. Let $\Delta^P = \cup_n \Sigma_n^P = \cup_n \Pi_n^P$. Since $\Sigma_1^P = \mathrm{NP}$ it is obvious that $\mathrm{NP} = \Delta^P \leftrightarrow \mathrm{NP} = \mathrm{co} - \mathrm{NP}$. Also as observed by Paris and Dimitracopoulos [28], the sets $S \subseteq \Delta^P$ are precisely those subsets of $\mathbb{N}$ with the property that $\mathrm{S} = \{x : \phi(x)\}$ for some $\phi(x)$ of the form:

$$Q_1 y_1 \leq x^{[\log_2 x]^m} Q_2 y_2 \leq x^{[\log_2 x]^m} \ldots Q_s y_s \leq x^{[\log_2 x]^m} \lambda(x, \vec{y}) \qquad (1)$$

where $Q_1, Q_2, \ldots, Q_s$ denotes quantifiers ($\forall$ or $\exists$), $m \in \mathbb{N}$, and λ is a quantifier free $L_{\leq,+,\cdot}$ formula.

Furthermore,the polynomial time hierarchy collapses at some point (that is, $\Delta^P = \Sigma_n^P = \Pi_n^P$ for some n) if and only if a fixed bound independent of S can be placed of the number of quantifiers required in (1).

An analogous open problem for Δ_0 formulas is whether there is some fixed k such that every $R \in \mathrm{RUD}$ can be defined by a formula of the form:

$$Q_1 \vec{v}^{(1)} \leq x Q_2 \vec{v}^{(2)} \leq x \ldots Q_k \vec{v}^{(k)} \leq x \rho(x, \vec{v}) \qquad (2)$$

where $Q_i \vec{v}^{(i)} \leq x$ denotes a block of similar bounded quantifiers of the form $\exists v_1^{(i)} \leq x \exists v_2^{(i)} \leq x \ldots \exists v_r^{(i)} \leq x$ or $\forall v_1^{(i)} \leq x \forall v_2^{(i)} \leq x \ldots \forall v_r^{(i)} \leq x$, and ρ is a quantifier free $L_{\leq,+,\cdot}$ formula. A bound k on the number of *changes* of quantifier does exist if RUD =*linear space*, as is shown in Harrow [19]. (Wilkie [48] has proved it is not possible to put a bound on the number of *individual* quantifiers).

On the other hand Paris and Dimitracopoulos [28] show that for every k there is some formula $\phi_k(x)$ of type (1) which cannot be defined by any bounded $L_{\leq,+,\cdot}$ formula with only k changes of quantifier. This is because it is possible to use numbers $u \leq x^{[\log_2 x]}$ as codes for sequences $v_1, \ldots, v_r \leq x$ and thus replace blocks of quantifiers of the form $Qv_i \leq x$ by single quantifiers $Qu \leq x^{[\log_2 x]}$, to obtain a sort of *universal formula* $T_k(x, n)$ with

$$T_k(x, \ulcorner\theta\urcorner) \leftrightarrow \theta(x)$$

for all $\theta(x)$ of form (2) having Gödel numbers $\ulcorner\theta\urcorner$ sufficiently small *compared with x.*

Theorem 1. $\mathrm{RUD} = S \Rightarrow \mathrm{NP} \neq \mathrm{co} - \mathrm{NP}$.

Proof. Suppose RUD $= S$. Then RUD =*linear space* so there exists some fixed k such that every $R \in \mathrm{RUD}$ can be defined by some formula of form (2). Thus there is some formula $\phi_k(x)$ of form (1) such that the set

$$\mathrm{S} = \{x : \phi_k(x)\} \notin \mathrm{RUD},$$

that is, there is some $\mathrm{S} \in \Delta^P$ such that $\mathrm{S} \notin \mathrm{RUD}$. But since $\mathrm{NP} \subseteq$

NEXP $= S =$ RUD, we see that S $\notin$ NP, so NP $\neq \Delta^P$, that is, NP $\neq$ co $-$ NP. $\dashv$

Not only does assuming RUD $= S$ ensure the existence of an upper bound on k in (2), but it also implies a lower bound, namely that some bounded $L_{\leq,+,\cdot}$ formula $\phi(x)$ is *not* equivalent to any *bounded diophantine formula*:

$$\exists v_1 \leq x \exists v_2 \leq x \ldots \exists v_r \leq x \lambda(x, \vec{v})$$

where λ is quantifier free and can be assumed without loss of generality to be an equation between two polynomials.

For all such formulas obviously define sets S $\in$ NP, and NP $\neq$ NEXP by a well known nondeterministic time hierarchy theorem.

Finally it should be remarked that *assuming the existence of an upper bound on k allows us to deduce $\exists n(\Sigma_n^P = \Delta^P)$.*

For suppose S $\in \Delta^P$. Then S is defined by some formula $\phi(x)$ of the form:

$$Q_1 y_1 \leq x^{[\log_2 x]^m} Q_2 y_2 \leq x^{[\log_2 x]^m} \ldots Q_s y_s \leq x^{[\log_2 x]^m} \lambda(x, \vec{y}).$$

We need only show that an upper bound can be placed on the s required here. Let $\theta(x, z)$ be the bounded $L_{\leq,+,\cdot}$ formula:

$$Q_1 y_1 \leq z Q_2 y_2 \leq z \ldots Q_s y_s \leq z \lambda(x, \vec{y}).$$

Choose a suitable *polynomial* $p(x, z)$ coding ordered pairs $< x, z >$ by the single numbers $p(x, z)$, with $p(x, z) \leq C z^2$ (C constant) whenever $x \leq z$. Obviously we can find a bounded formula $\theta^\star(w)$ such that:

$$\theta^\star(p(x, z)) \leftrightarrow \theta(x, z).$$

But *if* every bounded $L_{\leq,+,\cdot}$ formula is equivalent to one with only k changes of quantifier then $\theta^\star(w)$ can be assumed to be of the form:

$$Q_1 \vec{v}^{(1)} \leq w Q_2 \vec{v}^{(2)} \leq w \ldots Q_k \vec{v}^{(k)} \leq w \rho(w, \vec{v})$$

where $\rho(x, \vec{v})$ is quantifier free. Using coding, each block $Q_i \vec{v}^{(i)} \leq w$ of quantifiers of the form $Q_i v_1^{(i)} \leq w Q_i v_2^{(i)} \leq w \ldots Q_i v_r^{(i)} \leq w$ can be replaced by a single quantifier of the form $Q_i y_i \leq w^r$. Thus $\theta^\star(w)$ is equivalent to some formula of the form:

$$Q_1 y_1 \leq w^r Q_2 y_2 \leq w^r \ldots Q_k y_k \leq w^r \rho_1(w, \vec{v})$$

where $\rho_1(w, \vec{v})$ is a formula with all quantifiers bounded by $w^{[\log_2 w]^n}$ for some n, and *the number of these quantifiers is less than some bound independent of* S. (This is possible since there is a polynomial bound on the time required, given the y_i's, to compute the sequences $v_1^{(i)}, v_2^{(i)}, \ldots, v_r^{(i)}$ coded by the y_i's and then determine whether $\rho_1(w, \vec{v})$ is satisfied).

Clearly it follows that

$$\phi(x) \leftrightarrow \theta\left(x, x^{[\log_2 x]^m}\right) \leftrightarrow \theta^\star\left(p\left(x, x^{[\log_2 x]^m}\right)\right) \leftrightarrow$$

$$Q_1 y_1 \leq x^{[\log_2 x]^{m+1}} Q_2 y_2 \leq x^{[\log_2 x]^{m+1}} \ldots Q_k y_k \leq x^{[\log_2 x]^{m+1}} \rho_2(x, \vec{y})$$

where all the quantifiers in ρ_2 are bounded by $x^{[\log_2 x]^n}$ for some n, and their *number* does not depend on S.

In particular it follows from this that:

Theorem 2. RUD $= S \Rightarrow \exists n(\Sigma_n^P = \Delta^P)$, *(that is, the Stockmeyer polynomial time hierarchy collapses if* RUD $= S$*).*

Of course by theorem 1 we must have $n > 1$, but this could conceivably be consistent with the *conjecture* of Baker and Selman [1] that $\Sigma_3^P = \Delta^P \neq \Sigma_2^P$.

Bibliography

[1] T. P. Baker and A.L. Selman. A second step toward the polynomial hierarchy. *Theor. Comput. Sci.*, 8:177–187, 1979.

[2] A. P. Bel'tyukov. Decidability of the universal theory of natural numbers with addition and divisibility (in russian). *Zap. Naučn. Sem. Leningrad. Otdel. Mat. Inst. Steklov (LOMI)*, 60:15–28, 1976.

[3] J. H. Bennett. *On Spectra.* PhD thesis, Princeton University, 1962.

[4] G. D. Birkhoff and H. S. Vandiver. On the integral divisors of $a^n - b^n$. *Ann. of Math*, 5:173–180, 1904.

[5] V. Brun. Le crible d'Ératosthène et le théorème de Goldbach. *Skr. Norske Vid.-Akad. Kristiana I*, 3:36, 1920.

[6] S. A. Cook and R. A. Reckhow. The relative efficiency of propositional proof systems. *J. Symbolic Logic*, 44:36–50, 1979.

[7] M. Davis. Hilbert's tenth problem is unsolvable. *Amer. Math. Monthly*, 80:233–269, 1973.

[8] L. E. Dickson. A new extension of Dirichlet's theorem on prime numbers. *Messanger of Mathematics*, 33:155–161, 1904.

[9] C. Dimitracopoulos. *Matijasevič's Theorem and Fragments of Arithmetic.* PhD thesis, University of Manchester, 1980.

[10] P. Erdös. Über die kleinste quadratfreie Zahl einer arithmetischen Reihe. *Monatsh. Math.*, 64:314–316, 1960.

[11] P. Erdös. How many pairs of products of consecutive integers have the same prime factors? *Amer. Math. Monthly*, 87:391–394, 1980.

[12] P. Erdös and T. N. Shorey. On the greatest prime factor of $2^p - 1$ for a prime p and other expressions. *Acta Arith.*, 30:257–265, 1975.

[13] R. G. Fagin. A spectrum hierarchy. *Z. Math. Logik Grundlagen Math*, 21:123–134, 1975.

[14] A. O. Gel'fond and Yu. V. Linnik. Elementary Methods in the Analytic Theory of Numbers. In *International Series of Monographs in Pure and Applied Mathematics*, volume 92. Pergamon Press, 1962.

[15] C. A. Grimm. A conjecture on consecutive composite numbers. *Amer. Math. Monthly*, 76:1126–1128, 1969.

[16] A. Grzegorczyk. Some classes of recursive functions. *Rozprawy Matematyczne*, 4, 1953.

[17] W. Hanf. Model-theoretic methods in the study of elementary logic. In *Symposium on the Theory of Models*, pages 132–145. North Holand, Amsterdam, 1965.

[18] G. H. Hardy and E. M. Wright. *An Introduction to the Theory of Numbers*. Oxford University Press, Oxford, 5 edition, 1979.

[19] K. Harrow. The bounded arithmetic hierarchy. *Inform, and Control*, 36:102–117, 1978.

[20] J. E. Hopcroft and J. D. Ullman. *Introduction to Automata Theory Languages and Computation*. Addison-Wesley, 1979.

[21] N. D. Jones and A. L. Selman. Turing machines and the spectra of first order formulas. *J. Symbolic Logic*, 39:139–150, 1974.

[23] M. Langevin. Plus grand facteur premier d'entiers voisins. *C.R. Acad. Sci. Paris Sér. A*, 281:491–493, 1975.

[22] M. Langevin. Plus grand facteur premier d'entiers consécutifs. *C.R. Acad. Sci. Paris Sér. A*, 280:1567–1570, 1975.

[24] M. Langevin. Facteurs premiers des coefficients binomiaux. In *Séminaire Delange-Pisot-Poitou (Théorie des nombres) 20e année*, volume 27, page 15, 1978/79.

[25] D. H. Lehmer. On a problem of Störmer. *Illinois J. Math.*, 8:57–79, 1964.

[26] H. Lessan. *Models of Arithmetic*. PhD thesis, University of Manchester, 1978.[1]

[27] L. Lipshitz. The diophantine problem for addition and divisibility. *Trans. Amer. Math. Soc.*, 235:271–283, 1978.

[1]Published as part III of this volume.

[28] J. B. Paris and C. Dimiracopoulos. Truth definitions for Δ_0 formulae. In *Logic and algorithmic*, pages 317–329. Univ. Genève, Geneva, 1982.

[29] J. B. Paris and L. Harrington. A mathematical incompleteness in Peano Arithmetic. In J. Barwise, editor, *Handbook of Mathematical Logic*. North Holland, 1977.

[30] K. Prachar. *Primzahlverteilung*. Springer-Verlag, Berlin, 1957.

[31] K. Prachar. Über die kleinste quadratfreie zahl einer arithmetischen Reihe. *Monatsh. Math.*, 62:173–176, 1958.

[32] M. Presburger. Über die Vollständigkeit eines gewissen Systems der Arithmetik ganzer Zahlen, in welchem die Adition als einzige Operation hervortritt. In *Sprawozdanie z I Kongresu Matematykòw Krajòw Slowiańskich (Comptes-rendus du Ier Congrès des Mathématiciens des Pays Slaves), Warsaw*, pages 92–101, 395, 1930.[2]

[33] R. W. Ritchie. Classes of predictably computable functions. *Trans. Amer. Math. Soc.*, 106:139–173, 1963.

[34] C. Størmer. Quelques théorèmes sur l'équation de Pell $x^2 - Dy^2 = \pm 1$ et leurs applications. *Skrifter Videnskabs selskabet (Christiana) I*, 2:48, 1897.

[35] J. Robinson. Definability and decision problems in arithmetic. *J. Symbolic Logic*, 1949.

[36] A. Schinzel. Remarks on the paper "sur certaines hypothèses concernant les nombres premiers". *Acta Arith.*, 7:1–8, 1961.

[37] A. Schinzel and W. Sierpìnski. Sur certaines hypothèses concernant les nombres premiers. *Acta Arith.*, 4:185–208, 1958.

[38] L. Schnirelmann. Über additive Eigenschaften von Zahlen. *Math. Ann.*, 1933.

[39] H. Scholz. Problems. *J. Symbolic Logic*, 17:160, 1952.

[40] H. N. Shapiro. On primes in arithmetic progression (II). *Ann. of Math.*, 52:231–243, 1950.

[41] J. C. Shepherdson. Non standard Models for fragments of number theory. In L. Henkin J. W. Addison and A Tarski, editors,

[2] R. Stansifer, Presburger's Article on Integer Arithmetic: Remarks and Translation, Cornell tech reports,
`http://ecommons.library.cornell.edu/bitstream/1813/6478/1/84-639.pdf`.

the Theory of Models, pages 342–358. North-Hoiland, Amsterdam, 1965.

[42] H. M. Stark. Effective estimates of solutions of some diophantine equations. *Acta Arith.*, 24:251–259, 1973.

[43] L. J. Stockmeyer. The polynomial time hierarchy. *Theor. Comput. Sci.*, 3:1–22, 1976.

[44] Sylvester. On arithmetical series (part I). *Messenger of Mathematics*, 21:1–19, 1981.

[45] A. Tarski. Undecidability of group theory. (Abstract). *J. Symbolic Logic*, 1949.

[46] W. Thomas. The theory of successor with an extra predicate. *Math. Ann.*, 237:121–132, 1978.

[47] A. J. Wilkie. Some results and problems on weak systems of arithmetic. In L. Pacholski L. A. Macintyre and J. Paris, editors, *Logic Colloguium'77*. North-Holland, Amsterdam, 1978.

[48] A. J. Wilkie. Applications of complexity theory to Σ_0-definability problems in arithmetic. In *Lecture Notes in Mathematics*, volume 834. Springer-Verlag, Berlin, 1980. Model Theory of Algebra and Arithmetic. (Proceedings, Karpacz, Poland 1979.).

[49] A. J. Wilkie. Axioms for division. *To appear.*[3]

[50] A. R. Woods. Algebraic Properties of Non-Standard Numbers. Master's thesis, Monash University, Clayton, Victoria, 1977.

[3]Unpublished.

Part III

Models of Arithmetic

by H. Lessan

Thesis presented for the degree of Doctor of Philosophy
at Manchester University, Faculty of Science –February 1978.

Hamid Lesan (1951–2006)

Hamid Lesan[4] was born in Kashan, Iran, in 1951, the eldest of five children of a professor of literature at Teheran University. In his late teens Hamid won an Iranian government scholarship to study in the UK in preparation for entry to Manchester University. From 1971 to 1974 he was an undergraduate student in mathematics at Manchester University where he gained a first class honours degree in 1974. At this stage Hamid had already become passionate about mathematical logic, having been exposed to it in advanced undergraduate lecture courses. He stayed on in the mathematics department at Manchester University for a further four years, gaining first an MSc and then a PhD, during which time I was privileged to be his supervisor. This was an exciting time for logic at Manchester and for nonstandard model theorists in particular; Jeff Paris and his PhD students were then working on the first combinatorial independence results from first order Peano arithmetic. Hamid's focus was somewhat different, with an emphasis on definability and on the model theoretic properties of nonstandard models of PA arising from recursive saturation. There was in that period a lively competitive camaraderie amongst the group of Manchester logic graduate students, who included also D. Anapolitanos, C. Dimitracopoulos, L. Kirby, G. Koletsos, G. Leversha, and J. Väänänen; Hamid's sense of humour, wide cultural interests, and infectious enthusiasm contributed enormously to the cohesion of the group and also to its interaction with the then relatively youthful logic faculty.

Hamid married his wife Lynda in 1976 and, after completing his PhD in the spring of 1978, returned to Iran at the start of the Iranian revolution, accompanied by Lynda and their newborn daughter, and

[4]Lessan changed the Roman alphabet spelling of his surname to Lesan shortly after the submission of his thesis. However most citations of the thesis use the former spelling of his name.

full of hope for the future of his country. For the following two years of revolutionary turmoil Hamid worked as a temporary university lecturer in Kashan and Teheran until the situation became so desperate at the start of the Iran-Iraq war that the family had no realistic option but to return to the UK. Unfortunately, from the Iranian government's viewpoint this move was illegal since Hamid's study in the UK had been financed by 7 years of government scholarships, which formally required a commitment on Hamid's part to work in Iran for a minimum of 14 years. Hamid's father was promptly arrested on account of Hamid's "default" and Hamid was forced to borrow what was then the sizeable sum of 12,000 pounds in order to secure his father's release from prison. This new debt, combined with a shortage of academic posts in the UK and the necessity to support his growing family, sadly ended Hamid's hopes of pursuing an academic career.

For the remaining 26 years of Hamid's life he worked in the UK in the design of computer software, being employed successively by a number of large companies. He became an acknowledged expert on safety critical systems, and in this capacity worked on the design of such systems for several important projects including the Channel Tunnel rail link and the European Airbus. After some years of ill-health Hamid died suddenly from heart failure in August 2006 at the age of 54. He was survived by his wife and two daughters.

Hamid Lesan was one of only a very small number of Iranian mathematical logicians of the 20th century. Although he did not continue his academic research in logic after gaining his PhD, his pioneering study of the structure of definability in nonstandard models of arithmetic was of considerable originality and influence, with many of its key ideas being absorbed into the folklore of the subject, thus providing the building blocks for later research. It is therefore very much to be welcomed that this seminal 35-year old work is finally being published in the present volume.

George Wilmers
Manchester, March 2013

Acknowledgements

My thanks go first and foremost to my supervisor, George Wilmers, for his advice and constructive criticism during the preparation and presentation of this thesis and to whom is due the credit for many ideas contained herein.

My thanks also go to my wife Lynda, for her understanding during the last three years and for typing the manuscript.

I am also indebted to my friends, colleagues and lecturers in Manchester and elsewhere, especially to Jeff Paris and Ken McAloon, who have helped me with many fruitful discussions and suggested many of the problems here discussed.

My thanks also go to Mrs. O. Behrouzi, of Bank Markazi, Iran, for her help concerning financial matters.

Financial support came from The Iranian National Insurance Organization.

"For let Philosopher and Doctor preach
Of what they will, and what they will not - each
Is but one link in an eternal Chain
That none can slip, nor break, nor over-reach."
(Rubáiyát of Omar Khayyám - E. Fitzgerald trans.)

Abstract

A-saturated and recursively saturated structures were introduced and extensively studied by Wilmers [W_1] and Barwise-Schlipf [B&S] respectively. These structures appear naturally in the study of models of arithmetic and their special properties help us to solve some problems regarding these models. Let PA denote the set of Peano's axioms and M a model of PA.

In Chapter 2 we show the existence of end extensions of M which are A-saturated, using this we prove some results regarding the initial segments of M and answer a question of McAloon concerning the position of the class of Δ_n-definable elements of M.

In Chapter 3 we look at the substructure of M determined by its set of Σ_n-definable elements and the initial segment of M determined by this set. It turns out that these structures are models of substantial parts of PA and that the set of standard integers is definable in them. In the second part of this Chapter we give a 'partial' solution to a problem of Gaifman concerning arithmetical structures.

In Chapter 4 we consider the initial segments of a model K of the set of Π_1-consequences of PA determined by an element of K and show that in certain circumstances these initial segments are A-saturated; using this we generalize some results of Chapter 2 to models of weaker systems than PA.

In Chapter 5 we continue to exploit the versatility of the A-saturated structures introduced in Chapter 4, showing the existence of approximating chains of models of the set of Π_2-consequences of PA. In the last section of this Chapter we introduce a useful class of initial segments of M and prove some results concerning the number of isomorphic initial segments of an initial segment I of M and its relation to the number of automorphisms of I, when M is countable.

Certain results concerning the relative position of the classes of Σ_n-

definable elements of a model of PA can be found in 3.1 and 4.2.

Declaration

No portion of the work referred to in this thesis has been submitted in support of an application for another degree or qualification of this or any other university or institution of learning.

1

Notation, terminology and prerequisites

1.1 Notation

Throughout we shall assume familiarity with general mathematical logic and model theory as developed in [C& K]. Our logical symbols consist of $\wedge, \vee, \neg, \rightarrow, \forall, \exists, \exists!$ (there exists a unique -). Other symbols are:

$\vdash$ – 'is a consequence of';
$\models$ – 'is a model of";
$\cong$ – 'is isomorphic to";
$\equiv$ – 'is elementary equivalent to';
$\equiv_{\infty\omega}$ – 'is $\infty\omega$ elementary equivalent to';
$\subseteq$ – 'is a subset of' or 'is a substructure of', depending on the context;
$\prec$ – 'is an elementary substructure of';
$\cap(\cup)$ – 'is the intersection (resp. union) of' models or sets, again depending on the context.

Structures will be denoted by upper-case Roman letters, e.g. M, M', K etc. Generally our notation does not distinguish between a structure and its domain, or between an element of a structure and its name.

The vector symbol $\vec{x}$ will denote a sequence $x_1, x_2, \ldots$ of some finite length. For a set A, $(A)^{<\omega}$ will denote the set of all finite sequences of A.

Let M be a structure for a language $\mathcal{L}$:

(i) The language obtained by adding names for each element of M to $\mathcal{L}$ will be called the language of M and denoted by $\mathcal{L}(M)$.

(ii) If $\vec{a} \in (M)^{<\omega}$ and $\varphi(\vec{x})$ is a formula of $\mathcal{L}(M)$, we shall write $M \models \varphi(\vec{a})$ to denote that : the length of $\vec{x}$ = the length of $\vec{a}$ and

M is a model of $\varphi(\vec{a})$, where $\varphi(\vec{a})$ is the result of substituting $\vec{a}$ for free occurrences of $\vec{x}$ in $\varphi(\vec{x})$.

(iii) If $a_1, \ldots, a_n$ are elements of M, $(M, a_1, \ldots, a_n)$ $((M, \vec{a}))$, will denote the expansion of M to a structure for $\mathcal{L} \cup \{a_1, \ldots, a_n\}$ and $Th(M, a_1, \ldots, a_n)$ $(Th(M, \vec{a}))$, the set of sentences of $\mathcal{L} \cup \{a_1, \ldots, a_n\}$ which hold in M. Similarly we define $Th(M)$ and $Th(M, a)_{a \in M}$.

If φ is a formula of $\mathcal{L}$, the notation $\varphi(x_1, \ldots, x_n)$ will mean that the free variables of φ are among $x_1, \ldots, x_n$, and that no occurrence of x_i $(1 \leq i \leq n)$ is bound in φ.

For a language $\mathcal{L}$:

$F\mathcal{L}_n$=the set of formulae of $\mathcal{L}$ with n free variables
$F\mathcal{L}_n$=the set of formulae of $\mathcal{L}$ with n free variables
$F\mathcal{L}$ =the set of formulae of $\mathcal{L}$
$S\mathcal{L}$ $=F\mathcal{L}_0$

By a *theory in* $\mathcal{L}$ we shall mean a *consistent* set of sentences of $\mathcal{L}$.

The Gödel number of a formula φ is denoted by $\ulcorner\varphi\urcorner$. However, we shall sometimes identify formulae with their Gödel numbers and sets of formulae with the corresponding sets of Gödel numbers, if in doing so no confusion arises.

The arithmetization of the syntax and the notations regarding this are as in [Sh].

1.2 Arithmetic

Throughout we denote by L the first-order language with identity containing two binary function symbols $+, \cdot$, one unary function symbol S and constant symbol 0. PA is the theory consisting of:

PA_1

(i) $\forall x(S(x) \neq 0)$
(ii) $\forall x, y(S(x) = S(y) \rightarrow x = y)$

PA_2

(i) $\forall x(x + 0 = x)$
(ii) $\forall x, y(x + S(y) = S(x + y))$

PA_3

(i) $\forall x(x \cdot 0 = 0)$
(ii) $\forall x, y(x \cdot S(y) = x \cdot y + x))$

and the *Induction Principle*

PA_φ

$$\forall \vec{x}[(\varphi(\vec{x}, 0) \wedge \forall y(\varphi(\vec{x}, y) \rightarrow \varphi(\vec{x}, S(y)))) \rightarrow \forall y \varphi(\vec{x}, y)]$$

for every formula φ of L.

Define $x \leq y$ by $\exists z(x + z = y)$ and $x < y$ by $x \leq y \wedge x \neq y$. The following called the *Least Number Principle* is a straightforward consequence of PA.

$$\forall \vec{x}[\exists y \varphi(\vec{x}, y) \to \exists y(\varphi(\vec{x}, y) \wedge \forall z < y \neg \varphi(\vec{x}, z))]$$

for every formula φ of L.

By PA^- we will denote the theory consisting of PA_1, PA_2, PA_3 and the sentences of L asserting that: '$\leq$ is a linear ordering'; '$+$ and $\cdot$ are commutative and associative'; '$\cdot$ is distributive (in $+$)' and the sentence $\forall x, y(x + z = y + z \to x = y)$.

The standard model for PA^- will be denoted by $\mathbb{N}$ and its domain by N. Elements of N are called the standard integers and will be denoted by m, n, p etc. In general, we shall identify the standard part of the models of PA^- with N. If $M \models PA^-$ and $M \neq N$, M is said to be *non-standard* and elements of $M - N$ are called non-standard integers.

1.3 Definitions and prerequisites

Given $\varphi \in FL$, we write $\forall x \leq y\, \varphi$ and $\exists x \leq y\, \varphi$ for $\forall x(x \leq y \to \varphi)$ and $\exists x(x \leq y \wedge \varphi)$ respectively. Similarly, $\forall x < y \varphi$ and $\exists x < y \varphi$ are defined. Quantifiers of the form $\forall x \leq y\, \varphi$ and $\exists x \leq y\, \varphi$ are called *bounded quantifiers*.

Definition 1.3.1. *The sets Σ_n, Π_n of formulae of L are defined by induction on $n \in N$, as follows:*

$\Sigma_0 = \Pi_0 = \{\varphi \in FL :$ *every quantifier in φ is a bounded quantifier*$\}$.

$$\Sigma_{n+1} = \{\exists \vec{x} \varphi : \varphi \in \Pi_n\},$$

$$\Pi_{n+1} = \{\forall \vec{x} \varphi : \varphi \in \Sigma_n\}.$$

Let M be a structure for L and $\vec{a} \in (M)^{<\omega}$, let $\varphi(\vec{x}, \vec{y})$ be any formula of L. We say $\varphi(\vec{x}, \vec{a})$ is Σ_n (resp. Π_n) if $\varphi(\vec{x}, \vec{y})$ is Σ_n (resp. Π_n).

The following is a well-known theorem.

Theorem 1.3.2. *Modulo PA, the sets Σ_n, Π_n are closed under $\wedge, \vee$ and prefixing of bounded quantifiers.*

The following theorem is part of the folklore for models of PA and is essentially due to Kleene.

Theorem 1.3.3.

(i) *For every $m \in N$ there is a formula $\Delta_{1,m}(y, \vec{x})$ with $m + 1$ free variables which is equivalent to both a Σ_1 and a Π_1-formula modulo PA and such that for any Σ_0-formula $\varphi(\vec{x})$ with m free vari-*

ables:

$$PA \vdash \forall \vec{x}(\varphi(\vec{x}) \leftrightarrow \Delta_{1,m}(\ulcorner \varphi(\vec{x}) \urcorner, \vec{x})).$$

(ii) *For any $m, n \in N$ and $n \neq 0$, there is a Σ_n-formula denoted by $\Sigma_{n,m}(y, \vec{x})$ with $m+1$ free variables such that for and Σ_n-formula $\varphi(\vec{x})$ with m free variables:*

$$PA \vdash \forall \vec{x}(\varphi(\vec{x}) \leftrightarrow \Sigma_{n,m}(\ulcorner \varphi(\vec{x}) \urcorner, \vec{x})).$$

Generally if there is no ambiguity about the number of free variables, we shall drop the subscript m from $\Delta_{1,m}(y, \vec{x})$ and $\Sigma_{n,m}(y, \vec{x})$ and simple write $\Delta_1(y, \vec{x})$ and $\Sigma_n(y, \vec{x})$.

Definition 1.3.4.

(i) *Let M and M' be structures for L. By $M \prec_n M'$ we understand that M is a substructure of M' and for every Σ_n-formula $\varphi(\vec{x})$ and $\vec{a} \in (M)^{<\omega}$:*

$$M \models \varphi(\vec{a}) \Leftrightarrow M' \models \varphi(\vec{a}).$$

(ii) *For a theory T in L, $\Sigma_n(T) = \{\varphi \in \Sigma_n : T \vdash \varphi\}$. For a structure M for L, $\Sigma_n(M) = \Sigma_n(Th(M))$. Similarly we define $\Pi_n(T)$ and $\Pi_n(M)$.*

The following is a formalization of Matijasevič's Theorem checked by A. Pridor (see [G]).

Result 1.3.5. *For every Σ_1-formula $\varphi(\vec{x})$ there is a diophantine equation $\delta(\vec{x}, \vec{y})$ such that:*

$$PA \vdash \forall \vec{x}(\varphi(\vec{x} \leftrightarrow \exists \vec{y} \delta(\vec{x}, \vec{y})).$$

Result 1.3.6 (Preservation Theorems).

(i) (Gaifman[G]) *Let M and M' be models of PA such that $M \subset M'$. Then for every Σ_1-formula $\varphi(\vec{x})$ and for all $\vec{a} \in (M)^{<\omega}$:*

$$M \models \varphi(\vec{a}) \Leftrightarrow M' \models \varphi(\vec{a}).$$

(ii) *Let M and M' be structures for L such that $M \prec_n M'$. Then for every Σ_{n+1}-formula $\varphi(\vec{x})$ and for all $\vec{a} \in (M)^{<\omega}$:*

$$M \models \varphi(\vec{a}) \Leftrightarrow M' \models \varphi(\vec{a}).$$

The following is a slight modification of a well-known Theorem of Tarski.

Theorem 1.3.7. *Let M and M' be structures for L. $M \prec_n M'$ iff $M \subset M'$ and given any Π_{n-1} formula $\varphi(\vec{x}, \vec{y})$ and $\vec{a} \in (M)^{<\omega}$ such that $M' \models \exists \vec{x} \varphi(\vec{x}, \vec{a})$, there exists $\vec{b} \in (M)^{<\omega}$ such that $M' \models \varphi(\vec{b}, \vec{a})$.*

Definition 1.3.8. *Suppose M is a structure for L and A is any subset of M:*

(i) *A is said to be an* initial segment *of M (written $A \subseteq_e M$) if for all $x \in A$ and $y \in M$ such that $M \models y \leq x$ we have $y \in A$. A is said to be a* proper *initial segment of M (written $A \subset_e M$) if $\emptyset \neq A \neq M$.*

(ii) *A is said to be* cofinal *in M if for all $y \in M$ there is some $x \in A$ such that $M \models y \leq x$.*

(iii) *The set $\overline{A} = \{x \in M : \exists y \in A, M \models x \leq y\}$ is called the* initial segment of M determined by A.

Note. Let M and $M^{'}$ be structures for L. We write $M \prec_{n,e} M^{'}$ ($M^{'}$ is an *n-elementary end extension* of M) to denote that $M \prec_n M^{'}$ and $M \subset_e M^{'}$; similarly we define $M \prec_e M^{'}$.

Henceforth, unless otherwise stated, by 'an initial segment' we understand a *proper* initial segment. Generally if $M \models PA^-$, $I \models PA^-$ and $I \subset_e M$, we shall assume that I is a substructure of M.

Definition 1.3.9. *Let M and $M^{'}$ be structures for L:*

(i) *$M^{'}$ is an* end extension *of M if $M \subset_e M^{'}$.*

(ii) *$M^{'}$ is a* cofinal extension *of M if M is cofinal in $M^{'}$.*

Result 1.3.10 (Gaifman [G]). *Let M and $M^{'}$ be models of PA such that $M \subset M^{'}$. Let $\overline{M}$ denote the (proper or improper) initial segment of $M^{'}$ determined b M and with the structure inherited from M. then $M \prec \overline{M}$.*

Lemma 1.3.11 (Robinson's Overspill Lemma). *Let M be a model of PA, let $\varphi(x)$ be a formula of $L(M)$ such that $M \models \varphi(n)$ for all $n \in N$. Then there is some $\beta \in M - N$, such that $M \models \forall x \leq \beta \varphi(x)$.*

Throughout let p_i denote the *i'th prime.*

Definition 1.3.12. *Let M be a structure for L, $A \subseteq N$ is* coded *in M by $a \in M$ if $n \in A \Leftrightarrow M \models p_n | a$ for all $n \in N$, where $x|y$ is defined as $\exists z \leq y(x \cdot z = y)$.*

We define R_M (the Reals of M) as the set of all $A \subseteq N$ such that A is coded in M.

Lemma 1.3.13. *Let M be a non-standard model of PA. Then $\Sigma_n(M)$, $\Pi_n(M) \in R_M$, for all $n \in N$.*

Lemma 1.3.14. *Let T be any r.e. theory in L extending PA and suppose $A \subset N$ is non-recursive, then there is a non-standard model of M of T such that $A \notin R_M$.*

For a proof of the above see (e.g.) [J&E].

In the following definition we identify subsets of N with their characteristic functions (from N to $\{0,1\}$). We also assume a recursive Gödel numbering for finite sequences of 0's and 1's. For a set A, $P(A)$ denotes the power set of A.

Definition 1.3.15. (due to Scott) $A \subseteq P(N)$ *is called a Scott set if it satisfies the following conditions:*

(i) *If* $a \subseteq N$ *is recursive in a finite number of elements of* A *then* $a \in A$.

(ii) *Every infinite binary tree (of 0's and 1's) belonging to* A *has a path belonging to* A*, where a path of a binary tree* Y *(of 0's and 1's) is defined as a function* $f \in 2^N$ *such that* $f\lceil n \in Y$ *for all* $n \in N$.

Result 1.3.16. (essentially due to Scott [S]) *For a countable* $A \subset P(N)$ *the following are equivalent:*

(i) A *is a Scott set*

(ii) *There is some non-standard model* M *of* PA *such that* $A = R_M$.

For a proof of the above see (e.g.) [F].

Note that given *any* non-standard model M of PA, R_M is a Scott set (see [F]).

1.4 A-saturated and recursively saturated structures

A-saturated and recursively saturated structures were introduced at the same time by Wilmers [W_1] and Barwise-Schlipf [B&S] respectively. As we shall see, these structures appear naturally in the study of models of arithmetic.

Throughout this section let $\mathcal{L}$ denote some fixed arithmetizable language, T a theory in $\mathcal{L}$ and M a structure for $\mathcal{L}$.

Definitions 1.4.1.

(i) *A* partial n-type *for* T *is a set of formulae* $\Gamma(x_1, \ldots, x_n) \subseteq F\mathcal{L}_n$ *for which there is a model* M *of* T *and elements* $a_1, \ldots, a_n$ *of* M*, such that* $M \models \varphi(a_1, \ldots, a_n)$ *for all* $\varphi(x_1, \ldots, x_n) \in \Gamma(x_1, \ldots, x_n)$.

(ii) n*-types are defined as maximal partial* n*-types and a type of* T *is an* n*-type of* T *for some* $n \in N$.

(iii) M *realizes a set of formulae* $\Gamma(x_1, \ldots, x_n) \subseteq F\mathcal{L}_n$ *if for some elements* $a_1, \ldots, a_n$ *of* M, $M \models \varphi(a_1, \ldots, a_n)$ *for every* $\varphi(x_1, \ldots, x_n) \in \Gamma(x_1, \ldots, x_n)$.

(iv) *For elements* $a_1, \ldots, a_n$ *of* M*, the type of* $a_1, \ldots, a_n$ *in* M *is defined as the set* $\{\varphi(x_1, \ldots, x_n) \in F\mathcal{L}_n : M \models \varphi(a_1, \ldots, a_n)\}$.

Given a set of formulae $\Gamma(\vec{x}, \vec{y})$ of $\mathcal{L}$, $\vec{a} \in (M)^{<\omega}$ and a set $A \subseteq P(N)$ we write $\Gamma(\vec{x}, \vec{a}) \in A$ or $\Gamma \in A$ to denote that $\Gamma(\vec{x}, \vec{y}) \in A$.

Definitions 1.4.2.

(i) (Barwise-Schlipf) *M is recursively saturated if for all $\vec{a} \in (M)^{<\omega}$, M realizes every recursive partial type of $Th(M, \vec{a})$.*

(ii) (Wilmers) *M is A-saturated if A is a Scott set and for all $\vec{a} \in (M)^{<\omega}$, M realizes every partial type $\Gamma(x, \vec{a})$ of $Th(M, \vec{a})$, where $\Gamma(x, \vec{y}) \in A$ and does not realize any type $\Gamma'(x, \vec{a})$ (of $Th(M, \vec{a})$) such that $\Gamma'(x, \vec{y}) \notin A$.*

Remarks 1.4.3.

(i) Using 1.4.7(i) below, Wilmers [W_2] shows that "if M is countable and recursively saturated, then for some A, M is A-saturated".

(ii) Note that if M is recursively (resp. A-) saturated, then for all $\vec{a} \in (M)^{<\omega}$, $(M, \vec{a})$ is recursively (resp. A-) saturated.

In what follows we mention some of the properties of the A-saturated and recursively saturated structures which we will need later. For more information and proofs we refer the reader to [W_1] and [B&S].

The following results (1.4.4.) are proved by simple Back and Forth arguments, and do not use the fact that A (or B) is a Scott set.

Results 1.4.4 (Wilmers [W_1]).

(i) *If M, M' are A-saturated and $M \equiv M'$ then $M \equiv_{\infty\omega} M'$.*

(ii) *If M, M' are A-saturated, countable and $M \equiv M'$ then $M \cong M'$.*

(iii) *If M is countable, A-saturated and M' is B-saturated, $A \subseteq B$ and $M \equiv M'$, then M is isomorphic to an elementary substructure of M'.*

The following definitions are due to Wilmers.

Definitions 1.4.5.

(i) *By $\mathcal{L}^{II}$ we denote the full second-order language obtained from $\mathcal{L}$ by adding n-ary relation variables X_i^n for each $i, n \in N$ and allowing quantification over such variables. Satisfaction for $\mathcal{L}^{II}$ is defined in the obvious way. A formula of $\mathcal{L}^{II}$ is Δ_0^{II} if it contains no relation quantifiers.*

(ii) *We shall assume a recursive Gödel numbering of the formulae of $\mathcal{L}^{II}$ and as before we shall usually identify sets of formulae with the corresponding sets of Gödel numbers.*

(iii) *For $A \subseteq P(N)$, an infinitary sentence of the form $\exists X_1 \ldots X_n \bigwedge \Phi$ is called a* $\Sigma(A)$-sentence *of $\mathcal{L}$ if $\Phi = \{\varphi_i : i \in N\} \in A$ and Φ*

is a set of Δ_0^{II}*-formulae of* $\mathcal{L}^{II}$ *having* $X_1, \ldots, X_n$ *as their only relation variables and having no free first-order variables.*

Definitions 1.4.6.

(i) *A* $\Sigma(A)$*-sentence* Ψ *of* $\mathcal{L}$ *is* consistent with T, *if there is a model* M *of* T *such that* $M \models \Psi$.

(ii) M *is a* $\Sigma(A)$-structure *if for all* $\Sigma(A)$*-sentences* Ψ *of* $\mathcal{L} \cup \{a_1, \ldots, a_n\}$ *consistent with* $Th(M, a_1, \ldots, a_n)$ *we have* $M \models \Psi$ *(for all elements* $a_1, \ldots, a_n$ *of* M*).*

Throughout let: $rec = \{A \subseteq N : A \text{ is recursive }\}$.

Results 1.4.7.

(i) (Barwise-Schlipf [B&S]) *If* M *is a countable recursively saturated structure, then* M *is a* $\Sigma(rec)$*-structure.*

(ii) (Wilmers [W$_1$]) *If* M *is a countable* A*-saturated structure then* M *is a* $\Sigma(A)$*-structure.*

2

A-saturated models of Arithmetic

In the first part of this chapter we prove some results concerning A-saturated models of PA, some of which can be easily generalized to other A-saturated structures that will be introduced in later chapters. In the second part, using the Arithmetized Completeness Theorem we find n-elementary end extensions of models of PA which are A-saturated; using this we prove some results regarding initial segments and definable elements of models of PA. In the last section we look at the reducts of models of arithmetic to certain sublanguages of L.

2.1 Some basic results

Lemma 2.1.1. *Let M be an A-saturated model of PA. Then $A = R_M$.*

Proof. $A \subseteq R_M$: suppose $a \in A$, let $\Gamma(x) = \{p_n | x : n \in a\} \cup \{p_n \not| x : n \notin a\}$. $\Gamma(x)$ is a partial type of $Th(M)$ and as $\Gamma(x)$ is recursive in a: $\Gamma(x) \in A$. Therefore by the definition of A-saturatedness $\Gamma(x)$ is realized in M.

$R_M \subseteq A$: for $\vec{a} \in (M)^{<\omega}$ let Γ be a partial type of $Th(M, \vec{a})$ which is coded in M by some $\alpha \in M$; let Γ' be the type of α in M. Γ is recursive in Γ' ($\varphi \in \Gamma \Leftrightarrow p_\varphi | x \in \Gamma'$) and as $\Gamma' \in A$ hence $\Gamma \in A$. ⊣

Corollary 2.1.2. *If M is an A-saturated model of PA then for all $\vec{a} \in (M)^{<\omega}$, $Th(M, \vec{a}) \in R_M$.*

Proof. Follows from 2.1.1. ⊣

Definition and remark 2.1.3. *For elements $a_1, \ldots, a_n$ of a model M of PA, we denote by $Def(M, a_1, \ldots, a_n)$ the submodel of M, whose every element is definable in $(M, a_1, \ldots, a_n)$. It is not difficult to show that given any finite number of elements $a_1, \ldots, a_n$ of M:*

(i) $Def(M, a_1, \ldots, a_n) \prec M$.

(ii) *There is some* $a \in M$ *(e.g.* $a = 2^{a_1} \dots p_n^{a_n}$*) such that* $Def(M, a) = Def(M, a_1, \dots, a_n)$.

Lemma 2.1.4. *Let* M *be an A-saturated model of PA and* $a \in M$, *then there is no non-standard initial segment* I *of* M *such that* $I \subseteq Def(M, a)$.

Proof. Suppose not, then by 2.1.2., we can find some $\alpha \in Def(M, a)$ such that α codes $Th(M, a)$ in $Def(M, a)$; but this contradicts Tarski's well-known Theorem on the undefinability of truth. ⊣

Definition 2.1.5. *A model* M *of PA is called* tall *if for all* $a \in M$, $Def(M, a)$ *is not cofinal in* M.

Lemma 2.1.6. *Let* M *be an A-saturated model of PA. Then* M *is tall.*

Proof. Let a be any element of M and let Γ be the type of a in M. Define $\Gamma^{'}(x, a)$ to be the following partial type of $Th(M, a)$: $\Gamma^{'}(x, a) = \{\forall z(\varphi(a, z) \to z < x) : \exists! z \varphi(a, z) \in \Gamma\}$. $\Gamma^{'}(x, y)$ is recursive in Γ, hence $\Gamma^{'}(x, y) \in A$ and therefore $\Gamma^{'}(x, a)$ is realized in (M, a). ⊣

Proposition 2.1.7. *Suppose* M *is an A-saturated model of PA,* $a, b, c \in M$ *and* $M \models b \leq c$. *Then the following are equivalent:*

(i) *There is some* $d \in M$, $M \models d < b$ *such that* $(M, a, d) \equiv (M, a, c)$.
(ii) *There is no* $b^{'} \in Def(M, a)$ *such that* $M \models b \leq b^{'} \leq c$.

Proof.

(i) $\Rightarrow$(ii). Suppose not, then there is some $b^{'} \in Def(M, a)$ such that $M \models b \leq b^{'} \leq c$. Suppose $b^{'}$ is defined in M by a formula $\varphi(x)$ of $L \cup \{a\}$, then: $(M, a, c) \models \exists! x(\varphi(x) \wedge x \leq c)$ hence by (i) $(M, a, d) \models \exists! x(\varphi(x) \wedge x \leq d)$; i.e. $M \models b^{'} < d$ contradicting the fact that $M \models d < b^{'}$.

(ii) $\Rightarrow$ (i). Let $\Gamma(a, x)$ be the type of c in (M, a) and define $\Gamma^{'}(a, b, x)$ as $\{\varphi(x) \wedge x < b : \varphi(x) \in \Gamma(a, x)\}$; clearly $\Gamma^{'}(z, y, x) \in A$. We claim $\Gamma^{'}(a, b, x)$ is a partial type of $Th(M, a, b)$. Suppose not, then as (M, a, b) is A-saturated, by Compactness Theorem there is a finite subset $\{\varphi_1(x), \dots, \varphi_n(x)\}$ of $\Gamma(a, x)$ such that:

$$(M, a, b) \models \neg \exists x < b(\varphi_1(x) \wedge \dots \wedge \varphi_n(x)).$$

Define $\psi(x)$ to be $\varphi_1(x) \wedge \dots \wedge \varphi_n(x)$; $(M, a) \models \neg \exists x < b \psi(x)$ but $(M, a) \models \psi(c)$ therefore for some $b^{'} \in M$, $(M, a) \models \psi(b^{'}) \wedge \forall t < b^{'} \neg \psi(t)$, i.e. $b^{'} \in Def(M, a)$ and $M \models b \leq b^{'} \leq c$, contradicting (ii). Hence, $\Gamma^{'}(a, b, x)$ is a partial type of $Th(M, a, b)$ and as $\Gamma^{'}(z, y, x) \in A$, $\Gamma^{'}(a, b, x)$ is realized in (M, a, b), the result follows. ⊣

Corollaries 2.1.8.

(i) *Suppose M is an A-saturated model of PA and that $M' \prec_e M$ then M' is A-saturated if and only if M' is tall.*

(ii) *If M is a countable A-saturated model of PA, $\vec{a} \in (M)^{<\omega}$, $b \in M$ and $b \notin Def(M, \vec{a})$, then we can find $b' \in M$, $M \models b' < b$, such that $(M, \vec{a}, b) \cong (M, \vec{a}, b')$. Hence for every $\vec{a} \in (M)^{<\omega}$ and $b \notin Def(M, \vec{a})$, $(M, \vec{a})$ has an automorphism f such that $M \models f(b) < b$.*

Proof.

(i) '$\Rightarrow$' follows from 2.1.6. and '$\Leftarrow$' from 2.1.7. and 2.1.3.(ii).

(ii) follows from 2.1.7., 2.1.3.(ii) and 1.4.4.(ii). $\dashv$

The following is part of a theorem of Kueker which we will often use.

Result 2.1.9. (Kueker) *Let M be any countable structure and suppose for all $\vec{a} \in (M)^{<\omega}$, $(M, \vec{a})$ has a non-trivial automorphism. Then M has $2^{\aleph_0}$ automorphisms.*

Proof. See [Ku]. $\dashv$

Proposition 2.1.10. *Suppose M is a countable A-saturated model of PA, $M' \prec_e M$ and $M' \neq N$. Then M' has $2^{\aleph_0}$ automorphisms.*

Proof. Either M' is tall in which case by 2.1.8.(i) M' is A-saturated, or there is some $a \in M'$ such that $Def(M', a)$ ($=Def(M, a)$) is cofinal in M'. We will prove the proposition for the second case, proof of the first case is similar[1]. Using 2.1.9. it is enough to show that given any $\vec{b} \in (M')^{<\omega}$ $(M', a, \vec{b})$ has a non-trivial automorphism. By 2.1.4. and 2.1.3.(ii) we can find some $c \in M'$ such that $c \notin Def(M', a, \vec{b})$. Hence, by 2.1.8.(ii) there is an automorphism f of $(M, a, \vec{b})$ such that $M \models f(c) < c$, as $Def(M', a)$ is cofinal in M': $f''(M', a, \vec{b}) = (M', a, \vec{b})$. Let g be the restriction of f to M', then g is an automorphism of $(M', a, \vec{b})$. The proposition follows. $\dashv$

2.2 End extensions which are A-saturated

In the following two theorems let L' *denote an expansion of* L *by finitely many new constant symbols and* T *a theory in* L' *extending* PA. Also assume that *every structure appearing in the following two theorems is a structure for* L'. The Arithmetized Completeness Theorem has many

[1] It is not difficult to show that any countable, non-finite A-saturated structure has $2^{\aleph_0}$ automorphisms (see e.g. [W₂]).

claimants, but the credit for the following form of it must go mainly to K. McAloon ([Mc] Theorems 1.7 and 2.2).:

Theorem 2.2.1. *Let M be a model of PA such that T is definable in M (with or without parameters from M) and $\Pi_{n+1}(T) \subseteq \Pi_{n+1}(M)$ for some $n \in N$. Then we can construct a model M' of T such that:*

(i) *M' is isomorphic to an n-elementary end extension of M.*

(ii) *There is a formula $S_{M'}(x, y)$ of $L(M)$ such that for all elements $a_1, \ldots, a_n$ of M' and formulae $\varphi(x_1, \ldots, x_m)$ of L:*

$$M' \models \varphi(a_1, \ldots, a_m) \Leftrightarrow M \models S_{M'}(\ulcorner \varphi(x_1, \ldots, x_m) \urcorner, \langle a_1, \ldots, a_m \rangle).$$

Theorem 2.2.2. *Let M be a non-standard model of PA, $T \in R_M$ and $\Pi_{n+1}(T) \subseteq \Pi_{n+1}(M)$ for some $n \in N$. Then there is an R_M-saturated model M' of T such that $M \prec_{n,e} M'$.*

Proof. We show that if M is non-standard then M' constructed as in 2.2.1. is R_M-saturated.

For elements $a_1, \ldots, a_m$ of M', let $\Gamma(x, a_1, \ldots, a_m)$ be a partial type of $Th(M', a_1, \ldots, a_m)$ such that $\Gamma(x, y_1, \ldots, y_m)$ is coded in M by some $\alpha \in M$. As $\Gamma(x, a_1, \ldots, a_m)$ is finitely realizable in M' by (ii) of 2.2.1. for all $k \in N$:

$$M \models \exists x \forall \varphi \leq k[p_\varphi | x \to S_{M'}(\ulcorner \varphi(x_0, x_1, \ldots, x_m) \urcorner, \langle x, a_1, \ldots, a_m \rangle)].$$

Hence by overspill lemma there is some $\beta \in M - N$ such that:

$$M \models \exists x \forall \varphi \leq \beta[p_\varphi | x \to S_{M'}(\ulcorner \varphi(x_0, x_1, \ldots, x_m) \urcorner, \langle x, a_1, \ldots, a_m \rangle)],$$

i.e. there is some $a_0 \in M'$ such that for all $\varphi \in \Gamma$, $M' \models \varphi(a_0, a_1, \ldots, a_m)$. Hence, $\Gamma(x, a_1, \ldots, a_m)$ is realized in $(M', a_1, \ldots, a_m)$. Now assume $\Gamma(x, a_1, \ldots, a_m)$ is a type of $Th(M', a_1, \ldots, a_m)$ which is realized in $(M', a_1, \ldots, a_m)$, then for some $a_0 \in M'$

$$\varphi \in \Gamma \Leftrightarrow M' \models \varphi(a_0, a_1, \ldots, a_m).$$

Therefore for all $\varphi \in N$:

$$\varphi \in \Gamma \Leftrightarrow M \models S_{M'}(\ulcorner \varphi(x_0, x_1, \ldots, x_m) \urcorner, \langle a_0, a_1, \ldots, a_m \rangle).$$

It easily follows that $\Gamma \in R_M$. ⊣

Corollaries 2.2.3.

(i) *Let M and T be as in 2.2.2., but now assume that T is complete. Then there is an R_M-saturated model M' of T such that $M \prec_{n,e} M'$ iff $T \in R_M$.*

(ii) *For every model M of PA and for each $n \in N$, there is an R_M-saturated model M' of PA such that $M \prec_{n,e} M'$.*

Proof. (i) Follows immediately from 2.2.2. and 2.1.2. To prove (ii) assume M is non-standard and let $T = PA \cup \Pi_{n+1}(M)$ in 2.2.2. Clearly $T \in R_M$ and $\Pi_{n+1}(T) \subseteq \Pi_{n+1}(M)$. The case for when $M = N$ follows trivially. $\dashv$

Definition and remark 2.2.4. *Let M be a structure for some language $\mathcal{L}$ and $\theta(x)$ a formula in the language of M. If $M \models \exists x\theta(x)$ we can define a natural structure on $\theta^M(x) = \{a \in M : M \models \theta(a)\}$ by restricting the operations of $\mathcal{L}$ in M to $\theta^M(x)$, replacing any function symbol of $\mathcal{L}$ by a relation symbol if necessary. Henceforth, for given M and $\theta(x)$, $\theta^M(x)$ will denote this natural structure.*

Corollary 2.2.5. *Let M be a model of PA and $\theta(x)$ a formula of $L(M)$ such that $\theta^M(x)$ is non-empty and bounded above in M. Then $\theta^M(x)$ is R_M-saturated.*

Proof. For some $n \in N$, $\theta(x)$ is equivalent to a Σ_n-formula modulo PA. By 2.2.3.(ii) there is an R_M-saturated model M' of PA such that $M \prec_{n,e} M'$. As $\theta^M(x)$ is bounded above in M, $\theta^M(x) = \theta^{M'}(x)$. Now given elements $a_1, \ldots, a_m$ of $\theta^M(x)$, every partial type of $Th(\theta^M(x), a_1, \ldots, a_m)$ is the relativization of a partial type of $Th(M', a_1, \ldots, a_m)$ to $\theta(x)$. We leave the rest of the proof to the reader. $\dashv$

Corollary 2.2.6. (Wilmers) *If M is a model of PA then for all $a \in M$, $I_a^M = (x \leq a)^M$ is R_M-saturated.*

Proof. Immediate from 2.2.5. $\dashv$

We shall return to these structures in chapter 4 and give a new proof of their A-saturatedness.

Proposition 2.2.7.

(i) *Suppose M_1 and M_2 are countable non-standard models of PA. Then there are models M_1' and M_2' of PA such that $M_1 \prec_{n,e} M_1'$, $M_2 \prec_{n,e} M_2'$ and $M_1' \cong M_2'$ iff $R_{M_1} = R_{M_2}$ and $\Pi_{n+1}(M_1) \subseteq \Pi_{n+1}(M_2)$ or $\Pi_{n+1}(M_2) \subseteq \Pi_{n+1}(M_1)$.*

(ii) *As in (i) but drop the countability condition and replace $M_1' \cong M_2'$ by $M_1' \equiv_{\infty\omega} M_2'$.*

Proof. We prove (i) and leave the proof of (ii) to the reader.

'$\Rightarrow$' Let M_1, M_2, M_1', M_2' be as in the hypothesis, identify M_2' with M_1'. Hence either $M_1 \prec_{n,e} M_2 \prec_{n,e} M_1'$ or $M_2 \prec_{n,e} M_1 \prec_{n,e} M_1'$, which by preservation theorems imply that either $\Pi_{n+1}(M_2) \subseteq \Pi_{n+1}(M_1)$ or $\Pi_{n+1}(M_1) \subseteq \Pi_{n+1}(M_2)$ respectively and clearly $R_{M_1} = R_{M_1'} = R_{M_2}$.

'$\Leftarrow$' Suppose $\Pi_{n+1}(M_2) \subseteq \Pi_{n+1}(M_1)$. By 2.2.3.(ii) there is an R_{M_2}-saturated M_2' such that $M_2' \models PA$ and $M_2 \prec_{n,e} M_2'$. By 2.1.2., $T = Th(M_2') \in R_{M_2}$ $(=R_{M_1})$ and clearly $\Pi_{n+1}(T) \subseteq \Pi_{n+1}(M_2) \subseteq \Pi_{n+1}(M_1)$. Hence using 2.2.2. there is an R_{M_1}-saturated model M_1' of T such that $M_1 \prec_{n,e} M_1'$. As both M_1' and M_2' are R_{M_1} $(=R_{M_2})$-saturated, countable and $M_1' \equiv M_2'$ then by 1.4.4. $M_1' \cong M_2'$. The proof for the case when $\Pi_{n+1}(M_1) \subseteq \Pi_{n+1}(M_2)$ is symmetrical. $\dashv$

Definition 2.2.8. *For a formula θ and a relation R, the relativization of θ to R is denoted by θ^R.*

Theorem 2.2.9. *Let M be a countable non-standard model of PA such that $\mathbb{N} \prec_n M$, let A be any subset of $Th(\mathbb{N})$ which is coded in M. Then for all $a \in M - N$ there is a non-standard initial segment M' of M such that:*

(i) $a \notin M'$

(ii) $M' \prec_{n,e} M$

(iii) $M' \models A$.

Proof. By 2.2.3.(ii) there is an R_M-saturated model M^* of PA such that $M \prec_{n,e} M^*$; clearly $\mathbb{N} \prec_{n,e} M^*$. Let a be any non-standard element of M. Add to L a unary relation symbol R and a constant symbol c. Let $T(R,a,c)$ be the union of the following sets of sentences in the new language:

$S_1 = \{\theta^R : \theta \in A\}$, *i.e.*	'$R \models A$'
$S_2 = \{R(c) \wedge n \neq c : n \in N\}$	'$R \neq N$'
$S_3 = \{\forall x \forall y (R(x) \wedge y \leq x) \rightarrow R(y))\}$	'*R is an initial segment*'
$S_4 = \{\neg R(a)\}$	'$a \notin R$'
$S_5 = \{\forall x (R(x) \rightarrow (\theta(x) \leftrightarrow \theta^R(x))) : \theta(x) \in \Sigma_n \cap FL_1\}$	'*R is an n-elementary substructure*'.

Claim. $Th(M^*, a) \cup T(R,a,c)$ is consistent in the new language.

Proof of the claim. Let $T'(R,a,c)$ be any finite subset of $T(R,a,c)$ and let m be the largest $n \in N$ such that $R(c) \wedge n \neq c$ of S_2 occurs in $T'(R,a,c)$. Interpret R as the standard part of M^* and c as $m+1$ in M^*. Under this interpretation $(M^*, N, a, m+1)$ is a model of $T'(R,a,c)$. Therefore, by Compactness Theorem, $Th(M^*,a) \cup T(R,a,c)$ is consistent. Define Φ to be:

$$\exists X \exists x [\bigwedge_{\theta \in A} \theta^X \wedge X(x) \wedge \bigwedge_{n \in N} x \neq n \wedge \forall x \forall y (X(x) \wedge y \leq x \rightarrow X(y))$$

$$\wedge \neg X(a) \wedge \bigwedge_{\theta \in \Sigma_n \cap FL_1} \forall x(X(x) \to (\theta(x) \leftrightarrow \theta^X(x)))]$$

It is easy to check that Φ is equivalent to a $\Sigma(R_M)$-sentence of $L \cup \{a\}$. By the above claim, Φ is consistent with $Th(M^*, a)$, hence as M^* is countable and R_M-saturated, by 1.4.7.(ii) $M^* \models \Phi$. The theorem follows. ⊣

Corollary 2.2.10. *Let M be a countable non-standard model of PA such that $\mathbb{N} \prec_n M$. Then for all $a \in M - N$ there is some $M' \models PA$ such that $M' \prec_{n,e} M$, and $M' \neq N$, $a \in M'$.*

Proof. Immediate from 2.2.9. ⊣

Corollary 2.2.11. *Suppose M is* any *non-standard model of PA and A is a subset of $Th(\mathbb{N})$ which is coded in M. Then for all $a \in M - N$ there is some $M' \subset_e M$ such that $M' \models A$, $a \notin M'$ and $M' \neq N$.*

Proof. Let $M_1 = Def(M, a, \alpha)$, where α codes $PA \cup A$ in M. As M_1 is countable there is some $M_2 \models PA \cup A$ such that $a \notin M_2$, $M_2 \subset_e M_1$ and $M_2 \neq N$. Define M' to be the substructure of M whose domain is the initial segment of M determined by M_2. By Gaifman's Theorem (1.3.10.) $M_2 \prec M' \subset_e M$; it is easy to see that M' satisfies the required conditions. ⊣

Remark. Despite 2.2.11. we were unable to generalize 2.2.9. to uncountable models as we could not see any way of changing n to $n+1$ in the conclusion of the following proposition.

Proposition 2.2.12. *Let $M \prec_{n+1} M'$ where M and M' are models of PA, let $\overline{M}$ be the substructure of M' whose domain is the initial segment of M' determined by M. Then $\overline{M} \prec_{n,e} M'$.*

Proof. By 1.3.7. we need to show that for any $\varphi(\vec{x}, \vec{y}) \in \Pi_{n-1}$ and $\vec{a} \in (\overline{M})^{<\omega}$ if $M' \models \exists \vec{x} \varphi(\vec{x}, \vec{a})$ then for some $\vec{b} \in (\overline{M})^{<\omega}$ $M' \models \varphi(\vec{b}, \vec{a})$. For simplicity, let $\vec{x} = x$, $\vec{y} = y$ and $\vec{a} = a$. Suppose $M' \models \exists x \varphi(x, a)$ where $\varphi(x, y) \in \Pi_{n-1}$; as $a \in \overline{M}$ then for some $c \in M$, $M' \models a \leq c$. Define φ' and ψ as follows:

$$\varphi'(x, y) \equiv \varphi(x, y) \wedge \forall z < x \neg \varphi(z, y)$$

$$\psi(x, t) \equiv \exists y \leq t \varphi'(x, y) \wedge \forall y \leq t \forall z (\varphi'(z, y) \to z < x),$$

i.e. $\psi(x, t)$ holds in M' iff $x = max\{z \in M' : M \models \varphi'(z, y) \wedge y \leq t\}$. A little checking yields that $\psi(x, t)$ is equivalent to a Π_n-formula modulo PA and $M' \models \exists z \psi(z, c)$. As $M \prec_{n+1} M'$ then $M \models \exists z \psi(z, c)$, i.e. for some $c' \in M$ we have $M \models \psi(c', c)$, hence, $M' \models \exists x \leq c' \varphi(x, a)$. The proof of the general case is in exactly similar lines. ⊣

Definition 2.2.13. *(i) Let M be a model of PA, $a \in M$ is Δ_n-definable in M if there is a formula $\varphi(x) \in FL_1$ such that $\varphi(x)$ is equivalent to both a Σ_n and a Π_n-formula modulo PA and $M \models \exists! x\varphi(x) \wedge \varphi(a)$.*

For M and a as above, a is said to be Σ_n-definable in M if for some $\varphi(x) \in \Sigma_n$ $M \models \exists! x\varphi(x) \wedge \varphi(a)$. Similarly, we define Δ_n and Σ_n-definable elements of $(M, \vec{a})$ for $\vec{a} \in (M)^{<\omega}$

(ii) Let M be a non-standard model of PA^-. $A \subseteq M - N$ is co-initial *in M if $A \neq \emptyset$ and for all $a \in M - N$ there is some $b \in A$ such that $M \models b \leq a$.*

Proposition 2.2.14. *Let M be a non-standard model of PA, let $a \in M$ and $n \in N$. Then there is no non-standard initial segment of M whose every element is Σ_n-definable in (M, a).*

Proof. Supppose not, let I be a non-standard initial segment of M whose every element is Σ_n-definable in (M, a). Let M^* be an R_M-saturated model of PA such that $M \prec_{n,e} M^*$. Then $I \subset_e Def(M^*, a)$, but this contradicts 2.1.4. The proposition follows. ⊣

Lemma 2.2.15. *For any model M of PA and $n \in N - \{0\}$ the following are equivalent.*

(i) $\mathbb{N} \prec_n M$

(ii) $M \models \Pi_n(\mathbb{N})$

(iii) *M does not contain any non-standard Σ_n-definable elements*

(iv) *M does not contain any non-standard Δ_n-definable elements.*

Proof. Straightforward. ⊣

Lemma 2.2.16. *Suppose $M \prec_n M'$ where M and M' are models of PA. Then $a \in M$ is Δ_{n+1}-definable in M iff a is Δ_{n+1}-definable in M'.*

Proof. Left to the reader. ⊣

Theorem 2.2.17. *Let M be a non-standard model of PA such that $\mathbb{N} \prec_n M$. Then the non-standard Δ_{n+1}-definable elements of M are co-initial in M iff $\Pi_{n+1}(N) \in R_M$.*[2]

Proof. It is easy to see that we only need to prove the theorem for countable M. Hence assume M is countable.

[2] This theorem was proved as a solution to a problem of K. McAloon who independently showed the existence of models of PA satisfying the left side of the above theorem [Mc]. For $n = 0$ the problem was solved independently by Jensen and Ehrenfeucht [J&E].

'⇒' Suppose not, i.e. $\Pi_{n+1}(\mathbb{N}) \in R_M$; then by 2.2.9. there is a non-standard M', $M' \models PA \cup \Pi_{n+1}(\mathbb{N})$ such that $M' \prec_{n,e} M$. By 2.2.15. $M' - N$ does not contain any Δ_{n+1}-definable elements of M', hence by 2.2.16. it does not contain any Δ_{n+1}-definable elements of M. Contradiction, hence $\Pi_{n+1}(\mathbb{N}) \notin R_M$.

'⇐' Suppose not, let $a \in M - N$ be such that there are no non-standard Δ_{n+1}-definable elements of M less than a in M. By 2.2.9. there is some $M' \models PA$ such that $a \notin M'$, $M' \neq N$ and $M' \prec_{n,e} M$. By 2.2.16. M' does not contain any non-standard Δ_{n+1}-definable elements, hence by 2.2.15. $M' \models \Pi_{n+1}(\mathbb{N})$ therefore $\Pi_{n+1}(M') = \Pi_{n+1}(\mathbb{N})$ and by 1.3.13. $\Pi_{n+1}(\mathbb{N}) \in R_{M'}$ $(=R_M)$. Contradiction. The theorem follows. ⊣

Corollary 2.2.18. *Suppose M is a non-standard model of PA. Then the non-standard Σ_0-definable elements of M are co-initial in M iff $\Pi_1(\mathbb{N}) \notin R_M$.*

Proof. This follows from the proof of 2.2.17. and the fact that for any model M' of PA: $\Sigma_1(M') \neq \Sigma_1(\mathbb{N})$ iff M' contains a non-standard Σ_0-definable element. ⊣

Proposition 2.2.19. (Immortality of the Submodels)

Let M_1 and M_2 be countable models of PA such that for some $n \in N$, $M_1 \prec_n M_2$, let $a, b \in M_1$ and suppose $T \in R_{M_1}$ is any theory in L. Then the following are equivalent.

(i) *There is some $M_1' \models T$ such that $a \in M_1', b \notin M_1'$ and $M_1' \prec_{n,e} M_1$.*

(ii) *There is some $M_2' \models T$ such that $a \in M_2', b \notin M_2'$ and $M_2' \prec_{n,e} M_2$.*

Proof. Let $\overline{M}_1$ denote the (proper or improper) initial segment of M_2 determined by M_1 (and the structure inherited from M_2). By 1.3.10. we have $M_1 \prec \overline{M}_1$. As $T \in R_{M_1}$, Φ defined as below is a $\Sigma(R_{M_1})$-sentence of $L \cup \{a, b\}$:

$$\exists X[\bigwedge_{\varphi \in T} \varphi^X \wedge X(a) \wedge \neg X(b) \wedge \forall x \forall y (X(x) \wedge y \leq x \rightarrow X(y)) \wedge$$

$$\bigwedge_{\psi(x) \in \Sigma_n \cap FL_1} \forall x (X(x) \rightarrow (\psi(x) \leftrightarrow \psi^X(x)))].$$

It is enough to prove that $M_2 \models \Phi \Leftrightarrow M_1 \models \Phi$. To show this, we prove $M_2 \models \Phi \Leftrightarrow \overline{M}_1 \models \Phi$ and $\overline{M}_1 \models \Phi \Leftrightarrow M_1 \models \Phi$.

Clearly for every Π_{n+1}-formula $\varphi(x, y)$ we have:

$$M_2 \models \varphi(a, b) \Rightarrow M_1 \models \varphi(a, b) \Leftrightarrow \overline{M}_1 \models \varphi(a, b). \qquad (\dagger)$$

By 2.2.3.(ii) there is an R_{M_2}-saturated model M_2^* of PA such that $M_2 \prec_{n,e} M_2^*$. By 2.1.2. $Th(M_2^*, a, b) \in R_{M_2}$ $(= R_{\overline{M}_1})$, hence using (†) and 2.2.2. there is an $R_{M_1^*}$-saturated model $\overline{M}_1^*$ of PA such that $\overline{M}_1^* \prec_{n,e} \overline{M}_1^*$ and $(\overline{M}_1^*, a, b) \equiv (M_2^*, a, b)$. Therefore by 1.4.7.(ii):

$$\overline{M}_1^* \models \Phi \Leftrightarrow M_2^* \models \Phi \text{ and hence } \overline{M}_1 \models \Phi \Leftrightarrow M_2 \models \Phi.$$

Similarly, we can show the existence of an R_{M_1}-saturated model K of PA such that $M_1 \prec_{n,e} K$. Since $M_1 < \overline{M}_1$ and $Th(K, a, b) \in R_{M_1}$, as before there is an $R_{\overline{M}_1}$-saturated model $\overline{K}$ of PA such that $\overline{M}_1 \prec_{n,e} \overline{K}$ and $(\overline{K}, a, b) \equiv (K, a, b)$. Exactly as above we have: $M_1 \models \Phi \Leftrightarrow \overline{M}_1 \models \Phi$. The proposition follows. ⊣

Remark. The above proposition can be generalized to uncountable models if $PA \subseteq T$ and $n = 0$ exactly as in 2.2.11., or we may drop the condition that M_1', M_2' be initial segments in which case we must replace $b \notin M_1'$ (resp. $b \notin M_2'$) by "b is greater than all elements of M_1' (resp. M_2')".

We end this section with:

Proposition 2.2.20. *Let M be a non-standard model of PA and let T be a complete consistent extension of PA. Then there is some $M' \models T$ such that $M \prec_{n,e} M'$ iff $\Pi_{n+1}(T) \subseteq \Pi_{n+1}(M)$ and for all $m \in N$: $\Pi_m(T) \in R_M$.*

Proof. '⇒' is trivial.

'⇐' Using 2.2.1. we construct the chain:

$$M \prec_{n,e} M_1 \prec_{n+1,e} M_2 \ldots M_k \prec_{n+k,e} M_{k+1} \ldots$$

where M_k is a model of $\Pi_{n+k+1}(T) \cup PA$. Let $M' = \cup_{i \in N} M_i$. It is straightforward to check that M' satisfies the required conditions. ⊣

2.3 Reducts of Arithmetic

Throughout this section let $\mathcal{L}$ denote the language containing L and a name for each symbol defined from L; PA^* the theory in $\mathcal{L}$ containing PA and defining sentences for each new symbol. Let $\mathcal{L}'$ denote a reduct of $\mathcal{L}$, T a consistent extension of PA^* in $\mathcal{L}$ and T' the reduct of T to $\mathcal{L}'$. If M is a structure for $\mathcal{L}$ then $M\lceil\mathcal{L}'$ denotes the reduct of M to $\mathcal{L}'$.

Definition 2.3.1. T' *is* complete in T *if there is a formula* $\tau(x, y) \in FL_2$ *such that for all formulae* $\varphi(x_1, \ldots, x_n)$ *of* $\mathcal{L}'$

$$T \vdash \forall x_1 \ldots x_n(\varphi(x_1, \ldots, x_n) \leftrightarrow \tau(\ulcorner\varphi(x_1, \ldots, x_n)\urcorner, < x_1, \ldots, x_n >)).$$

Proposition 2.3.2. *Suppose M is a non-standard model of T and T' is complete in T. Then $M\lceil\mathcal{L}'$ is R_M-saturated.*

Proof. As T' is complete in T, then for all formulae $\varphi(x_1,\ldots,x_n)$ of $\mathcal{L}'$ and elements $a_1,\ldots,a_n$ of M:

$$M\lceil\mathcal{L}' \models \varphi(a_1,\ldots,a_n) \Leftrightarrow M \models \tau(\ulcorner\varphi(x_1,\ldots,x_n)\urcorner, \langle a_1,\ldots,a_n\rangle).$$

Now the proposition follows exactly as in the proof of 2.2.2. ⊣

Definition 2.3.3. T' *is* strongly complete in T *if* T' *is both complete and complete in* T.

Corollary 2.3.4. *Let* T' *be strongly complete in* T *and suppose* M_1 *and* M_2 *are models of* T *such that* $R_{M_1} = R_{M_2}$. *Then* $M_1\lceil\mathcal{L}' \equiv_{\infty\omega} M_2\lceil\mathcal{L}'$. *Hence if* M_1 *are* M_2 *are countable,* $M_1\lceil\mathcal{L}' \cong M_2\lceil\mathcal{L}'$.[3]

Proof. Immediate from 2.3.2. and 1.4.4. ⊣

Corollary 2.3.5. *Suppose* T' *is strongly complete in* T *then a countable non-standard model* M' *of* T' *is expandable to a model* M *of* T *iff* M' *is* R_M*-saturated.*

Proof. Directly from 2.3.2. and 1.4.4. ⊣

Corollary 2.3.6. *Let* L' *be any sub-language of* L *and* PA' *a reduct of* PA *which is strongly complete in* PA. *Then a countable non-standard model* M' *of* PA' *is expandable to a model of* PA *iff* M' *is recursively saturated.*

Proof. '$\Rightarrow$' follows from 2.3.2 and '$\Leftarrow$' from 1.4.3.(i) and 1.3.16. ⊣

Note. There are a few examples for L' such that PA' is strongly complete in PA, e.g. $L' = \{S, +\}$, $L' = \{\cdot\}$. For a discussion of these and other examples see [J&E].

We end this chapter with two open questions:

(i) What are the necessary and sufficient conditions for a non-standard model of PA to have an A-saturated elementary end extension?

(ii) Is there an extension T of PA^* and a reduct $\mathcal{L}'$ of $\mathcal{L}$ such that T' is complete in T but not strongly complete in T?

[3] A similar result was proved (independently) by Jensen and Ehrenfeucht ([J&E]).

3

Σ_n-definable elements and a problem of H. Gaifman

In this chapter we consider two kinds of structures which arise naturally in the study of the definable elements of models of PA. As we shall see these structures are models of substantial parts of PA and that N is first-order definable in them. In section 1 we prove some results concerning the model theory of these structures and the height of the Σ_n-definable elements of models of PA and in section 2 we look at a problem of H. Gaifman.

3.1 Σ_n-definable elements

Definitions and notations 3.1.1. i) *For $n \in N$, let B_n denote the set of Π_n-consequences of PA. The set of Π_2-consequences of PA is denoted by B, i.e. $B = B_2$.*

ii) *For $M \models PA^-$, $n \in N$ and $a \in M$, $K^n(M, a)$ is defined as the submodel of M with the domain:*

$$K^n(M, a) = \{b \in M : b \text{ is } \Sigma_n - definable \text{ in } (M, a)\}.$$

iii) *For M, n, a as above, $I^n(M, a)$ is defined as the submodel of M with the domain:*

$$I^n(M, a) = \{b \in M : \exists c \in K^n(M, a) \text{ such that } M \models b \leq c\}.$$

iv) *$K^n(M)$ and $I^n(M)$ denote $K^n(M, 0)$ and $I^n(M, 0)$ respectively.*

Remark. Most of the results of this chapter proved for $K^n(M)$ and $I^n(M)$ hold for $K^n(M, a)$ and $I^n(M, a)$ respectively when $a \notin N$ (if $M \neq N$) with perhaps some slight modifications, e.g. replacing $\Sigma_n(M)$ by $\Sigma_n(M, a)$ where $\Sigma_n(M, a) = \{\varphi(x) \in \Sigma_n : M \models \varphi(a)\}$ and/or replacing Σ_n-definable elements of M by Σ_n-definable elements of (M, a). We leave these generalizations to the reader.

Lemma 3.1.2. *For every $\varphi \in FL$ the universal closure of the following formula is provable in PA:*

$$\forall x \leq y \exists \vec{z} \varphi \rightarrow \exists t \forall x \leq y \exists \vec{z} \leq t \varphi.$$

Proof. Elementary. ⊣

Lemma 3.1.3. *Let M be a (countable) model of B_{n+1}. Then there exists a (countable) model M' of PA such that $M \prec_n M'$.*

Proof. Define $(n+1) - Diag(M)$ as $\{\varphi(\vec{x}) \in \Sigma_n \cup \Pi_n :$ for some $\vec{a} \in (M)^{<\omega}, M \models \varphi(\vec{a})\}$. To prove the lemma it is enough to show that $PA \cup (n+1) - Diag(M)$ is a consistent set of sentences of $L(M)$. This follows by a simple application of Compactness Theorem. If M is countable then by Löwenheim-Skolem we can ensure that M' will also be countable. ⊣

Lemma 3.1.4. *For any model M of PA and $n \in N - \{0\}$*

(i) $K^n(M) \prec_n M$ *and hence* $K^n(M) \models \Pi_{n+1}(M)$.

(ii) $I^n(M) \prec_{n-1} M$ *and* $I^n(M) \models \Pi_{n+1}(M)$[1].

Proof. (i) Let $\varphi(\vec{x}, \vec{y}) \in \Pi_{n-1}$ be such that for some $\vec{b} \in (M)^{<\omega}$ $M \models \exists \vec{y} \varphi(\vec{b}, \vec{y})$. We show that for some $\vec{c} \in (K^n(M))^{<\omega} M \models \varphi(\vec{b}, \vec{c})$. For simplicity, let $\vec{x} = x, \vec{y} = y$ and $\vec{b} = b$. Suppose b is defined in M by some formula $\theta(x) \in \Sigma_n$. The formula $\exists x(\theta(x) \wedge \varphi(x, y) \wedge \forall u < y \neg \varphi(x, u))$ is equivalent in PA to a Σ_n-formula $\sigma(y)$ and $M \models \exists! y \sigma(y)$. Let c be the element of $K^n(M)$ defined by $\sigma(y)$ in M. Then $M \models \varphi(b, c)$. The proof of the general case is similar. Therefore by 1.3.7. $K^n(M) \prec_n M$.

(ii) We omit the proof of $I^n(M) \prec_{n-1} M$, as except for some minor differences it is the same as the proof of 2.2.12.

To prove $I^n(M) \models \Pi_{n+1}(M)$, it is enough to show that given any $a \in K^n(M)$ and $\varphi(\vec{x}, \vec{y}) \in \Pi_{n-1}$ such that $M \models \forall \vec{x} \exists \vec{y} \varphi(\vec{x}, \vec{y})$ then $I^n(M) \models \forall \vec{x} \leq a \exists \vec{y} \varphi(\vec{x}, \vec{y})$. Let φ and a be as above. Then clearly $M \models \forall \vec{x} \leq a \exists \vec{y} \varphi(\vec{x}, \vec{y})$ and hence by 3.1.2, $M \models \exists t \forall \vec{x} \leq a \exists \vec{y} \leq t \varphi(\vec{x}, \vec{y})$. Suppose $a \in K^n(M)$ is defined in M by some Σ_n-formula $\theta(z)$. Define ψ, ψ' and σ as follows:

$$\psi(z, t) \equiv \forall \vec{x} \leq z \exists \vec{y} \leq t \varphi(\vec{x}, \vec{y})$$

$$\psi'(z, t) \equiv \psi(z, t) \wedge \forall v < t \neg \psi(z, v)$$

$$\sigma(t) \equiv \exists z(\psi'(z, t) \wedge \theta(z)).$$

$\sigma(t)$ is equivalent to a Σ_n-formula modulo PA and hence for some $b \in K^n(M)$ $M \models \sigma(b) \wedge \exists! t \sigma(t)$. Going through the proof it is easy

[1] A similar result was proved by Wilkie [Wk].

to check that $M \models \forall \vec{x} \leq a \exists \vec{y} \leq b\varphi(\vec{x}, \vec{y})$. Hence by the fact that $I^n(M) \prec_{n-1,e} M$ we have $I^n(M) \models \forall \vec{x} \leq a \exists \vec{y} \varphi(\vec{x}, \vec{y})$. The proposition follows. ⊣

Corollary 3.1.5. *If M is a model of B_{n+1} then $K^n(M) \prec_n M$.*

Proof. Follows from 3.1.3. and 3.1.4 (i). ⊣

Proposition 3.1.6. *For models M, M' of PA and $n \in N - \{0\}$*

(i) $K^n(M) \cong K^n(M')$ *iff* $\Sigma_n(M) = \Sigma^n(M')$.

(ii) $K^n(M)$ *is pointwise definable.*

(iii) *If M and M' are countable and $I^n(M) \neq N$, then $I^n(M) \cong I^n(M')$ iff $\Sigma_n(M) = \Sigma_n(M')$ and $R_M = R_{M'}$.*

(iv) *If M is countable and $I^n(M) \neq N$, then $I^n(M)$ has $2^{\aleph_0}$ automorphisms.*

Proof. The only ones deserving proofs are (iii) and (iv).

(iii) '⇒' Assume $I^n(M) \cong I^n(M')$. As $I^n(M)$ and $I^n(M')$ are initial segments of M and M' respectively, $R_M = R_{M'}$. By 3.1.4.(ii) $I^n(M) \models \Pi_{n+1}(M)$, therefore $\Sigma_n(M) = \Sigma_n(I^n(M))$. Hence $\Sigma_n(M) = \Sigma_n(M') = \Sigma_n(I^n(M))$.

'⇐' By 2.2.7. we can identify M or M' with an $(n-1)$-elementary end extension of the other, suppose $M \prec_{n-1,e} M'$. It is enough to show that $K^n(M) = K^n(M')$. If $a \in K^n(M)$ then for some some Σ_n-formula $\varphi(x)$ $M \models \varphi(a) \wedge \exists! x\varphi(x)$, as $M \prec_{n-1,e} M'$ then $M' \models \varphi(a)$ and as $\Sigma_n(M) = \Sigma_n(M')$, $M' \models \varphi(a) \wedge \exists! x\varphi(x)$, hence $K^n(M) \subseteq K^n(M')$. Similarly, $K^n(M') \subseteq K^n(M)$. Therefore $K^n(M) = K^n(M')$ and this yields: $I^n(M) = I^n(M')$.

(iv) We only give a hint as the proof is essentially the same as the proof of 2.1.10. By 2.2.3.(ii) there is some R_M-saturated model M^* of PA such that $M \prec_{n,e} M^*$. Clearly $I^n(M) = I^n(M^*)$. If f is any automorphism of M^* then $f''I^n(M) = I^n(M)$. The rest of the proof is left to the reader. ⊣

Strength of B_n. There is some difficulty concerning the complexity of formulae modulo B_n. Indeed, Hirschfeld and Wheeler (9.8 of [H&W]) essentially provide us with an example of a Σ_1-formula φ such that $\forall x \leq y\varphi$ is not equivalent to any Σ_1-formula modulo B, but clearly $\forall x \leq y\varphi$ is equivalent to a Σ_1-formula in PA. The following provides some relief:

Lemma 3.1.7. (i) *For $m \leq n \in N$, if φ is equivalent to a Σ_m (resp. Π_m)-formula in B_{n+2}, so is $\forall x \leq y\varphi$ (resp. $\exists x \leq y\varphi$).*

(ii) *For* $m \leq n \in N$, *if* φ *is equivalent to both a* Σ_{m+1} *and a* Π_{m+1}-*formula in* B_{n+2}, *so are* $\forall x \leq y\varphi$ *and* $\exists x \leq y\varphi$.

Proof. By 3.1.2. and induction on m and n $(0 \leq m \leq n)$. ⊣

Lemma 3.1.8. (i) *For* $m \leq n \in N$, $n \neq 0$ *and for any* Σ_m-*formula* $\varphi(\vec{x})$:

$$B_{n+1} \vdash \forall \vec{x}(\varphi(\vec{x}) \leftrightarrow \Sigma_n(\ulcorner \varphi(\vec{x}) \urcorner, \vec{x})).$$

(ii) *For any* Σ_1-*formula* $\varphi(\vec{x})$ *there is a diophantine equation* $\delta(\vec{x}, \vec{y})$ *such that:*

$$B \vdash \forall \vec{x}(\varphi(\vec{x}) \leftrightarrow \exists \vec{y}\delta(\vec{x}, \vec{y})).$$

(iii) *For* $n > 0$ *if* φ *is equivalent to both a* Σ_n *and a* Π_n-*formula in* B_{n+1}, *then:*

$$B_{n+1} \vdash \forall \vec{x}[\exists y\varphi(\vec{x}, y) \rightarrow \exists y(\varphi(\vec{x}, y) \wedge \forall z < y \neg\varphi(\vec{x}, z))].$$

Proof. (i) and (ii) follow from 1.3.3. and 1.3.5. respectively.
(iii) follows from 3.1.7.(ii). ⊣

Definition 3.1.9. $A \subseteq N^k$ *is* arithmetic *if there is a formula* $\varphi(\vec{x}) \in FL_k$ *such that:*

$$(n_1, \ldots, n_k) \in A \Leftrightarrow \mathbb{N} \models \varphi(n_1, \ldots, n_k).$$

Theorem 3.1.10. *Let* $n \in N - \{0\}$. *Suppose* M *is a model of* B_{n+1} *and* $K^n(M)$ *is cofinal in* M. *Then:*

(i) N *is definable in* M *by a formula of* L.
(ii) $Th(M)$ *is non-arithmetic.*

Proof. (i) By 3.1.8.(i) for any Σ_n-formula $\varphi(x)$:

$$M \models \forall x(\Sigma_n(\ulcorner \varphi(x) \urcorner, x) \leftrightarrow \varphi(x)).$$

Define $\Phi_n(i, x)$ as $\Sigma_n(i, x) \wedge \forall y(\Sigma_n(i, y) \rightarrow y = x)$. As Σ_n-definable elements of M are cofinal in M:

$m \in N \quad \Leftrightarrow \quad$ Every set of elements of M which are definable in M by formulae whose Gödel numbers are less than m has an upper-bound in M;

that is

$m \in N \quad \Leftrightarrow \quad M \models \exists y \forall x \forall i < m(\Phi_n(i, x) \rightarrow x \leq y)$.

(ii) Suppose $\theta_n(x)$ defines N in M. For a formula φ of L, let φ^{θ_n} denote the relativization of φ to $\theta_n(x)$. For every sentence φ of L:

$$\varphi \in Th(\mathbb{N}) \Leftrightarrow \varphi^{\theta_n} \in Th(M).$$

Therefore, $Th(\mathbb{N})$ is recursive in $Th(M)$, but by a well-known Theorem of Tarski, $Th(\mathbb{N})$ is non-arithmetic; hence, $Th(M)$ is non-arithmetic. ⊣

Corollary 3.1.11. *For any model M of PA and $n \in N - \{0\}$:*

(i) *N is definable in $K^n(M)$ and $I^n(M)$ by a formula of L.*
(ii) *$Th(K^n(M))$ and $Th(I^n(M))$ are non-arithmetic.*
(iii) *$Th(K^n(M))$ is recursive in $Th(K^{n+1}(M))$.*

Proof. (i) and (ii) follow directly from 3.1.4. and 3.1.10. noting that by 3.1.4. we have $K^n(M) \prec_n I^n(M)$. To show (iii) note that as N is definable in $K^{n+1}(M)$, then $K^n(M)$ is definable in $K^{n+1}(M)$. ⊣

The definition of N in 3.1.10 is not the best if the complexity of formulae are of concern (see notes at the end of this chapter). Throughout this section let $\Sigma_n^*(i, \vec{t})$ denote the Π_{n-1}-formula such that $\exists \vec{t}\, \Sigma_n^*(i, \vec{t}) \equiv \Sigma_n(i)$.

Lemma 3.1.12. *Let M and $n \in N$ be as in 3.1.10, then the following formula $\theta_n(v)$ defines N in M:*

$$\exists x \forall i < v(\exists \vec{t}\, \Sigma_n^*(i, \vec{t}) \rightarrow \exists \vec{t} \leq x \Sigma_n^*(i, \vec{t})).$$

Hence, by 3.1.7, N is defined in M by a Σ_{n+1}-formula.

Proof. Similar to 3.1.10.(i). ⊣

Proposition 3.1.13. *Let M be a model of B_{n+1} and let $n \in N - \{0\}$. Then $\Sigma_n(M) \notin R_{K^n(M)}$.*

Proof. Suppose not, i.e. $\Sigma_n(M)$ is coded in $K^n(M)$ by some $\alpha \in K^n(M)$. We can safely assume that $K^n(M) \neq N$ and by 3.1.3. we may assume that M is a model of PA.
Define $\varphi(x, y)$ as the formula:

$$\exists z \forall i < y(p_i | x \rightarrow \exists \vec{t} < z \Sigma_n^*(i, \vec{t})).$$

It is not difficult to prove by induction on y that $M \models \forall y \varphi(\alpha, y)$. By 3.1.7. $\varphi(x, y)$ is equivalent to a Σ_n-formula modulo B_{n+1}. Hence, as $K^n(M) \prec_n M$ we have: $K^n(M) \models \forall y \varphi(\alpha, y)$. Let β be any element of $K^n(M) - N$, then $K^n(M) \models \varphi(\alpha, \beta)$. Contradicting that Σ_n-definable elements of $K^n(M)$ are cofinal in $K^n(M)$, hence, $\Sigma_n(M) \notin R_{K^n(M)}$. ⊣

Remark. For a model M of PA, if $I^n(M) \neq N$ then $\Sigma_n(M)$ is coded in $I^n(M)$ and this helps us to find a Σ_n-formula with a parameter from $I^n(M)$ which defines N in $I^n(M)$. Namely:

$$m \in N \Leftrightarrow I^n(M) \models \exists z \forall i < m(p_i | \alpha \rightarrow \exists \vec{t} \leq z \Sigma_n^*(i, \vec{t})),$$

where α codes $\Sigma_n(M)$ in $I^n(M)$.

Corollary 3.1.14. *For any model M of PA and $n \in N - \{0\}$, if $I^n(M) \neq N$, then $I^n(M) \not\prec_n M$.*

Proof. Follows from the above remark and the proof of 3.1.13. ⊣

Theorem 3.1.15. *For any model M of PA:*

(i) $K^0(M)$ *is* always *cofinal in* $K^1(M)$.

(ii) *For* $n \in N - \{0\}$, *if* $K^{n+1}(M) \neq N$, *then* $K^n(M)$ *is* never *cofinal in* $K^{n+1}(M)$.

Proof. We leave the proof of (i) to the reader.

(ii) Suppose not, then $I^{n+1}(M) = I^n(M)$ for some $n \in N - \{0\}$. Hence by 3.1.4(ii) $I^n(M) \prec_n M$. As $I^n(M) \neq N$, this contradicts 3.1.14. Hence the corollary. ⊣

Proposition 3.1.16. *For any model M of PA and $n \in N$, if $K^n(M) \neq N$, then $K^n(M)$ is not an initial segment of $K^{n+1}(M)$.*

Proof. (Sketch) By 3.1.15.(i) we can assume $n \neq 0$. Using 3.1.15.(ii), it is rather tedious but straightforward to show that $\Sigma_n(M)$ is coded in $K^{n+1}(M)$, however, by 3.1.13. $\Sigma_n(M)$ is not coded in $K^n(M)$. Therefore $K^n(M)$ is not an initial segment of $K^{n+1}(M)$. ⊣

Theorem 3.1.17. *For any model M of PA and $n \in N - \{0\}$:*

(i) *$K^n(M)$ is the unique model of $Th(K^n(M))$ in which Σ_n-definable elements are cofinal.*

(ii) *$K^n(M)$ is the unique model of $Th(K^n(M))$ in which N is definable by a formula of L.*

(iii) *If $M' \equiv K^n(M)$ and $M' \neq K^n(M)$, then M' contains a non-standard initial segment which is a model of $Th(\mathbb{N})$.*

(iv) *Suppose $K^n(M) \prec M'$ and $K^n(M) \neq M'$, $a_1, a_2 \in K^n(M)$ and $K^n(M) \models a_1 < a_2$. Then if a_1 and a_2 are not finitely apart in $K^n(M)$ there is an element $c \in M' - K^n(M)$ such that $M' \models a_1 < c < a_2$.*

(v) *If $K^n(M) \neq N$, then $K^n(M)$ does not have a proper elementary cofinal or end extension.*

Proof. Throughout this proof let $\theta_n(x)$ be the formula which defines N in $K^n(M)$ as in the proof of 3.1.10.(i), and let $\Phi_n(x, y)$ denote the formula $\Sigma_n(x, y) \wedge \forall z(\Sigma_n(x, z) \rightarrow z = y)$.

(i) Suppose $M' \equiv K^n(M)$ and Σ_n-definable elements of M' are cofinal in M'. By 3.1.10.(i), $\theta_n(x)$ defines N in both M' and $K^n(M)$. As $K^n(M)$ is pointwise definable: $K^n(M) \models \forall y \exists x(\theta_n(x) \wedge \Phi_n(x, y))$. Hence: $M' \models \forall y \exists x(\theta_n(x) \wedge \Phi_n(x, y))$, i.e. M' is also pointwise definable, hence, as $K^n(M) \equiv M'$, we have $K^n(M) \cong M'$.

(ii) Suppose $\varphi(x)$ defines N in M' where $M' \equiv K^n(M)$. Clearly, $\varphi(x)$ defines N in $K^n(M)$. The proof follows as in (i) above.

(iii) We show that the submodel of M' with the domain: $M^* = \{a \in M' : M' \models \theta_n(a)\}$ satisfies the required conditions. By (ii) above, as $M' \neq K^n(M)$, M' is non-standard. It is also easy to see that M^* is an (perhaps improper) initial segment of M'. Now given any sentence φ of L, $M^* \models \varphi \Leftrightarrow M' \models \varphi^{\theta_n} \Leftrightarrow \mathbb{N} \models \varphi$. (If $K^n(M) = N$, then $M^* = M'$. However, by a result of Friedman (see Chapter 5) we can find a proper initial segment of M' which is isomorphic to M').

(iv) As $M' \neq K^n(M)$, by (ii) above there is some $\alpha \in M' - N$ such that $M' \models \theta_n(\alpha)$. Suppose that a_1 and a_2 are as in the hypothesis and that every element c of M' satisfying $M' \models a_1 < c < a_2$ is in $K^n(M)$, i.e. is Σ_n-definable. Then:

$$M' \models \forall y(a_1 \leq y \leq a_2 \rightarrow \exists z < \alpha \Phi_n(z, y)).$$

Hence, as $M' \models \theta_n(\alpha)$,

$$M' \models \exists x(\theta_n(x) \wedge \forall y(a_1 \leq y \leq a_2 \rightarrow \exists z < x \Phi_n(z, y))).$$

As $K^n(M) \prec M'$, the above sentence holds in $K^n(M)$ and as $\theta_n(x)$ defines N in $K^n(M)$, then for some $m \in N$:

$$K^n(M) \models \forall y(a_1 \leq y \leq a_2 \rightarrow \exists z < m \Phi_n(z, y)).$$

Contradicting the fact that a_1 and a_2 are not finitely apart. Therefore for some $c \in M' - K^n(M)$: $M' \models a_1 < c < a_2$.

(v) By (i), $K^n(M)$ does not have any proper elementary cofinal extensions and, by (iv), if $K^n(M) \neq N$, then $K^n(M)$ does not have any (proper) elementary end extensions. ⊣

3.2 A problem of H. Gaifman

Definition 3.2.1. *A structure M for L is said to be* arithmetical *if there is a bijection $f : M \rightarrow N$ and formulae of L which we denote by $S^M(x) = y$, $x +^M y = z$ and $x \cdot^M y = z$ such that:*

(i) $M \models S(a) = b \Leftrightarrow \mathbb{N} \models S^M(f(a)) = f(b)$

(ii) $M \models a + b = c \Leftrightarrow \mathbb{N} \models f(a) +^M f(b) = f(c)$

(iii) $M \models a \cdot b = c \Leftrightarrow \mathbb{N} \models f(a) \cdot^M f(b) = f(c)$,

for all elements $a, b, c \in M$; or, equivalently, up to isomorphism: the domain of M is an arithmetic subset of N and $S, +$ and $\cdot$ of M are arithmetic relations in N.

Clearly, N is arithmetical and by 2.2.1. any arithmetic extension T of PA has an arithmetical model. Gaifman asks whether there is a non-standard arithmetical model M of PA such that $Th(M)$ is non-arithmetic. Below we give a positive solution to a slightly weaker problem.

Proposition 3.2.2. *For any arithmetical M of PA and $n \in N - \{0\}$, $K^n(M)$ and $I^n(M)$ are arithmetical.*

Proof. We prove the proposition for $K^n(M)$, the proof for $I^n(M)$ is on similar lines.

Suppose $f, S^M(x) = y$ etc. are as in Definition 3.2.1. Let N_M denote the standard part of M. We first show that $f''N_M$ is an arithmetic subset of N. Let $m_0 = f(0^M)$, where 0^M denotes the interpretation of 0 in M.

Define $g : N \to N$ as follows:

$$g(0) = m_0$$
$$g(m+1) = S^M(g(m))$$

g is recursive in S^M and is therefore arithmetic, i.e. there is a formula of L denoted henceforth by $\varphi(x, y)$ such that:

$$\text{For all } m, p \in \mathbb{N} : \ f(m) = p \Leftrightarrow \mathbb{N} \models \varphi(m, p).$$

Therefore, for all $p \in N$: $p \in f''N_M \Leftrightarrow \mathbb{N} \models \exists x \varphi(x, p)$ (*).

As $S^M, +^M$ and $\cdot^M$ are arithmetic, we can show by induction on the complexity of formulae of L that for every formula $\psi(\vec{x})$ of L there is another formula of L denoted by $\psi^+(\vec{x})$ with the same number of variables such that:

$$\text{for all } \vec{a} \in (M)^{<\omega}, M \models \psi(\vec{a}) \Leftrightarrow \mathbb{N} \models \psi^+(\vec{f(a)}).$$

Let $\Phi_n(i, x)$ be as in the proof of 3.1.10.(i). Then:

$$\begin{aligned} a \in K^n(M) \quad &\Leftrightarrow \quad \text{there is some } p \in N_M \text{ such that } M \models \Phi_n(p, a) \\ &\Leftrightarrow \quad \text{there is some } p \in N_M \text{ such that } \mathbb{N} \models \Phi_n^+(f(p), f(a)) \\ &\Leftrightarrow \quad \mathbb{N} \models \exists x \exists y (\varphi(x, y) \wedge \Phi_n^+(y, f(a)) \text{ by } (*). \end{aligned}$$

Hence, the image of the domain of $K^n(M)$ under f is arithmetic. We leave the rest of the proof to the reader. ⊣

Remark. The above proposition contains an added bonus, namely:

If M is arithmetical then $Th(M)$ is hyperarithmetical; this is because for any sentence φ of L: $M \models \varphi \Leftrightarrow \mathbb{N} \models \varphi^+$.

Hence if M ($M \models PA$) is arithmetical and $n \in N - \{0\}$, then $Th(K^n(M))$ and $Th(I^n(M))$ are recursively equivalent with $Th(\mathbb{N})$.

The following is an interesting corollary of 3.2.2.

Corollary 3.2.3. *For any model M of PA and $n \in N - \{0\}$, the following are equivalent:*

(i) $\Sigma_n(M)$ *is arithmetic.*

(ii) *$K^n(M)$ is arithmetical.*

Proof. Left to the reader. ⊣

Theorem 3.2.4. *For any $n \in N$, there is a non-standard arithmetical model M^* of B_n such that $Th(M^*)$ is non-arithmetic.*

Proof. Let $T = PA \cup \{\neg ConPA\}$. By a Henkin-type argument we can construct an arithmetical model M of T (as in 2.2.1.). Let $M^* = K^{n-1}(M)$ $(n > 1)$. Then, by 3.1.4.(i), $M^* \models B_n$. By 3.2.1., M^* is arithmetical and, by 3.1.11., $Th(M^*)$ is non-arithmetic. ⊣

Notes. Using a different method, A. Robinson (ref. 85 of [H&W]) was the first to show that N is first-order definable in certain models of B. For more information regarding the definability of N in these models we refer the reader to [H&W] and a paper of Macintyre, Simmons and Goldrei (ref. 63 of [H&W]); also see [Mc] for formulae of least complexity defining N in models of B_n.

4

Internal A-saturated structures

In chapter 2 (Corollary 2.2.6.) we showed that for any model M of PA and $\alpha \in M$, I_α^M, as defined, is an A-saturated structure. In the first section of this chapter we generalize Corollary 2.2.6. to certain elements of models of some weaker systems and use this generalization to extend some results of chapter 2 to models of B_1 and B. We show that the class of small initial segments (see definition 4.1.3. below) of models of B_1 is equal to the class of initial segments of models of PA. In section two we solve some problems regarding the relative positions of the classes of Σ_n-definable elements of a model of PA.

4.1 Initial segments with a top element.

Definition and remarks 4.1.1. *Following 2.2.4. given a structure M for L and $\alpha \in M$, we define a natural structure on $I_\alpha^M = \{x \in M : M \models x \leq \alpha\}$. As was noted earlier, we have to think of $S, +, \cdot$ as relations on I_α^M rather than functions, however, for simplicity we shall use the same symbolism. We also note that if M is a model of PA^- plus the least number principle for Σ_0-formulae, then I_α^M would be a model of the least number principle for any formula in the language of I_α^M. This has the rather interesting consequence that in the above circumstances, for $\vec{a} \in (I_\alpha^M)^{<\omega}$, the set of definable elements of $(I_\alpha^M, \vec{a})$, denoted by $Def(I_\alpha^M, \vec{a})$, with the structure inherited from I_α^M, forms an elementary substructure of I_α^M.*

In what follows let K denote a non-standard model of PA, α an element of $K - N$ and J an elementary substructure of I_α^K.

Lemma 4.1.2. *Let $A \subseteq N$ be recursive. Then A is coded in J.*

Proof. As A is recursive, there is a Σ_1-formula $\varphi(x)$ such that $n \in A \Leftrightarrow PA \vdash \varphi(n)$ and $n \notin A \Leftrightarrow PA \vdash \neg\varphi(n)$ for all $n \in N$ (see [Sh]). Clearly, $n \in A \Leftrightarrow K \models \varphi(n)$ for all $n \in N$. Let $\psi(\vec{y}, x)$ denote the Σ_0-formula such that $\varphi(x) \equiv \exists\vec{y}\psi(\vec{y}, x)$. As $n \in A \Leftrightarrow \mathbb{N} \models \varphi(n)$ it is straightforward to show that $n \in A \Leftrightarrow K \models \exists\vec{y} \leq \alpha\psi(\vec{y}, n)$ for all $n \in N$. Therefore, for every $n \in N$, $n \in A \Leftrightarrow I_\alpha^K \models \exists\vec{y} \leq \alpha\psi(\vec{y}, n)$ and as $J \prec I_\alpha^K$, $n \in A \Leftrightarrow J \models \exists\vec{y} \leq \alpha\psi(\vec{y}, n)$. Now for every $m \in N$:

$$J \models \exists x \forall t < m(p_t | x \leftrightarrow \exists\vec{y} \leq \alpha\psi(\vec{y}, t)).$$

The lemma follows by a simple over-spill argument. ⊣

Definition 4.1.3.

(i) $\beta \in I_\alpha^K$ *is* small *in* I_α^K *(or* J*) if* $K \models 2^{(\beta^\gamma)} < \alpha$ *for some* $\gamma \in K - N$.

(ii) *An initial segment* I *of a model* M *of* B_1 *is* small *in* M *if* $2^{(\beta^\gamma)} \in M$ *for some* $\beta \in M - I$ *and* $\gamma \in M - N$.

Note. Definition (ii) above is meaningful, since by 3.1.3. for every model M of B_1 there is a model M' of PA such that $M \prec_0 M'$.

Lemma 4.1.4. *There is a formula* $S(x, y, z)$ *of* L *such that for all* $\beta \in J$*,* β *small in* J*, formulae* $\varphi(\vec{x})$ *of* L *and* $\vec{a} \in (I_\beta^J)^{<\omega}$

$$I_\beta^J \models \varphi(a_1, \ldots, a_n) \Leftrightarrow J \models S(\beta, \ulcorner\varphi(x_1, \ldots, x_n)\urcorner, \langle a_1, \ldots, a_n\rangle).$$

Proof. (Sketch) It is enough to show that we can define satisfaction for bounded formulae of the language of I_β^J in J. To do this for a (bounded) formula $\exists x \leq \beta\varphi(x)$, speaking informally inside J, we need 2^β elements of J, coding functions from $\{\varphi(i) : J \models i \leq \beta\}$ to $\{0, 1\}$ (exactly as in the case for propositional calculus, while considering the validity of a disjunction).

The proof follows by induction on the complexity of formulae and checking that as β is small in J, the above coding can be carried out in J. ⊣

Proposition 4.1.5. *Suppose* $\beta \in J$ *is small in* J*, then* I_β^J *is recursively saturated.*

Proof. Using 4.1.4., the proposition follows as in the proof of 2.2.2., noting that by 4.1.2. every recursive set is coded in J. ⊣

Lemma 4.1.6. *Suppose* M *is a substructure of* J *and that* M *is a model of* PA*. Let* $\overline{M}$ *denote the initial segment of* J *determined by* M *and the structure inherited from* J*. Then* $M < \overline{M}$*.*

Proof. This is a straightforward modification of Theorem 4 of Gaifman [G]. ⊣

Proposition 4.1.7. *Let A be any subset of $Th(\mathbb{N})$ such that A is coded in J. Then for all $a \in J - N$, there exists a non-standard initial segment M of J such that $M \models A$ and $a \notin M$.*

Proof. Using 4.1.6. and 4.1.1., it is enough to prove the proposition for the case when J is countable, hence assume J is countable. Choose $\beta \in J - N$ such that β is small in J. By 4.1.5., I^J_β is recursively saturated and hence by 1.4.7.(i) it is a $\Sigma(rec)$-structure. Let c be any element of J such that $J \models c < b$ and c codes A in J. Let Φ denote the following $\Sigma(rec)$-sentence of $L \cup \{c\}$:

$$\exists X \exists x (\bigwedge_{n \in N} x \neq n \wedge \forall x \forall y (X(x) \wedge y \leq x \rightarrow X(y)) \wedge \bigwedge_{\varphi \in SL} (p_\varphi | c \rightarrow \varphi^X)).$$

Φ is consistent with $Th(I^J_\beta, c)$ (the details are as in the proof of 2.2.9.). Hence, $I^J_\beta \models \Phi$ and the proposition follows. ⊣

Corollary 4.1.8.

(i) *R_J is a Scott set.*

(ii) *If β is small in J, then I^J_β is R_J-saturated.*

Proof. (i) By 4.1.7. there is a non-standard initial segment M of J, such that $M \models PA$. R_M is a Scott set and $R_M = R_J$, hence R_J is a Scott set.

(ii) Exactly as in the proof of 2.2.2., using (i) above. ⊣

Lemma 4.1.9. *Let M be a model of B_1, then there exists a model M' of PA such that $I^M_\alpha \prec I^{M'}_\alpha$ for all $\alpha \in M$.*

Proof. Follows directly from 3.1.3. ⊣

Corollary 4.1.10. *Let M be a non-standard model of B_1. Then*

(i) *R_M is a Scott set.*

(ii) *If I^M_β is small in M then I^M_β is R_M-saturated (where $\beta \in M$).*

(iii) *For all $a \in M - N$ and $A \subseteq Th(\mathbb{N})$ such that $A \in R_M$, there is a non-standard initial segment M' of M such that $M' \models A$ and $a \notin M'$.*

Proof. Using 4.1.9. the corollary follows from 4.1.7. and 4.1.8. ⊣

Theorem 4.1.11. *Let M be any countable structure for L. Suppose M has an end extension to a model M' of B_1 such that M is small in M'. Then M has an end extension to a model of PA.*[1]

[1] For a related result see [P&K], Theorem C.

Proof. First assume M' is countable, let β be any element of M' such that $I_\beta^{M'}$ is small in M'. By 4.1.9. there is a model K of PA such that $I_\beta^{M'} \prec I_\beta^K$. By 4.1.10. $I_\beta^{M'}$ is recursively saturated and hence by 1.4.7.(i) it is a $\Sigma(rec)$-structure. Let $PA^{(t,U_1,U_2,U_3)}$ denote the re-writing of PA in a new language derived from L by replacing 0 by a constant symbol t, S by a unary function symbol U_1 etc. Define the $\Sigma(rec)$-sentence Φ as the formalization of the following statement:

> $\exists X, U_1, U_2, U_3, F \exists x, t[X(x) \wedge X(t) \wedge U_1$ is a unary function from X into $X \wedge U_2, U_3$ are binary functions from X^2 into $\wedge$ F is an isomorphism between the universe and $I_x^X \wedge \bigwedge_{\varphi \in PA^{(t,U_1,U_2,U_3)}} \varphi^X]$.

Φ expresses the fact that the structure is end extendable to a model of PA. Clearly, $I_\beta^K \models \Phi$ and since $I_\beta^{M'} \prec I_\beta^K$, Φ is consistent with $Th(I_\beta^{M'})$. As $I_\beta^{M'}$ is a $\Sigma(rec)$-structure, $I_\beta^{M'} \models \Phi$ and the proposition follows for the case when M' is countable. If M' is uncountable, define J' as $J' = \bigcup_{a \in M} Def(I_\beta^{M'}, a)$. Clearly, $J' \prec I_\beta^{M'}$ and J' is countable; the proof follows exactly as in the countable case. ⊣

Remark. The following generalization of 4.1.11 can be proved exactly as above.

Theorem. *Let T be a consistent extension of B_1 such that $T \in R_{M'}$, where M' is a model of $\Pi_1(T)$. Let M be a small initial segment of M'. Then M is isomorphic to an initial segment of a model of T.*

Corollary 4.1.12. *Let M be any non-standard model of PA and let $a \in M - N$ then:*

(i) *$K^1(M, a)$ does not have any end extension $M' \models B_1$ such that $K^1(M, a)$ is small in M'.*

(ii) *$K^1(m, a)$ is not isomorphic to an initial segment of itself.*

Proof. (i) Suppose not, then by 4.1.11., $K^1(M, a)$ has an end extension to a model M_1 of PA. Therefore, M_1 has a non-standard initial segment whose every element is Σ_1-definable in (M_1, a), but this contradicts 2.2.14.

(ii) Follows from (i) above and noting that as in the proof of 3.1.4.(i) we can show that $K^1(M, a) \models B$. ⊣

Models of B

In what follows, we will prove some lemmas concerning models of B which will be used in the next chapter.

Definition 4.1.13. *A formula $\varphi(\vec{x})$ of L is a Δ_1^B-formula if $\varphi(\vec{x})$ is equivalent to both a Σ_1 and a Π_1-formula modulo B.*

Lemma 4.1.14. *Suppose M is a non-standard model of B and that $\varphi(x, \vec{y})$ is a Δ_1^B-formula such that for some $\vec{a} \in (M)^{<\omega}$ and for all $n \in N$, $M \models \varphi(n, \vec{a})$. Then there exists some $\alpha \in M - N$ such that $M \models \forall x \leq \alpha \varphi(x, \vec{a})$.*

Proof. Immediate from 3.1.8.(iii). ⊣

Lemma 4.1.15. *There is a Δ_1^B-formula $S(x, y, z)$ such that given any model M of B, $\alpha \in M$ and any formula $\varphi(\vec{x})$ of L, for all $\vec{a} \in (I_\alpha^M)^{<\omega}$:*

$$I_\alpha^M \models \varphi(a_1, \ldots, a_n) \Leftrightarrow M \models S(\alpha, \ulcorner\varphi(x_1, \ldots, x_n)\urcorner, \langle a_1, \ldots, a_n\rangle).$$

Proof. By 1.3.3.(i) there are a Σ_1-formula $\sigma(y, \vec{x})$ and a Π_1-formula $\Pi(y, \vec{x})$ such that $B \models \forall\vec{x}\forall y(\sigma(y, \vec{x}) \leftrightarrow \Pi(y, \vec{x}))$ (as the statement is a Π_2-statement) and for any Σ_0-formula $\varphi(\vec{x})$, $B \vdash \forall\vec{x}(\varphi(\vec{x}) \leftrightarrow \sigma(\ulcorner\varphi(\vec{x})\urcorner, \vec{x}))$. We leave the rest of the proof to the reader. ⊣

Note. Henceforth, instead of writing

$$M \models S(\alpha, \ulcorner\varphi(x_1, \ldots, x_n)\urcorner, \langle a_1, \ldots, a_n\rangle)$$

as in 4.1.15., we shall write $M \models Sat(\alpha, \varphi(\vec{a}))$, which, although misleading, is much more pleasant.

Remark 4.1.16. Let M be a model of B. Then I_α^M is R_M-saturated for all $\alpha \in M$.

Proof. Immediate from 4.1.10.(ii). ⊣

Lemma 4.1.17. *Let M_1 and M_2 be models of B such that $M_1 \subset M_2$. Then $I_\alpha^{M_1} \prec I_\alpha^{M_2}$ for all $\alpha \in M_1$.*

Proof. Let $\varphi(\vec{x})$ be any formula of L and let $\vec{a} \in (I_\alpha^{M_1})^{<\omega}$. Then $I_\alpha^{M_1} \models \varphi(\vec{a}) \Leftrightarrow M_1 \models \varphi^{y \leq \alpha}(\vec{a}) \Leftrightarrow$ by 3.1.8.(ii), $M_2 \models \varphi^{y\leq\alpha}(\vec{a}) \Leftrightarrow I_\alpha^{M_2} \models \varphi(\vec{a})$, where $\varphi^{y \leq x}$ denotes the relativization of φ to $y \leq x$. ⊣

Proposition 4.1.18. (Immortality of the submodels)
Suppose M_1 and M_2 are countable models of B such that $M_1 \subset M_2$, $a, b \in M_1$. Let T be any theory in L such that $T \in R_{M_1}$, then the following are equivalent.

(i) *There is some $M_1^{'} \models T$ such that $M_1^{'} \subset_e M_1, a \in M_1^{'}$ and $b \notin M_1^{'}$.*
(ii) *There is some $M_2^{'} \models T$ such that $M_2^{'} \subset_e M_2, a \in M_2^{'}$ and $b \notin M_2^{'}$.*

Proof. Similar to the proof of 2.2.19. using 4.1.17. ⊣

Remark. For the cases when the above proposition can be generalized to uncountable models, see the Remarks following 2.2.19.

4.2 The relative positions of Σ_n-definable elements of models of PA.

Definition 4.2.1. *Let M be a non-standard model of PA. For subsets A, B of M define:*

(i) $A \downarrow\downarrow B$ *if* $\forall a \in A - N \exists b \in B - N \; M \models b \leq a$ *and* $\forall b \in B - N \exists a \in A - N \; M \models a \leq b$.

(ii) $A| \downarrow B$ *if* $\exists b \in B - N \forall a \in A - N \; M \models b < a$.

Note that if $A \subset B$ then not $(A \downarrow\downarrow B) \Leftrightarrow A| \downarrow B$.

Lemma 4.2.2. *Let M be a model of B such that $K^0(M) \neq N$. Then $K^0(M)| \downarrow M$ iff $\Pi_1(\mathbb{N}) \in R_M$.*

Proof. Similar to the proof of 2.2.18. using 4.1.10.(iii). ⊣

Proposition 4.2.3. *Let M be a model of PA such that $K^0(M) \neq N$.*[2] *Then:*

(i) $K^0(M)| \downarrow K^1(M)$ *iff* $\Pi_1(\mathbb{N}) \in R_{K^1(M)}$.

(ii) *For $n \in N - \{0\}$, $K^n(M)| \downarrow K^{n+1}(M)$ and $K^0(M) \downarrow\downarrow K^n(M)$ iff $\Pi_1(\mathbb{N}) \in R_{K^{n+1}(M)} - R_{K^n(M)}$.*

Proof. (i) Follows directly from 4.2.2. remembering that $K^1(M) \models B$.

(ii) '$\Rightarrow$' as $K^0(M) \downarrow\downarrow K^n(M)$, then by 4.2.2. $\Pi_1(\mathbb{N}) \notin R_{K^n(M)}$ and as $K^n(M)| \downarrow K^{n+1}(M)$ then $K^0(M)| \downarrow K^{n+1}(M)$, hence again by 4.2.2. $\Pi_1(\mathbb{N}) \in R_{K^{n+1}(M)}$.

'$\Leftarrow$' Suppose $\Pi_1(\mathbb{N}) \notin R_{K^n(M)}$ and $\Pi_1(\mathbb{N}) \in R_{K^{n+1}(M)}$, then by 4.2.2. $K^0(M) \downarrow\downarrow K^n(M)$ and $K^0(M)| \downarrow K^{n+1}(M)$, hence $K^n(M)| \downarrow K^{n+1}(M)$. ⊣

Definition and remarks 4.2.4. *To show that the above proposition is non-void, we need some terminology and a result which is essentially due to Mostowski (for references see [Mc] p. 33). Following Scott [S], for any sentence φ of L let φ^1 denote φ and φ^0 denote $\neg\varphi$. For a theory T in L, a set of sentences $S = \{\varphi_i : i \in N\}$ of L is* independent of T *if $T \cup \{\varphi_i^{f(i)} : i \in N\}$ is consistent, no matter what $f : N \to \{0,1\}$ is chosen. For a model M of PA define $T_n(M) = \Sigma_n(M) \cup \Pi_n(M) \cup PA$.*

Result 4.2.5. (Mostowski-Friedman) *For any model M of PA and for each $n \in N$, there is a set S of Σ_{n+1}-sentences such that S is recursive in $T_n(M)$ and independent of $T_n(M)$.*

---

[2]We do not know if (ii) of 4.2.3. will hold for $n > 2$ if $K^0(M) = N$.

Proof. For an outline proof of a slightly different result see [Mc], pp. 33-34. ⊣

Proposition 4.2.6. *Let A be a non-recursive subset of N. Then for each $n \in N - \{0\}$, there is a model M of PA such that $K^1(M) \neq N$ and $A \in R_{K^{n+1}(M)} - R_{K^n(M)}$.*

Proof. Let $M^{'}$ be a model of PA such that $A \notin R_{M'}$ and $K^1(M) \neq N$ (possible by 1.3.14). By 4.2.5. there is a set $S = \{\varphi_i : i \in N\}$ of Σ_{n+1}-sentences which is recursive in $T_n(M^{'})$ and independent of $T_n(M^{'})$. Define $g : N \to N$ by $g(i) = \ulcorner\varphi_i\urcorner$. As S is recursive in $T_n(M^{'})$, it is tedious but elementary to show the existence of a Σ_{n+1}-formula $\gamma(x, y)$ such that $g(i) = \lceil\varphi_i\rceil \Leftrightarrow T_n(M^{'}) \models \gamma(i, \ulcorner\varphi_i\urcorner)$.

Define $\psi(x)$ as $\exists y(\gamma(x, y) \wedge \Sigma_{n+1}(y))$. $\psi(x)$ is Σ_{n+1} and it is easy to see that $T_n(M^{'}) \vdash \varphi_i \leftrightarrow \psi(i)$ for all $i \in N$. Let $C = \{2m : m \in A\} \cup \{2m+1 : m \notin A\}$ and define T to be the following set of sentences:

$$T = T_n(M^{'}) \cup \{\psi(m) : m \in C\} \cup \{\neg\psi(m) : m \notin C\}.$$

By 4.2.5. T is consistent, let M be any model of T, then for all $m \in N$, $m \in A \Leftrightarrow M \models \psi(2m)$ and $m \notin A \Leftrightarrow M \models \psi(2m + 1)$.

Let a be any non-standard element of $K^{n+1}(M)$ and define $\delta(a, y)$ to be the following formula of $L \cup \{a\}$

$$\forall t < a[(p_t | y \to \psi(2t)) \wedge (p_t \not| y \to \psi(2t + 1))].$$

It is routine to check that $\delta(x, y)$ is equivalent to a Σ_{n+1}-formula $\delta^{'}(x, y)$ modulo PA and that $M \models \exists y\delta^{'}(x, y)$. As $K^{n+1}(M) \prec_{n+1} M$, then $K^{n+1}(M) \models \exists y\delta^{'}(a, y)$. Let $b \in K^{n+1}(M)$ be such that $K^{n+1}(M) \models \delta^{'}(a, b)$; b codes A in $K^{n+1}(M)$, and as $K^n(M) \cong K^n(M^{'})$, $A \notin R_{K^n(M)}$. ⊣

Corollary 4.2.7. *Given any $A \subseteq N$, there is a model M of PA such that $A \in R_{K^1(M)}$.*

Proof. Similar to the proof of 4.2.6. ⊣

Theorem 4.2.8. *For each $n \in N$, there is a model M of PA such that $K^n(M) | \downarrow K^{n+1}(M)$.*

Proof. Immediate from 4.2.3., 4.2.6. and 4.2.7. ⊣

We end this chapter by posing two related open problems:

(i) Does there exist a structure M for L such that M is end extendable to a model of B_1 but not to any model of PA?

(ii) Suppose $\alpha \in M$, where M is a model of PA; let $J = Def(I_\alpha^M)$. What are the necessary and sufficient conditions on $\beta \in J$ so that I_β^J is recursively saturated?

Note that by 4.1.5., if β is small in J, then I_β^J is recursively saturated.

5

Initial segments of models of B

In this chapter we continue to exploit the versatility of the A-saturated structures introduced in chapter 4. The first section is devoted to a generalization of a Theorem of Friedman, also our proof simplifies Friedman's original argument. In the second section we show the existence of approximating chains of models of PA for certain models of B. In the last section we prove some results regarding the number of isomorphic initial segments of an initial segment I of a model of B and the number of automorphisms of I.

Remember that the word initial segment (resp. end extension) refers to a *proper* initial segment (resp. end extension) unless otherwise stated.

5.1 A Theorem of Friedman

Lemma 5.1.1. *Let M be a non-standard model of B, $\alpha \in M$ and $\vec{\beta} \in (I^M_\alpha)^{<\omega}$. Then $Th(I^M_\alpha, \vec{\beta}) \in R_M$.*

Proof. Left to the reader. ⊣

Definition 5.1.2. *Let M be a structure for L.*

(i) *M is* 1-tall, *if Σ_1-definable elements are not cofinal in M.*

(ii) *M is* 1-extendable, *if there is an end extension M' of M such that $M' \models B$ and $\Sigma_1 = \Sigma_1(M')$.*

Remarks. Let M be a non-standard model of PA. If $\Sigma_1(M) \neq \Sigma_1(\mathbb{N})$, then $I^1(M)$ is a non-standard 1-extendable model of B, but $I^1(M)$ is not 1-tall. If $\Sigma_1(M) = \Sigma_1(\mathbb{N})$, then given any $a \in M - N$, $K^1(M, a)$ is clearly 1-tall, but by 4.1.12.(i) $K^1(M, a)$ is not 1-extendable. Although both these examples give models of B, we note that every non-standard model of PA is both 1-tall and 1-extendable.

Result 5.1.3. (Friedman [F]) *Let M_1 and M_2 be countable non-standard models of PA. Then M_1 is isomorphic to an initial segment of M_2 iff $\Sigma_1(M_1) \subsetneq \Sigma_1(M_2)$ and $R_{M_1} = R_{M_2}$.*

The following is a generalization of the above result.

Proposition 5.1.4. *Suppose M_1 and M_2 are countable structures for L, M_1 is 1-extendable and M_2 is a 1-tall model of B. Then*

(i) *M_1 is isomorphic to an initial segment of M_2 iff $R_{M_1} = R_{M_2}$ and $\Sigma_1(M_1) \subseteq \Sigma_1(M_2)$.*

(ii) *M_1 is isomorphic to a submodel of M_2 iff $R_{M_1} \subseteq R_{M_2}$ and $\Sigma_1(M_1) \subseteq \Sigma_1(M_2)$.*

Proof. (i) '$\Rightarrow$' is obvious.
'$\Leftarrow$' as M_1 is 1-extendable, there is a model M of B such that $M_1 \subset_e M$ and $\Sigma_1(M) = \Sigma_1(M_1)$. Let α be any element of $M - M_1$; by 5.1.1., $Th(I_\alpha^M) \in R_{M_1}$ and as $R_{M_1} = R_{M_2}$, $Th(I_\alpha^M) \in R_{M_2}$. Let φ be any sentence of L such that $\varphi \in Th(I_\alpha^M)$, then $M \models \exists x \varphi^{t \leq x}$, where as usual, $\varphi^{t \leq x}$ denotes the relativization of φ to $t \leq x$. As $\Sigma_1(M) = \Sigma_1(M_1) \subseteq \Sigma_1(M_2)$, then $M_2 \models \exists x\ \varphi^{t \leq x}$. Hence, $M_2 \models \exists x\ Sat(x, \varphi)$ for all $\varphi \in Th(I_\alpha^M)$. Let $b \in M_2$ be an upper-bound for $K^1(M_2)$ in M_2 (since M_2 is 1-tall, such an element exists). Then:

$$\text{for all } n \in N,\ M_2 \models \exists x \leq b \forall \varphi \leq n(p_\varphi / a \rightarrow Sat(x, \varphi)),$$

where $a \in M_2$ codes $Th(I_\alpha^M)$ in M_2. By 3.1.7.(ii) the above formula is Δ_1^B and hence by 4.1.14., for some $\gamma \in M_2 - M$:

$$M_2 \models \exists x \forall \varphi \leq \gamma(p_\varphi \rightarrow Sat(x, \varphi)),$$

i.e. there exists some $\beta \in M_2$ such that $I_\beta^{M_2} \equiv I_\alpha^M$. As both of these structures are countable and R_{M_1} ($=R_{M_2}$)-saturated, $I_\beta^{M_2} \cong I_\alpha^M$ by 1.4.4.(ii). The isomorphism carries $M_1 \subset_e I_\alpha^M$ to an isomorphic copy $M_1' \subset_e I_\beta^{M_2}$.

(ii) Is proved similarly using 1.4.4.(iii). Note that here M_2 need not be countable. ⊣

Corollary 5.1.5. *Let M be a countable model of B. M is isomorphic to an initial segment of itself iff M is both 1-tall and 1-extendable.*

Proof. Immediate from 5.1.4. ⊣

See the remarks after 5.1.2. for some counter-examples when either of the above conditions on M fails to hold.

Lemma 5.1.6. *Let M be a countable model of B, $a, b \in M$ and $M \models a \leq b$. Then the following are equivalent:*

(i) *There is no* $c \in K^0(M)$ *such that* $M \models a \leq c \leq b$.

(ii) *There is some* $b' \in M$ *such that* $M \models b' < a$ *and* $I_b^M \cong I_{b'}^M$.

Proof. (i)⇒(ii) Since for all $d \in M$, I_d^M is R_M-saturated, by 1.4.4.(ii) it is enough to show the existence of some $b' \in M$ such that $I_b^M \cong I_{b'}^M$ and $M \models b' < a$. Suppose there is no such b'; then:

$$\text{For all } \gamma \in M - N,\ M \models \neg \exists x < a \forall \varphi \leq \gamma (p_\varphi | \beta \to Sat(x, \varphi)),$$

where β codes $Th(I_b^M)$ in M. By 3.1.7.(ii), the above formula is Δ_1^B; hence by 4.1.14. for some $m \in N$:

$$M \models \neg \exists x < a \forall \varphi < m(p_\varphi | \beta \to Sat(x, \varphi)).$$

Define φ as $\varphi_1 \wedge \ldots \varphi_n$, where $\ulcorner \varphi_i \urcorner < m$ and $\varphi_i \in Th(I_b^M)$ $(1 \leq i \leq n)$. Then $M \models \neg \exists x < a \varphi^{t \leq x}$. As M satisfies the least number principle for Σ_0-formulae and $M \models \exists x \varphi^{t \leq x}$ (since $I_b^M \models \varphi$) then $M \models \varphi^{t \leq c} \wedge \forall v < c \neg \varphi^{t \leq v}$ for some $c \in M$. Clearly, $M \models a \leq c \leq b$ and $c \in K^0(M)$, contradicting (i). Therefore for some $b' \in M$ we have $M \models b' < a$ and $I_b^M \cong I_{b'}^M$.

(ii)⇒(i) Obvious. ⊣

Corollary 5.1.7. *Let* M *be any countable non-standard model of* B. *Then for all* $\alpha \in M - N$, *the set* $\{\beta \in M : I_\alpha^M \cong I_\beta^M\}$ *is co-initial in* M *iff* $\Sigma_1(M) = \Sigma_1(N)$.

Proof. Noting that $\Sigma_1(M) = \Sigma_1(N)$ iff $K^0(M) = N$, the corollary follows immediately from 5.1.6. ⊣

Corollary 5.1.8. *Let* M *be a countable model of* B *such that* M *is isomorphic to an initial segment of itself. Then:*

$$I^1(M) = \bigcap \{M' : M' \subset_e M \text{ and } M' \cong M\}.$$

Proof. (Sketch) Suppose $M' \subset_e M$ and $M' \cong M$; choose $b \in M - M'$. Let a be any element of $M - I^1(M)$; then by 5.1.6. for some $b' \in M$ we have $M \models b' < a$ and $I_b^M \cong I_{b'}^M$. ⊣

5.2 Approximations to models of B.

Theorem 5.2.1. *Let* M, M_1 *and* M_2 *be countable models of* B *such that* $M \prec_{1,e} M_1 \prec_{1,e} M_2$, *where* $M \neq N$ *and* M_1 *could be equal to* M. *Then there are arbitrarily large initial segments of* M *which are isomorphic to* M_1.

Proof. Let γ be any element of M. We need to show the existence of an initial segment M_1' of M such that $M_1' \cong M_1$ and $\gamma \in M_1'$. Let α be any element of $M_2 - M_1$. By 5.1.1., $Th(I^{M_2}_{\alpha,\gamma})$ is coded in M_2, hence it is coded in M by some $a \in M$. Clearly:

$$M_2 \models \exists x \forall \varphi(y) < a(p_{\varphi(y)}|a \to Sat(x, \varphi(\gamma))).$$

By 3.1.7.(ii), the above formula is equivalent to a Σ_1-formula modulo B. Therefore, as $M <_1 M_2$:

$$M \models \exists x \forall \varphi(y) < a(p_{\varphi(y)}|a \to Sat(x, \varphi(\gamma))),$$

i.e. there is some $\beta \in M$ such that $(I^{M_2}_{\alpha}, \gamma) \equiv (I^{M}_{\beta}, \gamma)$. Hence, by 1.4.4.(ii), $(I^{M_2}_{\alpha}, \gamma) \cong (I^{M}_{\beta}, \gamma)$ (as both are R_M ($=R_{M_2}$)-saturated and countable). Now $M_1 \subset_e I^{M_2}_{\alpha}$ and $\gamma \in M_1$, hence there is an isomorphic copy M_1' of M_1 such that $M_1' \subset_e I^{M}_{\beta}$ and $\gamma \in M_1'$. ⊣

Corollary 5.2.2. *Let M be a countable non-standard model of PA. Then*

(i) *(Wilkie [Wk]) M is isomorphic to arbitrarily large initial segments of itself.*

(ii) *There are arbitrarily large initial segments M' of M such that $M' \models PA$ and M' is R_M-saturated.*

(iii) *There are arbitrarily large initial segments M' of M such that $M' \equiv M$ and M' is R_M-saturated iff $Th(M) \in R_M$.*

(iv) *Suppose T is a consistent extension of PA and $T \in R_M$. Then M has arbitrarily large initial segments which are models of T iff $\Pi_2(T) \subseteq \Pi_2(M)$.*

(v) *Let T be any complete consistent extension of PA. Then there are arbitrarily large initial segments of M which are models of T iff $\Pi_2(T) \subseteq \Pi_2(M)$ and $\Pi_n(T) \in R_M$ for all $n \in N$.*

Proof. (i) Let $M = M_1$ and let M_2 be any countable elementary end extension of M in 5.2.1. The result follows by 5.2.1.

(ii) By 2.2.3.(ii), M has a countable R_M-saturated end extension M_1 such that $M \prec_{1,e} M_1$. Again let M_2 be a countable elementary end extension of M_1. The result follows by 5.2.1.

(iii) Exactly as (ii) above but here we use 2.2.3.(i) instead of 2.2.3.(ii).

(iv) '⇒' Suppose M has arbitrarily large initial segments M_i which are models of T. Then $M = \bigcup M_i$. Therefore by a well-known result in Model Theory $M \models \Pi_2(T)$.

To show '⇐', we just note that if $T \in R_M$ and $\Pi_2(T) \subseteq \Pi_2(M)$, then by 2.2.1. there is a countable model M_1 of T such that $M \prec_{1,e} M_1$.

(v) Exactly as in (iv) above but here we use 2.2.20. ⊣

Remarks. (i) For an example of a model of B which is isomorphic to an initial segment of itself, but not to arbitrarily large initial segments of itself, see 5.2.4.(iv).

(ii) By 1.3.10.(Gaifman's Theorem) we can generalize (iv) and (v) of 5.2.2. to uncountable models M of PA.

(iii) Suppose M is a model of B such that for some M', $M' \models B$, we have $M \prec_{1,e} M'$, then M is both 1-tall and 1-extendable. This follows from 5.2.1. and 5.1.5.

Theorem 5.2.3. *Suppose $M_1 \subset_e M_2$, where M_1 and M_2 are models of B. Then M_1 is either the union or the intersection of some sequence of initial segments of M_2 which are models of PA.*[1]

Proof. We prove the theorem for the case when M_2 is countable. The generalization is left to the reader. We need to show that given $a \in M_2 - M$ and $b \in M_1$, there is an initial segment M_1' of M_2 such that $M_1' \models PA$, $b \in M_1'$ and $a \notin M_1'$. By 3.1.3. there is some countable model M_3 of PA such that $M_1 \prec_1 M_3$. We claim that the following is a consistent set of sentences in some suitable expansion of L:

$$Th(M_2, c)_{c \in M_2} \cup Diag(M_3) \cup \{a > b_i : i \in M_3\},$$

where $Diag(M_3)$ denotes the atomic diagram of M_3. The claim follows by a simple application of the Compactness Theorem. Hence by Löwenheim-Skolem Theorem, there is a countable elementary extension M_4 of M_2 such that $M_3 \subset M_4$ and $a \in M_4$ $(a \in M_2)$ is greater than all elements of M_3. Let $\overline{M}_3$ denote the initial segment of M_4, determined by M_3 and the structure inherited from M_4. Then $M_3 \prec \overline{M}_3$ by 4.1.6. Clearly, $a \in M_4 - \overline{M}_3$ and $b \in \overline{M}_3$. The theorem follows by 4.1.18. (Immortality of the Submodels). ⊣

Remarks 5.2.4. (i) The following generalization of 5.2.3. can be proved exactly as above.

Theorem. *Suppose T is a consistent extension of PA, $M_1 \subset_e M_2$, where $M_2 \models B$, $M_1 \models \Pi_2(T)$ and $T \in R_{M_1}$. Then M_1 is either the union or the intersection of some sequence of initial segments of M_2 which are models of T.*

(ii) Suppose T, M_1, M_2 are as in (i) above. If M_1 is the intersection (resp. union) of a sequence of initial segments of M_2 which are models of $\Pi_2(T)$, then M_1 is the intersection (resp. union) of some sequence of initial segments of M_2 which are models of T.

[1] A similar result was proved independently by J. B. Paris [Unpublished].

(iii) Given any non-standard model M of PA, we can find non-standard initial segments M_1, M_2 of M such that M_1 and M_2 are models of B and M_1 (resp. M_2) is not the intersection (resp. union) of any sequence of initial segments of M which are models of B; e.g. let A be any element of $M - N$ and let:

$$M_1 = \bigcup\{J : J \models B \text{ and } a \notin J\}$$
$$M_2 = \bigcap\{J : J \models B \text{ and } a \in J\}.$$

It is straightforward to check that M_1 and M_2 satisfy the required conditions.

(iv) Note that if a is an upper bound for Σ_1-definable elements of M, then by 5.1.5., M_2 constructed as in (iii) above, is isomorphic to an initial segment of itself, but clearly M_2 is not isomorphic to arbitrarily large initial segments of itself (assuming M is countable).

Proposition 5.2.5. *Suppose M is a model of B such that for all $a \in M$, $K^1(M, a)$ is not cofinal in M. Then M is the union of some sequence of initial segments of M which are models of PA.*

Proof. (Sketch) Let M be countable, the proof for the case when M is uncountable is left to the reader. By 3.1.3., there is a countable model M' of PA such that $M \prec_1 M'$. Let $\overline{M}$ denote the initial segment of M' determined by M and the structure inherited from M'. For each $a \in M$, we can show exactly as in the proof of 3.1.4.(ii) that $I^1(M, a) \models B$. Clearly $I^1(M', a) \subset_e \overline{M}$, i.e. $\overline{M}$ has arbitrarily large initial segments which are models of B. The proposition follows by 5.2.4. and 4.1.18. (Immortality of the Submodels). ⊣

Remarks 5.2.6. (i) The following generalization of 5.2.5. can be proved exactly as above.

Proposition. *Suppose T is a consistent extension of PA, $M \models \Pi_2(T)$ and $T \in R_M$. If for all $a \in M$, $K^1(M, a)$ is not cofinal in M, then M is the union of some sequence of initial segments of M which are models of T.*

(ii) Let M be any model of B such that $Th(M)$ is arithmetic. Then M is the union of some sequence of initial segments of M which are models of PA. This follows by a simple generalization of 3.1.10.(ii), namely, if M is a model of B such that for some $a \in M$, $K^1(M, a)$ is cofinal in M, then $Th(M)$ is non-arithmetic.

(iii) Let M_2 be as constructed in 5.2.4.(iii), then by (ii) above $Th(M_2)$ is non-arithmetic and by the proof of 5.2.5. $K^1(M, a) \cap M_2$ is cofinal in M_2.

Lemma 5.2.7. *Let M be a countable model of B and let $\alpha \in M$. Suppose $b, c \in I_\alpha^M$ are such that $I_\alpha^M \models c \leq b$. Then given any $\vec{a} \in (I_\alpha^M)^{<\omega}$, the following are equivalent:*

(i) *There is no $d \in Def(I_\alpha^M, \vec{a})$ such that $I_\alpha^M \models c \leq d \leq b$.*

(ii) *There is some $b' \in I_\alpha^M$, $I_\alpha^M \models b' < c$ such that $(I_\alpha^M, \vec{a}, b) \cong (I_\alpha^M, \vec{a}, b')$.*

(iii) *There is some $c' \in I_\alpha^M$, $I_\alpha^M \models b < c'$ such that $(I_\alpha^M, \vec{a}, c) \cong (I_\alpha^M, \vec{a}, c')$.*

Proof. We omit the proof as it is essentially the same as the proof of 2.1.7. ⊣

Proposition 5.2.8. *Suppose M is a non-standard model of B and $A \subseteq Th(\mathbb{N})$ is coded in M. Let α, β be elements of $M - N$ such that $M \models \beta < a$ for all $a \in Def(I_\alpha^M) - N$. Then there is an initial segment M' of M such that $M' \models A$, $\beta \in M'$ and $\alpha \notin M'$.*

Proof. Let M be countable. Again we leave the case for when M is uncountable to the reader. By 4.1.10.(iii), M has a non-standard initial segment M_1 such that $M_1 \models A$ and $\beta \notin M_1$. Let γ be any element of $M_1 - N$, by the hypothesis and 5.2.7. there exists some $\gamma' \in I_\alpha^M$ such that $M \models \beta \leq \gamma'$ and $(I_\alpha^M) \cong (I_\alpha^M, \gamma')$. The proposition follows. ⊣

Definition 5.2.9. *For a model M of B, define $J_0(M)$ as the initial segment of M with the domain:*

$$J_0(M) = \{a \in M : \forall b \in K^0(M) - N, M \models a < b\}$$

and the structure inherited from M.

Corollary 5.2.10. *Let M be a model of B such that $K^0(M) \downdownarrows K^1(M)$. Then $J_0(M) \models \Pi_2(\mathbb{N})$.*

Proof. To avoid triviality, assume $J_0(M) \neq N$. It is enough to show that given any $\beta \in J_0(M)$ and $\varphi \in Th(\mathbb{N})$, there is a model M' of $PA \cup \{\varphi\}$ such that $M' \subset_e J_0(M)$ and $\beta \in M'$. Let α be any element of $K^1(M) - N$. By the hypothesis and the definition of $J_0(M)$ we have $J_0(M) \cap Def(I_\alpha^M) = N$. Hence by 5.2.8. there is a model M' of $PA \cup \{\varphi\}$ such that $\beta \in M'$ and $M' \subset_e M$. Also the construction of M' in 5.2.8. ensures that $\Sigma_1(M') = \Sigma_1(\mathbb{N})$. Hence $M' \subset_e J_0(M)$ and the corollary follows. ⊣

5.3 Number of isomorphic initial segments

Throughout this section let M denote a *countable non-standard model of PA* and I a non-standard initial segment of M. Define $H(I)$ and $A(I)$ by:

$$H(I) = \{I' \subset_e I : I' \cong I\}$$
$$A(I) = \{f : f \text{ is an automorphism of } I\}.$$

Definition 5.3.1. *For $A \subseteq M$ and $f : A \to M$ we say $a \in M$ codes f in M if for all $i \in$ domain of f, $f(i) =$ the greatest x such that p_i^x divides a.*

The following definition is due to J. B. Paris.

Definition 5.3.2. *I is* semi-regular *in M if whenever $\alpha \in I$ and $f : I_\alpha^M \to M$ is coded in M, then $I \cap f''I_\alpha^M$ is bounded in I.*

Remarks 5.3.3. Kirby and Paris (see e.g. [K]) show that if M is a model of PA, then M has an elementary end extension M' such that M is semi-regular in M' (actually they prove a stronger result). They also prove that if I is semi-regular in M then I with the structure inherited from M satisfies the least number principle for Σ_1-formulae.

Result A. (Wilkie [Wk]) *If I is a rigid model of PA then $|H(I)| = 2^{\aleph_0}$.*

Result B. (Paris [Unpublished]) *If I is semi-regular in M then $|H(I)| = 2^{\aleph_0}$.*

In this section we generalize the above results. Also our proofs simplify the original arguments.

Given $\alpha, \beta \in M$, where $M \models \alpha \leq \beta$, the set $\{x \in M : M \models \alpha \leq x \leq \beta\}$ is denoted by $[\alpha, \beta]^M$.

Definition 5.3.4.

(i) *Let $\alpha \in M - I$. I is* α-clear *in M if for all $\vec{a} \in (I_\alpha^M)^{<\omega}$ there are $\beta \in I$ and $\gamma \in M - I$ such that*

$$Def(I_\alpha^M, \vec{a}) \cap [\beta, \gamma]^M = \emptyset.$$

(ii) *I is* clear *(resp.* strongly clear*) in M if for some (resp. for all) $\alpha \in M - I$, I is α-clear in M.*

Proposition 5.3.5. *Let I be semi-regular in M. Then I is strongly clear in M.*

Proof. Let α be any element of $M - I$ and let $\vec{a} \in (I_\alpha^M)^{<\omega}$. For $\beta \in I - N$, define the function $f : I_\beta^M \to M$ as follows:

$f(i) = b$ if i is the Gödel number of some formula $\varphi(x, \vec{y})$ and $I_\alpha^M \models \exists! x \varphi(x, \vec{a}) \wedge \varphi(b, \vec{a})$.

$f(i) = 0$ otherwise.

Clearly f is coded in M. Hence as I is semi-regular in M, $I \cap f''I_\beta^M$ is bounded in I; i.e. $Def(I_\alpha^M, \vec{a}) \cap I$ is not cofinal in I. By a simple overspill type argument it follows that I is α-clear in M. Hence as α was arbitrarily chosen, I is strongly clear in M. ⊣

Remark 5.3.6. There is a strongly clear initial segment J of M such that J is not closed under addition.

Proof. Suppose I is a semi-regular initial segment of M (for existence see [K]) and let β be any element of $M - I$. Define the initial segment J of M by:

$$J = \{x \in M : \exists i \in I, M \models x \leq \beta + i\}.$$

Obviously J is strongly clear in M and $\beta + \beta \notin J$. ⊣

Note that by 5.3.3. J is an example of a strongly clear initial segment of M which is not semi-regular in M.

Result 5.3.7. (Kueker-Reyes) *Let M be any countable structure and let P be any subset of M such that for all $\vec{a} \in (M)^{<\omega}$ there are $b \in P$ and $b' \in M - P$ for which $(M, \vec{a}, b) \cong (M, \vec{a}, b')$. Then:*

$$|\{P' : (M, P) \cong (M, P')\}| = 2^{\aleph_0}.$$

Proof. See e.g. reference 15 of [Ku] where a stronger result is proved. ⊣

Theorem 5.3.8.

(i) *Let I be clear in M. Then $|H(I)| = 2^{\aleph_0}$.*
(ii) *Suppose $|A(I)| < 2^{\aleph_0}$. Then I is strongly clear in M.*

Proof. (i) Suppose $\alpha \in M - I$ is such that I is α-clear in M. By 5.3.7 it is enough to show that given $\vec{a} \in (I_\alpha^M)^{<\omega}$ we can find $b \in I$ and $b' \in I_\alpha^M - I$ such that $(I_\alpha^M, \vec{a}, b) \cong (I_\alpha^M, \vec{a}, b')$. This follows from 5.2.7. and the fact that I is α-clear in M.

(ii) (Sketch) Suppose not; then there is some $\alpha \in M - I$ and $\vec{a} \in (I_\alpha^M)^{<\omega}$ such that for all $\beta \in I$ and $\gamma \in M - I$, $Def(I_\alpha^M, \vec{a}) \cap [\beta, \gamma]^M \neq \emptyset$. Let f be any automorphism of $(I_\alpha^M, \vec{a})$; then $f''I = I$. By 2.2.14., given $\vec{b} \in (I_\alpha^M)^{<\omega}$ there is no non-standard initial segment of I_α^M whose every element is definable in $(I_\alpha^M, \vec{b})$; using this and 5.2.7. it follows exactly as in the proof of 2.1.10. (for the case when M' is not tall) that $|A(I)| = 2^{\aleph_0}$, contradicting the hypothesis. Hence I is strongly clear in M. ⊣

Corollary 5.3.9. *Let I be any non-standard initial segment of M. Then either $|A(I)| = 2^{\aleph_0}$ or $|H(I)| = 2^{\aleph_0}$.*

Proof. Immediate from 5.3.8. ⊣

Proposition 5.3.10. *Suppose I is clear in M. Then there are sequences K_i and J_i of initial segments of M which are isomorphic to I and $I = \bigcap K_i = \bigcup J_i$.*

Proof. Immediate from 5.2.7. ⊣

Corollaries and remarks 5.3.11.

(i) By 5.3.6. and 5.3.8(i) there are initial segments J of M such that J is not closed under addition but $|H(J)| = 2^{\aleph_0}$. Clearly this is the best possible result as far as closure properties are concerned.

(ii) There are initial segments I of M such that $I \models B$ and $|A(I)| = 2^{\aleph_0}$, e.g. let $I = M_2$, where M_2 is constructed as in 5.2.4.(iv).

(iii) If $M \neq N$ is pointwise definable, then by 5.3.8.(ii) M is strongly clear in any end extension M' of M such that $M' \models PA$.

(iv) Let M' be an initial segment of M such that $M' \models B$. Suppose M' is not the union (resp. the intersection) of any sequence of initial segments of M which are models of PA, then $|A(M')| = 2^{\aleph_0}$. This follows by 5.3.10. and 5.2.3.

Proposition 5.3.12. *Suppose I is closed under exponentiation and I is clear in M. Then I is strongly clear in M.*

Proof. (Sketch) Suppose I is α-clear in M; then I is γ-clear in M for all $\gamma \in M - I$ such that $M \models \gamma \leq \alpha$. Suppose I is not strongly clear in M, then for some $\beta \in M$, $M \models \alpha < \beta$ and $\vec{a} \in (I_\beta^M)^{<\omega}$ we have:

$$Def(I_\beta^M, \vec{a}) \cap [b, c]^M \neq \emptyset \text{ for all } b \in I \text{ and } c \in M - I.$$

Let d be any element of $M - I$, there is some $d' \in M - I$ such that in M:

$$\text{“}p_i^a | d' \leftrightarrow a < d \wedge a \text{ is definable by a formula } \varphi(x, \vec{y}) \text{ in } (I_\beta^M, \vec{a}) \text{ and } \ulcorner \varphi(x, \vec{y}) \urcorner = i\text{”}.$$

It is obvious that I is not d'-clear in M. Now as I is closed under exponentiation, by choosing a small enough $d \in M - I$, we can ensure that $M \models d' < \alpha$; contradicting that I is α-clear in M. Hence I is strongly clear in M. ⊣

Note. We do not know whether every clear initial segment of M is strongly clear in M.

Proposition 5.3.13. *There is a model M of PA and a non-standard initial segment I of M such that $I \prec_e M$ and I is not clear in M.*

Proof. Let $M^{'}$ be any countable non-standard model of PA, enumerate elements of $M^{'}$ by an ω-sequence $(a_i)_{i \in N}$. Add to L a new constant symbol c. By the Compactness Theorem the following is a consistent set of sentences of $L(M^{'}) \cup \{c\}$:

$$T = Th(M^{'}, a)_{a \in M'} \cup \{p_i^{a_i} | c : a_i \in M^{'}\}.$$

Let M be any model of T and let I denote the initial segment of M determined by $M^{'}$ (and the structure inherited from M). By an obvious modification of 2.2.12. we have $I \prec_e M$. Clearly, I is not c-clear in M, hence by 5.3.11. I is not clear in M. ⊣

We end this chapter by posing some open problems.

(i) What is the weakest theory $T \supseteq B_1$ such that every model of T is isomorphic to an initial segment of itself?

(ii) Is there an initial segment I of M such that $I \models B$ and I is isomorphic to arbitrarily large initial segments of itself but $|H(I)| \neq 2^{\aleph_0}$?

(iii) What are the necessary and sufficient conditions for a model M of PA to have an end extension $M^{'}$ such that $M^{'} \models PA$ and M is not clear in $M^{'}$?

Bibliography

[B&S] J. Barwise and J. Schlipf, "An Introduction to Recursively Saturated and Resplendent Models", J.S.L. Vol. 41 (1976).

[C&K] C. C. Chang and H. J. Keisler, "Model Theory", North Holland (1973).

[F] H. Friedman, "Countable Models of Set Theories", Cambridge Summer School in Mathematical Logic, 1971, Springer Lecture Notes 337, pp. 539–573.

[G] H. Gaifman, "A Note on Models and Submodels of Arithmetic", Conference in Mathematical Logic, London'70, Springer Lecture Notes 255, pp. 128–144.

[H&W] J. Hirschfeld and W. Wheeler, "Forcing, Arithmetic, Division Rings", Springer Lecture Notes 454.

[J&E] D. Jensen and A. Ehrenfeucht, "Some Problems in Elementary Arithmetics", Fundamenta Mathematicae (1976).

[K] L. A. S. Kirby, Ph.D. Thesis, Manchester (1977).

[Ku] D. W. Kueker, "Back-and-Forth Arguments and Infinitary Logics", Infinitary Logic: In Memoriam Carol Karp, Springer Lecture Notes 492, pp. 17–73 (1975).

[Mc] K. McAloon, "Completeness Theorems, Incompleteness Theorems and Models of Arithmetic", Transactions of the American Mathematical Society 239 (1978) 253-277.

[P&K] J. B. Paris and L. A. S. Kirby, "Σ_n-Collection Schemas in Arithmetic" (mimeographed).

[S] D. Scott, "Algebras of Sets Binumerable in Complete Extensions of Arithmetic", Proc. Sympos. Pure Math. Vol. 5, A.M.S. (1962).

[Sh] J. R. Shoenfield, "Mathematical Logic", Addison-Wesley (1967).

[Sm] C. Smorynski, "Consistence and Related Mathematical Properties", University of Amsterdam Report 75-02.

[Wk] A. J. Wilkie, Ph.D. Thesis, Bedford College, London, 1971.
[W_1] G. M. Wilmers, Ph.D. Thesis, Oxford, 1975.
[W_2] G. M. Wilmers, Lecture Notes, Wrocław, Winter, 1976.